The Feynman

LECTURES ON

PHYSICS

QUANTUM MECHANICS

RICHARD P. FEYNMAN

Richard Chace Tolman Professor of Theoretical Physics
California Institute of Technology

ROBERT B. LEIGHTON

Professor of Physics
California Institute of Technology

MATTHEW SANDS

Professor
Stanford University

ADDISON-WESLEY PUBLISHING COMPANY

Reading, Massachusetts

Menlo Park, California · London · Amsterdam · Don Mills, Ontario · Sydney

Feynman's Preface

These are the lectures in physics that I gave last year and the year before to the freshman and sophomore classes at Caltech. The lectures are, of course, not verbatim—they have been edited, sometimes extensively and sometimes less so. The lectures form only part of the complete course. The whole group of 180 students gathered in a big lecture room twice a week to hear these lectures and then they broke up into small groups of 15 to 20 students in recitation sections under the guidance of a teaching assistant. In addition, there was a laboratory session once a week.

The special problem we tried to get at with these lectures was to maintain the interest of the very enthusiastic and rather smart students coming out of the high schools and into Caltech. They have heard a lot about how interesting and exciting physics is—the theory of relativity, quantum mechanics, and other modern ideas. By the end of two years of our previous course, many would be very discouraged because there were really very few grand, new, modern ideas presented to them. They were made to study inclined planes, electrostatics, and so forth, and after two years it was quite stultifying. The problem was whether or not we could make a course which would save the more advanced and excited student by maintaining his enthusiasm.

The lectures here are not in any way meant to be a survey course, but are very serious. I thought to address them to the most intelligent in the class and to make sure, if possible, that even the most intelligent student was unable to completely encompass everything that was in the lectures—by putting in suggestions of applications of the ideas and concepts in various directions outside the main line of attack. For this reason, though, I tried very hard to make all the statements as accurate as possible, to point out in every case where the equations and ideas fitted into the body of physics, and how—when they learned more—things would be modified. I also felt that for such students it is important to indicate what it is that they should—if they are sufficiently clever—be able to understand by deduction from what has been said before, and what is being put in as something new. When new ideas came in, I would try either to deduce them if they were deducible, or to explain that it *was* a new idea which hadn't any basis in terms of things they had already learned and which was not supposed to be provable—but was just added in.

At the start of these lectures, I assumed that the students knew something when they came out of high school—such things as geometrical optics, simple chemistry ideas, and so on. I also didn't see that there was any reason to make the lectures in a definite order, in the sense that I would not be allowed to mention something until I was ready to discuss it in detail. There was a great deal of mention of things to come, without complete discussions. These more complete discussions would come later when the preparation became more advanced. Examples are the discussions of inductance, and of energy levels, which are at first brought in in a very qualitative way and are later developed more completely.

At the same time that I was aiming at the more active student, I also wanted to take care of the fellow for whom the extra fireworks and side applications are merely disquieting and who cannot be expected to learn most of the material in the lecture at all. For such students I wanted there to be at least a central core or backbone of material which he *could* get. Even if he didn't understand everything in a lecture, I hoped he wouldn't get nervous. I didn't expect him to understand everything, but only the central and most direct features. It takes, of course, a certain intelligence on his part to see which are the central theorems and central ideas, and which are the more advanced side issues and applications which he may understand only in later years.

In giving these lectures there was one serious difficulty: in the way the course was given, there wasn't any feedback from the students to the lecturer to indicate how well the lectures were going over. This is indeed a very serious difficulty, and I don't know how good the lectures really are. The whole thing was essentially an experiment. And if I did it again I wouldn't do it the same way—I hope I *don't* have to do it again! I think, though, that things worked out—so far as the physics is concerned—quite satisfactorily in the first year.

In the second year I was not so satisfied. In the first part of the course, dealing with electricity and magnetism, I couldn't think of any really unique or different way of doing it—of any way that would be particularly more exciting than the usual way of presenting it. So I don't think I did very much in the lectures on electricity and magnetism. At the end of the second year I had originally intended to go on, after the electricity and magnetism, by giving some more lectures on the properties of materials, but mainly to take up things like fundamental modes, solutions of the diffusion equation, vibrating systems, orthogonal functions, ... developing the first stages of what are usually called "the mathematical methods of physics." In retrospect, I think that if I were doing it again I would go back to that original idea. But since it was not planned that I would be giving these lectures again, it was suggested that it might be a good idea to try to give an introduction to the quantum mechanics—what you will find in Volume III.

It is perfectly clear that students who will major in physics can wait until their third year for quantum mechanics. On the other hand, the argument was made that many of the students in our course study physics as a background for their primary interest in other fields. And the usual way of dealing with quantum mechanics makes that subject almost unavailable for the great majority of students because they have to take so long to learn it. Yet, in its real applications—especially in its more complex applications, such as in electrical engineering and chemistry—the full machinery of the differential equation approach is not actually used. So I tried to describe the principles of quantum mechanics in a way which wouldn't require that one first know the mathematics of partial differential equations. Even for a physicist I think that is an interesting thing to try to do—to present quantum mechanics in this reverse fashion—for several reasons which may be apparent in the lectures themselves. However, I think that the experiment in the quantum mechanics part was not completely successful—in large part because I really did not have enough time at the end (I should, for instance, have had three or four more lectures in order to deal more completely with such matters as energy bands and the spatial dependence of amplitudes). Also, I had never presented the subject this way before, so the lack of feedback was particularly serious. I now believe the quantum mechanics should be given at a later time. Maybe I'll have a chance to do it again someday. Then I'll do it right.

The reason there are no lectures on how to solve problems is because there were recitation sections. Although I did put in three lectures in the first year on how to solve problems, they are not included here. Also there was a lecture on inertial guidance which certainly belongs after the lecture on rotating systems, but which was, unfortunately, omitted. The fifth and sixth lectures are actually due to Matthew Sands, as I was out of town.

The question, of course, is how well this experiment has succeeded. My own point of view—which, however, does not seem to be shared by most of the people who worked with the students—is pessimistic. I don't think I did very well by the students. When I look at the way the majority of the students handled the problems on the examinations, I think that the system is a failure. Of course, my friends point out to me that there were one or two dozen students who—very surprisingly —understood almost everything in all of the lectures, and who were quite active in working with the material and worrying about the many points in an excited and interested way. These people have now, I believe, a first-rate background in physics—and they are, after all, the ones I was trying to get at. But then, "The power of instruction is seldom of much efficacy except in those happy dispositions where it is almost superfluous." (Gibbons)

Still, I didn't want to leave any student completely behind, as perhaps I did. I think one way we could help the students more would be by putting more hard work into developing a set of problems which would elucidate some of the ideas in the lectures. Problems give a good opportunity to fill out the material of the lectures and make more realistic, more complete, and more settled in the mind the ideas that have been exposed.

I think, however, that there isn't any solution to this problem of education other than to realize that the best teaching can be done only when there is a direct individual relationship between a student and a good teacher—a situation in which the student discusses the ideas, thinks about the things, and talks about the things. It's impossible to learn very much by simply sitting in a lecture, or even by simply doing problems that are assigned. But in our modern times we have so many students to teach that we have to try to find some substitute for the ideal. Perhaps my lectures can make some contribution. Perhaps in some small place where there are individual teachers and students, they may get some inspiration or some ideas from the lectures. Perhaps they will have fun thinking them through—or going on to develop some of the ideas further.

RICHARD P. FEYNMAN

June, 1963

Foreword

A great triumph of twentieth-century physics, the theory of quantum mechanics, is now nearly 40 years old, yet we have generally been giving our students their introductory course in physics (for many students, their last) with hardly more than a casual allusion to this central part of our knowledge of the physical world. We should do better by them. These lectures are an attempt to present them with the basic and essential ideas of the quantum mechanics in a way that would, hopefully, be comprehensible. The approach you will find here is novel, particularly at the level of a sophomore course, and was considered very much an experiment. After seeing how easily some of the students take to it, however, I believe that the experiment was a success. There is, of course, room for improvement, and it will come with more experience in the classroom. What you will find here is a record of that first experiment.

In the two-year sequence of the Feynman Lectures on Physics which were given from September 1961 through May 1963 for the introductory physics course at Caltech, the concepts of quantum physics were brought in whenever they were necessary for an understanding of the phenomena being described. In addition, the last twelve lectures of the second year were given over to a more coherent introduction to some of the concepts of quantum mechanics. It became clear as the lectures drew to a close, however, that not enough time had been left for the quantum mechanics. As the material was prepared, it was continually discovered that other important and interesting topics could be treated with the elementary tools that had been developed. There was also a fear that the too brief treatment of the Schrödinger wave function which had been included in the twelfth lecture would not provide a sufficient bridge to the more conventional treatments of many books the students might hope to read. It was therefore decided to extend the series with seven additional lectures; they were given to the sophomore class in May of 1964. These lectures rounded out and extended somewhat the material developed in the earlier lectures.

In this volume we have put together the lectures from both years with some adjustment of the sequence. In addition, two lectures originally given to the freshman class as an introduction to quantum physics have been lifted bodily from Volume I (where they were Chapters 37 and 38) and placed as the first two chapters here—to make this volume a self-contained unit, relatively independent of the first two. A few ideas about the quantization of angular momentum (including a discussion of the Stern-Gerlach experiment) had been introduced in Chapters 34 and 35 of Volume II, and familiarity with them is assumed; for the convenience of those who will not have that volume at hand, those two chapters are reproduced here as an Appendix.

This set of lectures tries to elucidate from the beginning those features of the quantum mechanics which are most basic and most general. The first lectures tackle head on the ideas of a probability amplitude, the interference of amplitudes, the abstract notion of a state, and the superposition and resolution of states—and the Dirac notation is used from the start. In each instance the ideas are introduced together with a detailed discussion of some specific examples—to try to make the physical ideas as real as possible. The time dependence of states including states of definite energy comes next, and the ideas are applied at once to the study of two-state systems. A detailed discussion of the ammonia maser provides the framework for the introduction to radiation absorption and induced transitions. The lectures then go on to consider more complex systems, leading to a discussion of the propagation of electrons in a crystal, and to a rather complete treatment of the quantum mechanics of angular momentum. Our introduction to quantum mechanics ends in Chapter 20 with a discussion of the Schrödinger wave function, its differential equation, and the solution for the hydrogen atom.

The last chapter of this volume is not intended to be a part of the "course." It is a "seminar" on superconductivity and was given in the spirit of some of the entertainment lectures of the first two volumes, with the intent of opening to the students a broader view of the relation of what they were learning to the general culture of physics. Feynman's "epilogue" serves as the period to the three-volume series.

As explained in the Foreword to Volume I, these lectures were but one aspect of a program for the development of a new introductory course carried out at the California Institute of Technology under the supervision of the Physics Course Revision Committee (Robert Leighton, Victor Neher, and Matthew Sands). The program was made possible by a grant from the Ford Foundation. Many people helped with the technical details of the preparation of this volume: Marylou Clayton, Julie Curcio, James Hartle, Tom Harvey, Martin Israel, Patricia Preuss, Fanny Warren, and Barbara Zimmerman. Professors Gerry Neugebauer and Charles Wilts contributed greatly to the accuracy and clarity of the material by reviewing carefully much of the manuscript.

But the story of quantum mechanics you will find here is Richard Feynman's. Our labors will have been well spent if we have been able to bring to others even some of the intellectual excitement we experienced as we saw the ideas unfold in his real-life Lectures on Physics.

December, 1964   MATTHEW SANDS

Contents

CHAPTER 1. QUANTUM BEHAVIOR

1–1 Atomic mechanics 1–1
1–2 An experiment with bullets 1–1
1–3 An experiment with waves 1–3
1–4 An experiment with electrons 1–4
1–5 The interference of electron waves 1–5
1–6 Watching the electrons 1–6
1–7 First principles of quantum mechanics 1–9
1–8 The uncertainty principle 1–11

CHAPTER 2. THE RELATION OF WAVE AND PARTICLE VIEWPOINTS

2–1 Probability wave amplitudes 2–1
2–2 Measurement of position and momentum 2–2
2–3 Crystal diffraction 2–4
2–4 The size of an atom 2–5
2–5 Energy levels 2–7
2–6 Philosophical implications 2–8

CHAPTER 3. PROBABILITY AMPLITUDES

3–1 The laws of combining amplitudes 3–1
3–2 The two-slit interference pattern 3–5
3–3 Scattering from a crystal 3–7
3–4 Identical particles 3–9

CHAPTER 4. IDENTICAL PARTICLES

4–1 Bose particles and Fermi particles 4–1
4–2 States with two Bose particles 4–3
4–3 States with n Bose particles 4–6
4–4 Emission and absorption of photons 4–7
4–5 The blackbody spectrum 4–8
4–6 Liquid helium 4–12
4–7 The exclusion principle 4–12

CHAPTER 5. SPIN ONE

5–1 Filtering atoms with a Stern-Gerlach apparatus 5–1
5–2 Experiments with filtered atoms 5–5
5–3 Stern-Gerlach filters in series 5–6
5–4 Base states 5–8
5–5 Interfering amplitudes 5–10
5–6 The machinery of quantum mechanics 5–12
5–7 Transforming to a different base 5–15
5–8 Other situations 5–16

CHAPTER 6. SPIN ONE-HALF

6–1 Transforming amplitudes 6–1
6–2 Transforming to a rotated coordinate system 6–3
6–3 Rotations about the z-axis 6–6
6–4 Rotations of 180° and 90° about y 6–9
6–5 Rotations about x 6–11
6–6 Arbitrary rotations 6–12

CHAPTER 7. THE DEPENDENCE OF AMPLITUDES ON TIME

7–1 Atoms at rest; stationary states 7–1
7–2 Uniform motion 7–3
7–3 Potential energy; energy conservation 7–6
7–4 Forces; the classical limit 7–9
7–5 The "precession" of a spin one-half particle 7–10

CHAPTER 8. THE HAMILTONIAN MATRIX

8–1 Amplitudes and vectors 8–1
8–2 Resolving state vectors 8–3
8–3 What are the base states of the world? 8–5
8–4 How states change with time 8–7
8–5 The Hamiltonian matrix 8–10
8–6 The ammonia molecule 8–11

CHAPTER 9. THE AMMONIA MASER

9–1 The states of an ammonia molecule 9–1
9–2 The molecule in a static electric field 9–5
9–3 Transitions in a time-dependent field 9–9
9–4 Transitions at resonance 9–11
9–5 Transitions off resonance 9–13
9–6 The absorption of light 9–14

CHAPTER 10. OTHER TWO-STATE SYSTEMS

10–1 The hydrogen molecular ion 10–1
10–2 Nuclear forces 10–6
10–3 The hydrogen molecule 10–8
10–4 The benzene molecule 10–10
10–5 Dyes 10–12
10–6 The Hamiltonian of a spin one-half particle in a magnetic field 10–12
10–7 The spinning electron in a magnetic field 10–15

CHAPTER 11.   MORE TWO-STATE SYSTEMS

11–1   The Pauli spin matrices 11–1
11–2   The spin matrices as operators 11–5
11–3   The solution of the two-state equations 11–8
11–4   The polarization states of the photon 11–9
11–5   The neutral K-meson 11–12
11–6   Generalization to N-state systems 11–20

CHAPTER 12.   THE HYPERFINE SPLITTING IN HYDROGEN

12–1   Base states for a system with two spin one-half particles 12–1
12–2   The Hamiltonian for the ground state of hydrogen 12–3
12–3   The energy levels 12–7
12–4   The Zeeman splitting 12–9
12–5   The states in a magnetic field 12–12
12–6   The projection matrix for spin one 12–14

CHAPTER 13.   PROPAGATION IN A CRYSTAL LATTICE

13–1   States for an electron in a one-dimensional lattice 13–1
13–2   States of definite energy 13–3
13–3   Time-dependent states 13–6
13–4   An electron in a three-dimensional lattice 13–7
13–5   Other states in a lattice 13–8
13–6   Scattering by imperfections in the lattice 13–10
13–7   Trapping by a lattice imperfection 13–12
13–8   Scattering amplitudes and bound states 13–13

CHAPTER 14.   SEMICONDUCTORS

14–1   Electrons and holes in semiconductors 14–1
14–2   Impure semiconductors 14–4
14–3   The Hall effect 14–7
14–4   Semiconductor junctions 14–8
14–5   Rectification at a semiconductor junction 14–10
14–6   The transistor 14–11

CHAPTER 15.   THE INDEPENDENT PARTICLE APPROXIMATION

15–1   Spin waves 15–1
15–2   Two spin waves 15–4
15–3   Independent particles 15–6
15–4   The benzene molecule 15–7
15–5   More organic chemistry 15–10
15–6   Other uses of the approximation 15–12

CHAPTER 16.   THE DEPENDENCE OF AMPLITUDES ON POSITION

16–1   Amplitudes on a line 16–1
16–2   The wave function 16–5
16–3   States of definite momentum 16–7
16–4   Normalization of the states in x 16–9
16–5   The Schrödinger equation 16–11
16–6   Quantized energy levels 16–14

CHAPTER 17.   SYMMETRY AND CONSERVATION LAWS

17–1   Symmetry 17–1
17–2   Symmetry and conservation 17–3
17–3   The conservation laws 17–7
17–4   Polarized light 17–9
17–5   The distintegration of the Λ° 17–11
17–6   Summary of the rotation matrices 17–15

CHAPTER 18.   ANGULAR MOMENTUM

18–1   Electric dipole radiation 18–1
18–2   Light scattering 18–3
18–3   The annihilation of positronium 18–5
18–4   Rotation matrix for any spin 18–9
18–5   Measuring a nuclear spin 18–13
18–6   Composition of angular momentum 18–14
         Added Note 1: Derivation of the rotation matrix 18–19
         Added Note 2: Conservation of parity in photon emission 18–22

CHAPTER 19.   THE HYDROGEN ATOM AND THE PERIODIC TABLE

19–1   Schrödinger's equation for the hydrogen atom 19–1
19–2   Spherically symmetric solutions 19–2
19–3   States with an angular dependence 19–6
19–4   The general solution for hydrogen 19–10
19–5   The hydrogen wave functions 19–12
19–6   The periodic table 19–13

CHAPTER 20.   OPERATORS

20–1   Operations and operators 20–1
20–2   Average energies 20–3
20–3   The average energy of an atom 20–6
20–4   The position operator 20–8
20–5   The momentum operator 20–9
20–6   Angular momentum 20–14
20–7   The change of averages with time 20–15

CHAPTER 21.   THE SCHRÖDINGER EQUATION IN A CLASSICAL CONTEXT: A SEMINAR ON SUPERCONDUCTIVITY

21–1   Schrödinger's equation in a magnetic field 21–1
21–2   The equation of continuity for probabilities 21–3
21–3   Two kinds of momentum 21–4
21–4   The meaning of the wave function 21–6
21–5   Superconductivity 21–7
21–6   The Meissner effect 21–8
21–7   Flux quantization 21–10
21–8   The dynamics of superconductivity 21–12
21–9   The Josephson junction 21–14

FEYNMAN'S EPILOGUE

APPENDIX

INDEX

1

Quantum Behavior

1–1 Atomic mechanics

"Quantum mechanics" is the description of the behavior of matter and light in all its details and, in particular, of the happenings on an atomic scale. Things on a very small scale behave like nothing that you have any direct experience about. They do not behave like waves, they do not behave like particles, they do not behave like clouds, or billiard balls, or weights on springs, or like anything that you have ever seen.

Newton thought that light was made up of particles, but then it was discovered that it behaves like a wave. Later, however (in the beginning of the twentieth century), it was found that light did indeed sometimes behave like a particle. Historically, the electron, for example, was thought to behave like a particle, and then it was found that in many respects it behaved like a wave. So it really behaves like neither. Now we have given up. We say: "It is like *neither*."

There is one lucky break, however—electrons behave just like light. The quantum behavior of atomic objects (electrons, protons, neutrons, photons, and so on) is the same for all, they are all "particle waves," or whatever you want to call them. So what we learn about the properties of electrons (which we shall use for our examples) will apply also to all "particles," including photons of light.

The gradual accumulation of information about atomic and small-scale behavior during the first quarter of this century, which gave some indications about how small things do behave, produced an increasing confusion which was finally resolved in 1926 and 1927 by Schrödinger, Heisenberg, and Born. They finally obtained a consistent description of the behavior of matter on a small scale. We take up the main features of that description in this chapter.

Because atomic behavior is so unlike ordinary experience, it is very difficult to get used to, and it appears peculiar and mysterious to everyone—both to the novice and to the experienced physicist. Even the experts do not understand it the way they would like to, and it is perfectly reasonable that they should not, because all of direct, human experience and of human intuition applies to large objects. We know how large objects will act, but things on a small scale just do not act that way. So we have to learn about them in a sort of abstract or imaginative fashion and not by connection with our direct experience.

In this chapter we shall tackle immediately the basic element of the mysterious behavior in its most strange form. We choose to examine a phenomenon which is impossible, *absolutely* impossible, to explain in any classical way, and which has in it the heart of quantum mechanics. In reality, it contains the *only* mystery. We cannot make the mystery go away by "explaining" how it works. We will just *tell* you how it works. In telling you how it works we will have told you about the basic peculiarities of all quantum mechanics.

1–2 An experiment with bullets

To try to understand the quantum behavior of electrons, we shall compare and contrast their behavior, in a particular experimental setup, with the more familiar behavior of particles like bullets, and with the behavior of waves like water waves. We consider first the behavior of bullets in the experimental setup shown diagrammatically in Fig. 1–1. We have a machine gun that shoots a stream of bullets. It is not a very good gun, in that it sprays the bullets (randomly) over a fairly large angular spread, as indicated in the figure. In front of the gun we have

1–1 Atomic mechanics

1–2 An experiment with bullets

1–3 An experiment with waves

1–4 An experiment with electrons

1–5 The interference of electron waves

1–6 Watching the electrons

1–7 First principles of quantum mechanics

1–8 The uncertainty principle

Note: This chapter is almost exactly the same as Chapter 37 of Volume I.

a wall (made of armor plate) that has in it two holes just about big enough to let a bullet through. Beyond the wall is a backstop (say a thick wall of wood) which will "absorb" the bullets when they hit it. In front of the wall we have an object which we shall call a "detector" of bullets. It might be a box containing sand. Any bullet that enters the detector will be stopped and accumulated. When we wish, we can empty the box and count the number of bullets that have been caught. The detector can be moved back and forth (in what we will call the x-direction). With this apparatus, we can find out experimentally the answer to the question: "What is the probability that a bullet which passes through the holes in the wall will arrive at the backstop at the distance x from the center?" First, you should realize that we should talk about probability, because we cannot say definitely where any particular bullet will go. A bullet which happens to hit one of the holes may bounce off the edges of the hole, and may end up anywhere at all. By "probability" we mean the chance that the bullet will arrive at the detector, which we can measure by counting the number which arrive at the detector in a certain time and then taking the ratio of this number to the *total* number that hit the backstop during that time. Or, if we assume that the gun always shoots at the same rate during the measurements, the probability we want is just proportional to the number that reach the detector in some standard time interval.

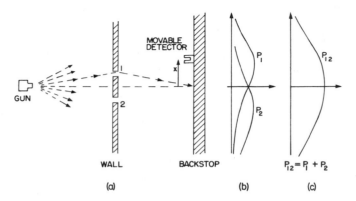

Fig. 1–1. Interference experiment with bullets.

For our present purposes we would like to imagine a somewhat idealized experiment in which the bullets are not real bullets, but are *indestructible* bullets— they cannot break in half. In our experiment we find that bullets always arrive in lumps, and when we find something in the detector, it is always one whole bullet. If the rate at which the machine gun fires is made very low, we find that at any given moment either nothing arrives, or one and only one—exactly one—bullet arrives at the backstop. Also, the size of the lump certainly does not depend on the rate of firing of the gun. We shall say: "Bullets *always* arrive in identical lumps." What we measure with our detector is the probability of arrival of a lump. And we measure the probability as a function of x. The result of such measurements with this apparatus (we have not yet done the experiment, so we are really imagining the result) are plotted in the graph drawn in part (c) of Fig. 1–1. In the graph we plot the probability to the right and x vertically, so that the x-scale fits the diagram of the apparatus. We call the probability P_{12} because the bullets may have come either through hole 1 or through hole 2. You will not be surprised that P_{12} is large near the middle of the graph but gets small if x is very large. You may wonder, however, why P_{12} has its maximum value at $x = 0$. We can understand this fact if we do our experiment again after covering up hole 2, and once more while covering up hole 1. When hole 2 is covered, bullets can pass only through hole 1, and we get the curve marked P_1 in part (b) of the figure. As you would expect, the maximum of P_1 occurs at the value of x which is on a straight line with the gun and hole 1. When hole 1 is closed, we get the symmetric curve P_2 drawn in the figure. P_2 is the probability distribution for bullets that pass through hole 2. Comparing parts (b) and (c) of Fig. 1–1, we find the important result that

$$P_{12} = P_1 + P_2. \tag{1.1}$$

The probabilities just add together. The effect with both holes open is the sum of the effects with each hole open alone. We shall call this result an observation of *"no interference,"* for a reason that you will see later. So much for bullets. They come in lumps, and their probability of arrival shows no interference.

1–3 An experiment with waves

Now we wish to consider an experiment with water waves. The apparatus is shown diagrammatically in Fig. 1–2. We have a shallow trough of water. A small object labeled the "wave source" is jiggled up and down by a motor and makes circular waves. To the right of the source we have again a wall with two holes, and beyond that is a second wall, which, to keep things simple, is an "absorber," so that there is no reflection of the waves that arrive there. This can be done by building a gradual sand "beach." In front of the beach we place a detector which can be moved back and forth in the x-direction, as before. The detector is now a device which measures the "intensity" of the wave motion. You can imagine a gadget which measures the height of the wave motion, but whose scale is calibrated in proportion to the *square* of the actual height, so that the reading is proportional to the intensity of the wave. Our detector reads, then, in proportion to the *energy* being carried by the wave—or rather, the rate at which energy is carried to the detector.

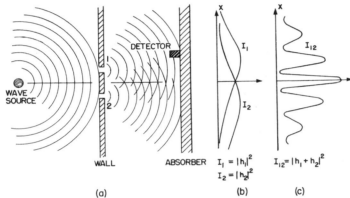

Fig. 1–2. Interference experiment with water waves.

With our wave apparatus, the first thing to notice is that the intensity can have *any* size. If the source just moves a very small amount, then there is just a little bit of wave motion at the detector. When there is more motion at the source, there is more intensity at the detector. The intensity of the wave can have any value at all. We would *not* say that there was any "lumpiness" in the wave intensity.

Now let us measure the wave intensity for various values of x (keeping the wave source operating always in the same way). We get the interesting-looking curve marked I_{12} in part (c) of the figure.

We have already worked out how such patterns can come about when we studied the interference of electric waves in Volume I. In this case we would observe that the original wave is diffracted at the holes, and new circular waves spread out from each hole. If we cover one hole at a time and measure the intensity distribution at the absorber we find the rather simple intensity curves shown in part (b) of the figure. I_1 is the intensity of the wave from hole 1 (which we find by measuring when hole 2 is blocked off) and I_2 is the intensity of the wave from hole 2 (seen when hole 1 is blocked).

The intensity I_{12} observed when both holes are open is certainly *not* the sum of I_1 and I_2. We say that there is "interference" of the two waves. At some places (where the curve I_{12} has its maxima) the waves are "in phase" and the wave peaks add together to give a large amplitude and, therefore, a large intensity. We say that the two waves are "interfering constructively" at such places. There will be such constructive interference wherever the distance from the detector to one hole is a whole number of wavelengths larger (or shorter) than the distance from the detector to the other hole.

At those places where the two waves arrive at the detector with a phase difference of π (where they are "out of phase") the resulting wave motion at the detector will be the difference of the two amplitudes. The waves "interfere destructively," and we get a low value for the wave intensity. We expect such low values wherever the distance between hole 1 and the detector is different from the distance between hole 2 and the detector by an odd number of half-wavelengths. The low values of I_{12} in Fig. 1–2 correspond to the places where the two waves interfere destructively.

You will remember that the quantitative relationship between I_1, I_2, and I_{12} can be expressed in the following way: The instantaneous height of the water wave at the detector for the wave from hole 1 can be written as (the real part of) $h_1 e^{i\omega t}$, where the "amplitude" h_1 is, in general, a complex number. The intensity is proportional to the mean squared height or, when we use the complex numbers, to the absolute value squared $|h_1|^2$. Similarly, for hole 2 the height is $h_2 e^{i\omega t}$ and the intensity is proportional to $|h_2|^2$. When both holes are open, the wave heights add to give the height $(h_1 + h_2)e^{i\omega t}$ and the intensity $|h_1 + h_2|^2$. Omitting the constant of proportionality for our present purposes, the proper relations for *interfering waves* are

$$I_1 = |h_1|^2, \qquad I_2 = |h_2|^2, \qquad I_{12} = |h_1 + h_2|^2. \tag{1.2}$$

You will notice that the result is quite different from that obtained with bullets (Eq. 1–1). If we expand $|h_1 + h_2|^2$ we see that

$$|h_1 + h_2|^2 = |h_1|^2 + |h_2|^2 + 2|h_1||h_2| \cos \delta, \tag{1.3}$$

where δ is the phase difference between h_1 and h_2. In terms of the intensities, we could write

$$I_{12} = I_1 + I_2 + 2\sqrt{I_1 I_2} \cos \delta. \tag{1.4}$$

The last term in (1.4) is the "interference term." So much for water waves. The intensity can have any value, and it shows interference.

1–4 An experiment with electrons

Now we imagine a similar experiment with electrons. It is shown diagrammatically in Fig. 1–3. We make an electron gun which consists of a tungsten wire heated by an electric current and surrounded by a metal box with a hole in it. If the wire is at a negative voltage with respect to the box, electrons emitted by the wire will be accelerated toward the walls and some will pass through the hole. All the electrons which come out of the gun will have (nearly) the same energy. In front of the gun is again a wall (just a thin metal plate) with two holes in it. Beyond the wall is another plate which will serve as a "backstop." In front of the backstop we place a movable detector. The detector might be a geiger counter or, perhaps better, an electron multiplier, which is connected to a loudspeaker.

We should say right away that you should not try to set up this experiment (as you could have done with the two we have already described). This experiment

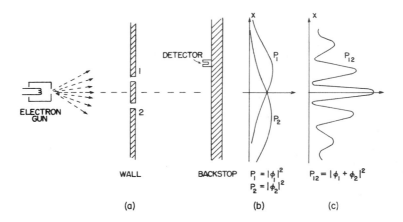

Fig. 1–3. Interference experiment with electrons.

has never been done in just this way. The trouble is that the apparatus would have to be made on an impossibly small scale to show the effects we are interested in. We are doing a "thought experiment," which we have chosen because it is easy to think about. We know the results that *would* be obtained because there *are* many experiments that have been done, in which the scale and the proportions have been chosen to show the effects we shall describe.

The first thing we notice with our electron experiment is that we hear sharp "clicks" from the detector (that is, from the loudspeaker). And all "clicks" are the same. There are *no* "half-clicks."

We would also notice that the "clicks" come very erratically. Something like: click . . . . . click-click . . . click . . . . click . . . . click-click . . . . . . click . . . , etc., just as you have, no doubt, heard a geiger counter operating. If we count the clicks which arrive in a sufficiently long time—say for many minutes—and then count again for another equal period, we find that the two numbers are very nearly the same. So we can speak of the *average rate* at which the clicks are heard (so-and-so-many clicks per minute on the average).

As we move the detector around, the *rate* at which the clicks appear is faster or slower, but the size (loudness) of each click is always the same. If we lower the temperature of the wire in the gun, the rate of clicking slows down, but still each click sounds the same. We would notice also that if we put two separate detectors at the backstop, one *or* the other would click, but never both at once. (Except that once in a while, if there were two clicks very close together in time, our ear might not sense the separation.) We conclude, therefore, that whatever arrives at the backstop arrives in "lumps." All the "lumps" are the same size: only whole "lumps" arrive, and they arrive one at a time at the backstop. We shall say: "Electrons always arrive in identical lumps."

Just as for our experiment with bullets, we can now proceed to find experimentally the answer to the question: "What is the relative probability that an electron 'lump' will arrive at the backstop at various distances x from the center?" As before, we obtain the relative probability by observing the rate of clicks, holding the operation of the gun constant. The probability that lumps will arrive at a particular x is proportional to the average rate of clicks at that x.

The result of our experiment is the interesting curve marked P_{12} in part (c) of Fig. 1–3. Yes! That is the way electrons go.

1–5 The interference of electron waves

Now let us try to analyze the curve of Fig. 1–3 to see whether we can understand the behavior of the electrons. The first thing we would say is that since they come in lumps, each lump, which we may as well call an electron, has come either through hole 1 or through hole 2. Let us write this in the form of a "Proposition":

Proposition A: Each electron *either* goes through hole 1 *or* it goes through hole 2.

Assuming Propositon A, all electrons that arrive at the backstop can be divided into two classes: (1) those that come through hole 1, and (2) those that come through hole 2. So our observed curve must be the sum of the effects of the electrons which come through hole 1 and the electrons which come through hole 2. Let us check this idea by experiment. First, we will make a measurement for those electrons that come through hole 1. We block off hole 2 and make our counts of the clicks from the detector. From the clicking rate, we get P_1. The result of the measurement is shown by the curve marked P_1 in part (b) of Fig. 1–3. The result seems quite reasonable. In a similar way, we measure P_2, the probability distribution for the electrons that come through hole 2. The result of this measurement is also drawn in the figure.

The result P_{12} obtained with *both* holes open is clearly not the sum of P_1 and P_2, the probabilities for each hole alone. In analogy with our water-wave experiment, we say: "There is interference."

$$\textit{For electrons:} \qquad P_{12} \neq P_1 + P_2. \qquad\qquad (1.5)$$

How can such an interference come about? Perhaps we should say: "Well, that means, presumably, that it is *not true* that the lumps go either through hole 1 or hole 2, because if they did, the probabilities should add. Perhaps they go in a more complicated way. They split in half and ..." But no! They cannot, they always arrive in lumps ... "Well, perhaps some of them go through 1, and then they go around through 2, and then around a few more times, or by some other complicated path ... then by closing hole 2, we changed the chance that an electron that *started out* through hole 1 would finally get to the backstop ..." But notice! There are some points at which very few electrons arrive when *both* holes are open, but which receive many electrons if we close one hole, so *closing* one hole *increased* the number from the other. Notice, however, that at the center of the pattern, P_{12} is more than twice as large as $P_1 + P_2$. It is as though closing one hole *decreased* the number of electrons which come through the other hole. It seems hard to explain *both* effects by proposing that the electrons travel in complicated paths.

It is all quite mysterious. And the more you look at it the more mysterious it seems. Many ideas have been concocted to try to explain the curve for P_{12} in terms of individual electrons going around in complicated ways through the holes. None of them has succeeded. None of them can get the right curve for P_{12} in terms of P_1 and P_2.

Yet, surprisingly enough, the *mathematics* for relating P_1 and P_2 to P_{12} is extremely simple. For P_{12} is just like the curve I_{12} of Fig. 1–2, and *that* was simple. What is going on at the backstop can be described by two complex numbers that we can call ϕ_1 and ϕ_2 (they are functions of x, of course). The absolute square of ϕ_1 gives the effect with only hole 1 open. That is, $P_1 = |\phi_1|^2$. The effect with only hole 2 open is given by ϕ_2 in the same way. That is, $P_2 = |\phi_2|^2$. And the combined effect of the two holes is just $P_{12} = |\phi_1 + \phi_2|^2$. The *mathematics* is the same as that we had for the water waves! (It is hard to see how one could get such a simple result from a complicated game of electrons going back and forth through the plate on some strange trajectory.)

We conclude the following: The electrons arrive in lumps, like particles, and the probability of arrival of these lumps is distributed like the distribution of intensity of a wave. It is in this sense that an electron behaves "sometimes like a particle and sometimes like a wave."

Incidentally, when we were dealing with classical waves we defined the intensity as the mean over time of the square of the wave amplitude, and we used complex numbers as a mathematical trick to simplify the analysis. But in quantum mechanics it turns out that the amplitudes *must* be represented by complex numbers. The real parts alone will not do. That is a technical point, for the moment, because the formulas look just the same.

Since the probability of arrival through both holes is given so simply, although it is not equal to $(P_1 + P_2)$, that is really all there is to say. But there are a large number of subtleties involved in the fact that nature does work this way. We would like to illustrate some of these subtleties for you now. First, since the number that arrives at a particular point is *not* equal to the number that arrives through 1 plus the number that arrives through 2, as we would have concluded from Proposition A, undoubtedly we should conclude that *Proposition A is false*. It is *not* true that the electrons go *either* through hole 1 or hole 2. But that conclusion can be tested by another experiment.

1–6 Watching the electrons

We shall now try the following experiment. To our electron apparatus we add a very strong light source, placed behind the wall and between the two holes, as shown in Fig. 1–4. We know that electric charges scatter light. So when an

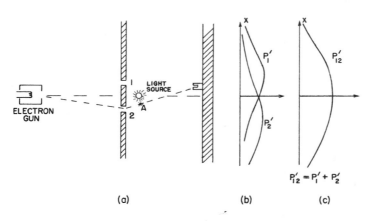

Fig. 1–4. A different electron experiment.

$$P'_{12} = P'_1 + P'_2$$

(a)   (b)   (c)

electron passes, however it does pass, on its way to the detector, it will scatter some light to our eye, and we can *see* where the electron goes. If, for instance, an electron were to take the path via hole 2 that is sketched in Fig. 1–4, we should see a flash of light coming from the vicinity of the place marked A in the figure. If an electron passes through hole 1, we would expect to see a flash from the vicinity of the upper hole. If it should happen that we get light from both places at the same time, because the electron divides in half . . . Let us just do the experiment!

Here is what we see: *every* time that we hear a "click" from our electron detector (at the backstop), we *also see* a flash of light *either* near hole 1 *or* near hole 2, but *never* both at once! And we observe the same result no matter where we put the detector. From this observation we conclude that when we look at the electrons we find that the electrons go either through one hole or the other. Experimentally, Proposition A is necessarily true.

What, then, is wrong with our argument *against* Proposition A? Why *isn't* P_{12} just equal to $P_1 + P_2$? Back to experiment! Let us keep track of the electrons and find out what they are doing. For each position (x-location) of the detector we will count the electrons that arrive and *also* keep track of which hole they went through, by watching for the flashes. We can keep track of things this way: whenever we hear a "click" we will put a count in Column 1 if we see the flash near hole 1, and if we see the flash near hole 2, we will record a count in Column 2. Every electron which arrives is recorded in one of two classes: those which come through 1 and those which come through 2. From the number recorded in Column 1 we get the probability P'_1 that an electron will arrive at the detector via hole 1; and from the number recorded in Column 2 we get P'_2, the probability that an electron will arrive at the detector via hole 2. If we now repeat such a measurement for many values of x, we get the curves for P'_1 and P'_2 shown in part (b) of Fig. 1–4.

Well, that is not too surprising! We get for P'_1 something quite similar to what we got before for P_1 by blocking off hole 2; and P'_2 is similar to what we got by blocking hole 1. So there is *not* any complicated business like going through both holes. When we watch them, the electrons come through just as we would expect them to come through. Whether the holes are closed or open, those which we see come through hole 1 are distributed in the same way whether hole 2 is open or closed.

But wait! What do we have *now* for the *total* probability, the probability that an electron will arrive at the detector by any route? We already have that information. We just pretend that we never looked at the light flashes, and we lump together the detector clicks which we have separated into the two columns. We *must* just *add* the numbers. For the probability that an electron will arrive at the backstop by passing through *either* hole, we do find $P'_{12} = P_1 + P_2$. That is, although we succeeded in watching which hole our electrons come through, we no longer get the old interference curve P_{12}, but a new one, P'_{12}, showing no interference! If we turn out the light P_{12} is restored.

We must conclude that *when we look at the electrons* the distribution of them on the screen is different than when we do not look. Perhaps it is turning on our light source that disturbs things? It must be that the electrons are very delicate, and the light, when it scatters off the electrons, gives them a jolt that changes their motion. We know that the electric field of the light acting on a charge will exert a force on it. So perhaps we *should* expect the motion to be changed. Anyway, the light exerts a big influence on the electrons. By trying to "watch" the electrons we have changed their motions. That is, the jolt given to the electron when the photon is scattered by it is such as to change the electron's motion enough so that if it *might* have gone to where P_{12} was at a maximum it will instead land where P_{12} was a minimum; that is why we no longer see the wavy interference effects.

You may be thinking: "Don't use such a bright source! Turn the brightness down! The light waves will then be weaker and will not disturb the electrons so much. Surely, by making the light dimmer and dimmer, eventually the wave will be weak enough that it will have a negligible effect." O.K. Let's try it. The first thing we observe is that the flashes of light scattered from the electrons as they pass by does *not* get weaker. *It is always the same-sized flash.* The only thing that happens as the light is made dimmer is that sometimes we hear a "click" from the detector but see *no flash at all*. The electron has gone by without being "seen." What we are observing is that light *also* acts like electrons, we *knew* that it was "wavy," but now we find that it is also "lumpy." It always arrives—or is scattered—in lumps that we call "photons." As we turn down the *intensity* of the light source we do not change the *size* of the photons, only the *rate* at which they are emitted. *That* explains why, when our source is dim, some electrons get by without being seen. There did not happen to be a photon around at the time the electron went through.

This is all a little discouraging. If it is true that whenever we "see" the electron we see the same-sized flash, then those electrons we see are *always* the disturbed ones. Let us try the experiment with a dim light anyway. Now whenever we hear a click in the detector we will keep a count in three columns: in Column (1) those electrons seen by hole 1, in Column (2) those electrons seen by hole 2, and in Column (3) those electrons not seen at all. When we work up our data (computing the probabilities) we find these results: Those "seen by hole 1" have a distribution like P'_1; those "seen by hole 2" have a distribution like P'_2 (so that those "seen by either hole 1 or 2" have a distribution like P'_{12}); and those "not seen at all" have a "wavy" distribution just like P_{12} of Fig. 1–3! *If the electrons are not seen, we have interference!*

That is understandable. When we do not see the electron, no photon disturbs it, and when we do see it, a photon has disturbed it. There is always the same amount of disturbance because the light photons all produce the same-sized effects and the effect of the photons being scattered is enough to smear out any interference effect.

Is there not *some* way we can see the electrons without disturbing them? We learned in an earlier chapter that the momentum carried by a "photon" is inversely proportional to its wavelength ($p = h/\lambda$). Certainly the jolt given to the electron when the photon is scattered toward our eye depends on the momentum that photon carries. Aha! If we want to disturb the electrons only slightly we should not have lowered the *intensity* of the light, we should have lowered its *frequency* (the same as increasing its wavelength). Let us use light of a redder color. We could even use infrared light, or radiowaves (like radar), and "see" where the electron went with the help of some equipment that can "see" light of these longer wavelengths. If we use "gentler" light perhaps we can avoid disturbing the electrons so much.

Let us try the experiment with longer waves. We shall keep repeating our experiment, each time with light of a longer wavelength. At first, nothing seems to change. The results are the same. Then a terrible thing happens. You remember that when we discussed the microscope we pointed out that, due to the *wave nature* of the light, there is a limitation on how close two spots can be and still be seen as two separate spots. This distance is of the order of the wavelength of light. So now, when we make the wavelength longer than the distance between our holes, we see a *big* fuzzy flash when the light is scattered by the electrons. We can no longer tell which hole the electron went through! We just know it went somewhere! And it is just with light of this color that we find that the jolts given to the electron

1–8

are small enough so that P'_{12} begins to look like P_{12}—that we begin to get some interference effect. And it is only for wavelengths much longer than the separation of the two holes (when we have no chance at all of telling where the electron went) that the disturbance due to the light gets sufficiently small that we again get the curve P_{12} shown in Fig. 1–3.

In our experiment we find that it is impossible to arrange the light in such a way that one can tell which hole the electron went through, and at the same time not disturb the pattern. It was suggested by Heisenberg that the then new laws of nature could only be consistent if there were some basic limitation on our experimental capabilities not previously recognized. He proposed, as a general principle, his *uncertainty principle*, which we can state in terms of our experiment as follows: "It is impossible to design an apparatus to determine which hole the electron passes through, that will not at the same time disturb the electrons enough to destroy the interference pattern." If an apparatus is capable of determining which hole the electron goes through, it *cannot* be so delicate that it does not disturb the pattern in an essential way. No one has ever found (or even thought of) a way around the uncertainty principle. So we must assume that it describes a basic characteristic of nature.

The complete theory of quantum mechanics which we now use to describe atoms and, in fact, all matter, depends on the correctness of the uncertainty principle. Since quantum mechanics is such a successful theory, our belief in the uncertainty principle is reinforced. But if a way to "beat" the uncertainty principle were ever discovered, quantum mechanics would give inconsistent results and would have to be discarded as a valid theory of nature.

"Well," you say, "what about Proposition A? Is it true, or is it *not* true, that the electron either goes through hole 1 or it goes through hole 2?" The only answer that can be given is that we have found from experiment that there is a certain special way that we have to think in order that we do not get into inconsistencies. What we must say (to avoid making wrong predictions) is the following. If one looks at the holes or, more accurately, if one has a piece of apparatus which is capable of determining whether the electrons go through hole 1 or hole 2, then one *can* say that it goes either through hole 1 or hole 2. *But*, when one does *not* try to tell which way the electron goes, when there is nothing in the experiment to disturb the electrons, then one may *not* say that an electron goes either through hole 1 or hole 2. If one does say that, and starts to make any deductions from the statement, he will make errors in the analysis. This is the logical tightrope on which we must walk if we wish to describe nature successfully.

If the motion of all matter—as well as electrons—must be described in terms of waves, what about the bullets in our first experiment? Why didn't we see an interference pattern there? It turns out that for the bullets the wavelengths were so tiny that the interference patterns became very fine. So fine, in fact, that with any detector of finite size one could not distinguish the separate maxima and minima. What we saw was only a kind of average, which is the classical curve. In Fig. 1–5 we have tried to indicate schematically what happens with large-scale objects. Part (a) of the figure shows the probability distribution one might predict for bullets, using quantum mechanics. The rapid wiggles are supposed to represent the interference pattern one gets for waves of very short wavelength. Any physical detector, however, straddles several wiggles of the probability curve, so that the measurements show the smooth curve drawn in part (b) of the figure.

1–7 First principles of quantum mechanics

We will now write a summary of the main conclusions of our experiments. We will, however, put the results in a form which makes them true for a general class of such experiments. We can write our summary more simply if we first define an "ideal experiment" as one in which there are no uncertain external influences, i.e., no jiggling or other things going on that we cannot take into ac-

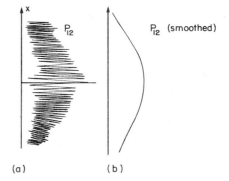

Fig. 1–5. Interference pattern with bullets: (a) actual (schematic), (b) observed.

count. We would be quite precise if we said: "An ideal experiment is one in which all of the initial and final conditions of the experiment are completely specified." What we will call "an event" is, in general, just a specific set of initial and final conditions. (For example: "an electron leaves the gun, arrives at the detector, and nothing else happens.") Now for our summary.

SUMMARY

(1) The probability of an event in an ideal experiment is given by the square of the absolute value of a complex number ϕ which is called the probability amplitude:

$$P = \text{probability},$$
$$\phi = \text{probability amplitude},$$
$$P = |\phi|^2. \tag{1.6}$$

(2) When an event can occur in several alternative ways, the probability amplitude for the event is the sum of the probability amplitudes for each way considered separately. There is interference:

$$\phi = \phi_1 + \phi_2,$$
$$P = |\phi_1 + \phi_2|^2. \tag{1.7}$$

(3) If an experiment is performed which is capable of determining whether one or another alternative is actually taken, the probability of the event is the sum of the probabilities for each alternative. The interference is lost:

$$P = P_1 + P_2. \tag{1.8}$$

One might still like to ask: "How does it work? What is the machinery behind the law?" No one has found any machinery behind the law. No one can "explain" any more than we have just "explained." No one will give you any deeper representation of the situation. We have no ideas about a more basic mechanism from which these results can be deduced.

We would like to emphasize a very important difference between classical and quantum mechanics. We have been talking about the probability that an electron will arrive in a given circumstance. We have implied that in our experimental arrangement (or even in the best possible one) it would be impossible to predict exactly what would happen. We can only predict the odds! This would mean, if it were true, that physics has given up on the problem of trying to predict exactly what will happen in a definite circumstance. Yes! physics *has* given up. *We do not know how to predict what would happen in a given circumstance,* and we believe now that it is impossible—that the only thing that can be predicted is the probability of different events. It must be recognized that this is a retrenchment in our earlier ideal of understanding nature. It may be a backward step, but no one has seen a way to avoid it.

We make now a few remarks on a suggestion that has sometimes been made to try to avoid the description we have given: "Perhaps the electron has some kind of internal works—some inner variables—that we do not yet know about. Perhaps that is why we cannot predict what will happen. If we could look more closely at the electron, we could be able to tell where it would end up." So far as we know, that is impossible. We would still be in difficulty. Suppose we were to assume that inside the electron there is some kind of machinery that determines where it is going to end up. That machine must *also* determine which hole it is going to go through on its way. But we must not forget that what is inside the electron should not be dependent on what *we* do, and in particular upon whether we open or close one of the holes. So if an electron, before it starts, has already made up its mind (a) which hole it is going to use, and (b) where it is going to land, we should find P_1 for those electrons that have chosen hole 1, P_2 for those that have chosen hole 2, *and necessarily* the sum $P_1 + P_2$ for those that arrive through the two holes. There seems to be no way around this. But we have verified experimentally that that is not the case. And no one has figured a way out of this puzzle. So at the present time we must limit ourselves to computing probabilities. We say "at the present time," but we suspect very strongly that it is something that will be with us forever—that it is impossible to beat that puzzle—that this is the way nature really *is*.

1–8 The uncertainty principle

This is the way Heisenberg stated the uncertainty principle originally: If you make the measurement on any object, and you can determine the *x*-component of its momentum with an uncertainty Δp, you cannot, at the same time, know its *x*-position more accurately than $\Delta x = h/\Delta p$, where h is a definite fixed number given by nature. It is called "Planck's constant," and is approximately 6.63×10^{-34} joule-seconds. The uncertainties in the position and momentum of a particle at any instant must have their product greater than Planck's constant. This is a special case of the uncertainty principle that was stated above more generally. The more general statement was that one cannot design equipment in any way to determine which of two alternatives is taken, without, at the same time, destroying the pattern of interference.

Let us show for one particular case that the kind of relation given by Heisenberg must be true in order to keep from getting into trouble. We imagine a modification of the experiment of Fig. 1–3, in which the wall with the holes consists of a plate mounted on rollers so that it can move freely up and down (in the *x*-direction), as shown in Fig. 1–6. By watching the motion of the plate carefully we can try to tell which hole an electron goes through. Imagine what happens when the detector is placed at $x = 0$. We would expect that an electron which passes through hole 1 must be deflected downward by the plate to reach the detector. Since the vertical component of the electron momentum is changed, the plate must recoil with an equal momentum in the opposite direction. The plate will get an upward kick. If the electron goes through the lower hole, the plate should feel a downward kick. It is clear that for every position of the detector, the momentum received by the plate will have a different value for a traversal via hole 1 than for a traversal via hole 2. So! Without disturbing the electrons *at all*, but just by watching the *plate*, we can tell which path the electron used.

Now in order to do this it is necessary to know what the momentum of the screen is, before the electron goes through. So when we measure the momentum after the electron goes by, we can figure out how much the plate's momentum has changed. But remember, according to the uncertainty principle we cannot at the same time know the position of the plate with an arbitrary accuracy. But if we do not know exactly *where* the plate is, we cannot say precisely where the two holes are. They will be in a different place for every electron that goes through. This means that the center of our interference pattern will have a different location for each electron. The wiggles of the interference pattern will be smeared out. We shall show quantitatively in the next chapter that if we determine the momentum of the plate sufficiently accurately to determine from the recoil measurement which hole was used, then the uncertainty in the *x*-position of the plate will, according to the uncertainty principle, be enough to shift the pattern observed at the detector up and down in the *x*-direction about the distance from a maximum to its nearest minimum. Such a random shift is just enough to smear out the pattern so that no interference is observed.

The uncertainty principle "protects" quantum mechanics. Heisenberg recognized that if it were possible to measure the momentum and the position simultaneously with a greater accuracy, the quantum mechanics would collapse. So he proposed that it must be impossible. Then people sat down and tried to figure out ways of doing it, and nobody could figure out a way to measure the position and the momentum of anything—a screen, an electron, a billiard ball, anything—with any greater accuracy. Quantum mechanics maintains its perilous but still correct existence.

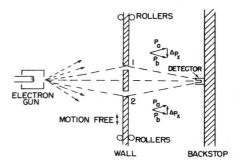

Fig. 1–6. An experiment in which the recoil of the wall is measured.

The Relation of Wave and Particle Viewpoints

2-1 Probability wave amplitudes

In this chapter we shall discuss the relationship of the wave and particle viewpoints. We already know, from the last chapter, that neither the wave viewpoint nor the particle viewpoint is correct. We would always like to present things accurately, or at least precisely enough that they will not have to be changed when we learn more—it may be extended, but it will not be changed! But when we try to talk about the wave picture or the particle picture, both are approximate, and both will change. Therefore what we learn in this chapter will not be accurate in a certain sense; we will deal with some half-intuitive arguments which will be made more precise later. But certain things will be changed a little bit when we interpret them correctly in quantum mechanics. We are doing this so that you can have some qualitative feeling for some quantum phenomena before we get into the mathematical details of quantum mechanics. Furthermore, all our experiences are with waves and with particles, and so it is rather handy to use the wave and particle ideas to get some understanding of what happens in given circumstances before we know the complete mathematics of the quantum-mechanical amplitudes. We shall try to indicate the weakest places as we go along, but most of it is very nearly correct—it is just a matter of interpretation.

First of all, we know that the new way of representing the world in quantum mechanics—the new framework—is to give an amplitude for every event that can occur, and if the event involves the reception of one particle, then we can give the amplitude to find that one particle at different places and at different times. The probability of finding the particle is then proportional to the absolute square of the amplitude. In general, the amplitude to find a particle in different places at different times varies with position and time.

In some special case it can be that the amplitude varies sinusoidally in space and time like $e^{i(\omega t - k \cdot r)}$, where r is the vector position from some origin. (Do not forget that these amplitudes are complex numbers, not real numbers.) Such an amplitude varies according to a definite frequency ω and wave number k. Then it turns out that this corresponds to a classical limiting situation where we would have believed that we have a particle whose energy E was known and is related to the frequency by

$$E = \hbar\omega, \tag{2.1}$$

and whose momentum p is also known and is related to the wave number by

$$p = \hbar k. \tag{2.2}$$

(The symbol $\hbar$ represents the number h divided by 2π; $\hbar = h/2\pi$.)

This means that the idea of a particle is limited. The idea of a particle—its location, its momentum, etc.—which we use so much, is in certain ways unsatisfactory. For instance, if an amplitude to find a particle at different places is given by $e^{i(\omega t - k \cdot r)}$, whose absolute square is a constant, that would mean that the probability of finding a particle is the same at all points. That means we do not know *where* it is—it can be anywhere—there is a great uncertainty in its location.

On the other hand, if the position of a particle is more or less well known and we can predict it fairly accurately, then the probability of finding it in different places must be confined to a certain region, whose length we call Δx. Outside this region, the probability is zero. Now this probability is the absolute square of an amplitude, and if the absolute square is zero, the amplitude is also zero, so that

2-1 **Probability wave amplitudes**

2-2 **Measurement of position and momentum**

2-3 **Crystal diffraction**

2-4 **The size of an atom**

2-5 **Energy levels**

2-6 **Philosophical implications**

Note: This chapter is almost exactly the same as Chapter 38 of Volume I.

Fig. 2-1. A wave packet of length Δx.

we have a wave train whose length is Δx (Fig. 2-1), and the wavelength (the distance between nodes of the waves in the train) of that wave train is what corresponds to the particle momentum.

Here we encounter a strange thing about waves; a very simple thing which has nothing to do with quantum mechanics strictly. It is something that anybody who works with waves, even if he knows no quantum mechanics, knows: namely, *we cannot define a unique wavelength for a short wave train.* Such a wave train does not *have* a definite wavelength; there is an indefiniteness in the wave number that is related to the finite length of the train, and thus there is an indefiniteness in the momentum.

2-2 Measurement of position and momentum

Let us consider two examples of this idea—to see the reason that there is an uncertainty in the position and/or the momentum, if quantum mechanics is right. We have also seen before that if there were not such a thing—if it were possible to measure the position and the momentum of anything simultaneously—we would have a paradox; it is fortunate that we do not have such a paradox, and the fact that such an uncertainty comes naturally from the wave picture shows that everything is mutually consistent.

Here is one example which shows the relationship between the position and the momentum in a circumstance that is easy to understand. Suppose we have a single slit, and particles are coming from very far away with a certain energy—so that they are all coming essentially horizontally (Fig. 2-2). We are going to concentrate on the vertical components of momentum. All of these particles have a certain horizontal momentum p_0, say, in a classical sense. So, in the classical sense, the vertical momentum p_y, before the particle goes through the hole, is definitely known. The particle is moving neither up nor down, because it came from a source that is far away—and so the vertical momentum is of course zero. But now let us suppose that it goes through a hole whose width is B. Then after it has come out through the hole, we know the position vertically—the y-position—with considerable accuracy—namely $\pm B$.† That is, the uncertainty in position, Δy, is of order B. Now we might also want to say, since we known the momentum is absolutely horizontal, that Δp_y is zero; but that is wrong. We *once* knew the momentum was horizontal, but we do not know it any more. Before the particles passed through the hole, we did not know their vertical positions. Now that we have found the vertical position by having the particle come through the hole, we have lost our information on the vertical momentum! Why? According to the wave theory, there is a spreading out, or diffraction, of the waves after they go through the slit, just as for light. Therefore there is a certain probability that particles coming out of the slit are not coming exactly straight. The pattern is spread out by the diffraction effect, and the angle of spread, which we can define as the angle of the first minimum, is a measure of the uncertainty in the final angle.

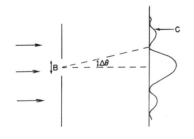

Fig. 2-2. Diffraction of particles passing through a slit.

How does the pattern become spread? To say it is spread means that there is some chance for the particle to be moving up or down, that is, to have a component of momentum up or down. We say *chance* and *particle* because we can detect this diffraction pattern with a particle counter, and when the counter receives the particle, say at C in Fig. 2-2, it receives the *entire* particle, so that, in a classical sense, the particle has a vertical momentum, in order to get from the slit up to C.

To get a rough idea of the spread of the momentum, the vertical momentum p_y has a spread which is equal to $p_0 \Delta\theta$, where p_0 is the horizontal momentum. And how big is $\Delta\theta$ in the spread-out pattern? We know that the first minimum occurs at an angle $\Delta\theta$ such that the waves from one edge of the slit have to travel one wavelength farther than the waves from the other side—we worked that out before (Chapter 30 of Vol. I). Therefore $\Delta\theta$ is λ/B, and so Δp_y in this experiment is $p_0\lambda/B$. Note that if we make B smaller and make a more accurate measurement

---

† More precisely, the error in our knowledge of y is $\pm B/2$. But we are now only interested in the general idea, so we won't worry about factors of 2.

of the position of the particle, the diffraction pattern gets wider. So the narrower we make the slit, the wider the pattern gets, and the more is the likelihood that we would find that the particle has sidewise momentum. Thus the uncertainty in the vertical momentum is inversely proportional to the uncertainty of y. In fact, we see that the product of the two is equal to $p_0\lambda$. But λ is the wavelength and p_0 is the momentum, and in accordance with quantum mechanics, the wavelength times the momentum is Planck's constant h. So we obtain the rule that the uncertainties in the vertical momentum and in the vertical position have a product of the order h:

$$\Delta y \, \Delta p_y \approx h. \tag{2.3}$$

We cannot prepare a system in which we know the vertical position of a particle and can predict how it will move vertically with greater certainty than given by (2.3). That is, the uncertainty in the vertical momentum must exceed $h/\Delta y$, where Δy is the uncertainty in our knowledge of the position.

Sometimes people say quantum mechanics is all wrong. When the particle arrived from the left, its vertical momentum was zero. And now that it has gone through the slit, its position is known. Both position and momentum seem to be known with arbitrary accuracy. It is quite true that we can receive a particle, and on reception determine what its position is and what its momentum would have had to have been to have gotten there. That is true, but that is not what the uncertainty relation (2.3) refers to. Equation (2.3) refers to the *predictability* of a situation, not remarks about the *past*. It does no good to say "I knew what the momentum was before it went through the slit, and now I know the position," because now the momentum knowledge is lost. The fact that it went through the slit no longer permits us to predict the vertical momentum. We are talking about a predictive theory, not just measurements after the fact. So we must talk about what we can predict.

Now let us take the thing the other way around. Let us take another example of the same phenomenon, a little more quantitatively. In the previous example we measured the momentum by a classical method. Namely, we considered the direction and the velocity and the angles, etc., so we got the momentum by classical analysis. But since momentum is related to wave number, there exists in nature still another way to measure the momentum of a particle—photon or otherwise—which has no classical analog, because it uses Eq. (2.2). We measure the *wavelengths of the waves*. Let us try to measure momentum in this way.

Suppose we have a grating with a large number of lines (Fig. 2–3), and send a beam of particles at the grating. We have often discussed this problem: if the particles have a definite momentum, then we get a very sharp pattern in a certain direction, because of the interference. And we have also talked about how accurately we can determine that momentum, that is to say, what the resolving power of such a grating is. Rather than derive it again, we refer to Chapter 30 of Volume I, where we found that the relative uncertainty in the wavelength that can be measured with a given grating is $1/Nm$, where N is the number of lines on the grating and m is the order of the diffraction pattern. That is,

$$\Delta\lambda/\lambda = 1/Nm. \tag{2.4}$$

Now formula (2.4) can be rewritten as

$$\Delta\lambda/\lambda^2 = 1/Nm\lambda = 1/L, \tag{2.5}$$

where L is the distance shown in Fig. 2–3. This distance is the difference between the total distance that the particle or wave or whatever it is has to travel if it is reflected from the bottom of the grating, and the distance that it has to travel if it is reflected from the top of the grating. That is, the waves which form the diffraction pattern are waves which come from different parts of the grating. The first ones that arrive come from the bottom end of the grating, from the beginning of the wave train, and the rest of them come from later parts of the wave train, coming from different parts of the grating, until the last one finally arrives, and that involves a point in the wave train a distance L behind the first point. So in order that we

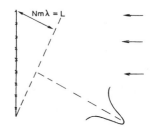

Fig. 2–3. Determination of momentum by using a diffraction grating.

shall have a sharp line in our spectrum corresponding to a definite momentum, with an uncertainty given by (2.4), we have to have a wave train of at least length L. If the wave train is too short, we are not using the entire grating. The waves which form the spectrum are being reflected from only a very short sector of the grating if the wave train is too short, and the grating will not work right—we will find a big angular spread. In order to get a narrower one, we need to use the whole grating, so that at least at some moment the whole wave train is scattering simultaneously from all parts of the grating. Thus the wave train must be of length L in order to have an uncertainty in the wavelength less than that given by (2.5). Incidentally,

$$\Delta\lambda/\lambda^2 = \Delta(1/\lambda) = \Delta k/2\pi. \qquad (2.6)$$

Therefore

$$\Delta k = 2\pi/L, \qquad (2.7)$$

where L is the length of the wave train.

This means that if we have a wave train whose length is less than L, the uncertainty in the wave number must exceed $2\pi/L$. Or the uncertainty in a wave number times the length of the wave train—we will call that for a moment Δx—exceeds 2π. We call it Δx because that is the uncertainty in the location of the particle. If the wave train exists only in a finite length, then that is where we could find the particle, within an uncertainty Δx. Now this property of waves, that the length of the wave train times the uncertainty of the wave number associated with it is at least 2π, is a property that is known to everyone who studies them. It has nothing to do with quantum mechanics. It is simply that if we have a finite train, we cannot count the waves in it very precisely.

Let us try another way to see the reason for that. Suppose that we have a finite train of length L; then because of the way it has to decrease at the ends, as in Fig. 2–1, the number of waves in the length L is uncertain by something like ± 1. But the number of waves in L is $kL/2\pi$. Thus k is uncertain, and we again get the result (2.7), a property merely of waves. The same thing works whether the waves are in space and k is the number of radians per centimeter and L is the length of the train, or the waves are in time and ω is the number of oscillations per second and T is the "length" in time that the wave train comes in. That is, if we have a wave train lasting only for a certain finite time T, then the uncertainty in the frequency is given by

$$\Delta\omega = 2\pi/T. \qquad (2.8)$$

We have tried to emphasize that these are properties of waves alone, and they are well known, for example, in the theory of sound.

The point is that in quantum mechanics we interpret the wave number as being a measure of the momentum of a particle, with the rule that $p = \hbar k$, so that relation (2.7) tells us that $\Delta p \approx h/\Delta x$. This, then, is a limitation of the classical idea of momentum. (Naturally, it has to be limited in some ways if we are going to represent particles by waves!) It is nice that we have found a rule that gives us some idea of when there is a failure of classical ideas.

2–3 Crystal diffraction

Next let us consider the reflection of particle waves from a crystal. A crystal is a thick thing which has a whole lot of similar atoms—we will include some complications later—in a nice array. The question is how to set the array so that we get a strong reflected maximum in a given direction for a given beam of, say, light (x-rays), electrons, neutrons, or anything else. In order to obtain a strong reflection, the scattering from all of the atoms must be in phase. There cannot be equal numbers in phase and out of phase, or the waves will cancel out. The way to arrange things is to find the regions of constant phase, as we have already explained; they are planes which make equal angles with the initial and final directions (Fig. 2–4).

If we consider two parallel planes, as in Fig. 2–4, the waves scattered from the two planes will be in phase, provided the difference in distance traveled by a wave

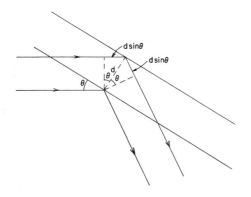

Fig. 2–4. Scattering of waves by crystal planes.

front is an integral number of wavelengths. This difference can be seen to be $2d \sin \theta$, where d is the perpendicular distance between the planes. Thus the condition for coherent reflection is

$$2d \sin \theta = n\lambda \qquad (n = 1, 2, \ldots). \qquad (2.9)$$

If, for example, the crystal is such that the atoms happen to lie on planes obeying condition (2.9) with $n = 1$, then there will be a strong reflection. If, on the other hand, there are other atoms of the same nature (equal in density) halfway between, then the intermediate planes will also scatter equally strongly and will interfere with the others and produce no effect. So d in (2.9) must refer to *adjacent* planes; we cannot take a plane five layers farther back and use this formula!

As a matter of interest, actual crystals are not usually as simple as a single kind of atom repeated in a certain way. Instead, if we make a two-dimensional analog, they are much like wallpaper, in which there is some kind of figure which repeats all over the wallpaper. By "figure" we mean, in the case of atoms, some arrangement—calcium and a carbon and three oxygens, etc., for calcium carbonate, and so on—which may involve a relatively large number of atoms. But whatever it is, the figure is repeated in a pattern. This basic figure is called a *unit cell*.

The basic pattern of repetition defines what we call the *lattice type*; the lattice type can be immediately determined by looking at the reflections and seeing what their symmetry is. In other words, where we find any reflections *at all* determines the lattice type, but in order to determine what is in each of the elements of the lattice one must take into account the *intensity* of the scattering at the various directions. *Which* directions scatter depends on the type of lattice, but *how strongly* each scatters is determined by what is inside each unit cell, and in that way the structure of crystals is worked out.

Two photographs of x-ray diffraction patterns are shown in Figs. 2–5 and 2–6; they illustrate scattering from rock salt and myoglobin, respectively.

Incidentally, an interesting thing happens if the spacings of the nearest planes are less than $\lambda/2$. In this case (2.9) has no solution for n. Thus if λ is bigger than twice the distance between adjacent planes, then there is no side diffraction pattern, and the light—or whatever it is—will go right through the material without bouncing off or getting lost. So in the case of light, where λ is much bigger than the spacing, of course it does go through and there is no pattern of reflection from the planes of the crystal.

This fact also has an interesting consequence in the case of piles which make neutrons (these are obviously particles, for anybody's money!). If we take these neutrons and let them into a long block of graphite, the neutrons diffuse and work their way along (Fig. 2–7). They diffuse because they are bounced by the atoms, but strictly, in the wave theory, they are bounced by the atoms because of diffraction from the crystal planes. It turns out that if we take a very long piece of graphite, the neutrons that come out the far end are all of long wavelength! In fact, if one plots the intensity as a function of wavelength, we get nothing except for wavelengths longer than a certain minimum (Fig. 2–8). In other words, we can get very slow neutrons that way. Only the slowest neutrons come through; they are not diffracted or scattered by the crystal planes of the graphite, but keep going right through like light through glass, and are not scattered out the sides. There are many other demonstrations of the reality of neutron waves and waves of other particles.

2–4 The size of an atom

We now consider another application of the uncertainty relation, Eq. (2.3). It must not be taken too seriously; the idea is right but the analysis is not very accurate. The idea has to do with the determination of the size of atoms, and the fact that, classically, the electrons would radiate light and spiral in until they settle down right on top of the nucleus. But that cannot be right quantum-mechanically because then we would know where each electron was and how fast it was moving.

Fig. 2–5. The pattern produced by the diffraction of a beam of x-rays in a crystal of sodium chloride.

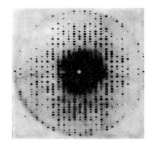

Fig. 2–6. The x-ray diffraction pattern of myoglobin.

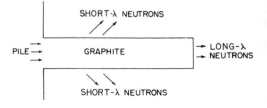

Fig. 2–7. Diffusion of pile neutrons through graphite block.

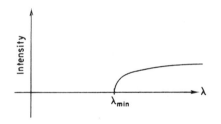

Fig. 2–8. Intensity of neutrons out of graphite rod as function of wavelength.

Suppose we have a hydrogen atom, and measure the position of the electron; we must not be able to predict exactly where the electron will be, or the momentum spread will then turn out to be infinite. Every time we look at the electron, it is somewhere, but it has an amplitude to be in different places so there is a probability of it being found in different places. These places cannot all be at the nucleus; we shall suppose there is a spread in position of order a. That is, the distance of the electron from the nucleus is usually about a. We shall determine a by minimizing the total energy of the atom.

The spread in momentum is roughly h/a because of the uncertainty relation, so that if we try to measure the momentum of the electron in some manner, such as by scattering x-rays off it and looking for the Doppler effect from a moving scatterer, we would expect not to get zero every time—the electron is not standing still—but the momenta must be of the order $p \approx h/a$. Then the kinetic energy is roughly $\frac{1}{2}mv^2 = p^2/2m = h^2/2ma^2$. (In a sense, this is a kind of dimensional analysis to find out in what way the kinetic energy depends upon Planck's constant, upon m, and upon the size of the atom. We need not trust our answer to within factors like 2, π, etc. We have not even defined a very precisely.) Now the potential energy is minus e^2 over the distance from the center, say $-e^2/a$, where, as defined in Volume I, e^2 is the charge of an electron squared, divided by $4\pi\epsilon_0$. Now the point is that the potential energy is reduced if a gets smaller, but the smaller a is, the higher the momentum required, because of the uncertainty principle, and therefore the higher the kinetic energy. The total energy is

$$E = h^2/2ma^2 - e^2/a. \tag{2.10}$$

We do not know what a is, but we know that the atom is going to arrange itself to make some kind of compromise so that the energy is as little as possible. In order to minimize E, we differentiate with respect to a, set the derivative equal to zero, and solve for a. The derivative of E is

$$dE/da = -h^2/ma^3 + e^2/a^2, \tag{2.11}$$

and setting $dE/da = 0$ gives for a the value

$$a_0 = h^2/me^2 = 0.528 \text{ angstrom}$$
$$= 0.528 \times 10^{-10} \text{ meter.} \tag{2.12}$$

This particular distance is called the *Bohr radius*, and we have thus learned that atomic dimensions are of the order of angstroms, which is right. This is pretty good—in fact, it is amazing, since until now we have had no basis for understanding the size of atoms! Atoms are completely impossible from the classical point of view, since the electrons would spiral into the nucleus.

Now if we put the value (2.12) for a_0 into (2.10) to find the energy, it comes out

$$E_0 = -e^2/2a_0 = -me^4/2h^2 = -13.6 \text{ ev.} \tag{2.13}$$

What does a negative energy mean? It means that the electron has less energy when it is in the atom than when it is free. It means it is bound. It means it takes energy to kick the electron out; it takes energy of the order of 13.6 ev to ionize a hydrogen atom. We have no reason to think that it is not two or three times this— or half of this—or $(1/\pi)$ times this, because we have used such a sloppy argument. However, we have cheated, we have used all the constants in such a way that it happens to come out the right number! This number, 13.6 electron volts, is called a Rydberg of energy; it is the ionization energy of hydrogen.

So we now understand why we do not fall through the floor. As we walk, our shoes with their masses of atoms push against the floor with *its* mass of atoms. In order to squash the atoms closer together, the electrons would be confined to a smaller space and, by the uncertainty principle, their momenta would have to be higher on the average, and that means high energy; the resistance to atomic compression is a quantum-mechanical effect and not a classical effect. Classically, we would expect that if we were to draw all the electrons and protons closer together, the energy would be reduced still further, and the best arrangement of positive and negative charges in classical physics is all on top of each other. This was well known in classical physics and was a puzzle because of the existence of the atom. Of course, the early scientists invented some ways out of the trouble—but never mind, we have the *right* way out, now!

Incidentally, although we have no reason to understand it at the moment, in a situation where there are many electrons it turns out that they try to keep away from each other. If one electron is occupying a certain space, then another does not occupy the same space. More precisely, there are two spin cases, so that two can sit on top of each other, one spinning one way and one the other way. But after that we cannot put any more there. We have to put others in another place, and that is the real reason that matter has strength. If we could put all the electrons in the same place, it would condense even more than it does. It is the fact that the electrons cannot all get on top of each other that makes tables and everything else solid.

Obviously, in order to understand the properties of matter, we will have to use quantum mechanics and not be satisfied with classical mechanics.

2–5 Energy levels

We have talked about the atom in its lowest possible energy condition, but it turns out that the electron can do other things. It can jiggle and wiggle in a more energetic manner, and so there are many different possible motions for the atom. According to quantum mechanics, in a stationary condition there can only be definite energies for an atom. We make a diagram (Fig. 2–9) in which we plot the energy vertically, and we make a horizontal line for each allowed value of the energy. When the electron is free, i.e., when its energy is positive, it can have any energy; it can be moving at any speed. But bound energies are not arbitrary. The atom must have one or another out of a set of allowed values, such as those in Fig. 2–9.

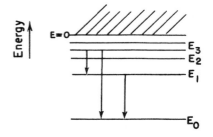

Fig. 2–9. Energy diagram for an atom, showing several possible transitions.

Now let us call the allowed values of the energy E_0, E_1, E_2, E_3. If an atom is initially in one of these "excited states," E_1, E_2, *etc.*, it does not remain in that state forever. Sooner or later it drops to a lower state and radiates energy in the form of light. The frequency of the light that is emitted is determined by conservation of energy plus the quantum-mechanical understanding that the frequency of the light is related to the energy of the light by (2.1). Therefore the frequency of the light which is liberated in a transition from energy E_3 to energy E_1 (for example) is

$$\omega_{31} = (E_3 - E_1)/\hbar. \qquad (2.14)$$

This, then, is a characteristic frequency of the atom and defines a spectral emission line. Another possible transition would be from E_3 to E_0. That would have a different frequency

$$\omega_{30} = (E_3 - E_0)/\hbar. \qquad (2.15)$$

Another possibility is that if the atom were excited to the state E_1 it could drop to the ground state E_0, emitting a photon of frequency

$$\omega_{10} = (E_1 - E_0)/\hbar. \qquad (2.16)$$

The reason we bring up three transitions is to point out an interesting relationship. It is easy to see from (2.14), (2.15), and (2.16) that

$$\omega_{30} = \omega_{31} + \omega_{10}. \qquad (2.17)$$

In general, if we find two spectral lines, we shall expect to find another line at the sum of the frequencies (or the difference in the frequencies), and that all the lines can be understood by finding a series of levels such that every line corresponds to the difference in energy of some pair of levels. This remarkable coincidence in spectral frequencies was noted before quantum mechanics was discovered, and it is called the *Ritz combination principle*. This is again a mystery from the point of view of classical mechanics. Let us not belabor the point that classical mechanics is a failure in the atomic domain; we seem to have demonstrated that pretty well.

We have already talked about quantum mechanics as being represented by amplitudes which behave like waves, with certain frequencies and wave numbers. Let us observe how it comes about from the point of view of amplitudes that the atom has definite energy states. This is something we cannot understand from what has been said so far, but we are all familiar with the fact that confined waves have definite frequencies. For instance, if sound is confined to an organ pipe, or anything like that, then there is more than one way that the sound can vibrate, but for each such way there is a definite frequency. Thus an object in which the waves are confined has certain resonance frequencies. It is therefore a property of waves in a confined space—a subject which we will discuss in detail with formulas later on—that they exist only at definite frequencies. And since the general relation exists between frequencies of the amplitude and energy, we are not surprised to find definite energies associated with electrons bound in atoms.

2–6 Philosophical implications

Let us consider briefly some philosophical implications of quantum mechanics. As always, there are two aspects of the problem: one is the philosophical implication for physics, and the other is the extrapolation of philosophical matters to other fields. When philosophical ideas associated with science are dragged into another field, they are usually completely distorted. Therefore we shall confine our remarks as much as possible to physics itself.

First of all, the most interesting aspect is the idea of the uncertainty principle; making an observation affects the phenomenon. It has always been known that making observations affects a phenomenon, but the point is that the effect cannot be disregarded or minimized or decreased arbitrarily by rearranging the apparatus. When we look for a certain phenomenon we cannot help but disturb it in a certain minimum way, and *the disturbance is necessary for the consistency of the viewpoint*. The observer was sometimes important in prequantum physics, but only in a trivial sense. The problem has been raised: if a tree falls in a forest and there is nobody there to hear it, does it make a noise? A *real* tree falling in a *real* forest makes a sound, of course, even if nobody is there. Even if no one is present to hear it, there are other traces left. The sound will shake some leaves, and if we were careful enough we might find somewhere that some thorn had rubbed against a leaf and made a tiny scratch that could not be explained unless we assumed the leaf were vibrating. So in a certain sense we would have to admit that there is a sound made. We might ask: was there a *sensation* of sound? No, sensations have to do, presumably, with consciousness. And whether ants are conscious and whether there were ants in the forest, or whether the tree was conscious, we do not know. Let us leave the problem in that form.

Another thing that people have emphasized since quantum mechanics was developed is the idea that we should not speak about those things which we cannot measure. (Actually relativity theory also said this.) Unless a thing can be defined by measurement, it has no place in a theory. And since an accurate value of the momentum of a localized particle cannot be defined by measurement it therefore has no place in the theory. The idea that this is what was the matter with classical theory *is a false position*. It is a careless analysis of the situation. Just because we cannot *measure* position and momentum precisely does not *a priori* mean that we *cannot* talk about them. It only means that we *need* not talk about them. The situation in the sciences is this: A concept or an idea which cannot be measured or cannot be referred directly to experiment may or may not be useful. It need not exist in a theory. In other words, suppose we compare the classical theory of the world with the quantum theory of the world, and suppose that it is true experimentally that we can measure position and momentum only imprecisely. The question is whether the *ideas* of the exact position of a particle and the exact momentum of a particle are valid or not. The classical theory admits the ideas; the quantum theory does not. This does not in itself mean that classical physics is wrong. When the new quantum mechanics was discovered, the classical people—which included everybody except Heisenberg, Schrödinger, and Born—said: "Look, your theory is not any good because you cannot answer certain questions like: what is the exact position of a particle?, which hole does it go through?, and some others." Heisenberg's answer was: "I do not need to answer such questions because you cannot ask such a question experimentally." It is that we do not *have* to. Consider two theories (a) and (b); (a) contains an idea that cannot be checked directly but which is used in the analysis, and the other, (b), does not contain the idea. If they disagree in their predictions, one could not claim that (b) is false because it cannot explain this idea that is in (a), because that idea is one of the things that cannot be checked directly. It is always good to know which ideas cannot be checked directly, but it is not necessary to remove them all. It is not true that we can pursue science completely by using only those concepts which are directly subject to experiment.

In quantum mechanics itself there is a probability amplitude, there is a potential, and there are many constructs that we cannot measure directly. The basis of a science is its ability to *predict*. To predict means to tell what will happen in an experiment that has never been done. How can we do that? By assuming that we know what is there, independent of the experiment. We must extrapolate the experiments to a region where they have not been done. We must take our concepts and extend them to places where they have not yet been checked. If we do not do that, we have no prediction. So it was perfectly sensible for the classical physicists to go happily along and suppose that the position—which obviously means something for a baseball—meant something also for an electron. It was not stupidity. It was a sensible procedure. Today we say that the law of relativity is supposed to be true at all energies, but someday somebody may come along and say how stupid we were. We do not know where we are "stupid" until we "stick our neck out," and so the whole idea is to put our neck out. And the only way to find out that we are wrong is to find out *what* our predictions are. It is absolutely necessary to make constructs.

We have already made a few remarks about the indeterminacy of quantum mechanics. That is, that we are unable now to predict what will happen in physics in a given physical circumstance which is arranged as carefully as possible. If we have an atom that is in an excited state and so is going to emit a photon, we cannot say *when* it will emit the photon. It has a certain amplitude to emit the photon at any time, and we can predict only a probability for emission; we cannot predict the future exactly. This has given rise to all kinds of nonsense and questions on the meaning of freedom of will, and of the idea that the world is uncertain.

Of course we must emphasize that classical physics is also indeterminate, in a sense. It is usually thought that this indeterminacy, that we cannot predict the future, is an important quantum-mechanical thing, and this is said to explain the behavior of the mind, feelings of free will, etc. But if the world *were* classical—if the laws of mechanics were classical—it is not quite obvious that the mind would not feel more or less the same. It is true classically that if we knew the position and the velocity of every particle in the world, or in a box of gas, we could predict exactly what would happen. And therefore the classical world is deterministic. Suppose, however, that we have a finite accuracy and do not know *exactly* where just one atom is, say to one part in a billion. Then as it goes along it hits another atom, and because we did not know the position better than to one part in a billion, we find an even larger error in the position after the collision. And that is amplified, of course, in the next collision, so that if we start with only a tiny error it rapidly magnifies to a very great uncertainty. To give an example: if water falls over a dam, it splashes. If we stand nearby, every now and then a drop will land on our nose. This appears to be completely random, yet such a behavior would be predicted by purely classical laws. The exact position of all the drops depends upon the precise wigglings of the water before it goes over the dam. How? The tiniest irregularities are magnified in falling, so that we get complete randomness. Obviously, we cannot really predict the position of the drops unless we know the motion of the water *absolutely exactly*.

Speaking more precisely, given an arbitrary accuracy, no matter how precise, one can find a time long enough that we cannot make predictions valid for that long a time. Now the point is that this length of time is not very large. It is not that the time is millions of years if the accuracy is one part in a billion. The time goes, in fact, only logarithmically with the error, and it turns out that in only a very, very tiny time we lose all our information. If the accuracy is taken to be one part in billions and billions and billions—no matter how many billions we wish, provided we do stop somewhere—then we can find a time less than the time it took to state the accuracy—after which we can no longer predict what is going to happen! It is therefore not fair to say that from the apparent freedom and indeterminacy of the human mind, we should have realized that classical "deterministic" physics could not ever hope to understand it, and to welcome quantum mechanics as a release from a "completely mechanistic" universe. For already in classical mechanics there was indeterminability from a practical point of view.

3

Probability Amplitudes

3-1 The laws for combining amplitudes

When Schrödinger first discovered the correct laws of quantum mechanics, he wrote an equation which described the amplitude to find a particle in various places. This equation was very similar to the equations that were already known to classical physicists—equations that they had used in describing the motion of air in a sound wave, the transmission of light, and so on. So most of the time at the beginning of quantum mechanics was spent in solving this equation. But at the same time an understanding was being developed, particularly by Born and Dirac, of the basically new physical ideas behind quantum mechanics. As quantum mechanics developed further, it turned out that there were a large number of things which were not directly encompassed in the Schrödinger equation—such as the spin of the electron, and various relativistic phenomena. Traditionally, all courses in quantum mechanics have begun in the same way, retracing the path followed in the historical development of the subject. One first learns a great deal about classical mechanics so that he will be able to understand how to solve the Schrödinger equation. Then he spends a long time working out various solutions. Only after a detailed study of this equation does he get to the "advanced" subject of the electron's spin.

We had also originally considered that the right way to conclude these lectures on physics was to show how to solve the equations of classical physics in complicated situations—such as the description of sound waves in enclosed regions, modes of electromagnetic radiation in cylindrical cavities, and so on. That was the original plan for this course. However, we have decided to abandon that plan and to give instead an introduction to the quantum mechanics. We have come to the conclusion that what are usually called the advanced parts of quantum mechanics are, in fact, quite simple. The mathematics that is involved is particularly simple, involving simple algebraic operations and no differential equations or at most only very simple ones. The only problem is that we must jump the gap of no longer being able to describe the behavior *in detail* of particles in space. So this is what we are going to try to do: to tell you about what conventionally would be called the "advanced" parts of quantum mechanics. But they are, we assure you, by all odds the simplest parts—in a deep sense of the word—as well as the most basic parts. This is frankly a pedagogical experiment; it has never been done before, as far as we know.

In this subject we have, of course, the difficulty that the quantum mechanical behavior of things is quite strange. Nobody has an everyday experience to lean on to get a rough, intuitive idea of what will happen. So there are two ways of presenting the subject: We could either describe what can happen in a rather rough physical way, telling you more or less what happens without giving the precise laws of everything; or we could, on the other hand, give the precise laws in their abstract form. But, then because of the abstractions, you wouldn't know what they were all about, physically. The latter method is unsatisfactory because it is completely abstract, and the first way leaves an uncomfortable feeling because one doesn't know exactly what is true and what is false. We are not sure how to overcome this difficulty. You will notice, in fact, that Chapters 1 and 2 showed this problem. The first chapter was relatively precise; but the second chapter was a rough description of the characteristics of different phenomena. Here, we will try to find a happy medium between the two extremes.

3-1 The laws for combining amplitudes
3-2 The two-slit interference pattern
3-3 Scattering from a crystal
3-4 Identical particles

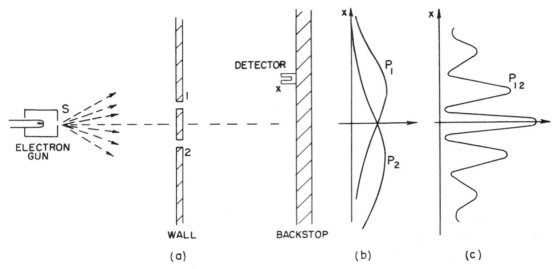

Fig 3–1. Interference experiment with electrons.

We will begin in this chapter by dealing with some general quantum mechanical ideas. Some of the statements will be quite precise, others only partially precise. It will be hard to tell you as we go along which is which, but by the time you have finished the rest of the book, you will understand in looking back which parts hold up and which parts were only explained roughly. The chapters which follow this one will not be so imprecise. In fact, one of the reasons we have tried carefully to be precise in the succeeding chapters is so that we can show you one of the most beautiful things about quantum mechanics—how much can be deduced from so little.

We begin by discussing again the superposition of *probability amplitudes*. As an example we will refer to the experiment described in Chapter 1, and shown again here in Fig. 3–1. There is a source *s* of particles, say electrons; then there is a wall with two slits in it; after the wall, there is a detector located at some position *x*. We ask for the probability that a particle will be found at *x*. Our *first general principle* in quantum mechanics is that the *probability* that a particle will arrive at *x*, when let out at the source *s*, can be represented quantitatively by the absolute square of a complex number called a *probability amplitude*—in this case, the "amplitude that a particle from *s* will arrive at *x*." We will use such amplitudes so frequently that we will use a shorthand notation—invented by Dirac and generally used in quantum mechanics—to represent this idea. We write the probability amplitude this way:

$$\langle \text{Particle arrives at } x \mid \text{particle leaves } s \rangle. \tag{3.1}$$

In other words, the two brackets $\langle \ \rangle$ are a sign equivalent to "the amplitude that"; the expression at the *right* of the vertical line always gives the *starting* condition, and the one at the left, the *final* condition. Sometimes it will also be convenient to abbreviate still more and describe the initial and final conditions by single letters. For example, we may on occasion write the amplitude (3.1) as

$$\langle x \mid s \rangle. \tag{3.2}$$

We want to emphasize that such an amplitude is, of course, just a single number— a *complex* number.

We have already seen in the discussion of Chapter 1 that when there are two ways for the particle to reach the detector, the resulting probability is not the sum of the two probabilities, but must be written as the absolute square of the sum of two amplitudes. We had that the probability that an electron arrives at the detector when both paths are open is

$$P_{12} = |\phi_1 + \phi_2|^2. \tag{3.3}$$

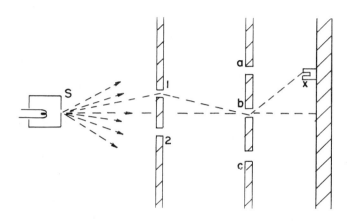

Fig. 3–2. A more complicated inter-ference experiment.

We wish now to put this result in terms of our new notation. First, however, we want to state our *second general principle* of quantum mechanics: When a particle can reach a given state by two possible routes, the total amplitude for the process is the *sum of the amplitudes* for the two routes considered separately. In our new notation we write that

$$\langle x \mid s \rangle_{\text{both holes open}} = \langle x \mid s \rangle_{\text{through 1}} + \langle x \mid s \rangle_{\text{through 2}}. \tag{3.4}$$

Incidentally, we are going to suppose that the holes 1 and 2 are small enough that when we say an electron goes through the hole, we don't have to discuss which part of the hole. We could, of course, split each hole into pieces with a certain amplitude that the electron goes to the top of the hole and the bottom of the hole and so on. We will suppose that the hole is small enough so that we don't have to worry about this detail. That is part of the roughness involved; the matter can be made more precise, but we don't want to do so at this stage.

Now we want to write out in more detail what we can say about the amplitude for the process in which the electron reaches the detector at x by way of hole 1. We can do that by using our *third general principle*: When a particle goes by some particular route the amplitude for that route can be written as the *product* of the *amplitude* to go part way with the *amplitude* to go the rest of the way. For the setup of Fig. 3–1 the amplitude to go from s to x by way of hole 1 is equal to the amplitude to go from s to 1, multiplied by the amplitude to go from 1 to x.

$$\langle x \mid s \rangle_{\text{via 1}} = \langle x \mid 1 \rangle \langle 1 \mid s \rangle. \tag{3.5}$$

Again this result is not completely precise. We should also include a factor for the amplitude that the electron will get through the hole at 1; but in the present case it is a simple hole, and we will take this factor to be unity.

You will note that Eq. (3.5) appears to be written in reverse order. It is to be read from right to left: The electron goes from s to 1 and then from 1 to x. In summary, if events occur in succession—that is, if you can analyze one of the routes of the particle by saying it does this, then it does this, then it does that—the resultant amplitude for that route is calculated by multiplying in succession the amplitude for each of the successive events. Using this law we can rewrite Eq. (3.4) as

$$\langle x \mid s \rangle_{\text{both}} = \langle x \mid 1 \rangle \langle 1 \mid s \rangle + \langle x \mid 2 \rangle \langle 2 \mid s \rangle.$$

Now we wish to show that just using these principles we can calculate a much more complicated problem like the one shown in Fig. 3–2. Here we have two walls, one with two holes, 1 and 2, and another which has three holes, a, b, and c. Behind the second wall there is a detector at x, and we want to know the amplitude for a particle to arrive there. Well, one way you can find this is by calculating the superposition, or interference, of the waves that go through; but you can also do it by saying that there are six possible routes and superposing an amplitude for each. The electron can go through hole 1, then through hole a, and then to x; or it could go through hole 1, then through hole b, and then to x; and so on. According-ing to our second principle, the amplitudes for alternative routes add, so we should be able to write the amplitude from s to x as a sum of six separate amplitudes. On the other hand, using the third principle, each of these separate amplitudes can be written as a product of three amplitudes. For example, one of them is the amplitude for s to 1, times the amplitude for 1 to a, times the amplitude for a to x. Using our shorthand notation, we can write the complete amplitude to go from s to x as

$$\langle x \mid s \rangle = \langle x \mid a \rangle \langle a \mid 1 \rangle \langle 1 \mid s \rangle + \langle x \mid b \rangle \langle b \mid 1 \rangle \langle 1 \mid s \rangle + \cdots + \langle x \mid c \rangle \langle c \mid 2 \rangle \langle 2 \mid s \rangle.$$

We can save writing by using the summation notation

$$\langle x \mid s \rangle = \sum_{\substack{i=1,2 \\ \alpha=a,b,c}} \langle x \mid \alpha \rangle \langle \alpha \mid i \rangle \langle i \mid s \rangle. \tag{3.6}$$

In order to make any calculations using these methods, it is, naturally, necessary to know the amplitude to get from one place to another. We will give a rough idea of a typical amplitude. It leaves out certain things like the polarization of light or the spin of the electron, but aside from such features it is quite accurate. We give it so that you can solve problems involving various combinations of slits. Suppose a particle with a definite energy is going in empty space from a location r_1 to a location r_2. In other words, it is a free particle with no forces on it. Except for a numerical factor in front, the amplitude to go from r_1 to r_2 is

$$\langle r_2 \mid r_1 \rangle = \frac{e^{ip \cdot r_{12}/\hbar}}{r_{12}}, \tag{3.7}$$

where $r_{12} = r_2 - r_1$, and p is the momentum which is related to the energy E by the relativistic equation

$$p^2 c^2 = E^2 - (m_0 c^2)^2,$$

or the nonrelativistic equation

$$\frac{p^2}{2m} = \text{Kinetic energy}.$$

Equation (3.7) says in effect that the particle has wavelike properties, the amplitude propagating as a wave with a wave number equal to the momentum divided by $\hbar$.

In the most general case, the amplitude and the corresponding probability will also involve the time. For most of these initial discussions we will suppose that the source always emits the particles with a given energy so we will not need to worry about the time. But we could, in the general case, be interested in some other questions. Suppose that a particle is liberated at a certain place P at a certain time, and you would like to know the amplitude for it to arrive at some location, say r, at some later time. This could be represented symbolically as the amplitude $\langle r, t = t_1 \mid P, t = 0 \rangle$. Clearly, this will depend upon both r and t. You will get different results if you put the detector in different places and measure at different times. This function of r and t, in general, satisfies a differential equation which is a wave equation. For example, in a nonrelativistic case it is the Schrödinger equation. One has then a wave equation analogous to the equation for electromagnetic waves or waves of sound in a gas. However, it must be emphasized that the wave function that satisfies the equation is not like a real wave in space; one cannot picture any kind of reality to this wave as one does for a sound wave.

Although one may be tempted to think in terms of "particle waves" when dealing with one particle, it is not a good idea, for if there are, say, two particles, the amplitude to find one at r_1 and the other at r_2 is not a simple wave in three-dimensional space, but depends on the *six* space variables r_1 and r_2. If we are, for example, dealing with two (or more) particles, we will need the following additional principle: Provided that the two particles do not interact, the amplitude that one particle will do one thing *and* the other one something else is the product of the two amplitudes that the two particles would do the two things separately. For example, if $\langle a \mid s_1 \rangle$ is the amplitude for particle 1 to go from s_1 to a, and $\langle b \mid s_2 \rangle$

is the amplitude for particle 2 to go from s_2 to b, the amplitude that *both* things will happen together is

$$\langle a \mid s_1 \rangle \langle b \mid s_2 \rangle.$$

There is one more point to emphasize. Suppose that we didn't know where the particles in Fig. 3–2 come from before arriving at holes 1 and 2 of the first wall. We can still make a prediction of what will happen beyond the wall (for example, the amplitude to arrive at x) provided that we are given two numbers: the amplitude to have arrived at 1 and the amplitude to have arrived at 2. In other words, because of the fact that the amplitude for successive events multiplies, as shown in Eq. (3.6), all you need to know to continue the analysis is two numbers—in this particular case $\langle 1 \mid s \rangle$ and $\langle 2 \mid s \rangle$. These two complex numbers are enough to predict all the future. That is what really makes quantum mechanics easy. It turns out that in later chapters we are going to do just such a thing when we specify a starting condition in terms of two (or a few) numbers. Of course, these numbers depend upon where the source is located and possibly other details about the apparatus, but given the two numbers, we do not need to know any more about such details.

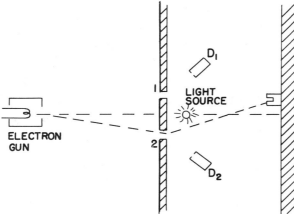

Fig. 3–3. An experiment to determine which hole the electron goes through.

3–2 The two-slit interference pattern

Now we would like to consider a matter which was discussed in some detail in Chapter 1. This time we will do it with the full glory of the amplitude idea to show you how it works out. We take the same experiment shown in Fig. 3–1, but now with the addition of a light source behind the two holes, as shown in Fig. 3–3. In Chapter 1, we discovered the following interesting result. If we looked behind slit 1 and saw a photon scattered from there, then the distribution obtained for the electrons at x in coincidence with these photons was the same as though slit 2 were closed. The total distribution for electrons that had been "seen" at either slit 1 or slit 2 was the sum of the separate distributions and was completely different from the distribution with the light turned off. This was true at least if we used light of short enough wavelength. If the wavelength was made longer so we could not be sure at which hole the scattering had occurred, the distribution became more like the one with the light turned off.

Let's examine what is happening by using our new notation and the principles of combining amplitudes. To simplify the writing, we can again let ϕ_1 stand for the amplitude that the electron will arrive at x by way of hole 1, that is,

$$\phi_1 = \langle x \mid 1 \rangle \langle 1 \mid s \rangle.$$

Similarly, we'll let ϕ_2 stand for the amplitude that the electron gets to the detector by way of hole 2:

$$\phi_2 = \langle x \mid 2 \rangle \langle 2 \mid s \rangle.$$

These are the amplitudes to go through the two holes and arrive at x if there is no light. Now if there is light, we ask ourselves the question: What is the amplitude for the process in which the electron starts at s and a photon is liberated by the light source L, ending with the electron at x and a photon seen behind slit 1? Suppose that we observe the photon behind slit 1 by means of a detector D_1, as shown in Fig. 3-3, and use a similar detector D_2 to count photons scattered behind hole 2. There will be an amplitude for a photon to arrive at D_1 and an electron at x, and also an amplitude for a photon to arrive at D_2 and an electron at x. Let's try to calculate them.

Although we don't have the correct mathematical formula for all the factors that go into this calculation, you will see the spirit of it in the following discussion. First, there is the amplitude $\langle 1 \mid s \rangle$ that an electron goes from the source to hole 1. Then we can suppose that there is a certain amplitude that while the electron is at hole 1 it scatters a photon into the detector D_1. Let us represent this amplitude by a. Then there is the amplitude $\langle x \mid 1 \rangle$ that the electron goes from slit 1 to the electron detector at x. The amplitude that the electron goes from s to x via slit 1 *and* scatters a photon into D_1 is then

$$\langle x \mid 1 \rangle a \langle 1 \mid s \rangle.$$

Or, in our previous notation, it is just $a\phi_1$.

There is also some amplitude that an electron going through slit 2 will scatter a photon into counter D_1. You say, "That's impossible; how can it scatter into counter D_1 if it is only looking at hole 1?" If the wavelength is long enough, there are diffraction effects, and it is certainly possible. If the apparatus is built well and if we use photons of short wavelength, then the amplitude that a photon will be scattered into detector 1, from an electron at 2 is very small. But to keep the discussion general we want to take into account that there is always some such amplitude, which we will call b. Then the amplitude that an electron goes via slit 2 *and* scatters a photon into D_1 is

$$\langle x \mid 2 \rangle b \langle 2 \mid s \rangle = b\phi_2.$$

The amplitude to find the electron at x and the photon in D_1 is the sum of two terms, one for each possible path for the electron. Each term is in turn made up of two factors: first, that the electron went through a hole, and second, that the photon is scattered by such an electron into detector 1; we have

$$\left\langle \begin{array}{c} \text{electron at } x \\ \text{photon at } D_1 \end{array} \middle| \begin{array}{c} \text{electron from } s \\ \text{photon from } L \end{array} \right\rangle = a\phi_1 + b\phi_2. \qquad (3.8)$$

We can get a similar expression when the photon is found in the other detector D_2. If we assume for simplicity that the system is symmetrical, then a is also the amplitude for a photon in D_2 when an electron passes through hole 2, and b is the amplitude for a photon in D_2 when the electron passes through hole 1. The corresponding total amplitude for a photon at D_2 and an electron at x is

$$\left\langle \begin{array}{c} \text{electron at } x \\ \text{photon at } D_2 \end{array} \middle| \begin{array}{c} \text{electron from } s \\ \text{photon from } L \end{array} \right\rangle = a\phi_2 + b\phi_1. \qquad (3.9)$$

Now we are finished. We can easily calculate the probability for various situations. Suppose that we want to know with what probability we get a count in D_1 and an electron at x. That will be the absolute square of the amplitude given in Eq. (3.8), namely, just $|a\phi_1 + b\phi_2|^2$. Let's look more carefully at this expression. First of all, if b is zero—which is the way we would like to design the apparatus—then the answer is simply $|\phi_1|^2$ diminished in total amplitude by the factor $|a|^2$. This is the probability distribution that you would get if there were only one hole—as shown in the graph of Fig. 3-4(a). On the other hand, if the wavelength is very long, the scattering behind hole 2 into D_1 may be just about the same as for hole 1. Although there may be some phases involved in a and b, we can ask about a simple case in which the two phases are equal. If a is practically equal to b, then the total probability becomes $|\phi_1 + \phi_2|^2$ multiplied by $|a|^2$, since the common factor a can be taken out. This, however, is just the probability

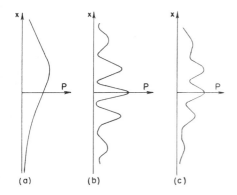

Fig. 3-4. The probability of counting an electron at x in coincidence with a photon at D in the experiment of Fig. 3-3: (a) for $b = 0$; (b) for $b = a$; (c) for $0 < b < a$.

distribution we would have gotten without the photons at all. Therefore, in the case that the wavelength is very long—and the photon detection ineffective—you return to the original distribution curve which shows interference effects, as shown in Fig. 3–4(b). In the case that the detection is partially effective, there is an interference between a lot of ϕ_1 and a little of ϕ_2, and you will get an intermediate distribution such as is sketched in Fig. 3–4(c). Needless to say, if we look for coincidence counts of photons at D_2 and electrons at x, we will get the same kinds of results. If you remember the discussion in Chapter 1, you will see that these results give a quantitative description of what was described there.

Now we would like to emphasize an important point so that you will avoid a common error. Suppose that you only want the amplitude that the electron arrives at x, *regardless* of whether the photon was counted at D_1 or D_2. Should you add the amplitudes given in Eqs. (3.8) and (3.9)? No! You must *never add amplitudes for different and distinct final states.* Once the photon is accepted by one of the photon counters, we can always determine which alternative occurred if we want, without any further disturbance to the system. Each alternative has a probability completely independent of the other. To repeat, do not add amplitudes for different *final* conditions, where by "final" we mean at that moment the *probability* is desired—that is, when the experiment is "finished." You do add the amplitudes for the different *indistinguishable* alternatives inside the experiment, before the complete process is finished. At the end of the process you may say that you "don't want to look at the photon." That's your business, but you still do not add the amplitudes. Nature does not know what you are looking at, and she behaves the way she is going to behave whether you bother to take down the data or not. So here we must not add the amplitudes. We first square the amplitudes for all possible different final events and then sum. The correct result for an electron at x and a photon at either D_1 or D_2 is

$$\left|\left\langle \begin{matrix} \text{e at } x \\ \text{ph at } D_1 \end{matrix} \middle| \begin{matrix} \text{e from } s \\ \text{ph from } L \end{matrix} \right\rangle\right|^2 + \left|\left\langle \begin{matrix} \text{e at } x \\ \text{ph at } D_2 \end{matrix} \middle| \begin{matrix} \text{e from } s \\ \text{ph from } L \end{matrix} \right\rangle\right|^2$$
$$= |a\phi_1 + b\phi_2|^2 + |a\phi_2 + b\phi_1|^2. \quad (3.10)$$

3–3 Scattering from a crystal

Our next example is a phenomenon in which we have to analyze the interference of probability amplitudes somewhat carefully. We look at the process of the scattering of neutrons from a crystal. Suppose we have a crystal which has a lot of atoms with nuclei at their centers, arranged in a periodic array, and a neutron beam that comes from far away. We can label the various nuclei in the crystal by an index i, where i runs over the integers $1, 2, 3, \ldots N$, with N equal to the total number of atoms. The problem is to calculate the probability of getting a neutron into a counter with the arrangement shown in Fig. 3–5. For any particular atom i, the amplitude that the neutron arrives at the counter C is the amplitude that the neutron gets from the source S to nucleus i, multiplied by the amplitude a that it gets scattered there, multiplied by the amplitude that it gets from i to the counter C. Let's write that down:

$$\langle \text{neutron at } C \mid \text{neutron from } S \rangle_{\text{via } i} = \langle C \mid i \rangle\, a\, \langle i \mid S \rangle. \quad (3.11)$$

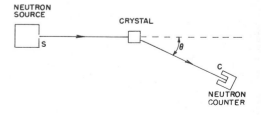

Fig. 3–5. Measuring the scattering of neutrons by a crystal.

In writing this equation we have assumed that the scattering amplitude a is the same for all atoms. We have here a large number of apparently indistinguishable routes. They are indistinguishable because a low-energy neutron is scattered from a nucleus without knocking the atom out of its place in the crystal—no "record" is left of the scattering. According to the earlier discussion, the total amplitude for a neutron at C involves a sum of Eq. (3.11) over all the atoms:

$$\langle \text{neutron at } C \mid \text{neutron from } S \rangle = \sum_{i=1}^{N} \langle C \mid i \rangle\, a\, \langle i \mid S \rangle. \quad (3.12)$$

Because we are adding amplitudes of scattering from atoms with different space positions, the amplitudes will have different phases giving the characteristic interference pattern that we have already analyzed in the case of the scattering of light from a grating.

The neutron intensity as a function of angle in such an experiment is indeed often found to show tremendous variations, with very sharp interference peaks and almost nothing in between—as shown in Fig. 3–6(a). However, for certain kinds of crystals it does not work this way, and there is—along with the interference peaks discussed above—a general background of scattering in all directions. We must try to understand the apparently mysterious reasons for this. Well, we have not considered one important property of the neutron. It has a spin of one-half, and so there are two conditions in which it can be: either spin "up" (say perpendicular to the page in Fig. 3–5) or spin "down." If the nuclei of the crystal have no spin, the neutron spin doesn't have any effect. But when the nuclei of the crystal also have a spin, say a spin of one-half, you will observe the background of smeared-out scattering described above. The explanation is as follows.

If the neutron has one direction of spin and the atomic nucleus has the same spin, then no change of spin can occur in the scattering process. If the neutron and atomic nucleus have opposite spin, then scattering can occur by two processes, one in which the spins are unchanged and another in which the spin directions are exchanged. This rule for no net change of the sum of the spins is analogous to our classical law of conservation of angular momentum. We can begin to understand the phenomenon if we assume that all the scattering nuclei are set up with spins in one direction. A neutron with the same spin will scatter with the expected sharp interference distribution. What about one with opposite spin? If it scatters without spin flip, then nothing is changed from the above; but if the two spins flip over in the scattering, we could, in principle, find out which nucleus had done the scattering, since it would be the only one with spin turned over. Well, if we can tell which atom did the scattering, what have the other atoms got to do with it? Nothing, of course. The scattering is exactly the same as that from a single atom.

To include this effect, the mathematical formulation of Eq. (3.12) must be modified since we haven't described the states completely in that analysis. Let's start with all neutrons from the source having spin up and all the nuclei of the crystal having spin down. First, we would like the amplitude that at the counter the spin of the neutron is up *and* all spins of the crystal are still down. This is not different from our previous discussion. We will let a be the amplitude to scatter with no flip or spin. The amplitude for scattering from the ith atom is, of course,

$$\langle C_{up}, \text{crystal all down} \mid S_{up}, \text{crystal all down} \rangle = \langle C \mid i \rangle a \langle i \mid S \rangle.$$

Since all the atomic spins are still down, the various alternatives (different values of i) cannot be distinguished. There is clearly no way to tell which atom did the scattering. For this process, all the amplitudes interfere.

We have another case, however, where the spin of the detected neutron is down although it started from S with spin up. In the crystal, one of the spins must be changed to the up direction—let's say that of the kth atom. We will assume that there is the same scattering amplitude with spin flip for every atom, namely b. (In a real crystal there is the disagreeable possibility that the reversed spin moves to some other atom, but let's take the case of a crystal for which this probability is very low.) The scattering amplitude is then

$$\langle C_{down}, \text{nucleus } k \text{ up} \mid S_{up}, \text{crystal all down} \rangle = \langle C \mid k \rangle b \langle k \mid S \rangle. \qquad (3.13)$$

If we ask for the probability of finding the neutron spin down and the kth nucleus spin up, it is equal to the absolute square of this amplitude, which is simply $|b|^2$ times $|\langle C \mid k \rangle \langle k \mid S \rangle|^2$. The second factor is almost independent of location in the crystal, and all phases have disappeared in taking the absolute square. The

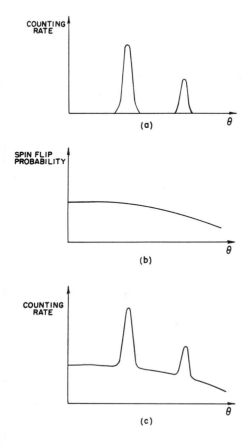

Fig. 3–6. The neutron counting rate as a function of angle: (a) for spin zero nuclei; (b) the probability of scattering with spin flip; (c) the observed counting rate with a spin one-half nucleus.

probability of scattering from *any nucleus* in the crystal with spin flip is now

$$|b|^2 \sum_{k=1}^{N} |\langle C \mid k \rangle \langle k \mid S \rangle|^2,$$

which will show a smooth distribution as in Fig. 3–6(b).

You may argue, "I don't care which atom is up." Perhaps you don't, but nature knows; and the probability is, in fact, what we gave above—there isn't any interference. On the other hand, if we ask for the probability that the spin is up at the detector and all the atoms still have spin down, then we must take the absolute square of

$$\sum_{i=1}^{N} \langle C \mid i \rangle \, a \, \langle i \mid S \rangle.$$

Since the terms in this sum have phases, they do interfere, and we get a sharp interference pattern. If we do an experiment in which we don't observe the spin of the detected neutron, then both kinds of events can occur; and the separate probabilities add. The total probability (or counting rate) as a function of angle then looks like the graph in Fig. 3–6(c).

Let's review the physics of this experiment. If you could, *in principle*, distinguish the alternative *final* states (even though you do not bother to do so), the total, final probability is obtained by calculating the *probability* for each state (not the amplitude) and then adding them together. If you *cannot* distinguish the final states *even in principle*, then the probability amplitudes must be summed before taking the absolute square to find the actual probability. The thing you should notice particularly is that if you were to try to represent the neutron by a wave alone, you would get the same kind of distribution for the scattering of a down-spinning neutron as for an up-spinning neutron. You would have to say that the "wave" would come from all the different atoms and interfere just as for the up-spinning one with the same wavelength. But we know that is not the way it works. So as we stated earlier, we must be careful not to attribute too much reality to the waves in space. They are useful for certain problems but not for all.

3–4 Identical particles

The next experiment we will describe is one which shows one of the beautiful consequences of quantum mechanics. It again involves a physical situation in which a thing can happen in two *indistinguishable* ways, so that there is an interference of amplitudes—as is *always* true in such circumstances. We are going to discuss the scattering, at relatively low energy, of nuclei on other nuclei. We start by thinking of α-particles (which, as you know, are helium nuclei) bombarding, say, oxygen. To make it easier for us to analyze the reaction, we will look at it in the center-of-mass system, in which the oxygen nucleus and the α-particle have their velocities in opposite directions before the collision and again in exactly opposite directions after the collision. See Fig. 3–7(a). (The magnitudes of the velocities are, of course, different, since the masses are different.) We will also suppose that there is conservation of energy and that the collision energy is low enough that neither particle is broken up or left in an excited state. The reason that the two particles deflect each other is, of course, that each particle carries a positive charge and, classically speaking, there is an electrical repulsion as they go by. The scattering will happen at different angles with different probabilities, and we would like to discuss something about the angle dependence of such scatterings. (It is possible, of course, to calculate this thing classically, and it is one of the most remarkable accidents of quantum mechanics that the answer to this problem comes out the same as it does classically. This is a curious point because it happens for no other force except the inverse square law—so it is indeed an accident.)

The probability of scattering in different directions can be measured by an experiment as shown in Fig. 3–7(a). The counter at position 1 could be designed to detect only α-particles; the counter at position 2 could be designed to detect

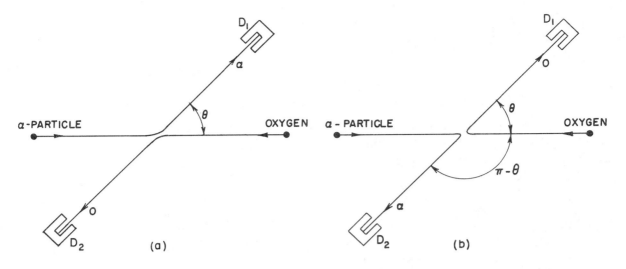

Fig. 3–7.   The scattering of α-particles from oxygen nuclei, as seen in the center-of-mass system.

only oxygen—just as a check. (In the laboratory system the detectors would not be opposite; but in the CM system they are.) Our experiment consists in measuring the probability of scattering in various directions. Let's call $f(\theta)$ the amplitude to scatter into the counters when they are at the angle θ; then $|f(\theta)|^2$ will be our experimentally determined probability.

Now we could set up another experiment in which our counters would respond to *either* the α-particle *or* the oxygen nucleus. Then we have to work out what happens when we do not bother to distinguish which particles are counted. Of course, if we are to get an oxygen in the position θ, there must be an α-particle on the opposite side at the angle $(\pi - \theta)$, as shown in Fig. 3–7(b). So if $f(\theta)$ is the amplitude for α-scattering through the angle θ, then $f(\pi - \theta)$ is the amplitude for oxygen scattering through the angle θ.† Thus, the probability for having *some* particle in the detector at position 1 is:

$$\text{Probability of } \textit{some} \text{ particle in } D_1 = |f(\theta)|^2 + |f(\pi - \theta)|^2. \quad (3.14)$$

Note that the two states are distinguishable in principle. Even though in this experiment we *do not* distinguish them, *we could.* According to the earlier discussion, then, we must add the probabilities, not the amplitudes.

The result given above is correct for a variety of target nuclei—for α-particles on oxygen, on carbon, on beryllium, on hydrogen. *But it is wrong for α-particles on α-particles.* For the one case in which both particles are exactly the same, the experimental data disagree with the prediction of (3.14). For example, the scattering probability at 90° is exactly twice what the above theory predicts and has nothing to do with the particles being "helium" nuclei. If the target is He^3, but the projectiles are α-particles (He^4), then there is agreement. Only when the target is He^4—so its nuclei are identical with the incoming α-particle—does the scattering vary in a peculiar way with angle.

Perhaps you can already see the explanation. There are two ways to get an α-particle into the counter: by scattering the bombarding α-particle at an angle θ, or by scattering it at an angle of $(\pi - \theta)$. How can we tell whether the bombarding particle or the target particle entered the counter? The answer is that we cannot. In the case of α-particles with α-particles there are two alternatives that cannot be distinguished. Here, we must let the probability *amplitudes* interfere by addition,

---

† In general, a scattering direction should, of course, be described by two angles, the polar angle ϕ, as well as the azimuthal angle θ. We would then say that an oxygen nucleus at (θ, ϕ) means that the α-particle is at $(\pi - \theta, \phi + \pi)$. However, for Coulomb scattering (and for many other cases), the scattering amplitude is independent of ϕ. Then the amplitude to get an oxygen at θ is the same as the amplitude to get the α-particle at $(\pi - \theta)$.

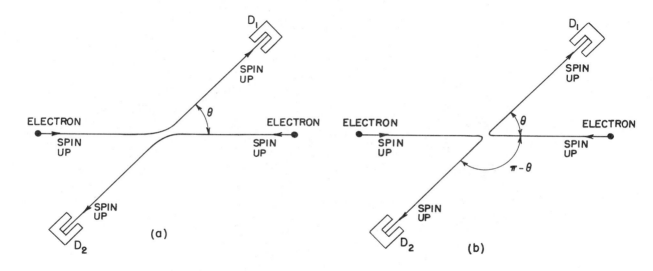

Fig. 3–8. The scattering of electrons on electrons. If the incoming electrons have parallel spins, the processes (a) and (b) are indistinguishable.

and the probability of finding an α-particle in the counter is the square of their sum:

$$\text{Probability of an } \alpha\text{-particle at } D_1 = |f(\theta) + f(\pi - \theta)|^2. \quad (3.15)$$

This is quite a different result than that in Eq. (3.14). We can take an angle of $\pi/2$ as an example, because it is easy to figure out. For $\theta = \pi/2$, we obviously have $f(\theta) = f(\pi - \theta)$, so the probability in Eq. (3.15) becomes $|f(\pi/2) + f(\pi/2)|^2 = 4|f(\pi/2)|^2$.

On the other hand, if they did not interfere, the result of Eq. (3.14) gives only $2|f(\pi/2)|^2$. So there is twice as much scattering at 90° as we might have expected. Of course, at other angles the results will also be different. And so you have the unusual result that when particles are identical, a certain new thing happens that doesn't happen when particles can be distinguished. In the mathematical description you must add the amplitudes for alternative process in which the two particles simply exchange roles and there is an interference.

An even more perplexing thing happens when we do the same kind of experiment by scattering electrons on electrons, or protons on protons. Neither of the above results is then correct! For these particles, we must invoke still a new rule, a most peculiar rule, which is the following: When you have a situation in which the identity of the electron which is arriving at a point is exchanged with another one, the new amplitude interferes with the old one with an *opposite phase*. It is interference all right, but with a minus sign. In the case of α-particles, when you exchange the α-particle entering the detector, the interfering amplitudes interfere with the positive sign. *In the case of electrons, the interfering amplitudes for exchange interfere with a negative sign.* Except for another detail to be discussed below, the proper equation for electrons in an experiment like the one shown in Fig. 3–8 is

$$\text{Probability of e at } D_1 = |f(\theta) - f(\pi - \theta)|^2. \quad (3.16)$$

The above statement must be qualified, because we have not considered the spin of the electron (α-particles have no spin). The electron spin may be considered to be either "up" or "down" with respect to the plane of the scattering. If the energy of the experiment is low enough, the magnetic forces due to the currents will be small and the spin will not be affected. We will assume that this is the case for the present analysis, so that there is no chance that the spins are changed during the collision. Whatever spin the electron has, it carries along with it. Now you see there are many possibilities. The bombarding and target particles can have both spins up, both down, or opposite spins. If both spins are up, as in Fig. 3–8 (or if both spins are down), the same will be true of the recoil particles and the *amplitude* for the process is the *difference* of the amplitudes for the two possibilities

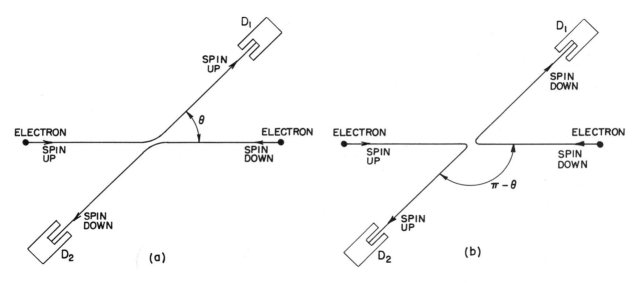

Fig. 3–9. The scattering of electrons with antiparallel spins.

shown in Fig. 3–8(a) and (b). The *probability* of detecting an electron in D_1 is then given by Eq. (3.16).

Suppose, however, the "bombarding" spin is up and the "target" spin is down. The electron entering counter 1 can have spin up or spin down, and by measuring this spin we can tell whether it came from the bombarding beam or from the target. The two possibilities are shown in Fig. 3–9(a) and (b); they are distinguishable in principle, and hence there will be no interference—merely an addition of the two probabilities. The same argument holds if both of the original spins are reversed— that is, if the left-hand spin is down and the right-hand spin is up.

Now if we take our electrons at random—as from a tungsten filament in which the electrons are completely unpolarized—then the odds are fifty-fifty that any particular electron comes out with spin up or spin down. If we don't bother to measure the spin of the electrons at any point in the experiment, we have what we call an unpolarized experiment. The results for this experiment are best calculated by listing all of the various possibilities as we have done in Table 3–1. A separate *probability* is computed for each distinguishable alternative. The total probability is then the sum of all the separate probabilities. Note that for unpolarized beams the result for $\theta = \pi/2$ is one-half that of the classical result with independent particles. The behavior of identical particles has many interesting consequences; we will discuss them in greater detail in the next chapter.

Table 3–1

Scattering of unpolarized spin one-half particles

Fraction of cases	Spin of particle 1	Spin of particle 2	Spin at D_1	Spin at D_2	Probability
$\frac{1}{4}$	up	up	up	up	$\lvert f(\theta) - f(\pi - \theta)\rvert^2$
$\frac{1}{4}$	down	down	down	down	$\lvert f(\theta) - f(\pi - \theta)\rvert^2$
$\frac{1}{4}$	up	down	up	down	$\lvert f(\theta)\rvert^2$
			down	up	$\lvert f(\pi - \theta)\rvert^2$
$\frac{1}{4}$	down	up	up	down	$\lvert f(\pi - \theta)\rvert^2$
			down	up	$\lvert f(\theta)\rvert^2$
Total probability $= \frac{1}{2}\lvert f(\theta) - f(\pi - \theta)\rvert^2 + \frac{1}{2}\lvert f(\theta)\rvert^2 + \frac{1}{2}\lvert f(\pi - \theta)\rvert^2$					

4

Identical Particles

4–1 Bose particles and Fermi particles

In the last chapter we began to consider the special rules for the interference that occurs in processes with two *identical* particles. By *identical particles* we mean things like electrons which can in no way be distinguished one from another. If a process involves two particles that are identical, reversing which one arrives at a counter is an alternative which cannot be distinguished and—like all cases of alternatives which cannot be distinguished—interferes with the original, un-exchanged case. The amplitude for an event is then the sum of the two interfering amplitudes; but, interestingly enough, the interference is in some cases with the *same* phase and, in others, with the *opposite* phase.

Suppose we have a collision of two particles a and b in which particle a scatters in the direction 1 and particle b scatters in the direction 2, as sketched in Fig. 4–1(a). Let's call $f(\theta)$ the amplitude for this process; then the probability P_1 of observing such an event is proportional to $|f(\theta)|^2$. Of course, it could also happen that particle b scattered into counter 1 and particle a went into counter 2, as shown in Fig. 4–1(b). Assuming that there are no special directions defined by spins or such, the probability P_2 for this process is just $|f(\pi - \theta)|^2$, because it is just equivalent to the first process with counter 1 moved over to the angle $\pi - \theta$. You might also think that the *amplitude* for the second process is just $f(\pi - \theta)$. But that is not necessarily so, because there could be an arbitrary phase factor. That is, the amplitude could be

$$e^{i\delta} f(\pi - \theta).$$

Such an amplitude still gives a probability P_2 equal to $|f(\pi - \theta)|^2$.

Now let's see what happens if a and b are identical particles. Then the two different processes shown in the two diagrams of Fig. 4–1 cannot be distinguished. There is an amplitude that *either* a or b goes into counter 1, while the other goes into counter 2. This amplitude is the sum of the amplitudes for the two processes shown in Fig. 4–1. If we call the first one $f(\theta)$, then the second one is $e^{i\delta} f(\pi - \theta)$, where now the phase factor is very important because we are going to be adding two amplitudes. Suppose we have to multiply the amplitude by a certain phase factor when we exchange the roles of the two particles. If we exchange them again we should get the same factor again. But we are then back to the first process.

4–1 Bose particles and Fermi particles

4–2 States with two Bose particles

4–3 States with n Bose particles

4–4 Emission and absorption of photons

4–5 The blackbody spectrum

4–6 Liquid helium

4–7 The exclusion principle

Review: Blackbody radiation in: Chapter 41, Vol. I, *The Brownian Movement*

Chapter 42, Vol. I, *Applications of Kinetic Theory*

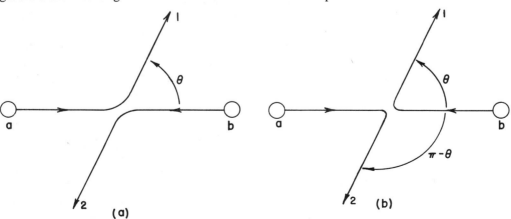

Fig. 4–1. In the scattering of two identical particles, the processes (a) and (b) are indistinguishable.

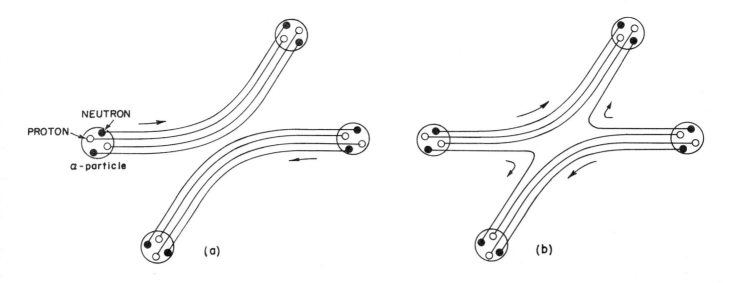

Fig. 4–2. The scattering of two α-particles. In (a) the two particles retain their identity; in (b) a neutron is exchanged during the collision.

The phase factor taken twice must bring us back where we started—its square must be equal to 1. There are only two possibilities: $e^{i\delta}$ is equal to $+1$, or is equal to -1. Either the exchanged case contributes with the *same* sign, or it contributes with the *opposite* sign. Both cases exist in nature, each for a different class of particles. Particles which interfere with a *positive* sign are called *Bose particles* and those which interfere with a *negative* sign are called *Fermi particles*. The Bose particles are the photon, the mesons, and the graviton. The Fermi particles are the electron, the muon, the neutrinos, the nucleons, and the baryons. We have, then, that the amplitude for the scattering of identical particles is:

Bose particles:
$$\text{(Amplitude direct)} + \text{(Amplitude exchanged)}. \tag{4.1}$$

Fermi particles:
$$\text{(Amplitude direct)} - \text{(Amplitude exchanged)}. \tag{4.2}$$

For particles with spin—like electrons—there is an additional complication. We must specify not only the location of the particles but the direction of their spins. It is only for identical particles *with identical spin states* that the amplitudes interfere when the particles are exchanged. If you think of the scattering of unpolarized beams—which are a mixture of different spin states—there is some extra arithmetic.

Now an interesting problem arises when there are two or more particles bound tightly together. For example, an α-particle has four particles in it—two neutrons and two protons. When two α-particles scatter, there are several possibilities. It may be that during the scattering there is a certain amplitude that one of the neutrons will leap across from one α-particle to the other, while a neutron from the other α-particle leaps the other way so that the two alphas which come out of the scattering are not the original ones—there has been an exchange of a pair of neutrons. See Fig. 4–2. The amplitude for scattering with an exchange of a pair of neutrons will interfere with the amplitude for scattering with no such exchange, and the interference must be with a minus sign because there has been an exchange of one pair of Fermi particles. On the other hand, if the relative energy of the two α-particles is so low that they stay fairly far apart—say, due to the Coulomb repulsion—and there is never any appreciable probability of exchanging any of the internal particles, we can consider the α-particle as a simple object, and we do not need to worry about its internal details. In such circumstances, there are only two contributions to the scattering amplitude. Either there is no exchange, or all four of the nucleons are exchanged in the scattering. Since the protons and the neutrons in the α-particle are all Fermi particles, an exchange of any pair reverses the sign of the scattering amplitude. So long as there are no internal changes in the α-particles, interchanging the two α-particles is the same as interchanging four pairs of Fermi particles. There is a change in sign for each pair, so the net result is that the amplitudes combine with a positive sign. The α-particle behaves like a Bose particle.

So the rule is that composite objects, in circumstances in which the composite object can be considered as a single object, behave like Fermi particles or Bose particles, depending on whether they contain an odd number or an even number of Fermi particles.

All the elementary Fermi particles we have mentioned—such as the electron, the proton, the neutron, and so on—have a spin $j = 1/2$. If several such Fermi particles are put together to form a composite object, the resulting spin may be either integral or half-integral. For example, the common isotope of helium, He^4, which has two neutrons and two protons, has a spin of zero, whereas Li^7, which has three protons and four neutrons, has a spin of $3/2$. We will learn later the rules for compounding angular momentum, and will just mention now that every composite object which has a *half-integral spin* imitates a *Fermi particle*, whereas every composite object with an *integral spin* imitates a *Bose particle*.

This brings up an interesting question: Why is it that particles with half-integral spin are Fermi particles whose amplitudes add with the minus sign, whereas particles with integral spin are Bose particles whose amplitudes add with the positive sign? We apologize for the fact that we cannot give you an elementary explanation. An explanation has been worked out by Pauli from complicated arguments of quantum field theory and relativity. He has shown that the two must necessarily go together, but we have not been able to find a way of reproducing his arguments on an elementary level. It appears to be one of the few places in physics where there is a rule which can be stated very simply, but for which no one has found a simple and easy explanation. The explanation is deep down in relativistic quantum mechanics. This probably means that we do not have a complete understanding of the fundamental principle involved. For the moment, you will just have to take it as one of the rules of the world.

4–2 States with two Bose particles

Now we would like to discuss an interesting consequence of the addition rule for Bose particles. It has to do with their behavior when there are several particles present. We begin by considering a situation in which two Bose particles are scattered from two different scatterers. We won't worry about the details of the scattering mechanism. We are interested only in what happens to the scattered particles. Suppose we have the situation shown in Fig. 4–3. The particle a is scattered into the state 1. By a *state* we mean a given direction and energy, or some other given condition. The particle b is scattered into the state 2. We want to assume that the two states 1 and 2 are nearly the same. (What we really want to find out eventually is the amplitude that the two particles are scattered into identical directions, or states; but it is best if we think first about what happens if the states are almost the same and then work out what happens when they become identical.)

Suppose that we had only particle a; then it would have a certain amplitude for scattering in direction 1, say $\langle 1 \mid a \rangle$. And particle b alone would have the amplitude $\langle 2 \mid b \rangle$ for landing in direction 2. If the two particles are not identical, the amplitude for the two scatterings to occur at the same time is just the product

$$\langle 1 \mid a \rangle \langle 2 \mid b \rangle.$$

The probability for such an event is then

$$|\langle 1 \mid a \rangle \langle 2 \mid b \rangle|^2,$$

which is also equal to

$$|\langle 1 \mid a \rangle|^2 |\langle 2 \mid b \rangle|^2.$$

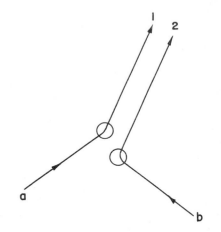

Fig. 4–3. A double scattering into nearby final states.

To save writing for the present arguments, we will sometimes set

$$\langle 1 \mid a \rangle = a_1, \qquad \langle 2 \mid b \rangle = b_2.$$

Then the probability of the double scattering is

$$|a_1|^2 |b_2|^2.$$

It could also happen that particle b is scattered into direction 1, while particle a goes into direction 2. The amplitude for this process is

$$\langle 2 \mid a \rangle \langle 1 \mid b \rangle,$$

and the probability of such an event is

$$|\langle 2 \mid a \rangle \langle 1 \mid b \rangle|^2 = |a_2|^2 |b_1|^2.$$

Imagine now that we have a pair of tiny counters that pick up the two scattered particles. The probability P_2 that they will pick up two particles together is just the sum

$$P_2 = |a_1|^2 |b_2|^2 + |a_2|^2 |b_1|^2. \tag{4.3}$$

Now let's suppose that the directions 1 and 2 are very close together. We expect that a should vary smoothly with direction, so a_1 and a_2 must approach each other as 1 and 2 get close together. If they are close enough, the amplitudes a_1 and a_2 will be equal. We can set $a_1 = a_2$ and call them both just a; similarly, we set $b_1 = b_2 = b$. Then we get that

$$P_2 = 2|a|^2 |b|^2. \tag{4.4}$$

Now suppose, however, that a and b are identical Bose particles. Then the process of a going into 1 and b going into 2 cannot be distinguished from the exchanged process in which a goes into 2 and b goes into 1. In this case the *amplitudes* for the two different processes can interfere. The *total* amplitude to obtain a particle in each of the two counters is

$$\langle 1 \mid a \rangle \langle 2 \mid b \rangle + \langle 2 \mid a \rangle \langle 1 \mid b \rangle. \tag{4.5}$$

And the probability that we get a pair is the absolute square of this amplitude,

$$P_2 = |a_1 b_2 + a_2 b_1|^2 = 4|a|^2 |b|^2. \tag{4.6}$$

We have the result that it is *twice as likely* to find two *identical* Bose particles scattered into the same state *as you would calculate assuming the particles were different*.

Although we have been considering that the two particles are observed in separate counters, this is not essential—as we can see in the following way. Let's imagine that both the directions 1 and 2 would bring the particles into a *single* small counter which is some distance away. We will let the direction 1 be defined by saying that it heads toward the element of area dS_1 of the counter. Direction 2 heads toward the surface element dS_2 of the counter. (We imagine that the counter presents a surface at right angles to the line from the scatterings.) Now we cannot give a probability that a particle will go into a precise direction or to a *particular* point in space. Such a thing is impossible—the chance for any exact direction is zero. When we want to be so specific, we shall have to define our amplitudes so that they give the probability of arriving *per unit area* of a counter. Suppose that we had only particle a; it would have a certain amplitude for scattering in direction 1. Let's define $\langle 1 \mid a \rangle = a_1$ to be the amplitude that a will scatter *into a unit area* of the counter in the direction 1. In other words, the scale of a_1 is chosen—we say it is "normalized" so that the probability that it will scatter *into an element of area dS_1* is

$$|\langle 1 \mid a \rangle|^2 \, dS_1 = |a_1|^2 \, dS_1. \tag{4.7}$$

If our counter has the total area ΔS, and we let dS_1 range over this area, the total probability that the particle a will be scattered into the counter is

$$\int_{\Delta S} |a_1|^2 \, dS_1. \tag{4.8}$$

As before, we want to assume that the counter is sufficiently small so that the amplitude a_1 doesn't vary significantly over the surface of the counter; a_1 is then a constant amplitude which we can call a. Then the probability that particle a is scattered somewhere into the counter is

$$p_a = |a|^2 \, \Delta S. \tag{4.9}$$

In the same way, we will have that the probability that particle b—when it is alone—scatters into some element of area, say dS_2, is

$$|b_2|^2 \, dS_2.$$

(We use dS_2 instead of dS_1 because we will later want a and b to go into different directions.) Again we set b_2 equal to the constant amplitude b; then the probability that particle b is counted in the detector is

$$p_b = |b|^2 \, \Delta S. \tag{4.10}$$

Now when both particles are present, the probability that a is scattered into dS_1 and b is scattered into dS_2 is

$$|a_1 b_2|^2 \, dS_1 \, dS_2 = |a|^2 |b|^2 \, dS_1 \, dS_2. \tag{4.11}$$

If we want the probability that *both* a and b get into the counter, we integrate both dS_1 and dS_2 over ΔS and find that

$$P_2 = |a|^2 |b|^2 \, (\Delta S)^2. \tag{4.12}$$

We notice, incidentally, that this is just equal to $p_a \cdot p_b$, just as you would suppose assuming that the particles a and b act independently of each other.

When the two particles are identical, however, there are two indistinguishable possibilities for each pair of surface elements dS_1 and dS_2. Particle a going into dS_2 and particle b going into dS_1 is indistinguishable from a into dS_1 and b into dS_2, so the amplitudes for these processes will interfere. (When we had two *different* particles above—although we did not *in fact* care which particle went where in the counter—we could, *in principle*, have found out; so there was no interference. For identical particles we cannot tell, even in principle.) We must write, then, that the probability that the two particles arrive at dS_1 and dS_2 is

$$|a_1 b_2 + a_2 b_1|^2 \, dS_1 \, dS_2. \tag{4.13}$$

Now, however, when we integrate over the area of the counter, we must be careful. If we let dS_1 and dS_2 range over the whole area ΔS, we would count each part of the area *twice* since (4.13) contains everything that can happen with any pair of surface elements dS_1 and dS_2† We can still do the integral that way, if we correct for the double counting by dividing the result by 2. We get then that P_2 for identical Bose particles is

$$P_2(\text{Bose}) = \tfrac{1}{2} \{4|a|^2 |b|^2 \, (\Delta S)^2\} = 2|a|^2 |b|^2 \, (\Delta S)^2. \tag{4.14}$$

Again, this is just twice what we got in Eq. (4.12) for distinguishable particles.

If we imagine for a moment that we knew that the b channel had already sent its particle into the particular direction, we can say that the *probability* that a second particle will go into the same direction is twice as great as we would have

---

† In (4.11) interchanging dS_1 and dS_2 gives a different event, so both surface elements should range over the whole area of the counter. In (4.13) we are treating dS_1 and dS_2 as a *pair* and including everything that can happen. If the integrals include again what happens when dS_1 and dS_2 are reversed, everything is counted twice.

expected if we had calculated it as an independent event. It is a property of Bose particles that if there is already one particle in a condition of some kind, the *probability* of getting a second one in the same condition is twice as great as it would be if the first one were not already there. This fact is often stated in the following way: If there is already one Bose particle in a given state, the amplitude for putting an identical one on top of it is $\sqrt{2}$ greater than if it weren't there. (This is not a proper way of stating the result from the physical point of view we have taken, but if it is used consistently as a rule, it will, of course, give the correct result.)

4–3 States with n Bose particles

Let's extend our result to a situation in which there are n particles present. We imagine the circumstance shown in Fig. 4–4. We have n particles $a, b, c, \ldots$, which are scattered and end up in the directions $1, 2, 3, \ldots, n$. All n directions are headed toward a small counter a long distance away. As in the last section, we choose to normalize all the amplitudes so that the probability that each particle acting alone would go into an element of surface dS of the counter is

$$|\langle\ \rangle|^2\, dS.$$

First, let's assume that the particles are all distinguishable; then the probability that n particles will be counted together in n different surface elements is

$$|a_1 b_2 c_3 \ldots|^2\, dS_1\, dS_2\, dS_3 \ldots \tag{4.15}$$

Again we take that the amplitudes don't depend on where dS is located in the counter (assumed small) and call them simply $a, b, c, \ldots$ The probability (4.15) becomes

$$|a|^2 |b|^2 |c|^2 \ldots dS_1\, dS_2\, dS_3 \ldots \tag{4.16}$$

Integrating each dS over the surface ΔS of the counter, we have that P_n (different), the probability of counting n different particles at once, is

$$P_n\ (\text{different}) = |a|^2 |b|^2 |c|^2 \ldots (\Delta S)^n. \tag{4.17}$$

This is just the product of the probabilities for each particle to enter the counter separately. They all act independently—the probability for one to enter does not depend on how many others are also entering.

Now suppose that all the particles are identical Bose particles. For each set of directions $1, 2, 3, \ldots$ there are many indistinguishable possibilities. If there were, for instance, just three particles, we would have the following possibilities:

$$
\begin{array}{ccc}
a \rightarrow 1 & a \rightarrow 1 & a \rightarrow 2 \\
b \rightarrow 2 & b \rightarrow 3 & b \rightarrow 1 \\
c \rightarrow 3 & c \rightarrow 2 & c \rightarrow 3 \\
\\
a \rightarrow 2 & a \rightarrow 3 & a \rightarrow 3 \\
b \rightarrow 3 & b \rightarrow 1 & b \rightarrow 2 \\
c \rightarrow 1 & c \rightarrow 2 & c \rightarrow 1 \\
\end{array}
$$

There are six different combinations. With n particles, there are $n!$ different, but *indistinguishable*, possibilities for which we must add amplitudes. The probability that n particles will be counted in n surface elements is then

$$|a_1 b_2 c_3 \ldots + a_1 b_3 c_2 \ldots + a_2 b_1 c_3 \ldots$$
$$+\ a_2 b_3 c_1 \ldots + \text{etc.} + \text{etc.}|^2\, dS_1\, dS_2\, dS_3 \ldots dS_n. \tag{4.18}$$

Once more we assume that all the directions are so close that we can set $a_1 = a_2 = \cdots = a = a_n$, and similarly for $b, c, \ldots$; the probability of (4.18) becomes

$$|n!\, abc \ldots|^2\, dS_1\, dS_2 \ldots dS_n. \tag{4.19}$$

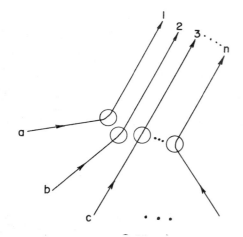

Fig. 4–4. The scattering of n particles into nearby final states.

When we integrate each dS over the area ΔS of the counter, each possible product of surface elements is counted $n!$ times; we correct for this by dividing by $n!$ and get

$$P_n(\text{Bose}) = \frac{1}{n!} \, |n!abc\ldots|^2 \, (\Delta S)^n$$

or

$$P_n(\text{Bose}) = n! \, |abc\ldots|^2 \, (\Delta S)^n. \qquad (4.20)$$

Comparing this result with Eq. (4.17), we see that the probability of counting n Bose particles together is $n!$ greater than we would calculate assuming that the particles were all distinguishable. We can summarize our result this way:

$$P_n(\text{Bose}) = n! \, P_n(\text{different}). \qquad (4.21)$$

Thus, the probability in the Bose case is larger by $n!$ than you would calculate assuming that the particles acted independently.

We can see better what this means if we ask the following question: What is the probability that a Bose particle will go into a particular state when there are *already n others* present? Let's call the newly added particle w. If we have $(n+1)$ particles, including w, Eq. (4.20) becomes

$$P_{n+1}(\text{Bose}) = (n+1)! \, |abc\ldots w|^2 \, (\Delta S)^{n+1}. \qquad (4.22)$$

We can write this as

$$P_{n+1}(\text{Bose}) = \{(n+1)|w|^2 \, \Delta S\} n! \, |abc\ldots|^2 \, \Delta S^n$$

or

$$P_{n+1}(\text{Bose}) = (n+1)|w|^2 \, \Delta S \, P_n(\text{Bose}). \qquad (4.23)$$

We can look at this result in the following way: The number $|w|^2 \, \Delta S$ is the probability for getting particle w into the detector if no other particles were present; $P_n(\text{Bose})$ is the chance that there are already n other Bose particles present. So Eq. (4.23) says that *when* there are n other identical Bose particles present, the probability that *one more* particle will enter the same state is *enhanced* by the factor $(n+1)$. The probability of getting a boson, where there are already n, is $(n+1)$ times stronger than it would be if there were none before. The *presence* of the other particles increases the probability of getting one more.

4–4 Emission and absorption of photons

Throughout our discussion we have talked about a process like the scattering of α-particles. But that is not essential; we could have been speaking of the creation of particles, as for instance the emission of light. When the light is emitted, a photon is "created." In such a case, we don't need the incoming lines in Fig. 4–4; we can consider merely that there are n atoms $a, b, c, \ldots$ emitting light, as in Fig. 4–5. So our result can also be stated: *The probability that an atom will emit a photon into a particular final state is increased by the factor $(n+1)$ if there are already n photons in that state.*

People like to summarize this result by saying that the *amplitude* to emit a photon is increased by the factor $\sqrt{n+1}$ when there are already n photons present. It is, of course, another way of saying the same thing if it is understood to mean that this amplitude is just to be squared to get the probability.

It is generally true in quantum mechanics that the amplitude to get from any condition ϕ to any other condition χ is the complex conjugate of the amplitude to get from χ to ϕ:

$$\langle \chi \mid \phi \rangle = \langle \phi \mid \chi \rangle^*. \qquad (4.24)$$

We will learn about this law a little later, but for the moment we will just assume it is true. We can use it to find out how photons are scattered or absorbed *out* of a given state. We have that the amplitude that a photon will be added to some state, say i, when there are already n photons present is, say,

$$\langle n+1 \mid n \rangle = \sqrt{n+1} \, a, \qquad (4.25)$$

Fig. 4–5. The creation of n photons in nearby states.

where $a = \langle i \mid a \rangle$ is the amplitude when there are no others present. Using Eq. (4.24), the amplitude to go the other way—from $(n + 1)$ photons to n—is

$$\langle n \mid n + 1 \rangle = \sqrt{n + 1} \, a^*. \qquad (4.26)$$

This isn't the way people usually say it; they don't like to think of going from $(n + 1)$ to n, but prefer always to start with n photons present. Then they say that the amplitude to absorb a photon when there are n present—in other words, to go from n to $(n - 1)$—is

$$\langle n - 1 \mid n \rangle = \sqrt{n} \, a^*. \qquad (4.27)$$

which is, of course, just the same as Eq. (4.26). Then they have trouble trying to remember when to use $\sqrt{n}$ or $\sqrt{n + 1}$. Here's the way to remember: The factor is always the square root of the largest number of photons present, whether it is before or after the reaction. Equations (4.25) and (4.26) show that the law is really symmetric—it only appears unsymmetric if you write it as Eq. (4.27).

There are many physical consequences of these new rules; we want to describe one of them having to do with the emission of light. Suppose we imagine a situation in which photons are contained in a box—you can imagine a box with mirrors for walls. Now say that in the box we have n photons, all of the same state—the same frequency, direction, and polarization—so they can't be distinguished, and that also there is an atom in the box that can emit another photon into the same state. Then the probability that it will emit a photon is

$$(n + 1)|a|^2, \qquad (4.28)$$

and the probability that it will absorb a photon is

$$n|a|^2, \qquad (4.29)$$

where $|a|^2$ is the probability it would emit if no photons were present. We have already discussed these rules in a somewhat different way in Chapter 42 of Vol. I. Equation (4.29) says that the probability that an atom will *absorb* a photon and make a transition to a higher energy state is proportional to the intensity of the light shining on it. But, as Einstein first pointed out, the rate at which an atom will make a transition *downward* has two parts. There is the probability that it will make a spontaneous transition $|a|^2$, plus the probability of an induced transition $n|a|^2$, which is proportional to the intensity of the light—that is, to the number of photons present. Furthermore, as Einstein said, the coefficients of absorption and of induced emission are equal and are related to the probability of spontaneous emission. What we learn here is that if the light intensity is measured in terms of the number of photons present (instead of as the energy per unit area, and per sec), the coefficients of absorption of induced emission and of spontaneous emission are all equal. This is the content of the relation between the Einstein coefficients A and B of Chapter 42, Vol. I, Eq. (42.18).

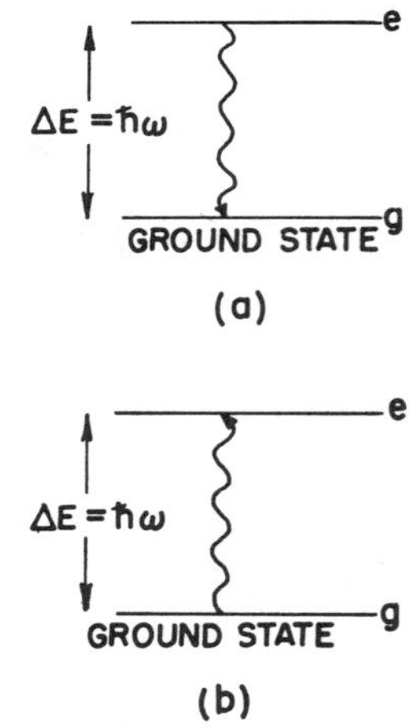

Fig. 4–6. Radiation and absorption of a photon with the frequency ω.

4–5 The blackbody spectrum

We would like to use our rules for Bose particles to discuss once more the spectrum of blackbody radiation (see Chapter 42, Vol. I). We will do it by finding out how many photons there are in a box if the radiation is in thermal equilibrium with some atoms in the box. Suppose that for each light frequency ω, there are a certain number N of atoms which have two energy states separated by the energy $\Delta E = \hbar\omega$. See Fig. 4–6. We'll call the lower-energy state the "ground" state and the upper state the "excited" state. Let N_g and N_e be the average numbers of atoms in the ground and excited states; then in thermal equilibrium at the temperature T, we have from statistical mechanics that

$$\frac{N_e}{N_g} = e^{-\Delta E / kT} = e^{-\hbar\omega / kT}. \qquad (4.30)$$

Each atom in the ground state can absorb a photon and go into the excited state, and each atom in the excited state can emit a photon and go to the ground state. In equilibrium, the rates for these two processes must be equal. The rates are proportional to the probability for the event and to the number of atoms present. Let's let $\bar{n}$ be the average number of photons present in a given state with the frequency ω. Then the absorption rate from that state is $N_g\bar{n}|a|^2$, and the emission rate into that state is $N_e(\bar{n}+1)|a|^2$. Setting the two rates equal, we have that

$$N_g\bar{n} = N_e(\bar{n}+1). \tag{4.31}$$

Combining this with Eq. (4.30), we have

$$\frac{\bar{n}}{\bar{n}+1} = e^{-\hbar\omega/kT}.$$

Solving for $\bar{n}$, we have

$$\bar{n} = \frac{1}{e^{\hbar\omega/kT}-1}, \tag{4.32}$$

which is the mean number of photons in any state with frequency ω, for a cavity in thermal equilibrium. Since each photon has the energy $\hbar\omega$, the energy in the photons of a given state is $\bar{n}\hbar\omega$, or

$$\frac{\hbar\omega}{e^{\hbar\omega/kT}-1}. \tag{4.33}$$

Incidentally, we once found a similar equation in another context [Chapter 41, Vol. I, Eq. (41.15)]. You remember that for any harmonic oscillator—such as a weight on a spring—the quantum mechanical energy levels are equally spaced with a separation $\hbar\omega$, as drawn in Fig. 4–7. If we call the energy of the nth level $n\hbar\omega$, we find that the mean energy of such an oscillator is also given by Eq. (4.33). Yet this equation was derived here for photons, by counting particles, and it gives the same results. That is one of the marvelous miracles of quantum mechanics. If one begins by considering a kind of state or condition for Bose particles which do not interact with each other (we have assumed that the photons do not interact with each other), and then considers that into this state there can be put either zero, or one, or two, . . . up to any number n of particles, one finds that this system behaves for all quantum mechanical purposes exactly like a harmonic oscillator. By such an oscillator we mean a dynamic system like a weight on a spring or a standing wave in a resonant cavity. And that is why it is possible to represent the electromagnetic field by photon particles. From one point of view, we can analyze the electromagnetic field in a box or cavity in terms of a lot of harmonic oscillators, treating each mode of oscillation according to quantum mechanics as a harmonic oscillator. From a different point of view, we can analyze the same physics in terms of identical Bose particles. And the results of both ways of working *are always in exact agreement*. There is no way to make up your mind whether the electromagnetic field is really to be described as a quantized harmonic oscillator or by giving how many photons there are in each condition. The two views turn out to be mathematically identical. So in the future we can speak either about the number of photons in a particular state in a box or the number of the energy level associated with a particular mode of oscillation of the electromagnetic field. They are two ways of saying the same thing. The same is true of photons in free space. They are equivalent to oscillations of a cavity whose walls have receded to infinity.

We have computed the mean energy in any particular mode in a box at the temperature T; we need only one more thing to get the blackbody radiation law: We need to know how many modes there are at each energy. (We assume that for every mode there are some atoms in the box—or in the walls—which have energy levels that can radiate into that mode, so that each mode can get into thermal equilibrium.) The blackbody radiation law is usually stated by giving the energy per unit volume carried by the light in a small frequency interval from ω to $\omega + \Delta\omega$. So we need to know how many modes there are in a box with frequencies in the

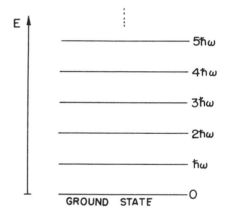

Fig. 4–7. The energy levels of a harmonic oscillator.

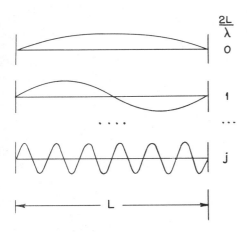

Fig. 4–8. The standing wave modes on a line.

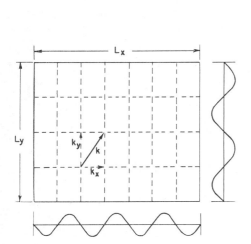

Fig. 4–9. Standing wave modes in two dimensions.

interval $\Delta\omega$. Although this question continually comes up in quantum mechanics, it is purely a classical question about standing waves.

We will get the answer only for a rectangular box. It comes out the same for a box of any shape, but it's very complicated to compute for the arbitrary case. Also, we are only interested in a box whose dimensions are very large compared with a wavelength of the light. Then there are billions and billions of modes; there will be many in any small frequency interval $\Delta\omega$, so we can speak of the "average number" in any $\Delta\omega$ at the frequency ω. Let's start by asking how many modes there are in a one-dimensional case—as for waves on a stretched string. You know that each mode is a sine wave that has to go to zero at both ends; in other words, there must be an integral number of half-wavelengths in the length of the line, as shown in Fig. 4–8. We prefer to use the wave number $k = 2\pi/\lambda$; calling k_j the wave number of the jth mode, we have that

$$k_j = \frac{j\pi}{L}, \tag{4.34}$$

where j is any integer. The separation δk between successive modes is

$$\delta k = k_{j+1} - k_j = \frac{\pi}{L}.$$

We want to assume that kL is so large that in a small interval Δk, there are many modes. Calling $\Delta\mathfrak{N}$ the number of modes in the interval Δk, we have

$$\Delta\mathfrak{N} = \frac{\Delta k}{\delta k} = \frac{L}{\pi}\,\Delta k. \tag{4.35}$$

Now theoretical physicists working in quantum mechanics usually prefer to say that there are one-half as many modes; they write

$$\Delta\mathfrak{N} = \frac{L}{2\pi}\,\Delta k. \tag{4.36}$$

We would like to explain why. They usually like to think in terms of travelling waves—some going to the right (with a positive k) and some going to the left (with a negative k). But a "mode" is a *standing* wave which is the sum of two waves, one going in each direction. In other words, they consider each standing wave as containing two distinct photon "states." So if by $\Delta\mathfrak{N}$, one prefers to mean the number of photon states of a given k (where now k ranges over positive and negative values), one should then take $\Delta\mathfrak{N}$ half as big. (All integrals must now go from $k = -\infty$ to $k = +\infty$, and the total number of states up to any given absolute value of k will come out O.K.) Of course, we are not then describing standing waves very well, but we are counting modes in a consistent way.

Now we want to extend the results to three dimensions. A standing wave in a rectangular box must have an integral number of half-waves *along each axis*. The situation for two of the dimensions is shown in Fig. 4–9. Each wave direction and frequency is described by a vector wave number $\boldsymbol{k}$, whose x, y, and z components must satisfy equations like Eq. (4.34). So we have that

$$k_x = \frac{j_x \pi}{L_x},$$

$$k_y = \frac{j_y \pi}{L_y},$$

$$k_z = \frac{j_z \pi}{L_z}.$$

The number of modes with k_x in an interval Δk_x is, as before,

$$\frac{L_x}{2\pi}\,\Delta k_x,$$

and similarly for Δk_y and Δk_z. If we call $\Delta\mathfrak{N}(\boldsymbol{k})$ the number of modes for a vector

4–10

wave number k whose x-component is between k_x and $k_x + \Delta k_x$, whose y-component is between k_y and $k_y + \Delta k_y$, and whose z-component is between k_z and $k_z + \Delta k_z$, then

$$\Delta \mathfrak{N}(\boldsymbol{k}) = \frac{L_x L_y L_z}{(2\pi)^3} \Delta k_x \, \Delta k_y \, \Delta k_z. \tag{4.37}$$

The product $L_x L_y L_z$ is equal to the volume V of the box. So we have the important result that for high frequencies (wavelengths small compared with the dimensions), the number of modes in a cavity is proportional to the volume V of the box and to the "volume in k-space" $\Delta k_x \, \Delta k_y \, \Delta k_z$. This result comes up again and again in many problems and should be memorized:

$$d\mathfrak{N}(\boldsymbol{k}) = V \frac{d^3 k}{(2\pi)^3}. \tag{4.38}$$

Although we have not proved it, the result is independent of the shape of the box.

We will now apply this result to find the number of photon modes for photons with frequencies in the range $\Delta \omega$. We are just interested in the energy in various modes—but not interested in the directions of the waves. We would like to know the number of modes in a given range of frequencies. In a vacuum the magnitude of k is related to the frequency by

$$|\boldsymbol{k}| = \frac{\omega}{c}. \tag{4.39}$$

So in a frequency interval $\Delta \omega$, these are all the modes which correspond to k's with a *magnitude* between k and $k + \Delta k$, independent of the direction. The "volume in k-space" between k and $k + \Delta k$ is a spherical shell of volume

$$4\pi k^2 \, \Delta k.$$

The number of modes is then

$$\Delta \mathfrak{N}(\omega) = \frac{V 4\pi k^2 \, \Delta k}{(2\pi)^3}. \tag{4.40}$$

However, since we are now interested in frequencies, we should substitute $k = \omega/c$, so we get

$$\Delta \mathfrak{N}(\omega) = \frac{V 4\pi \omega^2 \, \Delta \omega}{(2\pi)^3 c^3}. \tag{4.41}$$

There is one more complication. If we are talking about modes of an electromagnetic wave, for any given wave vector $\boldsymbol{k}$ there can be either of two polarizations (at right angles to each other). Since these modes are independent, we must—for light—double the number of modes. So we have

$$\Delta \mathfrak{N}(\omega) = \frac{V \omega^2 \, \Delta \omega}{\pi^2 c^3} \quad \text{(for light)}. \tag{4.42}$$

We have shown, Eq. (4.33), that each mode (or each "state") has on the average the energy

$$\bar{n} \hbar \omega = \frac{\hbar \omega}{e^{\hbar \omega / kT} - 1}.$$

Multiplying this by the number of modes, we get the energy ΔE in the modes that lie in the interval $\Delta \omega$:

$$\Delta E = \frac{\hbar \omega}{e^{\hbar \omega / kT} - 1} \frac{V \omega^2 \, \Delta \omega}{\pi^2 c^3}. \tag{4.43}$$

This is the law for the frequency spectrum of blackbody radiation, which we have already found in Chapter 41 of Vol. I. The spectrum is plotted in Fig. 4–10. You see now that the answer depends on the fact that photons are Bose particles, which

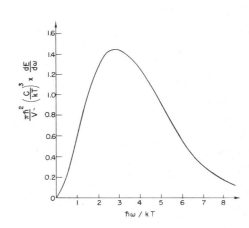

Fig. 4–10. The frequency spectrum of radiation in a cavity in thermal equilibrium, the "blackbody" spectrum.

have a tendency to try to get all into the same state (because the amplitude for doing so is large). You will remember, it was Planck's study of the blackbody spectrum (which was a mystery to classical physics), and his discovery of the formula in Eq. (4.43) that started the whole subject of quantum mechanics.

4–6 Liquid helium

Liquid helium has at low temperatures many odd properties which we cannot unfortunately take the time to describe in detail right now, but many of them arise from the fact that a helium atom is a Bose particle. One of the things is that liquid helium flows without any viscous resistance. It is, in fact, the ideal "dry" water we have been talking about in one of the earlier chapters—provided that the velocities are low enough. The reason is the following. In order for a liquid to have viscosity, there must be internal energy losses; there must be some way for one part of the liquid to have a motion that is different from that of the rest of the liquid. This means that it must be possible to knock some of the atoms into states that are different from the states occupied by other atoms. But at sufficiently low temperatures, when the thermal motions get very small, all the atoms try to get into the same condition. So, if some of them are moving along, then all the atoms try to move together in the same state. There is a kind of rigidity to the motion, and it is hard to break the motion up into irregular patterns of turbulence, as would happen, for example, with independent particles. So in a liquid of Bose particles, there is a strong tendency for all the atoms to go into the same state—which is represented by the $\sqrt{n + 1}$ factor we found earlier. (For a bottle of liquid helium n is, of course, a very large number!) This cooperative motion does not happen at high temperatures, because then there is sufficient thermal energy to put the various atoms into various different higher states. But at a sufficiently low temperature there suddenly comes a moment in which all the helium atoms try to go into the same state. The helium becomes a superfluid. Incidentally, this phenomenon only appears with the isotope of helium which has atomic weight 4. For the helium isotope of atomic weight 3, the individual atoms are Fermi particles, and the liquid is a normal fluid. Since superfluidity occurs only with He^4, it is evidently a quantum mechanical effect—due to the Bose nature of the α-particle.

4–7 The exclusion principle

Fermi particles act in a completely different way. Let's see what happens if we try to put two Fermi particles into the same state. We will go back to our original example and ask for the amplitude that two identical Fermi particles will be scattered into almost exactly the same direction. The amplitude that particle a will go in direction 1 and particle b will go in direction 2 is

$$\langle 1 \mid a \rangle \langle 2 \mid b \rangle,$$

whereas the amplitude that the outgoing directions will be interchanged is

$$\langle 2 \mid a \rangle \langle 1 \mid b \rangle.$$

Since we have Fermi particles, the amplitude for the process is the difference of these two amplitudes:

$$\langle 1 \mid a \rangle \langle 2 \mid b \rangle - \langle 2 \mid a \rangle \langle 1 \mid b \rangle. \tag{4.44}$$

Let's say that by "direction 1" we mean that the particle has not only a certain direction but also a given direction of its spin, and that "direction 2" is almost exactly the same as direction 1 and corresponds to the *same* spin direction. Then $\langle 1 \mid a \rangle$ and $\langle 2 \mid a \rangle$ are nearly equal. (This would not necessarily be true if the outgoing states 1 and 2 did not have the same spin, because there might be some reason why the amplitude would depend on the spin direction.) Now if we let

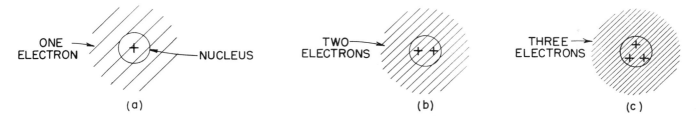

Fig. 4–11. How atoms might look if electrons behaved like Bose particles.

directions 1 and 2 approach each other, the total amplitude in Eq. (4.44) becomes zero. The result for Fermi particles is much simpler than for Bose particles. It just isn't possible at all for two Fermi particles—such as two electrons—to get into exactly the same state. You will never find two electrons in the same position with their two spins in the same direction. It is not possible for two electrons to have the same momentum and the same spin directions. If they are at the same location or with the same state of motion, the only possibility is that they must be spinning opposite to each other.

What are the consequences of this? There are a number of most remarkable effects which are a consequence of the fact that two Fermi particles cannot get into the same state. In fact, almost all the peculiarities of the material world hinge on this wonderful fact. The variety that is represented in the periodic table is basically a consequence of this one rule.

Of course, we cannot say what the world would be like if this one rule were changed, because it is just a part of the whole structure of quantum mechanics, and it is impossible to say what else would change if the rule about Fermi particles were different. Anyway, let's just try to see what would happen if only this one rule were changed. First, we can show that every atom would be more or less the same. Let's start with the hydrogen atom. It would not be noticeably affected. The proton of the nucleus would be surrounded by a spherically symmetric electron cloud, as shown in Fig. 4–11(a). As we have described in Chapter 2, the electron is attracted to the center, but the uncertainty principle requires that there be a balance between the concentration in space and in momentum. The balance means that there must be a certain energy and a certain spread in the electron distribution which determines the characteristic dimension of the hydrogen atom.

Now suppose that we have a nucleus with two units of charge, such as the helium nucleus. This nucleus would attract two electrons, and if they were Bose particles, they would—except for their electric repulsion—both crowd in as close as possible to the nucleus. A helium atom might look as shown in part (b) of the figure. Similarly, a lithium atom which has a triply charged nucleus would have an electron distribution like that shown in part (c) of Fig. 4–11. Every atom would look more or less the same—a little round ball with all the electrons sitting near the nucleus, nothing directional and nothing complicated.

Because electrons are Fermi particles, however, the actual situation is quite different. For the hydrogen atom the situation is essentially unchanged. The only difference is that the electron has a spin which we indicate by the little arrow in Fig. 4–12(a). In the case of a helium atom, however, we cannot put two electrons on top of each other. But wait, that is only true if their spins are the same. Two electrons *can* occupy the same state if their spins are opposite. So the helium atom does not look much different either. It would appear as shown in part (b) of Fig. 4–12. For lithium, however, the situation becomes quite different. Where can we put the third electron? The third electron cannot go on top of the other two because both spin directions are occupied. (You remember that for an electron or any particle with spin 1/2 there are only two possible directions for the spin.) The third electron can't go near the place occupied by the other two, so it must take up a special condition in a different kind of state farther away from the nucleus in part (c) of the figure. (We are speaking only in a rather rough way here, because in reality all three electrons are identical; since we cannot really distinguish which one is which, our picture is only an approximate one.)

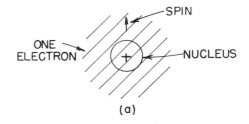

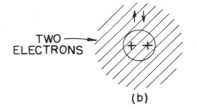

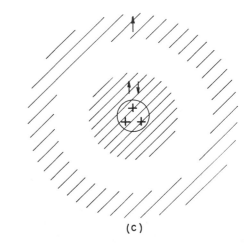

Fig. 4–12. Atomic configurations for real, Fermi-type, spin one-half electrons.

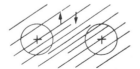

Fig. 4–13. The hydrogen molecule.

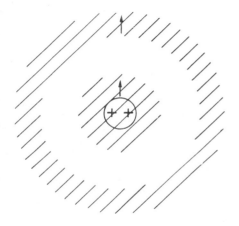

Fig. 4–14. Helium with one electron in a higher energy state.

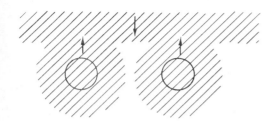

Fig. 4–15. The likely mechanism in a ferromagnetic crystal; the conduction electron is antiparallel to the unpaired inner electrons.

Now we can begin to see why different atoms will have different chemical properties. Because the third electron in lithium is farther out, it is relatively more loosely bound. It is much easier to remove an electron from lithium than from helium. (Experimentally, it takes 25 volts to ionize helium but only 5 volts to ionize lithium.) This accounts for the valence of the lithium atom. The directional properties of the valence have to do with the pattern of the waves of the outer electron, which we will not go into at the moment. But we can already see the importance of the so-called *exclusion principle*—which states that no two electrons can be found in exactly the same state (including spin).

The exclusion principle is also responsible for the stability of matter on a large scale. We explained earlier that the individual atoms in matter did not collapse because of the uncertainty principle; but this does not explain why it is that two hydrogen atoms can't be squeezed together as close as you want—why it is that all the protons don't get close together with one big smear of electrons around them. The answer is, of course, that since no more than two electrons—with opposite spins—can be in roughly the same place, the hydrogen atoms must keep away from each other. So the stability of matter on a large scale is really a consequence of the Fermi particle nature of the electrons.

Of course, if the outer electrons on two atoms have spins in opposite directions, they can get close to each other. This is, in fact, just the way that the chemical bond comes about. It turns out that two atoms together will generally have the lowest energy if there is an electron between them. It is a kind of an electrical attraction for the two positive nuclei toward the electron in the middle. It is possible to put two electrons more or less between the two nuclei so long as their spins are opposite, and the strongest chemical binding comes about this way. There is no stronger binding, because the exclusion principle does not allow there to be more than two electrons in the space between the atoms. We expect the hydrogen molecule to look more or less as shown in Fig. 4–13.

We want to mention one more consequence of the exclusion principle. You remember that if both electrons in the helium atom are to be close to the nucleus, their spins are necessarily opposite. Now suppose that we would like to try to arrange to have both electrons with the same spin—as we might consider doing by putting on a fantastically strong magnetic field that would try to line up the spins in the same direction. But then the two electrons could not occupy the same state in space. One of them would have to take on a different geometrical position, as indicated in Fig. 4–14. The electron which is located farther from the nucleus has less binding energy. The energy of the whole atom is therefore quite a bit higher. In other words, when the two spins are opposite, there is a much stronger total attraction.

So, there is an apparent, enormous force trying to line up spins opposite to each other when two electrons are close together. If two electrons are trying to go in the same place, there is a very strong tendency for the spins to become lined opposite. This apparent force trying to orient the two spins opposite to each other is much more powerful than the tiny force between the two magnetic moments of the electrons. You remember when we were speaking of ferromagnetism there was the mystery of why the electrons in different atoms had a strong tendency to line up parallel. Although there is still no quantitative explanation, it is believed that what happens is that the electrons around the core of one atom interact through the exclusion principle with the outer electrons which have become free to wander throughout the crystal. This interaction causes the spins of the free electrons and the inner electrons to take on opposite directions. But the free electrons and the inner atomic electrons can only be opposite provided all the inner electrons have the same spin direction, as indicated in Fig. 4–15. It seems probable that it is the effect of the exclusion principle acting indirectly through the free electrons that gives rise to the strong aligning forces responsible for ferromagnetism.

We will mention one further example of the influence of the exclusion principle. We have said earlier that the nuclear forces are the same between the neutron and the proton, between the proton and the proton, and between the proton and the neutron. Why is it then that a proton and a neutron can stick together to make a deuterium nucleus, whereas there is no nucleus with just two protons or with just two neutrons? The deuteron is, as a matter of fact, bound by an energy of about 2.2 million volts, yet, there is no corresponding binding between a pair of protons to make an isotope of helium with the atomic weight 2. Such nuclei do not exist. The combination of two protons does not make a bound state.

The answer is a result of two effects: first, the exclusion principle; and second, the fact that the nuclear forces are somewhat sensitive to the direction of spin. The force between a neutron and a proton is attractive and somewhat stronger when the spins are parallel than when they are opposite. It happens that these forces are just different enough that a deuteron can only be made if the neutron and proton have their spins parallel; when their spins are opposite, the attraction is not quite strong enough to bind them together. Since the spins of the neutron and proton are each one-half and are in the same direction, the deuteron has a spin of one. We know, however, that two protons are not allowed to sit on top of each other if their spins are parallel. If it were not for the exclusion principle, two protons would be bound, but since they cannot exist at the same place and with the same spin directions, the He^2 nucleus does not exist. The protons could come together with their spins opposite, but then there is not enough binding to make a stable nucleus, because the nuclear force for opposite spins is too weak to bind a pair of nucleons. The attractive force between neutrons and protons of opposite spins can be seen by scattering experiments. Similar scattering experiments with two protons with parallel spins show that there is the corresponding attraction. So it is the exclusion principle that helps explain why deuterium can exist when He^2 cannot.

Spin One

5–1 Filtering atoms with a Stern-Gerlach apparatus

In this chapter we really begin the quantum mechanics proper—in the sense that we are going to describe a quantum mechanical phenomenon in a completely quantum mechanical way. We will make no apologies and no attempt to find connections to classical mechanics. We want to talk about something new in a new language. The particular situation which we are going to describe is the behavior of the so-called quantization of the angular momentum, for a particle of *spin one*. But we won't use words like "angular momentum" or other concepts of classical mechanics until later. We have chosen this particular example because it is relatively simple, although not the simplest possible example. It is sufficiently complicated that it can stand as a prototype which can be generalized for the description of all quantum mechanical phenomena. Thus, although we are dealing with a particular example, all the laws which we mention are immediately generalizable, and we will give the generalizations so that you will see the general characteristics of a quantum mechanical description. We begin with the phenomenon of the splitting of a beam of atoms into three separate beams in a Stern-Gerlach experiment.

You remember that if we have an inhomogeneous magnetic field made by a magnet with a pointed pole tip and we send a beam through the apparatus, the beam of particles may be split into a number of beams—the number depending on the particular kind of atom and its state. We are going to take the case of an atom which gives three beams, and we are going to call that a particle of *spin one*. You can do for yourself the case of five beams, seven beams, two beams, etc.—you just copy everything down and where we have three terms, you will have five terms, seven terms, and so on.

Imagine the apparatus drawn schematically in Fig. 5–1. A beam of atoms (or particles of any kind) is collimated by some slits and passes through a non-uniform field. Let's say that the beam moves in the y-direction and that the magnetic field and its gradient are both in the z-direction. Then, looking from the side, we will see the beam split vertically into three beams, as shown in the figure. Now at the output end of the magnet we could put small counters which count the rate of arrival of particles in any one of the three beams. Or we can block off two of the beams and let the third one go on.

Suppose we block off the lower two beams and let the top-most beam go on and enter a second Stern-Gerlach apparatus of the same kind, as shown in Fig. 5–2. What happens? There are *not* three beams in the second apparatus; there is only the top beam.† This is what you would expect if you think of the second apparatus as simply an extension of the first. Those atoms which are being pushed upward continue to be pushed upward in the second magnet.

5–1 Filtering atoms with a Stern-Gerlach apparatus

5–2 Experiments with filtered atoms

5–3 Stern-Gerlach filters in series

5–4 Base states

5–5 Interfering amplitudes

5–6 The machinery of quantum mechanics

5–7 Transforming to a different base

5–8 Other situations

Review: Chapter 35, Vol. II, *Paramagnetism and Magnetic Resonance.* For your convenience this chapter is reproduced in the Appendix of this volume.

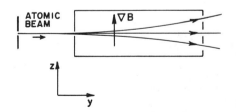

Fig. 5–1. In a Stern-Gerlach experiment, atoms of spin one are split into three beams.

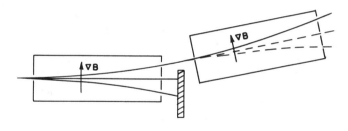

Fig. 5–2. The atoms from one of the beams are sent into a second identical apparatus.

---

† We are assuming that the deflection angles are very small.

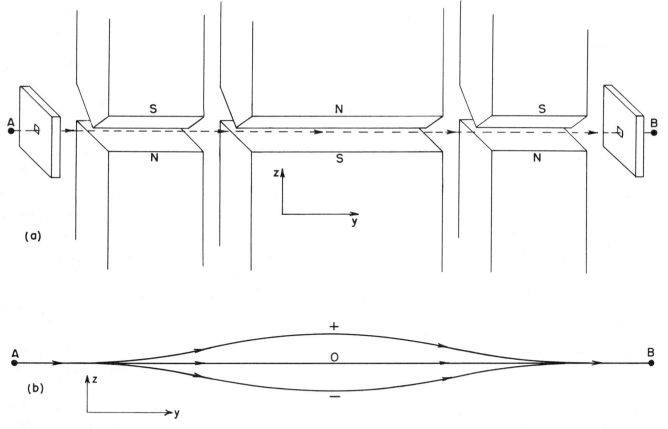

Fig. 5–3. (a) An imagined modification of a Stern-Gerlach apparatus. (b) The paths of spin-one atoms.

You can see then that the first apparatus has produced a beam of "purified" objects—atoms that get bent upward in the particular inhomogeneous field. The atoms, as they enter the original Stern-Gerlach apparatus, are of three "varieties," and the three kinds take different trajectories. By filtering out all but one of the varieties, we can make a beam whose future behavior in the same kind of apparatus is determined and predictable. We will call this a *filtered* beam, or a *polarized* beam, or a beam in which the atoms all are known to be in a *definite state*.

For the rest of our discussion, it will be more convenient if we consider a somewhat modified apparatus of the Stern-Gerlach type. The apparatus looks more complicated at first, but it will make all the arguments simpler. Anyway, since they are only "thought experiments," it doesn't cost anything to complicate the equipment. (Incidentally, no one has ever done all of the experiments we will describe in just this way, but we know what *would* happen from the laws of quantum mechanics, which are, of course, based on other similar experiments. These other experiments are harder to understand at the beginning, so we want to describe some idealized—but possible—experiments.)

Figure 5–3(a) shows a drawing of the "modified Stern-Gerlach apparatus" we would like to use. It consists of a sequence of three high-gradient magnets. The first one (on the left) is just the usual Stern-Gerlach magnet and splits the incoming beam of spin-one particles into three separate beams. The second magnet has the same cross section as the first, but is twice as long *and* the polarity of its magnetic field is opposite the field in magnet 1. The second magnet pushes in the opposite direction on the atomic magnets and bends their paths back toward the axis, as shown in the trajectories drawn in the lower part of the figure. The third magnet is just like the first, and brings the three beams back together again, so that leaves the exit hole along the axis. Finally, we would like to imagine that in front of the hole at A there is some mechanism which can get the atoms started from rest and that after the exit hole at B there is a decelerating mechanism that brings the atoms back to rest at B. That is not essential, but it will mean that in our analysis we won't have to worry about including any effects of the motion as the atoms come out, and can concentrate on those matters having only to do with the spin. The whole purpose of the "improved" apparatus is just to bring all the particles to the same place, and with zero speed.

Now if we want to do an experiment like the one in Fig. 5–2, we can first make a filtered beam by putting a plate in the middle of the apparatus that blocks two of the beams, as shown in Fig. 5–4. If we now put the polarized atoms through a second identical apparatus, all the atoms will take the upper path, as can be verified by putting similar plates in the way of the various beams of the second S filter and seeing whether particles get through.

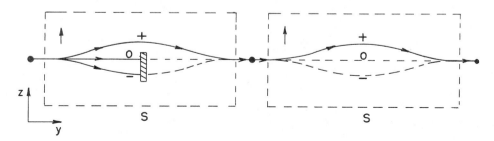

Fig. 5–4.   The "improved" Stern-Gerlach apparatus as a filter.

Suppose we call the first apparatus by the name S. (We are going to consider all sorts of combinations, and we will need labels to keep things straight.) We will say that the atoms which take the top path in S are in the "plus state with respect to S"; the ones which take the middle path are in the "zero state with respect to S"; and the ones which take the lowest path are in the "minus state with respect to S." (In the more usual language we would say that the z-component of the angular momentum was $+1\hbar$, 0, and $-1\hbar$, but we are not using that language now.) Now in Fig. 5–4 the second apparatus is oriented just like the first, so the filtered atoms will all go on the upper path. Or if we had blocked off the upper and lower beams in the first apparatus and let only the zero state through, all the filtered atoms would go through the middle path of the second apparatus. And if we had blocked off all but the lowest beam in the first, there would be only a low beam in the second. We can say that in each case our first apparatus has produced a filtered beam in a *pure* state with respect to S ($+$, 0, or $-$), and we can test which state is present by putting the atoms through a second, identical apparatus.

We can make our second apparatus so that it transmits only atoms of a particular state—by putting masks inside it as we did for the first one—and then we can test the state of the incoming beam just by seeing whether anything comes out the far end. For instance, if we block off the two lower paths in the second apparatus, 100 percent of the atoms will still come through; but if we block off the upper path, nothing will get through.

To make this kind of discussion easier, we are going to invent a shorthand symbol to represent one of our improved Stern-Gerlach apparatuses. We will let the symbol

$$\left\{\begin{array}{c}+\\0\\-\end{array}\right\}_S \tag{5.1}$$

stand for one complete apparatus. (This is *not* a symbol you will ever find used in quantum mechanics; we've just invented it for this chapter. It is simply meant to be a shorthand picture of the apparatus of Fig. 5–3.) Since we are going to want to use several apparatuses at once, and with various orientations, we will identify each with a letter underneath. So the symbol in (5.1) stands for the apparatus S. When we block off one or more of the beams inside, we will show that by some vertical bars indicating which beam is blocked, like this:

$$\left\{ \begin{matrix} + \\ 0 \\ - \end{matrix} \middle| \right\}_{S}. \tag{5.2}$$

The various possible combinations we will be using are shown in Fig. 5–5.

If we have two filters in succession (as in Fig. 5–4), we will put the two symbols next to each other, like this:

$$\left\{ \begin{matrix} + \\ 0 \\ - \end{matrix} \middle| \right\}_{S} \quad \left\{ \begin{matrix} + \\ 0 \\ - \end{matrix} \right\}_{S}. \tag{5.3}$$

For this setup, everything that comes through the first also gets through the second. In fact, even if we block off the "zero" and "minus" channels of the second apparatus, so that we have

$$\left\{ \begin{matrix} + \\ 0 \\ - \end{matrix} \middle| \right\}_{S} \quad \left\{ \begin{matrix} + \\ 0 \\ - \end{matrix} \middle| \right\}_{S}, \tag{5.4}$$

we still get 100 percent transmission through the second apparatus. On the other hand, if we have

$$\left\{ \begin{matrix} + \\ 0 \\ - \end{matrix} \middle| \right\}_{S} \quad \left\{ \begin{matrix} + \\ 0 \\ - \end{matrix} \middle| \right\}_{S}, \tag{5.5}$$

nothing at all comes out of the far end. Similarly,

$$\left\{ \begin{matrix} + \\ 0 \\ - \end{matrix} \middle| \right\}_{S} \quad \left\{ \begin{matrix} + \\ 0 \\ - \end{matrix} \middle| \right\}_{S} \tag{5.6}$$

would give nothing out. On the other hand,

$$\left\{ \begin{matrix} + \\ 0 \\ - \end{matrix} \middle| \right\}_{S} \quad \left\{ \begin{matrix} + \\ 0 \\ - \end{matrix} \middle| \right\}_{S} \tag{5.7}$$

would be just equivalent to

$$\left\{ \begin{matrix} + \\ 0 \\ - \end{matrix} \middle| \right\}_{S}$$

by itself.

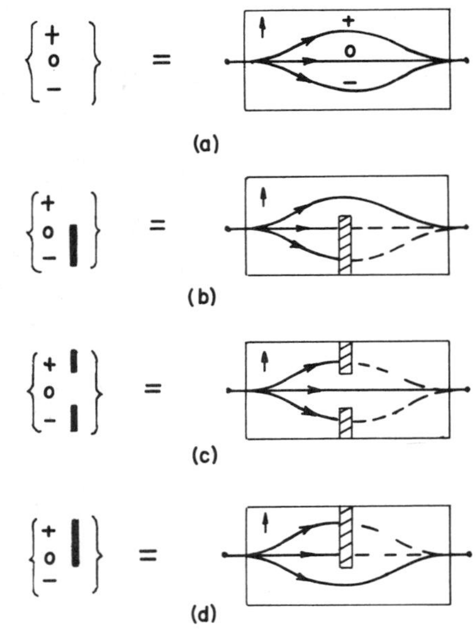

Fig. 5–5. Special shorthand symbols for Stern-Gerlach type filters.

Now we want to describe these experiments quantum mechanically. We will say that an atom is in the $(+S)$ state if it has gone through the apparatus of Fig. 5–5(b), that it is in a $(0\,S)$ state if it has gone through (c), and in a $(-S)$ state if it has gone through (d).† Then we let $\langle b \mid a \rangle$ be the *amplitude* that an atom which is in state a will get through an apparatus into the b state. We can say: $\langle b \mid a \rangle$ is the amplitude for an atom *in* the state a to *get into* the state b. The experiment (5.4) gives us that

$$\langle +S \mid +S \rangle = 1,$$

† Read: $(+S) = $ "plus-S"; $(0\,S) = $ "zero-S"; $(-S) = $ "minus-S."

5–4

whereas (5.5) gives us

$$\langle -S \mid +S \rangle = 0.$$

Similarly, the result of (5.6) is

$$\langle +S \mid -S \rangle = 0,$$

and of (5.7) is

$$\langle -S \mid -S \rangle = 1.$$

As long as we deal only with "pure" states—that is, we have only one channel open—there are nine such amplitudes, and we can write them in a table:

		from			
		$+S$	$0\,S$	$-S$	
to	$+S$	1	0	0	(5.8)
	$0\,S$	0	1	0	
	$-S$	0	0	1	

This array of nine numbers—called a *matrix*—summarizes the phenomena we've been describing.

5–2 Experiments with filtered atoms

Now comes the big question: What happens if the second apparatus is tipped to a different angle, so that its field axis is no longer parallel to the first? It could be not only tipped, but also pointed in a different direction—for instance, it could take the beam off at 90° with respect to the original direction. To take it easy at first, let's first think about an arrangement in which the second Stern-Gerlach experiment is tilted by some angle α about the y-axis, as shown in Fig. 5–6. We'll call the second apparatus T. Suppose that we now set up the following experiment:

$$\left\{ \begin{matrix} + \\ 0 \\ - \end{matrix} \Big| \right\}_{S} \quad \left\{ \begin{matrix} + \\ 0 \\ - \end{matrix} \Big| \right\}_{T},$$

or the experiment:

$$\left\{ \begin{matrix} + \\ 0 \\ - \end{matrix} \Big| \right\}_{S} \quad \left\{ \begin{matrix} + \\ 0 \\ - \end{matrix} \Big| \right\}_{T}.$$

What comes out at the far end in these cases?

The answer is this: If the atoms are in a definite state with respect to S, they are *not* in the same state with respect to T—a $(+S)$ state is *not* also a $(+T)$ state. There *is*, however, a certain *amplitude* to find the atom in a $(+T)$ state—or a $(0\,T)$ state or a $(-T)$ state.

In other words, as careful as we have been to make sure that we have the atoms in a definite condition, the fact of the matter is that if it goes through an apparatus which is tilted at a different angle it has, so to speak, to "reorient"

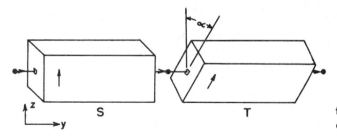

Fig. 5–6. Two Stern-Gerlach type filters in series; the second is tilted at the angle α with respect to the first.

itself—which it does, don't forget, by luck. We can put only one particle through at a time, and then we can only ask the question: What is the probability that it gets through? Some of the atoms that have gone through S will end in a $(+T)$ state, some of them will end in a $(0\,T)$, and some in a $(-T)$ state—all with different odds. These odds can be calculated by the absolute squares of complex amplitudes; what we want is some mathematical method, or quantum mechanical description, for these amplitudes. What we need to know are various quantities like

$$\langle -T \mid +S \rangle,$$

by which we mean the amplitude that an atom initially in the $(+S)$ state can get into the $(-T)$ condition (which is *not* zero unless T and S are lined up parallel to each other). There are other amplitudes like

$$\langle +T \mid 0\,S \rangle, \quad \text{or} \quad \langle 0\,T \mid -S \rangle, \quad \text{etc.}$$

There are, in fact, nine such amplitudes—another matrix—that a theory of particles should tell us how to calculate. Just as $F = ma$ tells us how to calculate what happens to a classical particle in any circumstance, the laws of quantum mechanics permit us to determine the amplitude that a particle will get through a particular apparatus. The central problem, then, is to be able to calculate—for any given tilt angle α, or in fact for any orientation whatever—the nine amplitudes:

$$
\begin{array}{lll}
\langle +T \mid +S \rangle, & \langle +T \mid 0\,S \rangle, & \langle +T \mid -S \rangle, \\
\langle 0\,T \mid +S \rangle, & \langle 0\,T \mid 0\,S \rangle, & \langle 0\,T \mid -S \rangle, \\
\langle -T \mid +S \rangle, & \langle -T \mid 0\,S \rangle, & \langle -T \mid -S \rangle.
\end{array}
\tag{5.9}
$$

We can already figure out some relations among these amplitudes. First, according to our definitions, the absolute square

$$|\langle +T \mid +S \rangle|^2$$

is the *probability* that an atom in a $(+S)$ state will enter a $(+T)$ state. We will often find it more convenient to write such squares in the equivalent form

$$\langle +T \mid +S \rangle \langle +T \mid +S \rangle^*.$$

In the same notation the number

$$\langle 0\,T \mid +S \rangle \langle 0\,T \mid +S \rangle^*$$

is the probability that a particle in the $(+S)$ state will enter the $(0\,T)$ state, and

$$\langle -T \mid +S \rangle \langle -T \mid +S \rangle^*$$

is the probability that it will enter the $(-T)$ state. But the way our apparatuses are made, every atom which enters the T apparatus must be found in *some* one of the three states of the T apparatus—there's nowhere else for a given kind of atom to go. So the sum of the three probabilities we've just written must be equal to 100 percent. We have the relation

$$
\langle +T \mid +S \rangle \langle +T \mid +S \rangle^* + \langle 0\,T \mid +S \rangle \langle 0\,T \mid +S \rangle^* \\
+ \langle -T \mid +S \rangle \langle -T \mid +S \rangle^* = 1.
\tag{5.10}
$$

There are, of course, two other such equations that we get if we start with a $(0\,S)$ or a $(-S)$ state. But they are all we can easily get, so we'll go on to some other general questions.

5–3 Stern-Gerlach filters in series

Here is an interesting question: Suppose we had atoms filtered into the $(+S)$ state, then we put them through a second filter, say into a $(0\,T)$ state, and *then* through *another* $+S$ filter. (We'll call the last filter S' just so we can distinguish it from the first S-fiter.) Do the atoms remember that they were once in a $(+S)$ state? In other words, we have the following experiment:

$$\left\{\begin{matrix}+\ \\ 0\ | \\ -\ | \end{matrix}\right\} \quad \left\{\begin{matrix}+\ | \\ 0\ \\ -\ | \end{matrix}\right\} \quad \left\{\begin{matrix}+\ \\ 0\ | \\ -\ | \end{matrix}\right\}. \tag{5.11}$$
$$STS'$$

We want to know whether all those that get through T also get through S'. *They do not.* Once they have been filtered by T, they *do not remember* in any way that they were in a $(+S)$ state when they entered T. Note that the second S apparatus in (5.11) is oriented exactly the same as the first, so it is still an S-type filter. The states filtered by S' are, of course, still $(+S)$, $(0\ S)$, and $(-S)$.

The important point is this: *If the T filter passes only one beam, the fraction* that gets through the second S filter depends only on the setup of the T filter, and is completely independent of what precedes it. The fact that the same atoms were once sorted by an S filter has no influence whatever on what they will do once they have been sorted again into a pure beam by a T apparatus. From then on, the probability for getting into different states is the same no matter what happened before they got into the T apparatus.

As an example, let's compare the experiment of (5.11) with the following experiment:

$$\left\{\begin{matrix}+\ | \\ 0\ \\ -\ | \end{matrix}\right\} \quad \left\{\begin{matrix}+\ | \\ 0\ \\ -\ | \end{matrix}\right\} \quad \left\{\begin{matrix}+\ \\ 0\ | \\ -\ | \end{matrix}\right\} \tag{5.12}$$
$$STS'$$

in which only the first S is changed. Let's say that the angle α (between S and T) is such that in experiment (5.11) one-third of the atoms that get through T also get through S'. In experiment (5.12), although there will, in general, be a different number of atoms coming through T, the *same fraction of these*—one-third—will also get through S'.

We can, in fact, show from what you have learned earlier that the fraction of the atoms that come out of T and get through any particular S' depends only on T and S', not on anything that happened earlier. Let's compare experiment (5.12) with

$$\left\{\begin{matrix}+\ | \\ 0\ \\ -\ | \end{matrix}\right\} \quad \left\{\begin{matrix}+\ | \\ 0\ \\ -\ | \end{matrix}\right\} \quad \left\{\begin{matrix}+\ | \\ 0\ \\ -\ | \end{matrix}\right\}. \tag{5.13}$$
$$STS'$$

The amplitude that an atom that comes out of S will also get through both T and S' is, for the experiments of (5.12),

$$\langle +S \mid 0\,T \rangle \langle 0\,T \mid 0\,S \rangle.$$

The corresponding probability is

$$|\langle +S \mid 0\,T \rangle \langle 0\,T \mid 0\,S \rangle|^2 = |\langle +S \mid 0\,T \rangle|^2\,|\langle 0\,T \mid 0\,S \rangle|^2.$$

The probability for experiment (5.13) is

$$|\langle 0\,S \mid 0\,T \rangle \langle 0\,T \mid 0\,S \rangle|^2 = |\langle 0\,S \mid 0\,T \rangle|^2\,|\langle 0\,T \mid 0\,S \rangle|^2.$$

The ratio is

$$\frac{|\langle 0\,S \mid 0\,T \rangle|^2}{|\langle +S \mid 0\,T \rangle|^2}$$

and depends only on T and S', and not at all on which beam $(+S)$, $(0\ S)$, or $(-S)$ is selected by S. (The absolute numbers may go up and down together depending on how much gets through T.) We would, of course, find the same result if we compared the probabilities that the atoms would go into the plus or the minus states with respect to S', or the ratio of the probabilities to go into the zero or minus states.

In fact, since these ratios depend only on which beam is allowed to pass through T, and not on the selection made by the first S filter, it is clear that we would get the same result even if the last apparatus were not an S filter. If we use for the third apparatus—which we will now call R—one rotated by some arbitrary angle with respect to T, we would find that a ratio such as $|\langle\, 0\, R\,|\, 0\, T\rangle|^2/|\langle +R\,|\, 0\, T\rangle|^2$ was independent of which beam was passed by the first filter S.

5–4 Base states

These results illustrate one of the basic principles of quantum mechanics: Any atomic system can be separated by a filtering process into a certain set of what we will call *base states*, and the future behavior of the atoms in any single given base state depends only on the nature of the base state—it is independent of any previous history.† The base states depend, of course, on the filter used; for instance, the three states $(+T)$, $(\,0\, T)$, and $(-T)$ are one set of base states; the three states $(+S)$, $(\,0\, S)$, and $(-S)$ are another. There are any number of possibilities each as good as any other.

We should be careful to say that we are considering *good* filters which do indeed produce "pure" beams. If, for instance, our Stern-Gerlach apparatus didn't produce a good separation of the three beams so that we could not separate them cleanly by our masks, then we could not make a complete separation into base states. We can tell if we have pure base states by seeing whether or not the beams can be split again in another filter of the same kind. If we have a pure $(+T)$ state, for instance, all the atoms will go through

$$\left\{\begin{matrix} + \\ 0 \\ - \end{matrix}\Big| \right\}_{T},$$

and none will go through

$$\left\{\begin{matrix} +\,| \\ 0 \\ -\,| \end{matrix}\right\}_{T},$$

or through

$$\left\{\begin{matrix} + \\ 0 \\ - \end{matrix}\Big|\right\}_{T}.$$

Our statement about base states means that it is *possible* to filter to some pure state, so that no further filtering by an identical apparatus is possible.

We must also point out that what we are saying is exactly true only in rather idealized situations. In any real Stern-Gerlach apparatus, we would have to worry about diffraction by the slits that could cause some atoms to go into states corresponding to different angles, or about whether the beams might contain atoms with different excitations of their internal states, and so on. We have idealized the situation so that we are talking only about the states that are split in a magnetic field; we are ignoring things having to do with position, momentum, internal excitations, and the like. In general, one would need to consider also base states which are sorted out with respect to such things also. But to keep the concepts simple, we are considering only our set of three states, which is sufficient for the exact treatment of the idealized situation in which the atoms don't get torn up in

---

† We do not intend the word "base state" to imply anything more than what is said here. They are not to be thought of as "basic" in any sense. We are using the word base with the thought of a *basis* for a description, somewhat in the sense that one speaks of "numbers to the *base* ten."

going through the apparatus, or otherwise badly treated, and come to rest when they leave the apparatus.

You will note that we always begin our thought experiments by taking a filter with only one channel open, so that we start with some definite base state. We do this because atoms come out of a furnace in various states determined at random by the accidental happenings inside the furnace. (It gives what is called an "unpolarized" beam.) This randomness involves probabilities of the "classical" kind—as in coin tossing—which are different from the quantum mechanical probabilities we are worrying about now. Dealing with an unpolarized beam would get us into additional complications that are better to avoid until after we understand the behavior of polarized beams. So don't try to consider at this point what happens if the *first* apparatus lets more than one beam through. (We will tell you how you can handle such cases at the end of the chapter.)

Let's now go back and see what happens when we go from a base state for one filter to a base state for a different filter. Suppose we start again with

$$\left\{ \begin{matrix} + \\ 0 \\ - \end{matrix} \right\}_{S} \quad \left\{ \begin{matrix} + \\ 0 \\ - \end{matrix} \right\}_{T} .$$

The atoms which come out of T are in the base state $(0\,T)$ and have no memory that they were once in the state $(+S)$. Some people would say that in the filtering by T we have "lost the information" about the previous state $(+S)$ because we have "disturbed" the atoms when we separated them into three beams in the apparatus T. But that is not true. The past information is not lost by the *separation* into three beams, but by the *blocking masks* that are put in—as we can see by the following set of experiments.

We start with a $+S$ filter and will call N the number of atoms that come through it. If we follow this by a $0\,T$ filter, the number of atoms that come out is some fraction of the original number, say αN. If we then put another $+\,S$ filter, only some fraction β of these atoms will get to the far end. We can indicate this in the following way:

$$\left\{ \begin{matrix} + \\ 0 \\ - \end{matrix} \right\}_{S} \xrightarrow{N} \left\{ \begin{matrix} + \\ 0 \\ - \end{matrix} \right\}_{T} \xrightarrow{\alpha N} \left\{ \begin{matrix} + \\ 0 \\ - \end{matrix} \right\}_{S'} \xrightarrow{\beta \alpha N} . \tag{5.14}$$

If our third apparatus S' selected a different state, say the $(0\,S)$ state, a different fraction, say γ, would get through.† We would have

$$\left\{ \begin{matrix} + \\ 0 \\ - \end{matrix} \right\}_{S} \xrightarrow{N} \left\{ \begin{matrix} + \\ 0 \\ - \end{matrix} \right\}_{T} \xrightarrow{\alpha N} \left\{ \begin{matrix} + \\ 0 \\ - \end{matrix} \right\}_{S'} \xrightarrow{\gamma \alpha N} . \tag{5.15}$$

Now suppose we repeat these two experiments but remove all the masks from T. We would then find the remarkable results as follows:

$$\left\{ \begin{matrix} + \\ 0 \\ - \end{matrix} \right\}_{S} \xrightarrow{N} \left\{ \begin{matrix} + \\ 0 \\ - \end{matrix} \right\}_{T} \xrightarrow{N} \left\{ \begin{matrix} + \\ 0 \\ - \end{matrix} \right\}_{S'} \xrightarrow{N} , \tag{5.16}$$

$$\left\{ \begin{matrix} + \\ 0 \\ - \end{matrix} \right\}_{S} \xrightarrow{N} \left\{ \begin{matrix} + \\ 0 \\ - \end{matrix} \right\}_{T} \xrightarrow{N} \left\{ \begin{matrix} + \\ 0 \\ - \end{matrix} \right\}_{S'} \xrightarrow{0} . \tag{5.17}$$

---

† In terms of our earlier notation $\alpha = |\langle 0\,T | + S\rangle|^2$, $\beta = |\langle +\,S | 0\,T\rangle|^2$, and $\gamma = |\langle 0\,S | 0\,T\rangle|^2$.

All the atoms get through S' in the first case, but *none* in the second case! This is one of the great laws of quantum mechanics. That nature works this way is not self-evident, but the results we have given correspond for our idealized situation to the quantum mechanical behavior observed in innumerable experiments.

5–5 Interfering amplitudes

How can it be that in going from (5.15) to (5.17)—by *opening more channels* —we let *fewer* atoms through? This is the old, deep mystery of quantum mechanics —the interference of amplitudes. It's the same kind of thing we first saw in the two-slit interference experiment with electrons. We saw that we could get fewer electrons at some places with both slits open than we got with one slit open. It works quantitatively this way. We can write the amplitude that an atom will get through T and S' in the apparatus of (5.17) as the sum of three amplitudes, one for each of the three beams in T; the sum is equal to zero:

$$\langle 0\,S \mid +T \rangle\langle +T \mid +S \rangle + \langle 0\,S \mid 0\,T \rangle\langle 0\,T \mid +S \rangle + \langle 0\,S \mid -T \rangle\langle -T \mid +S \rangle = 0.$$
(5.18)

None of the three individual amplitudes is zero—for example, the absolute square of the second amplitude is $\gamma\alpha$, see (5.15)—but the *sum is zero*. We would have also the same answer if S' were set to select the $(-S)$ state. However, in the setup of (5.16), the answer is different. If we call a the amplitude to get through T and S', in this case we have†

$$a = \langle +S \mid +T \rangle\langle +T \mid +S \rangle + \langle +S \mid 0\,T \rangle\langle 0\,T \mid +S \rangle$$
$$+ \langle +S \mid -T \rangle\langle -T \mid +S \rangle = 1. \qquad (5.19)$$

In the experiment (5.16) the beam has been split and recombined. Humpty Dumpty has been put back together again. The information about the original $(+S)$ state is retained—it is just as though the T apparatus were not there at all. This is true whatever is put after the "wide-open" T apparatus. We could follow it with an R filter—a filter at some odd angle—or anything we want. The answer will always be the same as if the atoms were taken directly from the first S filter.

So this is the important principle: A T filter—or any filter—with wide-open masks produces no change at all. We should make one additional condition. The wide-open filter must not only transmit all three beams, but it must also *not* produce unequal disturbances on the three beams. For instance, it should not have a strong electric field near one beam and not the others. The reason is that even if this extra disturbance would still let all the atoms through the filter, it could change the *phases* of some of the amplitudes. Then the interference would be changed, and the amplitudes in Eqs. (5.18) and (5.19) would be different. We will always assume that there are no such extra disturbances.

Let's rewrite Eqs. (5.18) and (5.19) in an improved notation. We will let i stand for any one of the three states $(+T)$, $(0\,T)$, or $(-T)$; then the equations can be written:

$$\sum_{\text{all } i} \langle 0\,S \mid i \rangle\langle i \mid +S \rangle = 0 \qquad (5.20)$$

and

$$\sum_{\text{all } i} \langle +S \mid i \rangle\langle i \mid +S \rangle = 1. \qquad (5.21)$$

Similarly, for an experiment where S' is replaced by a completely arbitrary filter R, we have

$$\left\{ \begin{matrix} + \\ 0 \\ - \end{matrix} \middle| \right\}_S \quad \left\{ \begin{matrix} + \\ 0 \\ - \end{matrix} \right\}_T \quad \left\{ \begin{matrix} + \\ 0 \\ - \end{matrix} \middle| \right\}_R. \qquad (5.22)$$

———

† We really cannot conclude from the experiment that $a = 1$, but only that $|a|^2 = 1$, so a might be $e^{i\delta}$, but it can be shown that the choice $\delta = 0$ represents no real loss of generality.

The results will always be the same as if the T apparatus were left out and we had only

$$\left\{ \begin{matrix} + \\ 0 \\ - \end{matrix} \middle| \right\}_S \qquad \left\{ \begin{matrix} + \\ 0 \\ - \end{matrix} \middle| \right\}_R .$$

Or, expressed mathematically,

$$\sum_{\text{all } i} \langle +R \mid i \rangle \langle i \mid +S \rangle = \langle +R \mid +S \rangle. \qquad (5.23)$$

This is our fundamental law, and it is generally true so long as i stands for the three base states of any filter.

You will notice that in the experiment (5.22) there is no special relation of S and R to T. Furthermore, the arguments would be the same no matter what states they selected. To write the equation in a general way, without having to refer to the specific states selected by S and R, let's call ϕ ("phi") the state prepared by the first filter (in our special example, $+S$) and χ ("khi") the state tested by the final filter (in our example, $+R$). Then we can state our fundamental law of Eq. (5.23) in the form

$$\langle \chi \mid \phi \rangle = \sum_{\text{all } i} \langle \chi \mid i \rangle \langle i \mid \phi \rangle, \qquad (5.24)$$

where i is to range over the three base states of some particular filter.

We want to emphasize again what we mean by base states. They are like the three states which can be selected by one of our Stern-Gerlach apparatuses. One condition is that if you have a base state, then the future is independent of the past. Another condition is that if you have a complete set of base states, Eq. (5.24) is true for any set of beginning and ending states ϕ and χ. There is, however, *no unique* set of base states. We began by considering base states *with respect to* a particular apparatus T. We could equally well consider a *different set* of base states with respect to an apparatus S, or with respect to R, etc.† We usually speak of the base states "in a certain representation."

Another condition on a set of base states in any particular representation is that they are all completely different. By that we mean that if we have a $(+T)$ state, there is no amplitude for it to go into a $(0T)$ or a $(-T)$ state. If we let i and j stand for any two base states of a particular set, the general rules discussed in connection with (5.8) are that

$$\langle j \mid i \rangle = 0$$

for all i and j that are not equal. Of course, we know that

$$\langle i \mid i \rangle = 1.$$

These two equations are usually written as

$$\langle j \mid i \rangle = \delta_{ji}, \qquad (5.25)$$

where δ_{ji} (the "Kronecker delta") is a symbol that is defined to be zero for $i \neq j$, and to be one for $i = j$.

Equation (5.25) is not independent of the other laws we have mentioned. It happens that we are not particularly interested in the mathematical problem of finding the minimum set of independent axioms that will give all the laws as consequences.‡ We are satisfied if we have a set that is complete and not apparently inconsistent. We can, however, show that Eqs. (5.25) and (5.24) are not independent. Suppose we let ϕ in Eq. (5.24) represent one of the base states of the

---

† In fact, for atomic systems with three or more base states, there exist other kinds of filters—quite different from a Stern-Gerlach apparatus—which can be used to get more choices for the set of base states (each set with the same *number* of states).

‡ Redundant *truth* doesn't bother us!

same set as i, say the jth state; then we have

$$\langle x \mid j \rangle = \sum_i \langle x \mid i \rangle \langle i \mid j \rangle.$$

But Eq. (5.25) says that $\langle i \mid j \rangle$ is zero unless $i = j$, so the sum becomes just $\langle x \mid j \rangle$ and we have an identity, which shows that the two laws are not independent.

We can see that there must be another relation among the amplitudes if both Eqs. (5.10) and (5.24) are true. Equation (5.10) is

$$\langle +T \mid +S \rangle \langle +T \mid +S \rangle^* + \langle 0\,T \mid +S \rangle \langle 0\,T \mid +S \rangle^* + \langle -T \mid +S \rangle \langle -T \mid +S \rangle^* = 1.$$

If we write Eq. (5.24), letting both ϕ and x be the state $(+S)$, the left-hand side is $\langle +S \mid +S \rangle$, which is clearly $=1$; so we get once more Eq. (5.19),

$$\langle +S \mid +T \rangle \langle +T \mid +S \rangle + \langle +S \mid 0\,T \rangle \langle 0\,T \mid +S \rangle + \langle +S \mid -T \rangle \langle -T \mid +S \rangle = 1.$$

These two equations are consistent (for all relative orientations of the T and S apparatuses) only if

$$\langle +S \mid +T \rangle = \langle +T \mid +S \rangle^*,$$
$$\langle +S \mid 0\,T \rangle = \langle 0\,T \mid +S \rangle^*,$$
$$\langle +S \mid -T \rangle = \langle -T \mid +S \rangle^*.$$

And it follows that for any states ϕ and x,

$$\langle \phi \mid x \rangle = \langle x \mid \phi \rangle^*. \tag{5.26}$$

If this were not true, probability wouldn't be "conserved," and particles would get "lost."

Before going on, we want to summarize the three important general laws about amplitudes. They are Eqs. (5.24), (5.25), and (5.26):

$$\text{I} \quad \langle j \mid i \rangle = \delta_{ji},$$
$$\text{II} \quad \langle x \mid \phi \rangle = \sum_{\text{all } i} \langle x \mid i \rangle \langle i \mid \phi \rangle, \tag{5.27}$$
$$\text{III} \quad \langle \phi \mid x \rangle = \langle x \mid \phi \rangle^*.$$

In these equations the i and j refer to *all* the base states of some *one* representation, while ϕ and x represent any possible states of the atom. It is important to note that II is valid only if the sum is carried out over *all* the base states of the system (in our case, three: $+T$, $0\,T$, $-T$). These laws say nothing about what we should choose for a base for our set of base states. We began by using a T apparatus, which is a Stern-Gerlach experiment with some arbitrary orientation; but any other orientation, say W, would be just as good. We would have a different set of states to use for i and j, but all the laws would still be good—there is no unique set. One of the great games of quantum mechanics is to make use of the fact that things can be calculated in more than one way.

5–6 The machinery of quantum mechanics

We want to show you why these laws are useful. Suppose we have an atom in a given condition (by which we mean that it was prepared in a certain way), and we want to know what will happen to it in some experiment. In other words, we start with our atom in the state ϕ and want to know what are the *odds* that it will go through some apparatus which accepts atoms only in the condition x. The laws say that we can describe the apparatus completely in terms of three complex numbers $\langle x \mid i \rangle$, the amplitudes for each base state to be in the condition x; and that we can tell what will happen if an atom is put into the apparatus if we describe the state of the atom by giving three numbers $\langle i \mid \phi \rangle$, the amplitudes for the atom in its original condition to be found in each of the three base states. This is an important idea.

Let's consider another illustration. Think of the following problem: We start with an S apparatus; then we have a complicated mess of junk, which we can call A, and then an R apparatus—like this:

$$\left\{ \begin{matrix} + \\ 0 \\ - \end{matrix} \Big| \right\}_S \quad \left\{ A \right\} \quad \left\{ \begin{matrix} + \\ 0 \\ - \end{matrix} \Big| \right\}_R . \tag{5.28}$$

By A we mean any complicated arrangement of Stern-Gerlach apparatuses with masks or half-masks, oriented at peculiar angles, with odd electric and magnetic fields . . . almost anything you want to put. (It's nice to do thought experiments— you don't have to go to all the trouble of actually *building* the apparatus!) The problem then is: With what amplitude does a particle that enters the section A in a $(+S)$ state come out of A in the $(0R)$ state, so that it will get through the last R filter? There is a regular notation for such an amplitude; it is

$$\langle 0R \mid A \mid +S \rangle.$$

As usual, it is to be read from right to left (like Hebrew):

$$\langle \text{finish} \mid \text{through} \mid \text{start} \rangle.$$

If by chance A doesn't do anything—but is just an open channel—then we write

$$\langle 0R \mid 1 \mid +S \rangle = \langle 0R \mid +S \rangle; \tag{5.29}$$

the two symbols are equivalent. For a more general problem, we might replace $(+S)$ by a general starting state ϕ and $(0R)$ by a general finishing state χ, and we would want to know the amplitude

$$\langle \chi \mid A \mid \phi \rangle.$$

A complete analysis of the apparatus A would have to give the amplitude $\langle \chi \mid A \mid \phi \rangle$ for every possible pair of states ϕ and χ—an infinite number of combinations! How then can we give a concise description of the behavior of the apparatus A? We can do it in the following way. Imagine that the apparatus of (5.28) is modified to be

$$\left\{ \begin{matrix} + \\ 0 \\ - \end{matrix} \Big| \right\}_S \quad \left\{ \begin{matrix} + \\ 0 \\ - \end{matrix} \right\}_T \quad \left\{ A \right\} \quad \left\{ \begin{matrix} + \\ 0 \\ - \end{matrix} \right\}_T \quad \left\{ \begin{matrix} + \\ 0 \\ - \end{matrix} \Big| \right\}_R . \tag{5.30}$$

This is really no modification at all since the wide-open T apparatuses don't do anything. But they do suggest how we can analyze the problem. There is a certain set of amplitudes $\langle i \mid +S \rangle$ that the atoms from S will get into the i state of T. Then there is another set of amplitudes that an i state (with respect to T) entering A will come out as a j state (with respect to T). And finally there is an amplitude that each j state will get through the last filter as a $(0R)$ state. For each possible alternative path, there is an amplitude of the form

$$\langle 0R \mid j \rangle \langle j \mid A \mid i \rangle \langle i \mid +S \rangle,$$

and the total amplitude is the sum of the terms we can get with all possible combinations of i and j. The amplitude we want is

$$\sum_{ij} \langle 0R \mid j \rangle \langle j \mid A \mid i \rangle \langle i \mid +S \rangle. \tag{5.31}$$

If $(0R)$ and $(+S)$ are replaced by general states χ and ϕ, we would have the same kind of expression; so we have the general result

$$\langle \chi \mid A \mid \phi \rangle = \sum_{ij} \langle \chi \mid j \rangle \langle j \mid A \mid i \rangle \langle i \mid \phi \rangle. \tag{5.32}$$

Now notice that the right-hand side of Eq. (5.32) is really "simpler" than the left-hand side. The apparatus A is completely described by the *nine* numbers $\langle j\,|\,A\,|\,i\rangle$ which tell the response of A with respect to the three base states of the apparatus T. Once we know these nine numbers, we can handle any two incoming and outgoing states ϕ and χ if we define each in terms of the three amplitudes for going into, or from, each of the three base states. The result of an experiment is predicted using Eq. (5.32).

This then is the machinery of quantum mechanics for a spin-one particle. Every *state* is described by three numbers which are the amplitudes to be in each of some selected set of base states. Every apparatus is described by nine numbers which are the amplitudes to go from one base state to another in the apparatus. From these numbers anything can be calculated.

The nine amplitudes which describe the apparatus are often written as a square matrix—called the matrix $\langle j\,|\,A\,|\,i\rangle$:

$$
\begin{array}{cccc}
 & & \text{from} & \\
 & + & 0 & - \\
\text{to} \quad + & \langle +\,|\,A\,|\,+\rangle & \langle +\,|\,A\,|\,0\rangle & \langle +\,|\,A\,|\,-\rangle \\
0 & \langle 0\,|\,A\,|\,+\rangle & \langle 0\,|\,A\,|\,0\rangle & \langle 0\,|\,A\,|\,-\rangle \\
- & \langle -\,|\,A\,|\,+\rangle & \langle -\,|\,A\,|\,0\rangle & \langle -\,|\,A\,|\,-\rangle
\end{array}
\tag{5.33}
$$

The mathematics of quantum mechanics is just an extension of this idea. We will give you a simple illustration. Suppose we have an apparatus C that we wish to analyze—that is, we want to calculate the various $\langle j\,|\,C\,|\,i\rangle$. For instance, we might want to know what happens in an experiment like

$$
\left\{ \begin{matrix} + \\ 0 \\ - \end{matrix} \Bigg| \right\}_S \quad \left\{ C \right\} \quad \left\{ \begin{matrix} + \\ 0 \\ - \end{matrix} \Bigg| \right\}_R .
\tag{5.34}
$$

But then we notice that C is just built of two pieces of apparatus A and B in series— the particles go through A and then through B—so we can write symbolically

$$
\left\{ C \right\} = \left\{ A \right\} \cdot \left\{ B \right\} .
\tag{5.35}
$$

We can call the C apparatus the "product" of A and B. Suppose also that we already know how to analyze the two parts; so we can get the matrices (with respect to T) of A and B. Our problem is then solved. We can easily find

$$
\langle \chi\,|\,C\,|\,\phi\rangle
$$

for any input and output states. First we write that

$$
\langle \chi\,|\,C\,|\,\phi\rangle = \sum_k \langle \chi\,|\,B\,|\,k\rangle\langle k\,|\,A\,|\,\phi\rangle .
$$

Do you see why? (*Hint:* Imagine putting a T apparatus between A and B.) Then if we consider the special case in which ϕ and χ are also base states (of T), say i and j, we have

$$
\langle j\,|\,C\,|\,i\rangle = \sum_k \langle j\,|\,B\,|\,k\rangle\langle k\,|\,A\,|\,i\rangle .
\tag{5.36}
$$

This equation gives the matrix for the "product" apparatus C in terms of the two matrices of the apparatuses A and B. Mathematicians call the new matrix $\langle j\,|\,C\,|\,i\rangle$ —formed from two matrices $\langle j\,|\,B\,|\,i\rangle$ and $\langle j\,|\,A\,|\,i\rangle$ according to the sum specified in Eq. (5.36)—the "product" matrix BA of the two matrices B and A. (Note that the *order* is important, $AB \neq BA$.) Thus, we can say that the matrix for a succession of two pieces of apparatus is the matrix product of the matrices for the two apparatuses (putting the *first* apparatus on the *right* in the product). Anyone who knows matrix algebra then understands that we mean just Eq. (5.36).

5–7 Transforming to a different base

We want to make one final point about the base states used in the calculations. Suppose we have chosen to work with some particular base—say the S base—and another fellow decides to do the same calculations with a different base—say the T base. To keep things straight let's call our base states the (iS) states, where $i = +, 0, -$. Similarly, we can call his base states (jT). How can we compare our work with his? The final answers for the result of any measurement should come out the same, but in the calculations the various amplitudes and matrices used will be different. How are they related? For instance, if we both start with the same ϕ, we will describe it in terms of the three amplitudes $\langle iS \mid \phi \rangle$ that ϕ goes into our base states in the S representation, whereas he will describe it by the amplitudes $\langle jT \mid \phi \rangle$ that the state ϕ goes into the base states is his T representation. How can we check that we are really both describing the same state ϕ? We can do it with the general rule II in (5.27). Replacing χ by any one of his states jT, we have

$$\langle jT \mid \phi \rangle = \sum_j \langle jT \mid iS \rangle \langle iS \mid \phi \rangle. \tag{5.37}$$

To relate the two representations, we need only give the nine complex numbers of the matrix $\langle jT \mid iS \rangle$. This matrix can then be used to convert all of his equations to our form. It tells us how to *transform* from one set of base states to another. (For this reason $\langle jT \mid iS \rangle$ is sometimes called "the transformation matrix from representation S to representation T." Big words!)

For the case of spin-one particles for which we have only three base states (for higher spins, there are more) the mathematical situation is analogous to what we have seen in vector algebra. Every vector can be represented by giving three numbers—the components along the axes x, y, and z. That is, every vector can be resolved into three "base" vectors which are vectors along the three axes. But suppose someone else chooses to use a different set of axes—x', y', and z'. He will be using different numbers to represent any particular vector. His calculations will look different, but the final results will be the same. We have considered this before and know the rules for transforming vectors from one set of axes to another.

You may want to see how the quantum mechanical transformations work by trying some out; so we will give here, without proof, the transformation matrices for converting the spin-one amplitudes in one representation S to another representation T, for various special relative orientations of the S and T filters. (We will show you in a later chapter how to derive these same results.)

First case: The T apparatus has the same y-axis (along which the particles move) as the S apparatus, but is rotated about the common y-axis by the angle α (as in Fig. 5–6). (To be specific, a set of coordinates x', y', z' is fixed in the T apparatus, related to the x, y, z coordinates of the S apparatus by: $z' = z \cos \alpha + x \sin \alpha, x' = x \cos \alpha - z \sin \alpha, y' = y$.) Then the transformation amplitudes are:

$$\langle +T \mid +S \rangle = \tfrac{1}{2}(1 + \cos \alpha),$$

$$\langle 0T \mid +S \rangle = -\frac{1}{\sqrt{2}} \sin \alpha,$$

$$\langle -T \mid +S \rangle = \tfrac{1}{2}(1 - \cos \alpha),$$

$$\langle +T \mid 0\,S \rangle = +\frac{1}{\sqrt{2}} \sin \alpha,$$

$$\langle 0T \mid 0\,S \rangle = \cos \alpha, \tag{5.38}$$

$$\langle -T \mid 0\,S \rangle = -\frac{1}{\sqrt{2}} \sin \alpha,$$

$$\langle +T \mid -S \rangle = \tfrac{1}{2}(1 - \cos \alpha),$$

$$\langle 0T \mid -S \rangle = +\frac{1}{\sqrt{2}} \sin \alpha,$$

$$\langle -T \mid -S \rangle = \tfrac{1}{2}(1 + \cos \alpha).$$

Second Case: The T apparatus has the same z-axis as S, but is rotated around the z-axis by the angle β. (The coordinate transformation is $z' = z$, $x' = x \cos \beta + y \sin \beta$, $y' = y \cos \beta - x \sin \beta$.) Then the transformation amplitudes are:

$$\langle +T \mid +S \rangle = e^{+i\beta},$$
$$\langle 0\,T \mid 0\,S \rangle = 1,$$
$$\langle -T \mid -S \rangle = e^{-i\beta}, \tag{5.39}$$
$$\text{all others} = 0.$$

Note that any rotations of T whatever can be made up of the two rotations described.

If a state ϕ is defined by the three numbers

$$C_+ = \langle +S \mid \phi \rangle, \qquad C_0 = \langle 0\,S \mid \phi \rangle, \qquad C_- = \langle -S \mid \phi \rangle, \tag{5.40}$$

and the same state is described from the point of view of T by the three numbers

$$C'_+ = \langle +T \mid \phi \rangle, \qquad C'_0 = \langle 0\,T \mid \phi \rangle, \qquad C'_- = \langle -T \mid \phi \rangle, \tag{5.41}$$

then the coefficients $\langle jT \mid iS \rangle$ of (5.38) or (5.39) give the transformation connecting C'_i and C'_i. In other words, the C_i are very much like the components of a vector that appear different from the point of view of S and T.

For a spin-one particle *only*—because it requires *three* amplitudes—the correspondence with a vector is very close. In each case, there are three numbers that must transform with coordinate changes in a certain definite way. In fact, there is a set of base states *which transform just like the three components of a vector*. The three combinations

$$C_x = -\frac{1}{\sqrt{2}}\,(C_+ - C_-), \qquad C_y = -\frac{i}{\sqrt{2}}\,(C_+ + C_-), \qquad C_z = C_0 \tag{5.42}$$

transform to C'_x, C'_y, and C'_z just the way that x, y, z transform to x', y', z'. [You can check that this is so by using the transformation laws (5.38) and (5.39).] Now you see why a spin-one particle is often called a "vector particle."

5–8 Other situations

We began by pointing out that our discussion of spin-one particles would be a prototype for any quantum mechanical problem. The generalization has only to do with the numbers of states. Instead of only three base states, any particular situation may involve n base states.† Our basic laws in Eq. (5.27) have exactly the same form—with the understanding that i and j must range over all n base states. Any phenomenon can be analyzed by giving the amplitudes that it starts in each one of the base states and ends in any other one of the base states, and then summing over the complete set of base states. Any proper set of base states can be used, and if someone wishes to use a different set, it is just as good; the two can be connected by using an n by n transformation matrix. We will have more to say later about such transformations.

Finally, we promised to remark on what to do if atoms come directly from a furnace, go through some apparatus, say A, and are then analyzed by a filter which selects the state χ. You do not know what the state ϕ is that they start out in. It is perhaps best if you don't worry about this problem just yet, but instead concentrate on problems that always start out with pure states. But if you insist, here is how the problem can be handled.

First, you have to be able to make some reasonable guess about the way the states are distributed in the atoms that come from the furnace. For example, if

---

† The number of base states n may be, and generally is, infinite.

there were nothing "special" about the furnace, you might reasonably guess that atoms would leave the furnace with random "orientations." Quantum mechanically, that corresponds to saying that you don't know anything about the states, but that one-third are in the $(+S)$ state, one-third are in the $(0\ S)$ state, and one-third are in the $(-S)$ state. For those that are in the $(+S)$ state the amplitude to get through is $\langle x \mid A \mid +S\rangle$ and the probability is $|\langle x \mid A \mid +S\rangle|^2$, and similarly for the others. The overall probability is then

$$\tfrac{1}{3}|\langle x \mid A \mid +S\rangle|^2 + \tfrac{1}{3}|\langle x \mid A \mid 0\ S\rangle|^2 + \tfrac{1}{3}|\langle x \mid A \mid -S\rangle|^2.$$

Why did we use S rather than, say, T? The answer is, surprisingly, the same no matter what we choose for our initial resolution—so long as we are dealing with completely random orientations. It comes about in the same way that

$$\sum_i |\langle x \mid iS\rangle|^2 = \sum_j |\langle x \mid jT\rangle|^2$$

for any x. (We leave it for you to prove.)

Note that it is *not* correct to say that the input state has the amplitudes $\sqrt{1/3}$ to be in $(+S)$, $\sqrt{1/3}$ to be in $(0\ S)$, and $\sqrt{1/3}$ to be in $(-S)$; that would imply that certain interferences might be possible. It is simply that you do not *know* what the initial state is; you have to think in terms of the probability that the system starts out in the various possible initial states, and then you have to take a weighted average over the various possibilities.

6

Spin One-Half†

6–1 Transforming amplitudes

In the last chapter, using a system of spin one as an example, we outlined the general principles of quantum mechanics:

Any state ψ can be described in terms of a set of base states by giving the amplitudes to be in each of the base states.

The amplitude to go from any state to another can, in general, be written as a sum of products, each product being the amplitude to go into one of the base states times the amplitude to go from that base state to the final condition, with the sum including a term for each base state:

$$\langle \chi \mid \psi \rangle = \sum_i \langle \chi \mid i \rangle \langle i \mid \psi \rangle. \tag{6.1}$$

The base states are orthogonal—the amplitude to be in one if you are in the other is zero:

$$\langle i \mid j \rangle = \delta_{ij}. \tag{6.2}$$

The amplitude to get from one state to another directly is the complex conjugate of the reverse:

$$\langle \chi \mid \psi \rangle^* = \langle \psi \mid \chi \rangle. \tag{6.3}$$

We also discussed a little bit about the fact that there can be more than one base for the states and that we can use Eq. (6.1) to convert from one base to another. Suppose, for example, that we have the amplitudes $\langle iS \mid \psi \rangle$ to find the state ψ in every one of the base states i of a base system S, but that we then decide that we would prefer to describe the state in terms of another set of base states, say the states j belonging to the base T. In the general formula, Eq. (6.1), we could substitute jT for χ and obtain this formula:

$$\langle jT \mid \psi \rangle = \sum_i \langle jT \mid iS \rangle \langle iS \mid \psi \rangle. \tag{6.4}$$

The amplitudes for the state (ψ) to be in the base states (iT) are related to the amplitudes to be in the base states (iS) by the set of coefficients $\langle jT \mid iS \rangle$. If there are N base states, there are N^2 such coefficients. Such a set of coefficients is often called the "*transformation matrix* to go from the *S-representation* to the *T-representation.*" This looks rather formidable mathematically, but with a little renaming we can see that it is really not so bad. If we call C_i the amplitude that the state ψ is in the base state iS—that is, $C_i = \langle iS \mid \psi \rangle$—and call C_j' the corresponding amplitudes for the base system T—that is, $C_j' = \langle jT \mid \psi \rangle$, then Eq. (6.4) can be written as

$$C_j' = \sum_i R_{ji} C_i, \tag{6.5}$$

where R_{ji} means the same thing as $\langle jT \mid iS \rangle$. Each amplitude C_j' is equal to a sum

6–1 Transforming amplitudes

6–2 Transforming to a rotated coordinate system

6–3 Rotations about the z-axis

6–4 Rotations of 180° and 90° about y

6–5 Rotations about x

6–6 Arbitrary rotations

---

† This chapter is a rather long and abstract side tour, and it does not introduce any idea which we will not also come to by a different route in later chapters. You can, therefore, skip over it, and come back later if you are interested.

over all i of one of the coefficients R_{ji} times each amplitude C_i. It has the same form as the transformation of a vector from one coordinate system to another.

In order to avoid being too abstract for too long, we have given you some examples of these coefficients for the spin-one case, so you can see how to use them in practice. On the other hand, there is a very beautiful thing in quantum mechanics—that from the sheer fact that there are three states and from the symmetry properties of space under rotations, these coefficients can be found purely by abstract reasoning. Showing you such arguments at this early stage has a disadvantage in that you are immersed in another set of abstractions before we get "down to earth." However, the thing is so beautiful that we are going to do it anyway.

We will show you in this chapter how the transformation coefficients can be derived for spin one-half particles. We pick this case, rather than spin one, because it is somewhat easier. Our problem is to determine the coefficients R_{ji} for a particle—an atomic system—which is split into two beams in a Stern-Gerlach apparatus. We are going to derive all the coefficients for the transformation from one representation to another by pure reasoning—plus a few assumptions. *Some assumptions are always necessary in order to use "pure" reasoning!* Although the arguments will be abstract and somewhat involved, the result we get will be relatively simple to state and easy to understand—and the result is the most important thing. You may, if you wish, consider this as a sort of cultural excursion. We have, in fact, arranged that all the essential results derived here are also derived in some other way when they are needed in later chapters. So you need have no fear of losing the thread of our study of quantum mechanics if you omit this chapter entirely, or study it at some later time. The excursion is "cultural" in the sense that it is intended to show that the principles of quantum mechanics are not only interesting, but are so deep that by adding only a few extra hypotheses about the structure of space, we can deduce a great many properties of physical systems. Also, it is important that we know where the different consequences of quantum mechanics come from, because so long as our laws of physics are incomplete—as we know they are—it is interesting to find out whether the places where our theories fail to agree with experiment is where our logic is the best or where our logic is the worst. Until now, it appears that where our logic is the most abstract it always gives correct results—it agrees with experiment. Only when we try to make specific models of the internal machinery of the fundamental particles and their interactions are we unable to find a theory that agrees with experiment. The theory then that we are about to describe agrees with experiment wherever it has been tested—for the strange particles as well as for electrons, protons, and so on.

One remark on an annoying, but interesting, point before we proceed: It is not possible to determine the coefficients R_{ji} uniquely, because there is always some arbitrariness in the probability amplitudes. If you have a set of amplitudes of any kind, say the amplitudes to arrive at some place by a whole lot of different routes, and if you multiply every single amplitude by the same phase factor—say by $e^{i\delta}$—you have another set that is just as good. So, it is always possible to make an arbitrary change in phase of all the amplitudes in any given problem if you want to.

Suppose you calculate some probability by writing a sum of several amplitudes, say $(A + B + C + \cdots)$ and taking the absolute square. Then somebody else calculates the same thing by using the sum of the amplitudes $(A' + B' + C' + \cdots)$ and taking the absolute square. If all the A', B', C', etc., are equal to the A, B, C, etc., except for a factor $e^{i\delta}$, all probabilities obtained by taking the absolute squares will be exactly the same, since $(A' + B' + C' + \cdots)$ is then equal to $e^{i\delta}(A + B + C + \cdots)$. Or suppose, for instance, that we were computing something with Eq. (6.1), but then we suddenly change all of the phases of a certain base system. Every one of the amplitudes $\langle i \mid \psi \rangle$ would be multiplied by the same factor $e^{i\delta}$. Similarly, the amplitudes $\langle i \mid \chi \rangle$ would also be changed by $e^{i\delta}$, but the amplitudes $\langle \chi \mid i \rangle$ are the complex conjugates of the amplitudes $\langle i \mid \chi \rangle$; therefore, the former gets changed by the factor $e^{-i\delta}$. The plus and minus $i\delta$'s in the exponents cancel out, and we would have the same expression we had before. So it is a general rule that if we change all the amplitudes with respect to a given base system by the same phase—or even if we just change *all* the amplitudes in any problem by the same phase—it makes no difference. There is, therefore, some freedom to choose the phases in our transformation matrix. Every now and then we will make such an arbitrary choice—usually following the conventions that are in general use.

6–2 Transforming to a rotated coordinate system

We consider again the "improved" Stern-Gerlach apparatus described in the last chapter. A beam of spin one-half particles, entering at the left, would, in general, be split into *two* beams, as shown schematically in Fig. 6–1. (There were three beams for spin *one*.) As before, the beams are put back together again unless one or the other of them is blocked off by a "stop" which intercepts the beam at its half-way point. In the figure we show an arrow which points in the direction of the increase of the *magnitude* of the field—say toward the magnet pole with the sharp edges. This arrow we take to represent *the "up" axis* of any particular apparatus. It is fixed relative to the apparatus and will allow us to indicate the relative orientations when we use several apparatuses together. We also assume that the direction of the magnetic field in each magnet is always the same with respect to the arrow.

We will say that those atoms which go in the "upper" beam are in the $(+)$ state *with respect to that apparatus* and that those in the "lower" beam are in the $(-)$ state. (There is no "zero" state for spin one-half particles.)

Now suppose we put two of our modified Stern-Gerlach apparatuses in sequence, as shown in Fig. 6–2(a). The first one, which we call S, can be used to prepare a pure $(+S)$ or a pure $(-S)$ state by blocking one beam or the other. [As shown it prepares a pure $(+S)$ state.] For each condition, there is some amplitude for a particle that comes out of S to be in either the $(+T)$ or the $(-T)$ beam of the second apparatus. There are, in fact, just four amplitudes: the amplitude to go from $(+S)$ to $(+T)$, from $(+S)$ to $(-T)$, from $(-S)$ to $(+T)$, from $(-S)$ to $(-T)$. These amplitudes are just the four coefficients of the transformation matrix R_{ji} to go from the S-representation to the T-representation. We can consider that the first apparatus "prepares" a particular state in one representation and that the second apparatus "analyzes" that state in terms of the second representation. The kind of question we want to answer, then, is this: If an atom has been prepared in a given condition—say the $(+S)$ state—by blocking one of the beams in the apparatus S, what is the chance that it will get through the second apparatus T if this is set for, say, the $(-T)$ state. The result will depend, of course, on the angles between the two systems S and T.

We should explain why it is that we could have any hope of finding the coefficients R_{ji} by deduction. You know that it is almost impossible to believe that if a particle has its spin lined up in the $+z$-direction, that there is some chance of finding the same particle with its spin pointing in the $+x$-direction—or in any other direction at all. In fact, it *is* almost impossible, but not quite. It is so nearly impossible that there is *only one way* it can be done, and that is the reason we can find out what that unique way is.

The first kind of argument we can make is this. Suppose we have a setup like the one in Fig. 6–2(a), in which we have the two apparatuses S and T, with T cocked at the angle α with respect to S, and we let only the $(+)$ beam through S and the $(-)$ beam through T. We would observe a certain number for the probability that the particles coming out of S get through T. Now suppose we make another measurement with the apparatus of Fig. 6–2(b). The *relative* orientation of S and T is the same, but the whole system sits at a different angle in space. We want to *assume* that both of these experiments give the same number for the chance that a particle in a pure state with respect to S will get into some particular state with respect to T. We are assuming, in other words, that the result of any experiment of this type is the same—that the *physics* is the same—no matter

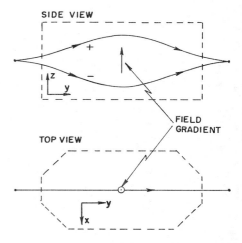

Fig. 6–1. Top and side views of an "improved" Stern-Gerlach apparatus with beams of a spin one-half particle.

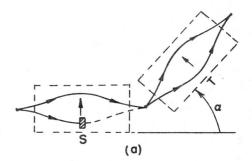

(a)

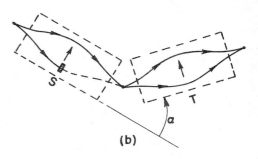

(b)

Fig. 6–2. Two equivalent experiments.

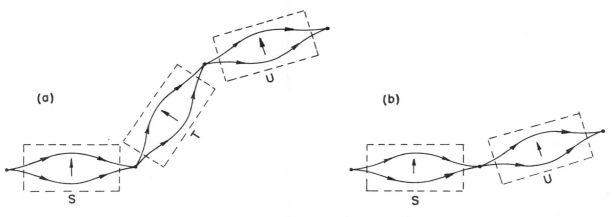

Fig. 6–3. If T is "wide open," (b) is equivalent to (a).

how the *whole* apparatus is oriented in space. (You say, "That's obvious." But it *is* an assumption, and it is "right" only if it is actually what happens.) That means that the coefficients R_{ji} depend only on the relation in space of S and T, and not on the absolute situation of S and T. To say this in another way, R_{ji} depends only on the *rotation* which carries S to T, for evidently what is the same in Fig. 6–2(a) and Fig. 6–2(b) is the three-dimensional rotation which would carry apparatus S into the orientation of apparatus T. When the transformation matrix R_{ji} depends only on a rotation, as it does here, it is called a *rotation matrix*.

For our next step we will need one more piece of information. Suppose we add a third apparatus which we can call U, which follows T at some arbitrary angle, as in Fig. 6–3(a). (It's beginning to look horrible, but that's the fun of abstract thinking—you can make the most weird experiments just by drawing lines!) Now what is the $S \to T \to U$ transformation? What we really want to ask for is the amplitude to go from some state with respect to S to some other state with respect to U, when we know the transformation from S to T and from T to U. We are then asking about an experiment in which both channels of T are open. We can get the answer by applying Eq. (6.5) twice in succession. For going from the S-representation to the T-representation, we have

$$C'_j = \sum_i R_{ji}^{TS} C_i, \tag{6.6}$$

where we put the superscripts TS on the R, so that we can distinguish it from the coefficients R^{UT} we will have for going from T to U.

Assuming the amplitudes to be in the base states of the U-representation C''_k, we can relate them to the T-amplitudes by using Eq. (6.5) once more; we get

$$C''_k = \sum_j R_{kj}^{UT} C'_j. \tag{6.7}$$

Now we can combine Eqs. (6.6) and (6.7) to get the transformation to U directly from S. Substituting C'_j from Eq. (6.6) in Eq. (6.7), we have

$$C''_k = \sum_j R_{kj}^{UT} \sum_i R_{ji}^{TS} C_i. \tag{6.8}$$

Or, since i does not appear in R_{kj}^{UT}, we can put the i-summation also in front, and write

$$C''_k = \sum_i \sum_j R_{kj}^{UT} R_{ji}^{TS} C_i. \tag{6.9}$$

This is the formula for a double transformation.

Notice, however, that so long as all the beams in T are unblocked, the state coming out of T is the same as the one that went in. We could just as well have made a transformation from the S-representation directly to the U-representation. It should be the same as putting the U apparatus right after S, as in Fig.

6–3(b). In that case, we would have written

$$C''_k = \sum_i R^{US}_{ki} C_i, \tag{6.10}$$

with the coefficients R^{US}_{ki} belonging to this transformation. Now, clearly, Eqs. (6.9) and (6.10) should give the same amplitudes C''_k, and this should be true no matter what the original state ϕ was which gave us the amplitudes C_i. So it must be that

$$R^{US}_{ki} = \sum_j R^{UT}_{kj} R^{TS}_{ji}. \tag{6.11}$$

In other words, for any rotation $S \rightarrow U$ of a reference base, which is viewed as a compounding of two successive rotations $S \rightarrow T$ and $T \rightarrow U$, the rotation matrix R^{US}_{ki} can be obtained from the matrices of the two partial rotations by Eq. (6.11). If you wish, you can find Eq. (6.11) directly from Eq. (6.1), for it is only a different notation for $\langle kU \mid iS \rangle = \sum_j \langle kU \mid jT \rangle \langle jT \mid iS \rangle$.

To be thorough, we should add the following parenthetical remarks. They are not terribly important, however, so you can skip to the next section if you want. What we have said is not quite right. We cannot really say that Eq. (6.9) and Eq. (6.10) must give *exactly* the same amplitudes. Only the *physics* should be the same; all the amplitudes could be different by some common phase factor like $e^{i\delta}$ without changing the result of any calculation about the real world. So, instead of Eq. (6.11), all we can say, really, is that

$$e^{i\delta} R^{US}_{ki} = \sum_j R^{UT}_{kj} R^{TS}_{ji}, \tag{6.12}$$

where δ is *some* real constant. What this extra factor of $e^{i\delta}$ means, of course, is that the amplitudes we get if we use the matrix R^{US} might all differ by the same phase $(e^{-i\delta})$ from the amplitude we would get using the two rotations R^{UT} and R^{TS}. We know that it doesn't matter if all amplitudes are changed by the same phase, so we could just ignore this phase factor if we wanted to. It turns out, however, that if we define all of our rotation matrices in a particular way, this extra phase factor will never appear—the δ in Eq. (6.12) will always be zero. Although it is not important for the rest of our arguments, we can give a quick proof by using a mathematical theorem about determinants. [If you don't yet know much about determinants, don't worry about the proof and just skip to the definition of Eq. (6.15).]

First, we should say that Eq. (6.11) is the mathematical definition of a "product" of two matrices. (It is just convenient to be able to say: "R^{US} is the product of R^{UT} and R^{TS}.") Second, there is a theorem of mathematics—which you can easily prove for the two-by-two matrices we have here—which says that the determinant of a "product" of two matrices is the product of their determinants. Applying this theorem to Eq. (6.12), we get

$$e^{i2\delta} (\text{Det } R^{US}) = (\text{Det } R^{UT}) \cdot (\text{Det } R^{TS}). \tag{6.13}$$

(We leave off the subscripts, because they don't tell us anything useful.) Yes, the 2δ is right. Remember that we are dealing with two-by-two matrices; every term in the matrix R^{US}_{ki} is multiplied by $e^{i\delta}$, so each product in the determinant—which has *two* factors—gets multiplied by $e^{i2\delta}$. Now let's take the square root of Eq. (6.13) and divide it into Eq. (6.12); we get

$$\frac{R^{US}_{ki}}{\sqrt{\text{Det } R^{US}}} = \sum_i \frac{R^{UT}_{kj}}{\sqrt{\text{Det } R^{UT}}} \frac{R^{TS}_{ji}}{\sqrt{\text{Det } R^{TS}}}. \tag{6.14}$$

The extra phase factor has disappeared.

Now it turns out that if we want all of our amplitudes in any given representation to be normalized (which means, you remember, that $\sum_i \langle \phi|i \rangle \langle i|\phi \rangle = 1$), the rotation matrices will all have determinants that are pure imaginary exponentials, like $e^{i\alpha}$. (We won't prove it; you will see that it always comes out that way.) So we can, if we wish, choose to make all our rotation matrices R have a unique phase by making Det $R = 1$. It is done like this. Suppose we find a rotation matrix R in some arbitrary way. We make it a rule to "convert" it to "standard form" by defining

$$R_{\text{standard}} = \frac{R}{\sqrt{\text{Det } R}}. \tag{6.15}$$

We can do this because we are just multiplying each term of R by the same phase factor, to get the phases we want. In what follows, we will always assume that our matrices have been put in the "standard form"; then we can use Eq. (6.11) without having any extra phase factors.

6–3 Rotations about the z-axis

We are now ready to find the transformation matrix R_{ji} between two different representations. With our rule for compounding rotations and our assumption that space has no preferred direction, we have the keys we need for finding the matrix of any arbitrary rotation. There is only *one* solution. We begin with the transformation which corresponds to a rotation about the z-axis. Suppose we have two apparatuses S and T placed in series along a straight line with their axes parallel and pointing out of the page, as shown in Fig. 6–4(a). We take our "z-axis" in this direction. Surely, if the beam goes "up" (toward $+z$) in the S apparatus, it will do the same in the T apparatus. Similarly, if it goes down in S, it will go down in T. Suppose, however, that the T apparatus were placed at some other angle, but still with its axis parallel to the axis of S, as in Fig. 6–4(b). Intuitively, you would say that a $(+)$ beam in S would still go with a $(+)$ beam in T, because the fields and field gradients are still in the same physical direction. And that would be quite right. Also, a $(-)$ beam in S would still go into a $(-)$ beam in T. The same result would apply for any orientation of T in the xy-plane of S. What does this tell us about the relation between $C'_+ = \langle +T \mid \psi \rangle$, $C'_- = \langle -T \mid \psi \rangle$ and $C_+ = \langle +S \mid \psi \rangle$, $C_- = \langle -S \mid \psi \rangle$? You might conclude that any rotation about the z-axis of the "frame of reference" for base states leaves the amplitudes C_+ to be "up" and "down," the same as before. We could write $C'_+ = C_+$ and $C'_- = C_-$ —but that is *wrong*. All we *can* conclude is that for such rotations the probabilities to be in the "up" beam are the same for the S and T apparatuses. That is,

$$|C'_+| = |C_+| \qquad \text{and} \qquad |C'_-| = |C_-|.$$

We cannot say that the *phases* of the amplitudes referred to the T apparatus may not be different for the two different orientations in (a) and (b) of Fig. 6–4.

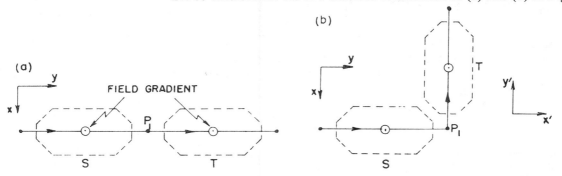

Fig. 6–4. Rotating 90° about the z-axis.

The two apparatuses in (a) and (b) of Fig. 6–4 are, in fact, different, as we can see in the following way. Suppose that we put an apparatus in front of S which produces a pure $(+x)$ state. (The x-axis points toward the bottom of the figure.) Such particles would be split into $(+z)$ and $(-z)$ beams in S, but the two beams would be recombined to give a $(+x)$ state again at P_1—the exit of S. The same thing happens again in T. If we follow T by a third apparatus U, whose axis is in the $(+x)$ direction and, as shown in Fig. 6–5(a), all the particles would go into the $(+)$ beam of U. Now imagine what happens if T and U are swung around *together* by 90° to the positions shown in Fig. 6–5(b). Again, the T apparatus puts out just what it takes in, so the particles that enter U are in a $(+x)$ state with respect to S. But U now analyzes for the $(+y)$ state with respect to S, which is different. (By symmetry, we would now expect only one-half of the particles to get through.)

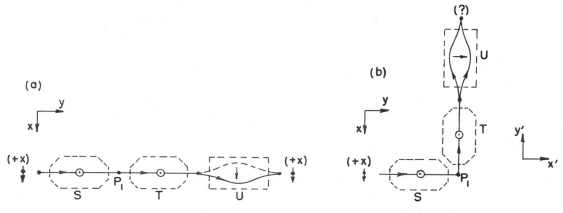

Fig. 6–5. Particle in a (+x) state behaves differently in (a) and (b).

What could have changed? The apparatuses T and U are still in the same *physical* relationship to each other. Can the *physics* be changed just because T and U are in a different orientation? Our original assumption is that it should not. It must be that the *amplitudes* with respect to T are different in the two cases shown in Fig. 6–5—and, therefore, also in Fig. 6–4. There must be some way for a particle to know that it has turned the corner at P_1. How could it tell? Well, all we have decided is that the *magnitudes* of C_1' and C_2' are the same in the two cases, but they could—in fact, *must*—have different *phases*. We conclude that C_+' and C_+ must be related by

$$C_+' = e^{i\lambda}C_+,$$

and that C_-' and C_- must be related by

$$C_-' = e^{i\mu}C_-,$$

where λ and μ are real numbers which must be related in some way to the angle between S and T.

The only thing we can say at the moment about λ and μ is that they must not be equal [except for the special case shown in Fig. 6–5(a), when T is in the same orientation as S]. We have seen that equal phase changes in all amplitudes have no physical consequence. For the same reason, we can always add the same arbitrary amount to both λ and μ without changing anything. So we are permitted to *choose* to make λ and μ equal to plus and minus the same number. That is, we can always take

$$\lambda' = \lambda - \frac{(\lambda + \mu)}{2}, \qquad \mu' = \mu - \frac{(\lambda + \mu)}{2}.$$

Then

$$\lambda' = \frac{\lambda}{2} - \frac{\mu}{2} = -\mu'.$$

So we adopt the convention† that $\mu = -\lambda$. We have then the general rule that for a rotation of the reference apparatus by some angle about the z-axis, the transformation is

$$C_+' = e^{+i\lambda}C_+, \qquad C_-' = e^{-i\lambda}C_-. \tag{6.16}$$

The absolute values are the same, only the phases are different. These phase factors are responsible for the different results in the two experiments of Fig. 6–5.

Now we would like to know the law that relates λ to the angle between S and T. We already know the answer for one case. If the angle is zero, λ is zero. Now we will *assume* that the phase shift λ is a continuous function of angle ϕ between S and T (see Fig. 6–4) as ϕ goes to zero—as only seems reasonable. In

———

† Looking at it another way, we are just putting the transformation in the "standard form" described in Section 6–2 by using Eq. (6.15).

ther words, if we rotate T from the straight line through S by the small angle ϵ, the λ is also a small quantity, say $m\epsilon$, where m is some number. We write it this way because we can show that λ must be proportional to ϵ. Suppose we were to put after T another apparatus T' which makes the angle ϵ with T, and, therefore, the angle 2ϵ with S. Then, with respect to T, we have

$$C'_+ = e^{i\lambda}C_+,$$

and with respect to T', we have

$$C''_+ = e^{i\lambda}C'_+ = e^{i2\lambda}C_+.$$

But we know that we should get the same result if we put T' right after S. Thus, when the angle is doubled, the phase is doubled. We can evidently extend the argument and build up any rotation at all by a sequence of infinitesimal rotations. We conclude that for *any* angle ϕ, λ is proportional to the angle. We can, therefore, write $\lambda = m\phi$.

The general result we get, then, is that for T rotated about the z-axis by the angle ϕ with respect to S

$$C'_+ = e^{im\phi}C_+, \qquad C'_- = e^{-im\phi}C_-. \tag{6.17}$$

For the angle ϕ, and for all rotations we speak of in the future, we adopt the standard convention that a *positive* rotation is a *right-handed* rotation about the positive direction of the reference axis. A positive ϕ has the sense of rotation of a right-handed screw advancing in the positive z-direction.

Now we have to find what m must be. First, we might try this argument: Suppose T is rotated by 360°; then, clearly, it is right back at zero degrees, and we should have $C'_+ = C_+$ and $C'_- = C_-$, or, what is the same thing, $e^{im\,2\pi} = 1$. We get $m = 1$. *This argument is wrong!* To see that it is, consider that T is rotated by 180°. If m were equal to 1, we would have $C'_+ = e^{i\pi}C_+ = -C_+$ and $C'_- = e^{-i\pi}C_- = -C_-$. However, this is just the *original* state all over again. *Both* amplitudes are just multiplied by -1 which gives back the original physical system. (It is again a case of a common phase change.) This means that if the angle between T and S in Fig. 6–5(b) is increased to 180°, the system (with respect to T) would be indistinguishable from the zero-degree situation, and the particles would again go through the $(+)$ state of the U apparatus. At 180°, though, the $(+)$ state of the U apparatus is the $(-x)$ state of the original S apparatus. So a $(+x)$ state would become a $(-x)$ state. But we have done nothing to *change* the original state; the answer is wrong. We cannot have $m = 1$.

We must have the situation that a rotation by 360° and *no smaller angle* reproduces the same physical state. This will happen if $m = \frac{1}{2}$. Then, and only then, will the first angle that reproduces the same *physical* state be $\phi = 360°$.† It gives

$$\left.\begin{aligned}C'_+ &= -C_+ \\ C'_- &= -C_-\end{aligned}\right\} 360° \text{ about } z\text{-axis.} \tag{6.18}$$

It is very curious to say that if you turn the apparatus 360° you get new amplitudes. They aren't really new, though, because the common change of sign doesn't give any different physics. If someone else had decided to change all the signs of the amplitudes because he thought he had turned 360°, that's all right; he gets the same physics.‡ So our final answer is that if we know the amplitudes C_+ and C_- for spin one-half particles with respect to a reference frame S, and we then use a base

---

† It appears that $m = -\frac{1}{2}$ would also work. However, we see in (6.17) that the change in sign merely redefines the notation for a spin-up particle.

‡ Also, if something has been rotated by a sequence of small rotations whose net result is to return it to the original orientation, it is possible to define the idea that it has been rotated 360°—as distinct from zero net rotation—if you have kept track of the whole history. (Interestingly enough, this is *not* true for a net rotation of 720°.)

system referred to T which is obtained from S by a rotation of ϕ around the z-axis, the new amplitudes are given in terms of the old by

$$\left.\begin{aligned} C'_+ &= e^{i\phi/2}C_+ \\ C'_- &= e^{-i\phi/2}C_- \end{aligned}\right\} \phi \text{ about } z. \tag{6.19}$$

6–4 Rotations of 180° and 90° about y

Next, we will try to guess the transformation for a rotation of T with respect to S of 180° around an axis *perpendicular* to the z-axis—say, about the y-axis. (We have defined the coordinate axes in Fig. 6–1.) In other words, we start with two identical Stern-Gerlach equipments, with the second one, T, turned "upside down" with respect to the first one, S, as in Fig. 6–6. Now if we think of our particles as little magnetic dipoles, a particle that is the $(+S)$ state—so that it goes on the "upper" path in the first apparatus—will also take the "upper" path in the second, so that it will be in the *minus* state with respect to T. (In the inverted T apparatus, both the gradients *and* the field direction are reversed; for a particle with its magnetic moment in a given direction, the force is unchanged.) Anyway, what is "up" with respect to S will be "down" with respect to T. For these relative positions of S and T, then, we know that the transformation must give

$$|C'_+| = |C_-|, \qquad |C'_-| = |C_+|.$$

As before, we cannot rule out some additional phase factors; we could have (for 180° about the y-axis)

$$C'_+ = e^{i\beta}C_- \qquad \text{and} \qquad C'_- = e^{i\gamma}C_+, \tag{6.20}$$

where β and γ are still to be determined.

What about a rotation of 360° about the y-axis? Well, we already know the answer for a rotation of 360° about the z-axis—the amplitude to be in any state changes sign. A rotation of 360° around any axis always brings us back to the original position. It must be that for *any* 360° rotation, the result is the same as a 360° rotation about the z-axis—all amplitudes simply change sign. Now suppose we imagine two successive rotations of 180° about y—using Eq. (6.20)—we should get the result of Eq. (6.18). In other words,

and
$$\begin{aligned} C''_+ &= e^{i\beta}C'_- = e^{i\beta}e^{i\gamma}C_+ = -C_+ \\[4pt] C''_- &= e^{i\gamma}C'_+ = e^{i\gamma}e^{i\beta}C_- = -C_-. \end{aligned} \tag{6.21}$$

This means that

$$e^{i\beta}e^{i\gamma} = -1 \qquad \text{or} \qquad e^{i\gamma} = -e^{-i\beta}.$$

So the transformation for a rotation of 180° about the y-axis can be written

$$C'_+ = e^{i\beta}C_-, \qquad C'_- = -e^{-i\beta}C_+. \tag{6.22}$$

The arguments we have just used would apply equally well to a rotation of 180° about *any* axis in the xy-plane, although different axes can, of course, give different numbers for β. However, that is the only way they can differ. Now there is a certain amount of arbitrariness in the number β, but once it is specified for one axis of rotation in the xy-plane it is determined for any other axis. It is *conventional* to choose to set $\beta = 0$ for a 180° rotation about the y-axis.

To show that we have this choice, suppose we imagine that β was not equal to zero for a rotation about the y-axis; then we can show that there is *some other* axis in the xy-plane, for which the corresponding phase factor *will* be zero. Let's find the phase factor β_A for an axis A that makes the angle α with the y-axis, as shown in Fig. 6–7(a). (For clarity, the figure is drawn with α equal to a negative number, but that doesn't matter.) Now if we take a T apparatus which is initially lined up with the S apparatus and is then rotated 180° about the axis A, its axes—which we will call x'', y'', and z''—will be as shown in Fig. 6–7(a). The amplitudes

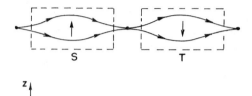

Fig. 6–6. A rotation of 180° about the y-axis.

with respect to T will then be

$$C''_+ = e^{i\beta_A} C_-, \qquad C''_- = -e^{-i\beta_A} C_+. \qquad (6.23)$$

We can now think of getting to the same orientation by the two successive rotations shown in (b) and (c) of the figure. First, we imagine an apparatus U which is rotated with respect to S by 180° about the y-axis. The axes x', y', and z' of U will be as shown in Fig. 6–7(b), and the amplitudes *with respect to U* are given by (6.22).

Now notice that we can go from U to T by a rotation about the "z-axis" of U, namely about z', as shown in Fig. 6–7(c). From the figure you can see that the angle required is two times the angle α but in the opposite direction (with respect to z'). Using the transformation of (6.19) with $\phi = -2\alpha$, we get

$$C''_+ = e^{-i\alpha} C'_+, \qquad C''_- = e^{+i\alpha} C'_-. \qquad (6.24)$$

Combining Eqs. (6.24) and (6.22), we get that

$$C''_+ = e^{i(\beta-\alpha)} C_-, \qquad C''_- = - e^{-i(\beta-\alpha)} C_+. \qquad (6.25)$$

These amplitudes must, of course, be the same as we got in 6.23). So β_A must be related to α and β by

$$\beta_A = \beta - \alpha. \qquad (6.26)$$

This means that if the angle α between the A-axis and the y-axis (of S) is equal to β, the transformation for a rotation of 180° about A will have $\beta_A = 0$.

Now so long as *some* axis perpendicular to the z-axis is going to have $\beta = 0$, we may as well take it to be the y-axis. It is purely a matter of *convention*, and we adopt the one in general use. Our result: For a rotation of 180° about the y-axis, we have

$$\left. \begin{array}{l} C'_+ = C_- \\[2mm] C'_- = -C_+ \end{array} \right\} 180° \text{ about } y. \qquad (6.27)$$

While we are thinking about the y-axis, let's next ask for the transformation matrix for a rotation of 90° about y. We can find it because we know that two successive 90° rotations about the same axis must equal one 180° rotation. We start by writing the transformation for 90° in the most general form:

$$C'_+ = aC_+ + bC_-, \qquad C'_- = cC_+ + dC_-. \qquad (6.28)$$

A second rotation of 90° about the same axis would have the same coefficients:

$$C''_+ = aC'_+ + bC'_-, \qquad C''_- = cC'_+ + dC'_-. \qquad (6.29)$$

Combining Eqs. (6.28) and (6.29), we have

$$C''_+ = a(aC_+ + bC_-) + b(cC_+ + dC_-), \\[2mm] C''_- = c(aC_+ + bC_-) + d(cC_+ + dC_-). \qquad (6.30)$$

However, from (6.27) we know that

$$C''_+ = C_-, \qquad C''_- = -C_+,$$

so that we must have that

$$\begin{array}{l} ab + bd = 1, \\ a^2 + bc = 0, \\ ac + cd = -1, \\ bc + d^2 = 0. \end{array} \qquad (6.31)$$

These four equations are enough to determine all our unknowns: a, b, c, and d.

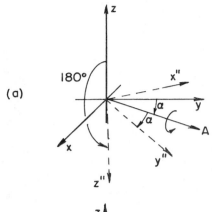

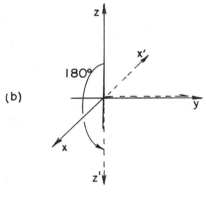

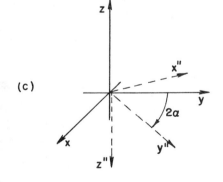

Fig. 6–7. A 180° rotation about the axis A is equivalent to a rotation of 180° about y, followed by a rotation about z'.

6–10

It is not hard to do. Look at the second and fourth equations. Deduce that $a^2 = d^2$, which means that $a = d$ or else that $a = -d$. But $a = -d$ is out, because then the first equation wouldn't be right. So $d = a$. Using this, we have immediately that $b = 1/2a$ and that $c = -1/2a$. Now we have everything in terms of a. Putting, say, the second equation all in terms of a, we have

$$a^2 - \frac{1}{4a^2} = 0 \qquad \text{or} \qquad a^4 = \frac{1}{4}.$$

This equation has four different solutions, but only two of them give the standard value for the determinant. We might as well take $a = 1/\sqrt{2}$; then†

$$a = 1/\sqrt{2}, \qquad b = 1/\sqrt{2},$$
$$c = -1/\sqrt{2}, \qquad d = 1/\sqrt{2}.$$

In other words, for two apparatuses S and T, with T rotated with respect to S by 90° about the y-axis, the transformation is

$$\left.\begin{aligned} C'_+ &= \frac{1}{\sqrt{2}}\,(C_+ + C_-) \\[2em] C'_- &= \frac{1}{\sqrt{2}}\,(-C_+ + C_-) \end{aligned}\right\} \quad 90^\circ \text{ about } y. \qquad (6.32)$$

We can, of course, solve these equations for C_+ and C_-, which will give us the transformation for a rotation of *minus* 90° about y. Changing the primes around, we would conclude that

$$\left.\begin{aligned} C'_+ &= \frac{1}{\sqrt{2}}\,(C_+ - C_-) \\[2em] C'_- &= \frac{1}{\sqrt{2}}\,(C_+ + C_-) \end{aligned}\right\} \quad -90^\circ \text{ about } y. \qquad (6.33)$$

6–5 Rotations about x

You may be thinking: "This is getting ridiculous. What are they going to do next, 47° around y, then 33° about x, and so on, forever?" No, we are almost finished. With just two of the transformations we have—90° about y, and an arbitrary angle about z (which we did first if you remember)—we can generate any rotation at all.

As an illustration, suppose that we want the angle α around x. We know how to deal with the angle α around z, but now we want it around x. How do we get it? First, we turn the axis z down onto x—which is a rotation of $+90^\circ$ about y, as shown in Fig. 6–8. Then we turn through the angle α around z'. Then we rotate -90° about y''. The net result of the three rotations is the same as turning around x by the angle α. It is a property of space.

(These facts of the combinations of rotations, and what they produce, are hard to grasp intuitively. It is rather strange, because we live in three dimensions, but it is hard for us to appreciate what happens if we turn this way and then that way. Perhaps, if we were fish or birds and had a real appreciation of what happens when we turn somersaults in space, we could more easily appreciate such things.)

Anyway, let's work out the transformation for a rotation by α around the x-axis by using what we know. From the first rotation by $+90^\circ$ around y the amplitudes go according to Eq. (6.32). Calling the rotated axes x', y', and z', the

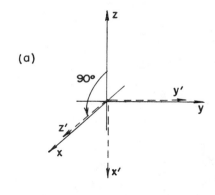

(a)

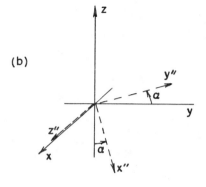

(b)

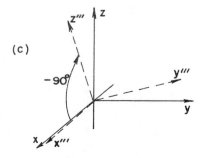

(c)

Fig. 6–8. A rotation by α about the x-axis is equivalent to: (a) a rotation by $+90^\circ$ about y, followed by (b) a rotation by α about z', followed by (c) a rotation of -90° about y''.

---

† The other solution changes all signs of a, b, c, and d and corresponds to a -270° rotation.

next rotation by the angle α around z' takes us to a frame x'', y'', z'', for which

$$C''_+ = e^{i\alpha/2}C'_+, \qquad C''_- = e^{-i\alpha/2}C'_-.$$

The last rotation of $-90°$ about y'' takes us to x''', y''', z'''; by (6.33),

$$C'''_+ = \frac{1}{\sqrt{2}}(C''_+ - C''_-), \qquad C'''_- = \frac{1}{\sqrt{2}}(C''_+ + C''_-).$$

Combining these last two transformations, we get

$$C'''_+ = \frac{1}{\sqrt{2}}(e^{+i\alpha/2}C'_+ - e^{-i\alpha/2}C'_-),$$

$$C'''_- = \frac{1}{\sqrt{2}}(e^{+i\alpha/2}C'_+ + e^{-i\alpha/2}C'_-).$$

Using Eqs. (6.32) for C'_+ and C'_-, we get the complete transformation:

$$C'''_+ = \tfrac{1}{2}\{e^{+i\alpha/2}(C_+ + C_-) - e^{-i\alpha/2}(-C_+ + C_-)\},$$
$$C'''_- = \tfrac{1}{2}\{e^{+i\alpha/2}(C_+ + C_-) + e^{-i\alpha/2}(-C_+ + C_-)\}.$$

We can put these formulas in a simpler form by remembering that

$$e^{i\theta} + e^{-i\theta} = 2\cos\theta, \qquad \text{and} \qquad e^{i\theta} - e^{i\theta} = 2i\sin\theta.$$

We get

$$\left.\begin{aligned} C'''_+ &= \left(\cos\frac{\alpha}{2}\right)C_+ + i\left(\sin\frac{\alpha}{2}\right)C_- \\ C'''_- &= i\left(\sin\frac{\alpha}{2}\right)C_+ + \left(\cos\frac{\alpha}{2}\right)C_- \end{aligned}\right\} \; \alpha \text{ about } x. \tag{6.34}$$

Here is our transformation for a rotation about the x-axis by *any* angle α. It is only a little more complicated than the others.

6–6 Arbitrary rotations

Now we can see how to do *any* angle at all. First, notice that any relative orientation of two coordinate frames can be described in terms of three angles, as shown in Fig. 6–9. If we have a set of axes x', y', and z' oriented in any way at all with respect to x, y, and z, we can describe the relationship between the two frames by means of the three Euler angles α, β, and γ, which define three successive rotations that will bring the x, y, z frame into the x', y', z' frame. Starting at x, y, z, we rotate our frame through the angle β about the z-axis, bringing the x-axis to the line x_1. Then, we rotate by α about this temporary x-axis, to bring z down to z'. Finally, a rotation about the new z-axis (that is, z') by the angle γ will bring the x-axis into x' and the y-axis into y'.† We know the transformations for each of the three rotations—they are given in (6.19) and (6.34). Combining them in the proper order, we get

$$C'_+ = \cos\frac{\alpha}{2}\, e^{i(\beta+\gamma)/2}C_+ + i\sin\frac{\alpha}{2}\, e^{-i(\beta-\gamma)/2}C_-,$$

$$C'_- = i\sin\frac{\alpha}{2}\, e^{i(\beta-\gamma)/2}C_+ + \cos\frac{\alpha}{2}\, e^{-i(\beta+\gamma)/2}C_-. \tag{6.35}$$

So just starting from some assumptions about the properties of space, we have derived the amplitude transformation for any rotation at all. That means that if

---

† With a little work you can show that the frame x, y, z can also be brought into the frame x', y', z' by the following three rotations about the *original* axes: (1) rotate by the angle γ around the original z-axis; (2) rotate by the angle α around the original x-axis; (3) rotate by the angle β around the original z-axis.

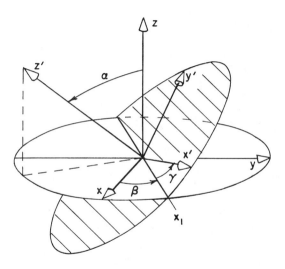

Fig. 6–9. The orientation of any coordinate frame x', y', z' relative to another frame x, y, z can be defined in terms of Euler's angles α, β, γ.

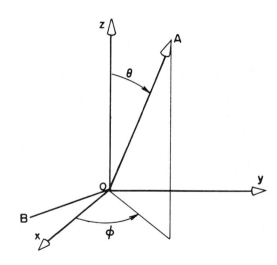

Fig. 6–10. An axis A defined by the polar angles θ and ϕ.

we know the amplitudes for any state of a spin one-half particle to go into the two beams of a Stern-Gerlach apparatus S, whose axes are x, y, and z, we can calculate what fraction would go into either beam of an apparatus T with the axes x', y', and z'. In other words, if we have a state ψ of a spin one-half particle, whose amplitudes are $C_+ = \langle + \mid \psi \rangle$ and $C_- = \langle - \mid \psi \rangle$ to be "up" and "down" with respect to the z-axis of the x, y, z frame, we also know the amplitudes C'_+ and C'_- to be "up" and "down" with respect to the z'-axis of any other frame x', y', z'. The four coefficients in Eqs. (6.35) are the terms of the "transformation matrix" with which we can project the amplitudes of a spin one-half particle into any other coordinate system.

We will now work out a few examples to show you how it all works. Let's take the following simple question. We put a spin one-half atom through a Stern-Gerlach apparatus that transmits only the $(+z)$ state. What is the amplitude that it will be in the $(+x)$ state? The $+x$ axis is the same as the $+z'$ axis of a system rotated 90° about the y-axis. For this problem, then, it is simplest to use Eqs. (6.32)—although you could, of course, use the complete equations of (6.35). Since $C_+ = 1$ and $C_- = 0$, we get $C'_+ = 1/\sqrt{2}$. The probabilities are the absolute square of these amplitudes; there is a 50 percent chance that the particle will go through an apparatus that selects the $(+x)$ state. If we had asked about the $(-x)$ state the amplitude would have been $-1/\sqrt{2}$, which also gives a probability $1/2$—as you would expect from the symmetry of space. So if a particle is in the $(+z)$ state, it is equally likely to be in $(+x)$ or $(-x)$, but with opposite phase.

There's no prejudice in y either. A particle in the $(+z)$ state has a 50–50 chance of being in $(+y)$ or in $(-y)$. However, for these (using the formula for rotating $-90°$ about x), the amplitudes are $1/\sqrt{2}$ and $-i/\sqrt{2}$. In this case, the two amplitudes have a phase difference of 90° instead of 180°, as they did for the $(+x)$ and $(-x)$. In fact, that's how the distinction between x and y shows up.

As our final example, suppose that we know that a spin one-half particle is in a state ψ such that it is polarized "up" along some axis A, defined by the angles θ and ϕ in Fig. 6–10. We want to know the amplitude $\langle C_+ \mid \psi \rangle$ that the particle is "up" along z and the amplitude $\langle C_- \mid \psi \rangle$ that it is "down" along z. We can find these amplitudes by imagining that A is the z-axis of a system whose x-axis lies in some arbitrary direction—say in the plane formed by A and z. We can then bring the frame of A into x, y, z by three rotations. First, we make a rotation by $-\pi/2$ about the axis A, which brings the x-axis into the line B in the figure. Then we rotate by θ about line B (the new x-axis of frame A) to bring A to the z-axis. Finally, we rotate by the angle $(\pi/2 - \phi)$ about x. Remembering that we have only a $(+)$

6–13

state with respect to A, we get

$$C_+ = \cos \frac{\theta}{2}\, e^{-i\phi/2}, \qquad C_- = \sin \frac{\theta}{2}\, e^{+i\phi/2}. \tag{6.36}$$

We would like, finally, to summarize the results of this chapter in a form that will be useful for our later work. First, we remind you that our primary result in Eqs. (6.35) can be written in another notation. Note that Eqs. (6.35) mean just the same thing as Eq. (6.4). That is, in Eqs. (6.35) the coefficients of $C_+ = \langle +S \mid \psi \rangle$ and $C_- = \langle -S \mid \psi \rangle$ are just the amplitudes $\langle jT \mid iS \rangle$ of Eq. (6.4)—the amplitudes that a particle in the i-state with respect to S will be in the j-state with respect to T (when the orientation of T with respect to S is given in terms of the angles α, β, and γ). We also called them R_{ji}^{TS} in Eq. (6.6). (We have a plethora of notations!) For example, $R_{-+}^{TS} = \langle -T \mid +S \rangle$ is the coefficient of C_+ in the formula for C'_-, namely, $i \sin(\alpha/2)\, e^{i(\beta-\gamma)/2}$. We can, therefore, make a summary of our results in the form of a table, as we have done in Table 6–1.

It will occasionally be handy to have these amplitudes already worked out for some simple special cases. Let's let $R_z(\phi)$ stand for a rotation by the angle ϕ about the z-axis. We can also let it stand for the corresponding rotation matrix (omitting the subscripts i and j, which are to be implicitly understood). In the same spirit $R_x(\phi)$ and $R_y(\phi)$ will stand for rotations by the angle ϕ about the x-axis or the y-axis. We give in Table 6–2 the matrices—the tables of amplitudes $\langle jT \mid iS \rangle$—which project the amplitudes from the S-frame into the T-frame, where T is obtained from S by the rotation specified.

Table 6–1

The amplitudes $\langle jT \mid iS \rangle$ for a rotation defined by the Euler angles α, β, γ of Fig. 6–9

$$R_{ji}(\alpha, \beta, \gamma)$$

$\langle jT \mid iS \rangle$	$+S$	$-S$
$+T$	$\cos \dfrac{\alpha}{2}\, e^{i(\beta+\gamma)/2}$	$i \sin \dfrac{\alpha}{2}\, e^{-i(\beta-\gamma)/2}$
$-T$	$i \sin \dfrac{\alpha}{2}\, e^{i(\beta-\gamma)/2}$	$\cos \dfrac{\alpha}{2}\, e^{-i(\beta+\gamma)/2}$

Table 6–2

The amplitudes $\langle jT \mid iS \rangle$ for a rotation $R(\phi)$ by the angle ϕ about the z-axis, x-axis, or y-axis

$$R_z(\phi)$$

$\langle jT \mid iS \rangle$	$+S$	$-S$
$+T$	$e^{i\phi/2}$	0
$-T$	0	$e^{-i\phi/2}$

$$R_x(\phi)$$

$\langle jT \mid iS \rangle$	$+S$	$-S$
$+T$	$\cos \phi/2$	$i \sin \phi/2$
$-T$	$i \sin \phi/2$	$\cos \phi/2$

$$R_y(\phi)$$

$\langle jT \mid iS \rangle$	$+S$	$-S$
$+T$	$\cos \phi/2$	$\sin \phi/2$
$-T$	$-\sin \phi/2$	$\cos \phi/2$

The Dependence of Amplitudes on Time

7-1 Atoms at rest; stationary states

We want now to talk a little bit about the behavior of probability amplitudes in time. We say a "little bit," because the actual behavior in time necessarily involves the behavior in space as well. Thus, we get immediately into the most complicated possible situation if we are to do it correctly and in detail. We are always in the difficulty that we can either treat something in a logically rigorous but quite abstract way, or we can do something which is not at all rigorous but which gives us some idea of a real situation—postponing until later a more careful treatment. With regard to energy dependence, we are going to take the second course. We will make a number of statements. We will not try to be rigorous—but will just be telling you things that have been found out, to give you some feeling for the behavior of amplitudes as a function of time. As we go along, the precision of the description will increase, so don't get nervous that we seem to be picking things out of the air. It is, of course, all out of the air—the air of experiment and of the imagination of people. But it would take us too long to go over the historical development, so we have to plunge in somewhere. We could plunge into the abstract and deduce everything—which you would not understand—or we could go through a large number of experiments to justify each statement. We choose to do something in between.

An electron alone in empty space can, under certain circumstances, have a certain definite energy. For example, if it is standing still (so it has no translational motion, no momentum, or kinetic energy), it has its rest energy. A more complicated object like an atom can also have a definite energy when standing still, but it could also be internally excited to another energy level. (We will describe later the machinery of this.) We can often think of an atom in an excited state as having a definite energy, but this is really only approximately true. An atom doesn't stay excited forever because it manages to discharge its energy by its interaction with the electromagnetic field. So there is some amplitude that a new state is generated—with the atom in a lower state, and the electromagnetic field in a higher state, of excitation. The total energy of the system is the same before and after, but the energy of the *atom* is reduced. So it is not precise to say an excited atom has a *definite* energy; but it will often be convenient and not too wrong to say that it does.

[Incidentally, why does it go one way instead of the other way? Why does an atom radiate light? The answer has to do with entropy. When the energy is in the electromagnetic field, there are so many different ways it can be—so many different places where it can wander—that if we look for the equilibrium condition, we find that in the most probable situation the field is excited with a photon, and the atom is de-excited. It takes a very long time for the photon to come back and find that it can knock the atom back up again. It's quite analogous to the classical problem: Why does an accelerating charge radiate? It isn't that it "wants" to lose energy, because, in fact, when it radiates, the energy of the world is the same as it was before. Radiation or absorption goes in the direction of increasing *entropy*.]

Nuclei can also exist in different energy levels, and in an approximation which disregards the electromagnetic effects, we can say that a nucleus in an excited state stays there. Although we know that it doesn't stay there forever, it is often useful to start out with an approximation which is somewhat idealized and easier to think about. Also it is often a legitimate approximation under certain circumstances. (When we first introduced the classical laws of a falling body, we did not include friction, but there is almost never a case in which there isn't *some* friction.)

7-1 **Atoms at rest; stationary states**

7-2 **Uniform motion**

7-3 **Potential energy; energy conservation**

7-4 **Forces; the classical limit**

7-5 **The "precession" of a spin one-half particle**

Review: Chapter 17, Vol. I, *Space-Time*
Chapter 48, Vol. I, *Beats*

Then there are the subnuclear "strange particles," which have various masses. But the heavier ones disintegrate into other light particles, so again it is not correct to say that they have a precisely definite energy. That would be true only if they lasted forever. So when we make the approximation that they have a definite energy, we are forgetting the fact that they must blow up. For the moment, then, we will intentionally forget about such processes and learn later how to take them into account.

Suppose we have an atom—or an electron, or any particle—which at rest would have a definite energy E_0. By the energy E_0 we mean the mass of the whole thing times c^2. This mass includes any internal energy; so an excited atom has a mass which is different from the mass of the same atom in the ground state. (The *ground* state means the state of lowest energy.) We will call E_0 the "energy at rest."

For an atom *at rest*, the quantum mechanical *amplitude* to find an atom at a place is the *same everywhere*; it does *not* depend on position. This means, of course, that the *probability* of *finding* the atom anywhere is the same. But it means even more. The *probability* could be independent of position, and still the *phase* of the *amplitude* could vary from point to point. But for a particle at rest, the complete amplitude is identical everywhere. It does, however, depend on the *time*. For a particle in a state of definite energy E_0, the amplitude to find the particle at (x, y, z) at the time t is

$$ae^{-i(E_0/\hbar)t}, \tag{7.1}$$

where a is some constant. The amplitude to be at any point in space is the same for all points, but depends on time according to (7.1). We shall simply assume this rule to be true.

Of course, we could also write (7.1) as

$$ae^{-i\omega t}, \tag{7.2}$$

with

$$\hbar\omega = E_0 = Mc^2,$$

where M is the rest mass of the atomic state, or particle. There are three different ways of specifying the energy: by the frequency of an amplitude, by the energy in the classical sense, or by the inertia. They are all equivalent; they are just different ways of saying the same thing.

You may be thinking that it is strange to think of a "particle" which has equal amplitudes to be found throughout all space. After all, we usually imagine a "particle" as a small object located "somewhere." But don't forget the uncertainty principle. If a particle has a definite energy, it has also a definite momentum. If the uncertainty in momentum is zero, the uncertainty relation, $\Delta p \, \Delta x = \hbar$, tells us that the uncertainty in the position must be infinite, and that is just what we are saying when we say that there is the same amplitude to find the particle at all points in space.

If the internal parts of an atom are in a different state with a different total energy, then the variation of the amplitude with time is different. If you don't know in which state it is, there will be a certain amplitude to be in one state and a certain amplitude to be in another—and each of these amplitudes will have a different frequency. There will be an interference between these different components —like a beat-note—which can show up as a varying probability. Something will be "going on" inside of the atom—even though it is "at rest" in the sense that its center of mass is not drifting. However, if the atom has one definite energy, the amplitude is given by (7.1), and the absolute square of this amplitude does not depend on time. You see, then, that if a thing has a definite energy and if you ask any *probability* question about it, the answer is independent of time. Although the *amplitudes* vary with time, if the energy is *definite* they vary as an imaginary exponential, and the absolute value doesn't change.

That's why we often say that an atom in a definite energy level is in a *stationary state*. If you make any measurements of the things inside, you'll find that nothing (in probability) will change in time. In order to have the probabilities change in time, we have to have the interference of two amplitudes at two different frequencies, and that means that we cannot know what the energy is. The object will have one amplitude to be in a state of one energy and another amplitude to be in a state of another energy. That's the quantum mechanical description of something when its *behavior* depends on time.

If we have a "condition" which is a mixture of two different states with different energies, then the amplitude for each of the two states varies with time according to Eq. (7.2), for instance, as

$$e^{-i(E_1/\hbar)t} \quad \text{and} \quad e^{-i(E_2/\hbar)t}. \tag{7.3}$$

And if we have some combination of the two, we will have an interference. But notice that if we added a constant to both energies, it wouldn't make any difference. If somebody else were to use a different scale of energy in which all the energies were increased (or decreased) by a constant amount—say, by the amount A—then the amplitudes in the two states would, from his point of view, be

$$e^{-i(E_1+A)t/\hbar} \quad \text{and} \quad e^{-i(E_2+A)t/\hbar}. \tag{7.4}$$

All of his amplitudes would be multiplied by the same factor $e^{-i(A/\hbar)t}$, and all linear combinations, or interferences, would have the same factor. When we take the absolute squares to find the probabilities, all the answers would be the same. The choice of an origin for our energy scale makes no difference; we can measure energy from any zero we want. For relativistic purposes it is nice to measure the energy so that the rest mass is included, but for many purposes that aren't relativistic it is often nice to subtract some standard amount from all energies that appear. For instance, in the case of an atom, it is usually convenient to subtract the energy $M_s c^2$, where M_s is the mass of all the *separate* pieces—the nucleus and the electrons—which is, of course, different from the mass of the atom. For other problems it may be useful to subtract from all energies the amount $M_g c^2$, where M_g is the mass of the whole atom *in the ground* state; then the energy that appears is just the excitation energy of the atom. So, sometimes we may shift our zero of energy by some very large constant, but it doesn't make any difference, provided we shift all the energies in a particular calculation by the same constant. So much for a particle standing still.

7-2 Uniform motion

If we suppose that the relativity theory is right, a particle at rest in one inertial system can be in uniform motion in another inertial system. In the rest frame of the particle, the probability amplitude is the same for all x, y, and z but varies with t. The *magnitude* of the amplitude is the same for all t, but the *phase* depends on t. We can get a kind of a picture of the behavior of the amplitude if we plot lines of equal phase—say, lines of zero phase—as a function of x and t. For a particle at rest, these equal-phase lines are parallel to the x-axis and are equally spaced in the t-coordinate, as shown by the dashed lines in Fig. 7–1.

In a different frame—x', y', z', t'—that is moving with respect to the particle in, say, the x-direction, the x' and t' coordinates of any particular point in space are related to x and t by the Lorentz transformation. This transformation can be represented graphically by drawing x' and t' axes, as is done in Fig. 7–1. (See Chapter 17, Vol. I, Fig. 17–2.) You can see that in the x'-t' system, points of equal phase† have a different spacing along the t'-axis, so the frequency of the time variation is different. Also there is a variation of the phase with x', so the probability amplitude must be a function of x'.

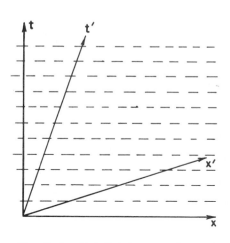

Fig. 7–1. Relativistic transformation of the amplitude of a particle at rest in the x-t systems.

---

† We are assuming that the phase should have the same value at corresponding points in the two systems. This is a subtle point, however, since the phase of a quantum mechanical amplitude is, to a large extent, arbitrary. A complete justification of this assumption requires a more detailed discussion involving interferences of two or more amplitudes.

Under a Lorentz transformation for the velocity v, say along the negative x-direction, the time t is related to the time t' by

$$t = \frac{t' - x'v/c^2}{\sqrt{1 - v^2/c^2}},$$

so our amplitude now varies as

$$e^{-(i/\hbar)E_0 t} = e^{-(i/\hbar)(E_0 t'/\sqrt{1-v^2/c^2} - E_0 v x'/c^2 \sqrt{1-v^2/c^2})}.$$

In the prime system it varies in space as well as in time. If we write the amplitude as

$$e^{-(i/\hbar)(E_p' t' - p' x')},$$

we see that $E_p' = E_0/\sqrt{1 - v^2/c^2}$ is the energy computed classically for a particle of rest energy E_0 travelling at the velocity v, and $p' = E_p'v/c^2$ is the corresponding particle momentum.

You know that $x_\mu = (t, x, y, z)$ and $p_\mu = (E, p_x, p_y, p_z)$ are four-vectors, and that $p_\mu x_\mu = Et - \mathbf{p} \cdot \mathbf{x}$ is a scalar invariant. In the rest frame of the particle, $p_\mu x_\mu$ is just Et; so if we transform to another frame, Et will be replaced by

$$E't' - \mathbf{p}' \cdot \mathbf{x}'.$$

Thus, the probability amplitude of a particle which has the momentum $\mathbf{p}$ will be proportional to

$$e^{-(i/\hbar)(E_p t - \mathbf{p} \cdot \mathbf{x})}, \tag{7.5}$$

where E_p is the energy of the particle whose momentum is p, that is,

$$E_p = \sqrt{(pc)^2 + E_0^2}, \tag{7.6}$$

where E_0 is, as before, the rest energy. For nonrelativistic problems, we can write

$$E_p = M_s c^2 + W_p, \tag{7.7}$$

where W_p is the energy over and above the rest energy $M_s c^2$ of the parts of the atom. In general, W_p would include both the kinetic energy of the atom as well as its binding or excitation energy, which we can call the "internal" energy. We would write

$$W_p = W_{\text{int}} + \frac{p^2}{2M}, \tag{7.8}$$

and the amplitudes would be

$$e^{-(i/\hbar)(W_p t - \mathbf{p} \cdot \mathbf{x})}. \tag{7.9}$$

Because we will generally be doing nonrelativistic calculations, we will use this form for the probability amplitudes.

Note that our relativistic transformation has given us the variation of the amplitude of an atom which moves in space without any additional assumptions. The wave number of the space variations is, from (7.9),

$$k = \frac{p}{\hbar}; \tag{7.10}$$

so the wavelength is

$$\lambda = \frac{2\pi}{k} = \frac{h}{p}. \tag{7.11}$$

This is the same wavelength we have used before for particles with the momentum p. This formula was first arrived at by de Broglie in just this way. For a moving particle, the *frequency* of the amplitude variations is still given by

$$\hbar\omega = W_p. \tag{7.12}$$

The absolute square of (7.9) is just 1, so for a particle in motion with a *definite energy*, the probability of finding it is the same everywhere and does not change with time. (It is important to notice that the amplitude is a *complex* wave. If we used a real sine wave, the square would vary from point to point, which would not be right.)

We know, of course, that there are situations in which particles move from place to place so that the probability depends on position and changes with time. How do we describe such situations? We can do that by considering amplitudes which are a superposition of two or more amplitudes for states of definite energy. We have already discussed this situation in Chapter 48 of Vol. I—even for probability amplitudes! We found that the sum of two amplitudes with different wave numbers k (that is, momenta) and frequencies ω (that is, energies) gives interference humps, or beats, so that the square of the amplitude varies with space and time. We also found that these beats move with the so-called "group velocity" given by

$$v_g = \frac{\Delta\omega}{\Delta k},$$

where Δk and $\Delta\omega$ are the differences between the wave numbers and frequencies for the two waves. For more complicated waves—made up of the sum of many amplitudes all near the same frequency—the group velocity is

$$v_g = \frac{d\omega}{dk}. \tag{7.13}$$

Taking $\omega = E_p/\hbar$ and $k = p/\hbar$, we see that

$$v_g = \frac{dE_p}{dp}. \tag{7.14}$$

Using Eq. (7.6), we have

$$\frac{dE_p}{dp} = c^2\,\frac{p}{E_p}. \tag{7.15}$$

But $E_p = Mc^2$, so

$$\frac{dE_p}{dp} = \frac{p}{M}, \tag{7.16}$$

which is just the classical velocity of the particle. Alternatively, if we use the non-relativistic expressions, we have

$$\omega = \frac{W_p}{\hbar} \quad\text{and}\quad k = \frac{p}{\hbar},$$

and

$$\frac{d\omega}{dk} = \frac{dW}{dp} = \frac{d}{dp}\left(\frac{p^2}{2M}\right) = \frac{p}{M}, \tag{7.17}$$

which is again the classical velocity.

Our result, then, is that if we have several amplitudes for pure energy states of nearly the same energy, their interference gives "lumps" in the probability that move through space with a velocity equal to the velocity of a classical particle of that energy. We should remark, however, that when we say we can add two amplitudes of different wave number together to get a beat-note that will correspond to a moving particle, we have introduced something new—something that we cannot deduce from the theory of relativity. We said what the amplitude did for a particle standing still and then deduced what it would do if the particle were moving. But we *cannot* deduce from these arguments what would happen when there are *two* waves moving with different speeds. If we stop one, we cannot stop the other. So we have added tacitly the *extra* hypothesis that not only is (7.9) a *possible* solution, but that there can also be solutions with all kinds of p's for the same system, and that the different terms will interfere.

7–3 Potential energy; energy conservation

Now we would like to discuss what happens when the energy of a particle can change. We begin by thinking of a particle which moves in a force field described by a potential. We discuss first the effect of a constant potential. Suppose that we have a large metal can which we have raised to some electrostatic potential ϕ, as in Fig. 7–2. If there are charged objects inside the can, their potential energy will be $q\phi$, which we will call V, and will be absolutely independent of position. Then there can be no change in the physics inside, because the constant potential doesn't make any difference so far as anything going on inside the can is concerned. Now there is no way we can deduce what the answer should be, so we must make a guess. The guess which works is more or less what you might expect: For the energy, we must use the sum of the potential energy V and the energy E_p—which is itself the sum of the internal and kinetic energies. The amplitude is proportional to

$$e^{-(i/\hbar)[(E_p+V)t-\boldsymbol{p}\cdot\boldsymbol{x}]}. \tag{7.18}$$

The *general principle* is that the coefficient of t, which we may call ω, is always given by the *total energy* of the system: internal (or "mass") energy, plus kinetic energy, plus potential energy:

$$\hbar\omega = E_p + V. \tag{7.19}$$

Or, for nonrelativistic situations,

$$\hbar\omega = W_{\text{int}} + \frac{p^2}{2M} + V. \tag{7.20}$$

Now what about physical phenomena inside the box? If there are several different energy states, what will we get? The amplitude for each state has the same additional factor

$$e^{-(i/\hbar)Vt}$$

over what it would have with $V = 0$. That is just like a change in the zero of our energy scale. It produces an equal phase change in all amplitudes, but as we have seen before, this doesn't change any of the probabilities. All the physical phenomena are the same. (We have assumed that we are talking about different states of the same charged object, so that $q\phi$ is the same for all. If an object could change its charge in going from one state to another, we would have quite another result, but conservation of charge prevents this.)

So far, our assumption agrees with what we would expect for a change of energy reference level. But if it is really right, it should hold for a potential energy that is not just a constant. In general, V could vary in any arbitrary way with both time and space, and the complete result for the amplitude must be given in terms of a differential equation. We don't want to get concerned with the general case right now, but only want to get some idea about how some things happen, so we will think only of a potential that is constant in time and varies very slowly in space. Then we can make a comparison between the classical and quantum ideas.

Suppose we think of the situation in Fig. 7–3, which has two boxes held at the constant potentials ϕ_1 and ϕ_2 and a region in between where we will assume that the potential varies smoothly from one to the other. We imagine that some particle has an amplitude to be found in any one of the regions. We also assume that the momentum is large enough so that in any small region in which there are many wavelengths, the potential is nearly constant. We would then think that in any part of the space the amplitude ought to look like (7.18) with the appropriate V for that part of the space.

Let's think of a special case in which $\phi_1 = 0$, so that the potential energy there is zero, but in which $q\phi_2$ is negative, so that classically the particle would have more energy in the second box. Classically, it would be going faster in the second box—it would have more energy and, therefore, more momentum. Let's see how that might come out of quantum mechanics.

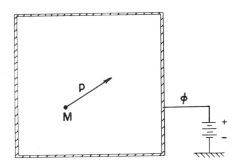

Fig. 7–2. A particle of mass M and momentum p in a region of constant potential.

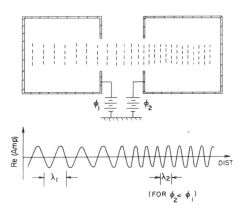

Fig. 7–3. The amplitude for a particle in transit from one potential to another.

With our assumption, the amplitude in the first box would be proportional to

$$e^{-(i/\hbar)[(W_{\text{int}}+p_1^2/2M+V_1)t-\mathbf{p}_1\cdot\mathbf{x}]},\tag{7.21}$$

and the amplitude in the second box would be proportional to

$$e^{-(i/\hbar)[(W_{\text{int}}+p_2^2/2M+V_2)t-\mathbf{p}_2\cdot\mathbf{x}]}.\tag{7.22}$$

(Let's say that the internal energy is not being changed, but remains the same in both regions.) The question is: How do these two amplitudes match together through the region between the boxes?

We are going to suppose that the potentials are all constant in time—so that nothing in the conditions varies. We will then suppose that the variations of the amplitude (that is, its phase) have the same *frequency* everywhere—because, so to speak, there is nothing in the "medium" that depends on time. If nothing in the space is changing, we can consider that the wave in one region "generates" subsidiary waves all over space which will all oscillate at the same frequency—just as light waves going through materials at rest do not change their frequency. If the frequencies in (7.21) and (7.22) are the same, we must have that

$$W_{\text{int}} + \frac{p_1^2}{2M} + V_1 = W_{\text{int}} + \frac{p_2^2}{2M} + V_2.\tag{7.23}$$

Both sides are just the classical total energies, so Eq. (7.23) is a statement of the conservation of energy. In other words, the classical statement of the conservation of energy is equivalent to the quantum mechanical statement that the frequencies for a particle are everywhere the same if the conditions are not changing with time. It all fits with the idea that $\hbar\omega = E$.

In the special example that $V_1 = 0$ and V_2 is negative, Eq. (7.23) gives that p_2 is greater than p_1, so the wavelength of the waves is shorter in region 2. The surfaces of equal phase are shown by the dashed lines in Fig. 7–3. We have also drawn a graph of the real part of the amplitude, which shows again how the wavelength decreases in going from region 1 to region 2. The group velocity of the waves, which is p/M, also increases in the way one would expect from the classical energy conservation, since it is just the same as Eq. (7.23).

There is an interesting special case where V_2 gets so large that $V_2 - V_1$ is greater than $p_1^2/2M$. Then p_2^2, which is given by

$$p_2^2 = 2M\left[\frac{p_1^2}{2M} - V_2 + V_1\right],\tag{7.24}$$

is *negative*. That means that p_2 is an imaginary number, say, ip'. Classically, we would say that the particle never gets into region 2—it doesn't have enough energy to climb the potential hill. Quantum mechanically, however, the amplitude is still given by Eq. (7.22); its space variation still goes as

$$e^{(i/\hbar)\mathbf{p}_2\cdot\mathbf{x}}.$$

But if p_2 is imaginary, the space dependence becomes a real exponential. Say that the particle was initially going in the $+x$-direction; then the amplitude would vary as

$$e^{-p'x/\hbar}.\tag{7.25}$$

The amplitude decreases rapidly with increasing x.

Imagine that the two regions at different potentials were very close together, so that the potential energy changed suddenly from V_1 to V_2, as shown in Fig. 7–4(a). If we plot the real part of the probability amplitude, we get the dependence shown in part (b) of the figure. The wave in the first region corresponds to a particle trying to get into the second region, but the amplitude there falls off

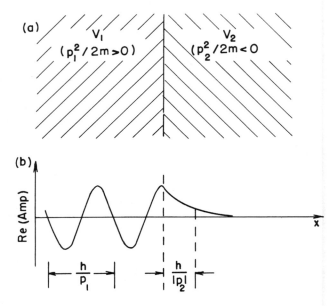

Fig. 7–4. The amplitude for a particle approaching a strongly repulsive potential.

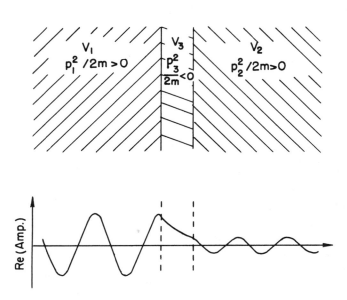

Fig. 7–5. The penetration of the amplitude through a potential barrier.

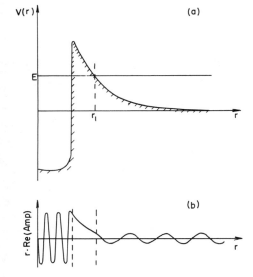

Fig. 7–6. (a) The potential function for an α-particle in a uranium nucleus. (b) The qualitative form of the probability amplitude.

rapidly. There is some chance that it will be observed in the second region—where it could *never* get classically—but the amplitude is very small except right near the boundary. The situation is very much like what we found for the total internal reflection of light. The light doesn't normally get out, but we can observe it if we put something within a wavelength or two of the surface.

You will remember that if we put a second surface close to the boundary where light was totally reflected, we could get some light transmitted into the second piece of material. The corresponding thing happens to particles in quantum mechanics. If there is a narrow region with a potential V, so great that the classical kinetic energy would be negative, the particle would classically never get past. But quantum mechanically, the exponentially decaying amplitude can reach across the region and give a small probability that the particle will be found on the other side where the kinetic energy is again positive. The situation is illustrated in Fig. 7–5. This effect is called the quantum mechanical "penetration of a barrier."

The barrier penetration by a quantum mechanical amplitude gives the explanation—or description—of the α-particle decay of a uranium nucleus. The potential energy of an α-particle, as a function of the distance from the center, is shown in Fig. 7–6(a). If one tried to shoot an α-particle with the energy E *into* the nucleus, it would feel an electrostatic repulsion from the nuclear charge z and would, classically, get no closer than the distance r_1 where its total energy is equal to the potential energy V. Closer in, however, the potential energy is much lower because of the strong attraction of the short-range nuclear forces. How is it then that in radioactive decay we find α-particles which started out inside the nucleus coming out with the energy E? Because they start out with the energy E inside the nucleus and "leak" through the potential barrier. The probability amplitude is roughly as sketched in part (b) of Fig. 7–6, although actually the exponential decay is much larger than shown. It is, in fact, quite remarkable that the mean life of an α-particle in the uranium nucleus is as long as $4\frac{1}{2}$ billion years, when the natural oscillations inside the nucleus are so extremely rapid—about 10^{22} per sec! How can one get a number like 10^9 years from 10^{-22} sec? The answer is that the exponential gives the tremendously small factor of about e^{-45}—which gives the very small, though definite, probability of leakage. Once the α-particle is in the nucleus, there is almost no amplitude at all for finding it outside; however, if you take many nuclei and wait long enough, you may be lucky and find one that has come out.

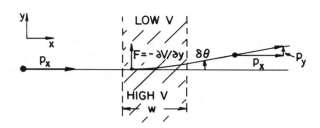

Fig. 7–7. The deflection of a particle by a transverse potential gradient.

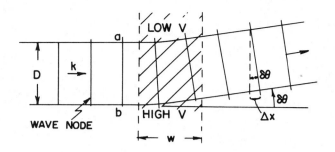

Fig. 7–8. The probability amplitude in a region with a transverse potential gradient.

7–4 Forces; the classical limit

Suppose that we have a particle moving along and passing through a region where there is a potential that varies at right angles to the motion. Classically, we would describe the situation as sketched in Fig. 7–7. If the particle is moving along the x-direction and enters a region where there is a potential that varies with y, the particle will get a transverse acceleration from the force $F = -\partial V/\partial y$. If the force is present only in a limited region of width w, the force will act only for the time w/v. The particle will be given the transverse momentum

$$p_y = F\frac{w}{v}.$$

The angle of deflection $\delta\theta$ is then

$$\delta\theta = \frac{p_y}{p} = \frac{Fw}{pv},$$

where p is the initial momentum. Using $-\partial V/\partial y$ for F, we get

$$\delta\theta = -\frac{w}{pv}\frac{\partial V}{\partial y}. \tag{7.26}$$

It is now up to us to see if our idea that the waves go as (7.20) will explain the same result. We look at the same thing quantum mechanically, assuming that everything is on a very large scale compared with a wavelength of our probability amplitudes. In any small region we can say that the amplitude varies as

$$e^{-(i/\hbar)[(W+p^2/2M+V)t-\boldsymbol{p}\cdot\boldsymbol{x}]}. \tag{7.27}$$

Can we see that this will also give rise to a deflection of the particle when V has a transverse gradient? We have sketched in Fig. 7–8 what the waves of probability amplitude will look like. We have drawn a set of "wave nodes" which you can think of as surfaces where the phase of the amplitude is zero. In every small region, the wavelength—the distance between successive nodes—is

$$\lambda = \frac{h}{p},$$

where p is related to V through

$$W + \frac{p^2}{2M} + V = \text{const.} \tag{7.28}$$

In the region where V is larger, p is smaller, and the wavelength is longer. So the angle of the wave nodes gets changed as shown in the figure.

To find the change in angle of the wave nodes we notice that for the two paths a and b in Fig. 7–8 there is a difference of potential $\Delta V = (\partial V/\partial y)D$, so there is a difference Δp in the momentum along the two tracks which can be obtained from (7.28):

$$\Delta\left(\frac{p^2}{2M}\right) = \frac{p}{M}\,\Delta p = -\Delta V. \tag{7.29}$$

The wave number $p/\hbar$ is, therefore, different along the two paths, which means that the phase is advancing at a different rate. The difference in the rate of increase of phase is $\Delta k = \Delta p/\hbar$, so the accumulated phase difference in the total distance w is

$$\Delta(\text{phase}) = \Delta k \cdot w = \frac{\Delta p}{\hbar} \cdot w = -\frac{M}{p\hbar}\,\Delta V \cdot w. \tag{7.30}$$

This is the amount by which the phase on path b is "ahead" of the phase on path a as the wave leaves the strip. But outside the strip, a phase advance of this amount corresponds to the wave node being ahead by the amount

$$\Delta x = \frac{\lambda}{2\pi}\,\Delta(\text{phase}) = \frac{\hbar}{p}\,\Delta(\text{phase})$$

or

$$\Delta x = -\frac{M}{p^2}\,\Delta V \cdot w. \tag{7.31}$$

Referring to Fig. 7–8, we see that the new wavefronts will be at the angle $\delta\theta$ given by

$$\Delta x = D\,\delta\theta; \tag{7.32}$$

so we have

$$D\,\delta\theta = -\frac{M}{p^2}\,\Delta V \cdot w. \tag{7.33}$$

This is identical to Eq. (7.26) if we replace p/m by v and $\Delta V/D$ by $\partial V/\partial y$.

The result we have just got is correct only if the potential variations are slow and smooth—in what we call the *classical limit*. We have shown that under these conditions we will get the same particle motions we get from $F = ma$, provided we assume that a potential contributes a phase to the probability amplitude equal to $Vt/\hbar$. *In the classical limit, the quantum mechanics will agree with Newtonian mechanics.*

7–5 The "precession" of a spin one-half particle

Notice that we have not assumed anything special about the potential energy—it is just that energy whose derivative gives a force. For instance, in the Stern-Gerlach experiment we had the energy $U = -\boldsymbol{\mu} \cdot \boldsymbol{B}$, which gives a force if $\boldsymbol{B}$ has a spatial variation. If we wanted to give a quantum mechanical description, we would have said that the particles in one beam had an energy that varied one way and that those in the other beam had an opposite energy variation. (We could put the magnetic energy U into the potential energy V or into the "internal" energy W; it doesn't matter.) Because of the energy variation, the waves are refracted, and the beams are bent up or down. (We see now that quantum mechanics would give us the same bending as we would compute from the classical mechanics.)

From the dependence of the amplitude on potential energy we would also expect that if a particle sits in a uniform magnetic field along the z-direction, its probability amplitude must be changing with time according to

$$e^{-(i/\hbar)(-\mu_z B)t}.$$

(We can consider that this is, in effect, a definition of μ_z.) In other words, if we place a particle in a uniform field B for a time τ, its probability amplitude will be multiplied by

$$e^{-(i/\hbar)(-\mu_z B)\tau}$$

over what it would be in no field. Since for a spin one-half particle, μ_z can be either plus or minus some number, say μ, the two possible states in a uniform field would have their phases changing at the same rate but in opposite directions. The two amplitudes get multiplied by

$$e^{\pm(i/\hbar)\mu B\tau}. \tag{7.34}$$

This result has some interesting consequences. Suppose we have a spin one-half particle in some state that is not purely spin up or spin down. We can describe its condition in terms of the amplitudes to be in the pure up and pure down states. But in a magnetic field, these two states will have phases changing at a different rate. So if we ask some question about the amplitudes, the answer will depend on how long it has been in the field.

As an example, we consider the disintegration of the muon in a magnetic field. When muons are produced as disintegration products of π-mesons, they are polarized (in other words, they have a preferred spin direction). The muons, in turn, disintegrate—in about 2.2 microseconds on the average—emitting an electron and two neutrinos:

$$\mu \to e + \nu + \bar{\nu}.$$

In this disintegration it turns out that (for at least the highest energies) the electrons are emitted preferentially in the direction opposite to the spin direction of the muon.

Suppose then that we consider the experimental arrangement shown in Fig. 7-9. If polarized muons enter from the left and are brought to rest in a block of material at A, they will, a little while later, disintegrate. The electrons emitted will, in general, go off in all possible directions. Suppose, however, that the muons all enter the stopping block at A with their spins in the x-direction. Without a magnetic field there would be some angular distribution of decay directions; we would like to know how this distribution is changed by the presence of the magnetic field. We expect that it may vary in some way with time. We can find out what happens by asking, for any moment, what the amplitude is that the muon will be found in the $(+x)$ state.

We can state the problem in the following way: A muon is known to have its spin in the $+x$-direction at $t = 0$; what is the amplitude that it will be in the same state at the time τ? Now we do not have any rule for the behavior of a spin one-half particle in a magnetic field at right angles to the spin, but we do know what happens to the spin up and spin down states with respect to the field—their amplitudes get multiplied by the factor (7.34). Our procedure then is to choose the representation in which the base states are spin up and spin down with respect to the z-direction (the field direction). Any question can then be expressed with reference to the amplitudes for these states.

Let's say that $\psi(t)$ represents the muon state. When it enters the block A, its state is $\psi(0)$, and we want to know $\psi(\tau)$ at the later time τ. If we represent the two base states by $(+z)$ and $(-z)$ we know the two amplitudes $\langle +z \mid \psi(0) \rangle$ and $\langle -z \mid \psi(0) \rangle$—we know these amplitudes because we know that $\psi(0)$ represents a state with the spin in the $(+x)$ state. From the results of the last chapter, these amplitudes are†

$$\langle +z \mid +x \rangle = C_+ = \frac{1}{\sqrt{2}}$$

and

$$\langle -z \mid +x \rangle = C_- = \frac{1}{\sqrt{2}}. \tag{7.35}$$

They happen to be equal. Since these amplitudes refer to the condition at $t = 0$, let's call them $C_+(0)$ and $C_-(0)$.

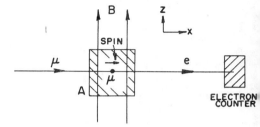

Fig. 7-9. A muon-decay experiment.

---

† If you skipped Chapter 6, you can just take (7.35) as an underived rule for now. We will give later (in Chapter 10) a more complete discussion of spin precession, including a derivation of these amplitudes.

Now we know what happens to these two amplitudes with time. Using (7.34), we have

$$C_+(t) = C_+(0)e^{-(i/\hbar)\mu Bt}$$

and

$$C_-(t) = C_-(0)e^{+(i/\hbar)\mu Bt}.$$

(7.36)

But if we know $C_+(t)$ and $C_-(t)$, we have all there is to know about the condition at t. The only trouble is that what we *want* to know is the probability that at t the spin will be in the $+x$-direction. Our general rules can, however, take care of this problem. We write that the amplitude to be in the $(+x)$ state at time t, which we may call $A_+(t)$, is

$$A_+(t) = \langle +x \mid \psi(t)\rangle = \langle +x \mid +z\rangle\langle +z \mid \psi(t)\rangle + \langle +x \mid -z\rangle\langle -z \mid \psi(t)\rangle$$

or

$$A_+(t) = \langle +x \mid +z\rangle C_+(t) + \langle +x \mid -z\rangle C_-(t).$$

(7.37)

Again using the results of the last chapter—or better the equality $\langle \phi \mid \chi\rangle = \langle \chi \mid \phi\rangle^*$ from Chapter 5—we know that

$$\langle +x \mid +z\rangle = \frac{1}{\sqrt{2}}, \qquad \langle +x \mid -z\rangle = \frac{1}{\sqrt{2}}.$$

So we know all the quantities in Eq. (7.37). We get

$$A_+(t) = \tfrac{1}{2}e^{(i/\hbar)\mu Bt} + \tfrac{1}{2}e^{-(i/\hbar)\mu Bt},$$

or

$$A_+(t) = \cos\frac{\mu B}{\hbar}t.$$

A particularly simple result! Notice that the answer agrees with what we expect for $t = 0$. We get $A_+(0) = 1$, which is right, because we assumed that the muon was in the $(+x)$ state at $t = 0$.

The probability P_+ that the muon will be found in the $(+x)$ state at t is $(A_+)^2$ or

$$P_+ = \cos^2\frac{\mu Bt}{\hbar}.$$

The probability oscillates between zero and one, as shown in Fig. 7–10. Note that the probability returns to one for $\mu Bt/\hbar = \pi$ (*not* 2π). Because we have squared the cosine function, the probability repeats itself with the frequency $2\mu B/\hbar$.

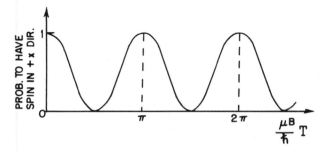

Fig. 7–10. Time dependence of the probability that a spin one-half particle will be in a $(+)$ state with respect to the x-axis.

Thus, we find that the chance of catching the decay electron in the electron counter of Fig. 7–9 varies periodically with the length of time the muon has been sitting in the magnetic field. The frequency depends on the magnetic moment μ. The magnetic moment of the muon has, in fact, been measured in just this way.

We can, of course, use the same method to answer any other questions about the muon decay. For example, how does the chance of detecting a decay electron in the y-direction at $90°$ to the x-direction but still at right angles to the field depend on t? If you work it out, the amplitude to be in the $(+y)$ state varies as $\cos^2\{(\mu Bt/\hbar) - \pi/4\}$, which oscillates with the same period but reaches its maximum one-quarter cycle later, when $\mu Bt/\hbar = \pi/4$. In fact, what is happening is that as time goes on, the muon goes through a succession of states which correspond to complete polarization in a direction that is continually rotating about the z-axis. We can describe this by saying that *the spin is precessing* at the frequency

$$\omega_p = \frac{2\mu B}{\hbar}. \tag{7.38}$$

You can begin to see the form that our quantum mechanical description will take when we are describing how things behave in time.

The Hamiltonian Matrix

8-1 Amplitudes and vectors

Before we begin the main topic of this chapter, we would like to describe a number of mathematical ideas that are used a lot in the literature of quantum mechanics. Knowing them will make it easier for you to read other books or papers on the subject. The first idea is the close mathematical resemblance between the equations of quantum mechanics and those of the scalar product of two vectors. You remember that if χ and ϕ are two states, the amplitude to start in ϕ and end up in χ can be written as a sum over a complete set of base states of the amplitude to go from ϕ into one of the base states and then from that base state out again into χ:

$$\langle \chi \mid \phi \rangle = \sum_{\text{all } i} \langle \chi \mid i \rangle \langle i \mid \phi \rangle. \tag{8.1}$$

We explained this in terms of a Stern-Gerlach apparatus, but we remind you that there is no need to have the apparatus. Equation (8.1) is a mathematical law that is just as true whether we put the filtering equipment in or not—it is not always necessary to imagine that the apparatus is there. We can think of it simply as a formula for the amplitude $\langle \chi \mid \phi \rangle$.

We would like to compare Eq. (8.1) to the formula for the dot product of two vectors $\boldsymbol{B}$ and $\boldsymbol{A}$. If $\boldsymbol{B}$ and $\boldsymbol{A}$ are ordinary vectors in three dimensions, we can write the dot product this way:

$$\sum_{\text{all } i} (\boldsymbol{B} \cdot \boldsymbol{e}_i)(\boldsymbol{e}_i \cdot \boldsymbol{A}), \tag{8.2}$$

with the understanding that the symbol $\boldsymbol{e}_i$ stands for the three unit vectors in the x, y, and z-directions. Then $\boldsymbol{B} \cdot \boldsymbol{e}_1$ is what we ordinarily call B_x; $\boldsymbol{B} \cdot \boldsymbol{e}_2$ is what we ordinarily call B_y; and so on. So Eq. (8.2) is equivalent to

$$B_x A_x + B_y A_y + B_z A_z,$$

which is the dot product $\boldsymbol{B} \cdot \boldsymbol{A}$.

Comparing Eqs. (8.1) and (8.2), we can see the following analogy: The states χ and ϕ correspond to the two vectors A and B. The base states i correspond to the special vectors $\boldsymbol{e}_i$ to which we refer all other vectors. Any vector can be represented as a linear combination of the three "base vectors" $\boldsymbol{e}_i$. Furthermore, if you know the coefficients of each "base vector" in this combination—that is, its three components—you know everything about a vector. In a similar way, any quantum mechanical state can be described completely by the amplitude $\langle i \mid \phi \rangle$ to go into the base states; and if you know these coefficients, you know everything there is to know about the state. Because of this close analogy, what we have called a "state" is often also called a "state vector."

Since the base vectors $\boldsymbol{e}_i$ are all at right angles, we have the relation

$$\boldsymbol{e}_i \cdot \boldsymbol{e}_j = \delta_{ij}. \tag{8.3}$$

This corresponds to the relations (5.25) among the base states i,

$$\langle i \mid j \rangle = \delta_{ij}. \tag{8.4}$$

You see now why one says that the base states i are all "orthogonal."

8-1 Amplitudes and vectors

8-2 Resolving state vectors

8-3 What are the base states of the world?

8-4 How states change with time

8-5 The Hamiltonian matrix

8-6 The ammonia molecule

Review: Chapter 49, Vol. I, *Modes*

There is one minor difference between Eq. (8.1) and the dot product. We have that

$$\langle \phi \mid \chi \rangle = \langle \chi \mid \phi \rangle^*. \qquad (8.5)$$

But in vector algebra,

$$\boldsymbol{A} \cdot \boldsymbol{B} = \boldsymbol{B} \cdot \boldsymbol{A}.$$

With the complex numbers of quantum mechanics we have to keep straight the order of the terms, whereas in the dot product, the order doesn't matter.

Now consider the following vector equation:

$$\boldsymbol{A} = \sum_i \boldsymbol{e}_i (\boldsymbol{e}_i \cdot \boldsymbol{A}). \qquad (8.6)$$

It's a little unusual, but correct. It means the same thing as

$$\boldsymbol{A} = \sum_i A_i \boldsymbol{e}_i = A_x \boldsymbol{e}_x + A_y \boldsymbol{e}_y + A_z \boldsymbol{e}_z. \qquad (8.7)$$

Notice, though, that Eq. (8.6) involves a quantity which is *different* from a dot product. A dot product is just a *number*, whereas Eq. (8.6) is a *vector* equation. One of the great tricks of vector analysis was to abstract away from the equations the idea of a *vector* itself. One might be similarly inclined to abstract a thing that is the analog of a "vector" from the quantum mechanical formula Eq. (8.1)—and one can indeed. We remove the $\langle \chi \mid$ from both sides Eq. (8.1) and write the following equation (don't get frightened—it's just a notation and in a few minutes you will find out what the symbols mean):

$$\mid \phi \rangle = \sum_i \mid i \rangle \langle i \mid \phi \rangle. \qquad (8.8)$$

One thinks of the bracket $\langle \chi \mid \phi \rangle$ as being divided into two pieces. The second piece $\mid \phi \rangle$ is often called a *ket*, and the first piece $\langle \chi \mid$ is called a *bra* (put together, they make a "bra-ket"—a notation proposed by Dirac); the half-symbols $\langle \chi \mid$ and $\mid \phi \rangle$ are also called *state vectors*. In any case, they are *not* numbers, and, in general, we want the results of our calculations to come out as numbers; so such "unfinished" quantities are only part-way steps in our calculations.

It happens that until now we have written all our results in terms of numbers. How have we managed to avoid vectors? It is amusing to note that even in ordinary vector algebra we *could* make all equations involve only numbers. For instance, instead of a vector equation like

$$\boldsymbol{F} = m\boldsymbol{a},$$

we could always have written

$$\boldsymbol{C} \cdot \boldsymbol{F} = \boldsymbol{C} \cdot (m\boldsymbol{a}).$$

We have then an equation between dot products that is true for *any* vector $\boldsymbol{C}$. But if it is true for any $\boldsymbol{C}$, it hardly makes sense at all to keep writing the $\boldsymbol{C}$!

Now look at Eq. (8.1). It is an equation that is true for *any* χ. So to save writing, we should just leave *out* the χ and write Eq. (8.8) instead. It has the same information *provided* we understand that it should always be "finished" by "multiplying on the left by"—which simply means reinserting—some $\langle \chi \mid$ on both sides. So Eq. (8.8) means exactly the same thing as Eq. (8.1)—no more, no less. When you want numbers, you put in the $\langle \chi \mid$ you want.

Maybe you have already wondered about the ϕ in Eq. (8.8). Since the equation is true for *any* ϕ, why do we keep *it*? Indeed, Dirac suggests that the ϕ also can just as well be abstracted away, so that we have only

$$\mid = \sum_i \mid i \rangle \langle i \mid. \qquad (8.9)$$

And this is the great law of quantum mechanics! (There is no analog in vector analysis.) It says that if you put *in* any two states χ and ϕ on the left and right of both sides, you *get back* Eq. (8.1). It is not really very useful, but it's a nice reminder that the equation is true for any two states.

8–2 Resolving state vectors

Let's look at Eq. (8.8) again; we can think of it in the following way. Any state vector $|\phi\rangle$ can be represented as a linear combination with suitable coefficients of a set of base "vectors"—or, if you prefer, as a superposition of "unit vectors" in suitable proportions. To emphasize that the coefficients $\langle i \mid \phi \rangle$ are just ordinary (complex) numbers, suppose we write

$$\langle i \mid \phi \rangle = C_i.$$

Then Eq. (8.8) is the same as

$$|\phi\rangle = \sum_i |i\rangle C_i. \tag{8.10}$$

We can write a similar equation for any other state vector, say $|\chi\rangle$, with, of course, different coefficients—say D_i. Then we have

$$|\chi\rangle = \sum_i |i\rangle D_i. \tag{8.11}$$

The D_i are just the amplitudes $\langle i \mid \chi \rangle$.

Suppose we had started by abstracting the ϕ from Eq. (8.1). We would have had

$$\langle \chi | = \sum_i \langle \chi \mid i \rangle \langle i |. \tag{8.12}$$

Remembering that $\langle \chi \mid i \rangle = \langle i \mid \chi \rangle^*$, we can write this as

$$\langle \chi | = \sum_i D_i^* \langle i |. \tag{8.13}$$

Now the interesting thing is that we can just *multiply* Eq. (8.13) and Eq. (8.10) to get back $\langle \chi \mid \phi \rangle$. When we do that, we have to be careful of the summation indices, because they are quite distinct in the two equations. Let's first rewrite Eq. (8.13) as

$$\langle \chi | = \sum_j D_j^* \langle j |,$$

which changes nothing. Then putting it together with Eq. (8.10), we have

$$\langle \chi \mid \phi \rangle = \sum_{ij} D_j^* \langle j \mid i \rangle C_i. \tag{8.14}$$

Remember, though, that $\langle j \mid i \rangle = \delta_{ij}$, so that in the sum we have left only the terms with $j = i$. We get

$$\langle \chi \mid \phi \rangle = \sum_i D_i^* C_i, \tag{8.15}$$

where, of course, $D_i^* = \langle i \mid \chi \rangle^* = \langle \chi \mid i \rangle$, and $C_i = \langle i \mid \phi \rangle$. Again we see the close analogy with the dot product

$$\boldsymbol{A} \cdot \boldsymbol{B} = \sum_i A_i B_i.$$

The only difference is the complex conjugate on D_i. So Eq. (8.15) says that if the state vectors $\langle \chi |$ and $|\phi\rangle$ are expanded in terms of the base vectors $\langle i |$ or $|i\rangle$, the amplitude to go from ϕ to χ is given by the kind of dot product in Eq. (8.15). This equation is, of course, just Eq. (8.1) written with different symbols. So we have just gone in a circle to get used to the new symbols.

We should perhaps emphasize again that while space vectors in three dimensions are described in terms of *three* orthogonal unit vectors, the base vectors $|i\rangle$ of the quantum mechanical states must range over the complete set applicable to any particular problem. Depending on the situation, two, or three, or five, or an infinite number of base states may be involved.

We have also talked about what happens when particles go through an apparatus. If we start the particles out in a certain state ϕ, then send them through an apparatus, and afterward make a measurement to see if they are in state χ, the result is described by the amplitude

$$\langle \chi \mid A \mid \phi \rangle. \tag{8.16}$$

Such a symbol doesn't have a close analog in vector algebra. (It is closer to tensor algebra, but the analogy is not particularly useful.) We saw in Chapter 5, Eq. (5.32), that we could write (8.16) as

$$\langle \chi \mid A \mid \phi \rangle = \sum_{ij} \langle \chi \mid i \rangle \langle i \mid A \mid j \rangle \langle j \mid \phi \rangle. \tag{8.17}$$

This is just an example of the fundamental rule Eq. (8.9), used twice.

We also found that if another apparatus B was added in series with A, then we could write

$$\langle \chi \mid BA \mid \phi \rangle = \sum_{ijk} \langle \chi \mid i \rangle \langle i \mid B \mid j \rangle \langle j \mid A \mid k \rangle \langle k \mid \phi \rangle. \tag{8.18}$$

Again, this comes directly from Dirac's method of writing Eq. (8.9)—remember that we can always place a bar ($\mid$), which is just like the factor 1, between B and A.

Incidentally, we can think of Eq. (8.17) in another way. Suppose we think of the particle entering apparatus A in the state ϕ and coming out of A in the state ψ ("psi"). In other words, we could ask ourselves this question: Can we find a ψ such that the amplitude to get from ψ to χ is always identically and everywhere the same as the amplitude $\langle \chi \mid A \mid \phi \rangle$? The answer is yes. We want Eq. (8.17) to be replaced by

$$\langle \chi \mid \psi \rangle = \sum_{i} \langle \chi \mid i \rangle \langle i \mid \psi \rangle. \tag{8.19}$$

We can clearly do this if

$$\langle i \mid \psi \rangle = \sum_{j} \langle i \mid A \mid j \rangle \langle j \mid \phi \rangle = \langle i \mid A \mid \phi \rangle, \tag{8.20}$$

which determines ψ. "But it doesn't determine ψ," you say; "it only determines $\langle i \mid \psi \rangle$." However, $\langle i \mid \psi \rangle$ *does* determine ψ, because if you have all the coefficients that relate ψ to the base states i, then ψ is uniquely defined. In fact, we can play with our notation and write the last term of Eq. (8.20) as

$$\langle i \mid \psi \rangle = \sum_{j} \langle i \mid j \rangle \langle j \mid A \mid \phi \rangle. \tag{8.21}$$

Then, since this equation is true for all i, we can write simply

$$\mid \psi \rangle = \sum_{j} \mid j \rangle \langle j \mid A \mid \phi \rangle. \tag{8.22}$$

Then we can say: "The state ψ is what we get if we start with ϕ and go through the apparatus A."

One final example of the tricks of the trade. We start again with Eq. (8.17). Since it is true for any χ and ϕ, we can drop them both! We then get†

$$A = \sum_{ij} \mid i \rangle \langle i \mid A \mid j \rangle \langle j \mid. \tag{8.23}$$

What does it mean? It means no more, no less, than what you get if you put back the ϕ and χ. As it stands, it is an "open" equation and incomplete. If we multiply it "on the left" by $\mid \phi \rangle$, it becomes

$$A \mid \phi \rangle = \sum_{ij} \mid i \rangle \langle i \mid A \mid j \rangle \langle j \mid \phi \rangle, \tag{8.24}$$

---

† You might think we should write $\mid A \mid$ instead of just A. But then it would look like the symbol for "absolute value of A," so the bars are usually dropped. In general, the bar ($\mid$) behaves much like the factor one.

which is just Eq. (8.22) all over again. In fact, we could have just dropped the j's from that equation and written

$$| \psi \rangle = A | \phi \rangle. \qquad (8.25)$$

The symbol A is neither an amplitude, nor a vector; it is a new kind of thing called an *operator*. It is something which "operates on" a state to produce a new state—Eq. (8.25) says that $| \psi \rangle$ is what results if A operates on $| \phi \rangle$. Again, it is still an open equation until it is completed with some bra like $\langle \chi |$ to give

$$\langle \chi | \psi \rangle = \langle \chi | A | \phi \rangle. \qquad (8.26)$$

The operator A is, of course, described completely if we give the matrix of amplitudes $\langle i | A | j \rangle$—also written A_{ij}—in terms of any set of base vectors.

We have really added nothing new with all of this new mathematical notation. One reason for bringing it all up was to show you the way of writing pieces of equations, because in many books you will find the equations written in the incomplete forms, and there's no reason for you to be paralyzed when you come across them. If you prefer, you can always add the missing pieces to make an equation between numbers that will look like something more familiar.

Also, as you will see, the "bra" and "ket" notation is a very convenient one. For one thing, we can from now on identify a state by giving its state vector. When we want to refer to a state of definite momentum p we can say: "the state $| p \rangle$". Or we may speak of some arbitrary state $| \psi \rangle$. For consistency we will always use the ket, writing $| \psi \rangle$, to identify a state. (It is, of course an arbitrary choice; we could equally well have chosen to use the bra, $\langle \psi |$.)

8–3 What are the base states of the world?

We have discovered that any state in the world can be represented as a superposition—a linear combination with suitable coefficients—of base states. You may ask, first of all, *what* base states? Well, there are many different possibilities. You can, for instance, project a spin in the z-direction or in some other direction. There are many, many different *representations*, which are the analogs of the different *coordinate systems* one can use to represent ordinary vectors. Next, *what* coefficients? Well, that depends on the physical circumstances. Different sets of coefficients correspond to different physical conditions. The important thing to know about is the "space" in which you are working—in other words, what the base states mean physically. So the first thing you have to know about, in general, is what the base states are like. Then you can understand how to describe a situation in terms of these base states.

We would like to look ahead a little and speak a bit about what the general quantum mechanical description of nature is going to be—in terms of the now current ideas of physics, anyway. First, one decides on a particular representation for the base states—different representations are always possible. For example, for a spin one-half particle we can use the plus and minus states with respect to the z-axis. But there's nothing special about the z-axis—you can take any other axis you like. For consistency we'll always pick the z-axis, however. Suppose we begin with a situation with one electron. In addition to the two possibilities for the spin ("up" and "down" along the z-direction), there is also the momentum of the electron. We pick a set of base states, each corresponding to one value of the momentum. What if the electron doesn't have a definite momentum? That's all right; we're just saying what the *base* states are. If the electron hasn't got a definite momentum, it has some amplitude to have one momentum and another amplitude to have another momentum, and so on. And if it is not necessarily spinning up, it has some amplitude to be spinning up going at this momentum, and some amplitude to be spinning down going at that momentum, and so on. The complete description of an electron, *so far as we know*, requires only that the base states be described by the *momentum* and the *spin*. So one acceptable set of base states $| i \rangle$ for a single electron refer to different values of the momentum and whether the spin is up or down. Different mixtures of amplitudes—that is, different combinations of the C's describe different circumstances. What any particular electron is doing is described by telling with what amplitude it has an up-spin or a down-spin and one momentum or another—for all possible momenta. So you can see what is involved in a complete quantum mechanical description of a single electron.

What about systems with more than one electron? Then the base states get more complicated. Let's suppose that we have two electrons. We have, first of all, four possible states with respect to spin: both electrons spinning up, the first one down and the second one up, the first one up and the second one down, or both down. Also we have to specify that the first electron has the momentum p_1, and the second electron, the momentum p_2. The base states for two electrons require the specification of two momenta and two spin characters. With seven electrons, we have to specify seven of each.

If we have a proton and an electron, we have to specify the spin direction of the proton and its momentum, and the spin direction of the electron and its momentum. At least that's approximately true. *We do not really know* what the correct representation is for the world. It is all very well to start out by supposing that if you specify the spin in the electron and its momentum, and likewise for a proton, you will have the base states; but what about the "guts" of the proton? Let's look at it this way. In a hydrogen atom which has one proton and one electron, we have many different base states to describe—up and down spins of the proton and electron and the various possible momenta of the proton and electron. Then there are different combinations of amplitudes C_i which together describe the character of the hydrogen atom in different states. But suppose we look at the whole hydrogen atom as a "particle." If we didn't know that the hydrogen atom was made out of a proton and an electron, we might have started out and said: "Oh, I know what the base states are—they correspond to a particular momentum of the hydrogen atom." No, because the hydrogen atom has internal parts. It may, therefore, have various states of different internal energy, and describing the real nature requires more detail.

The question is: Does a proton have internal parts? Do we have to describe a proton by giving all possible states of protons, and mesons, and strange particles? We don't know. And even though we suppose that the electron is simple, so that all we have to tell about it is its momentum and its spin, maybe tomorrow we will discover that the electron also has inner gears and wheels. It would mean that our representation is incomplete, or wrong, or approximate—in the same way that a representation of the hydrogen atom which describes only its momentum would be incomplete, because it disregarded the fact that the hydrogen atom could have become excited inside. If an electron could become excited inside and turn into something else like, for instance, a muon, then it would be described not just by giving the states of the new particle, but presumably in terms of some more complicated internal wheels. The *main problem in the study of the fundamental particles today* is to discover what are the correct representations for the description of nature. At the present time, we *guess* that for the electron it is enough to specify its momentum and spin. We also guess that there is an idealized proton which has its π-mesons, and k-mesons, and so on, that all have to be specified. Several dozen particles—that's crazy! The question of what *is* a fundamental particle and what *is not* a fundamental particle—a subject you hear so much about these days—is the question of what is the final *representation* going to look like in the ultimate quantum mechanical description of the world. Will the electron's momentum still be the right thing with which to describe nature? Or even, should the whole question be put this way at all! This question must always come up in any scientific investigation. At any rate, we see a problem—how to find a representation. We don't know the answer. We don't even know whether we have the "right" problem, but if we do, we must first attempt to find out whether any particular particle is "fundamental" or not.

In the nonrelativistic quantum mechanics—if the energies are not too high, so that you don't disturb the inner workings of the strange particles and so forth—

you can do a pretty good job without worrying about these details. You can just decide to specify the momenta and spins of the electrons and of the nuclei; then everything will be all right. In most chemical reactions and other low-energy happenings, nothing goes on in the nuclei; they don't get excited. Furthermore, if a hydrogen atom is moving slowly and bumping quietly against other hydrogen atoms—never getting excited inside, or radiating, or anything complicated like that, but staying always in the ground state of energy for internal motion—you can use an approximation in which you talk about the hydrogen atom as one object, or particle, and not worry about the fact that it *can* do something inside. This will be a good approximation as long as the kinetic energy in any collision is well below 10 electron volts—the energy required to excite the hydrogen atom to a different internal state. We will often be making an approximation in which we do not include the possibility of inner motion, thereby decreasing the number of details that we have to put into our base states. Of course, we then omit some phenomena which would appear (usually) at some higher energy, but by making such approximations we can simplify very much the analysis of physical problems. For example, we can discuss the collision of two hydrogen atoms at low energy—or any chemical process—without worrying about the fact that the atomic nuclei could be excited. To summarize, then, when we can neglect the effects of any internal excited states of a particle we can choose a base set which are the states of definite momentum and z-component of angular momentum.

One problem then in describing nature is to find a suitable representation for the base states. But that's only the beginning. We still want to be able to say what "happens." If we know the "condition" of the world at one moment, we would like to know the condition at a later moment. So we also have to find the laws that determine how things change with time. We now address ourselves to this second part of the framework of quantum mechanics—how states change with time.

8–4 How states change with time

We have already talked about how we can represent a situation in which we put something through an apparatus. Now one convenient, delightful "apparatus" to consider is merely a wait of a few minutes; that is, you prepare a state ϕ, and then before you analyze it, you just let it sit. Perhaps you let it sit in some particular electric or magnetic field—it depends on the physical circumstances in the world. At any rate, whatever the conditions are, you let the object sit from time t_1 to time t_2. Suppose that it is let out of your first apparatus in the condition ϕ at t_1. And then it goes through an "apparatus," but the "apparatus" consists of just delay until t_2. During the delay, various things could be going on—external forces applied or other shenanigans—so that something is happening. At the end of the delay, the amplitude to find the thing in some state χ is no longer exactly the same as it would have been without the delay. Since "waiting" is just a special case of an "apparatus," we can describe what happens by giving an amplitude with the same form as Eq. (8.17). Because the operation of "waiting" is especially important, we'll call it U instead of A, and to specify the starting and finishing times t_1 and t_2, we'll write $U(t_2, t_1)$. The amplitude we want is

$$\langle \chi \mid U(t_2, t_1) \mid \phi \rangle. \tag{8.27}$$

Like any other such amplitude, it can be represented in some base system or other by writing it

$$\sum_{ij} \langle \chi \mid i \rangle \langle i \mid U(t_2, t_1) \mid j \rangle \langle j \mid \phi \rangle. \tag{8.28}$$

Then U is completely described by giving the whole set of amplitudes—the matrix

$$\langle i \mid U(t_2, t_1) \mid j \rangle. \tag{8.29}$$

We can point out, incidentally, that the matrix $\langle i \mid U(t_2, t_1) \mid j \rangle$ gives much more detail than may be needed. The high-class theoretical physicist working in high-energy physics considers problems of the following general nature (because it's the way experiments are usually done). He starts with a couple of particles, like a proton and a proton, coming together from infinity. (In the lab, usually one particle is standing still, and the other comes from an accelerator that is practically at infinity on atomic level.) The things go crash and out come, say, two k-mesons, six π-mesons, and two neutrons in certain directions with certain momenta. What's the amplitude for this to happen? The mathematics looks like this: The ϕ-state specifies the spins and momenta of the incoming particles. The χ would be the question about what comes out. For instance, with what amplitude do you get the six mesons going in such-and-such directions, and the two neutrons going off in these directions, with their spins so-and-so. In other words, χ would be specified by giving all the momenta, and spins, and so on of the final products. Then the job of the theorist is to calculate the amplitude (8.27). However, he is really only interested in the special case that t_1 is $-\infty$ and t_2 is $+\infty$. (There is no experimental evidence on the details of the process, only on what comes in and what goes out.) The limiting case of $U(t_2, t_1)$ as $t_1 \rightarrow -\infty$ and $t_2 \rightarrow +\infty$ is called S, and what he wants is

$$\langle \chi \mid S \mid \phi \rangle.$$

Or, using the form (8.28), he would calculate the matrix

$$\langle i \mid S \mid j \rangle,$$

which is called the *S-matrix*. So if you see a theoretical physicist pacing the floor and saying, "All I have to do is calculate the S-matrix," you will know what he is worried about.

How to analyze—how to specify the laws for—the S-matrix is an interesting question. In relativistic quantum mechanics for high energies, it is done one way, but in nonrelativistic quantum mechanics it can be done another way, which is very convenient. (This other way can also be done in the relativistic case, but then it is not so convenient.) It is to work out the U-matrix for a small interval of time—in other words for t_2 and t_1 close together. If we can find a sequence of such U's for successive intervals of time we can watch how things go as a function of time. You can appreciate immediately that this way is not so good for relativity, because you don't want to have to specify how everything looks "simultaneously" everywhere. But we won't worry about that—we're just going to worry about nonrelativistic mechanics.

Suppose we think of the matrix U for a delay from t_1 until t_3 which is greater than t_2. In other words, let's take three successive times: t_1 less than t_2 less than t_3. Then we claim that the matrix that goes between t_1 and t_3 is the *product* in succession of what happens when you delay from t_1 until t_2 and then from t_2 until t_3. It's just like the situation when we had two apparatuses B and A in series. We can then write, following the notation of Section 5–6,

$$U(t_3, t_1) = U(t_3, t_2) \cdot U(t_2, t_1). \tag{8.30}$$

In other words, we can analyze any time interval if we can analyze a sequence of short time intervals in between. We just multiply together all the pieces; that's the way that quantum mechanics is analyzed nonrelativistically.

Our problem, then, is to understand the matrix $U(t_2, t_1)$ for an infinitesimal time interval—for $t_2 = t_1 + \Delta t$. We ask ourselves this: If we have a state ϕ now, what does the state look like an infinitesimal time Δt later? Let's see how we write that out. Call the state at the time t, $\mid \psi(t) \rangle$ (we show the time dependence of ψ to be perfectly clear that we mean the condition at the time t). Now we ask the question: What is the condition after the small interval of time Δt later? The answer is

$$\mid \psi(t + \Delta t) \rangle = U(t + \Delta t, t) \mid \psi(t) \rangle. \tag{8.31}$$

This means the same as we meant by (8.25), namely, that the amplitude to find x at the time $t + \Delta t$, is

$$\langle x \mid \psi(t + \Delta t)\rangle = \langle x \mid U(t + \Delta t, t) \mid \psi(t)\rangle. \tag{8.32}$$

Since we're not yet too good at these abstract things, let's project our amplitudes into a definite representation. If we multiply both sides of Eq. (8.31) by $\langle i \mid$, we get

$$\langle i \mid \psi(t + \Delta t)\rangle = \langle i \mid U(t + \Delta t, t) \mid \psi(t)\rangle. \tag{8.33}$$

We can also resolve the $\mid \psi(t)\rangle$ into base states and write

$$\langle i \mid \psi(t + \Delta t)\rangle = \sum_j \langle i \mid U(t + \Delta t, t) \mid j\rangle\langle j \mid \psi(t)\rangle. \tag{8.34}$$

We can understand Eq. (8.34) in the following way. If we let $C_i(t) = \langle i \mid \psi(t)\rangle$ stand for the amplitude to be in the base state i at the time t, then we can think of this amplitude (just a *number*, remember!) varying with time. Each C_i becomes a function of t. And we also have some information on *how* the amplitudes C_i vary with time. Each amplitude at $(t + \Delta t)$ is proportional to *all of the other* amplitudes at t multiplied by a set of coefficients. Let's call the U-matrix U_{ij}, by which we mean

$$U_{ij} = \langle i \mid U \mid j\rangle.$$

Then we can write Eq. (8.34) as

$$C_i(t + \Delta t) = \sum_j U_{ij}(t + \Delta t, t)C_j(t). \tag{8.35}$$

This, then, is how the dynamics of quantum mechanics is going to look.

We don't know much about the U_{ij} yet, except for one thing. We know that if Δt goes to zero, nothing can happen—we should get just the original state. So, $U_{ii} \to 1$ and $U_{ij} \to 0$, if $i \neq j$. In other words, $U_{ij} \to \delta_{ij}$ for $\Delta t \to 0$. Also, we can suppose that for small Δt, each of the coefficients U_{ij} should differ from δ_{ij} by amounts proportional to Δt; so we can write

$$U_{ij} = \delta_{ij} + K_{ij}\,\Delta t. \tag{8.36}$$

However, it is usual to take the factor $(-i/\hbar)$† out of the coefficients K_{ij}, for historical and other reasons; we prefer to write

$$U_{ij}(t + \Delta t, t) = \delta_{ij} - \frac{i}{\hbar} H_{ij}(t)\,\Delta t. \tag{8.37}$$

It is, of course, the same as Eq. (8.36) and, if you wish, just defines the coefficients $H_{ij}(t)$. The terms H_{ij} are just the derivatives with respect to t_2 of the coefficients $U_{ij}(t_2, t_1)$, evaluated at $t_2 = t_1 = t$.

Using this form for U in Eq. (8.35), we have

$$C_i(t + \Delta t) = \sum_j \left[\delta_{ij} - \frac{i}{\hbar} H_{ij}(t)\,\Delta t\right] C_j(t). \tag{8.38}$$

Taking the sum over the δ_{ij} term, we get just $C_i(t)$, which we can put on the other side of the equation. Then dividing by Δt, we have what we recognize as a derivative

$$\frac{C_i(t + \Delta t) - C_i(t)}{\Delta t} = -\frac{i}{\hbar} \sum_j H_{ij}(t)C_j(t)$$

or

$$i\hbar \frac{dC_i(t)}{dt} = \sum_j H_{ij}(t)C_j(t). \tag{8.39}$$

---

† We are in a bit of trouble here with notation. In the factor $(-i/\hbar)$, the i means the imaginary unit $\sqrt{-1}$, and *not* the *index* i that refers to the ith base state! We hope that you won't find it too confusing.

You remember that $C_i(t)$ is the amplitude $\langle i \mid \psi \rangle$ to find the state ψ in one of the base states i (at the time t). So Eq. (8.39) tells us how each of the coefficients $\langle i \mid \psi \rangle$ varies with time. But that is the same as saying that Eq. (8.39) tells us how the state ψ varies with time, since we are describing ψ in terms of the amplitudes $\langle i \mid \psi \rangle$. The variation of ψ in time is described in terms of the matrix H_{ij}, which has to include, of course, the things we are doing to the system to cause it to change. If we know the H_{ij}—which contains the physics of the situation and can, in general, depend on the time—we have a complete description of the behavior in time of the system. Equation (8.39) is then the quantum mechanical law for the dynamics of the world.

(We should say that we will always take a set of base states which are fixed and do not vary with time. There are people who use base states that also vary. However, that's like using a rotating coordinate system in mechanics, and we don't want to get involved in such complications.)

8–5 The Hamiltonian matrix

The idea, then, is that to describe the quantum mechanical world we need to pick a set of base states i and to write the physical laws by giving the matrix of coefficients H_{ij}. Then we have everything—we can answer any question about what will happen. So we have to learn what the rules are for finding the H's to go with any physical situation—what corresponds to a magnetic field, or an electric field, and so on. And that's the hardest part. For instance, for the new strange particles, we have no idea what H_{ij}'s to use. In other words, no one knows the *complete* H_{ij} for the whole world. (Part of the difficulty is that one can hardly hope to discover the H_{ij} when no one even knows what the base states are!) We do have excellent approximations for nonrelativistic phenomena and for some other special cases. In particular, we have the forms that are needed for the motions of electrons in atoms—to describe chemistry. But we don't know the full true H for the whole universe.

The coefficients H_{ij} are called *the Hamiltonian matrix* or, for short, just *the Hamiltonian*. (How Hamilton, who worked in the 1830's, got his name on a quantum mechanical matrix is a tale of history.) It would be much better called the *energy matrix*, for reasons that will become apparent as we work with it. So *the* problem is: Know your Hamiltonian!

The Hamiltonian has one property that can be deduced right away, namely, that

$$H_{ij}^* = H_{ji}. \qquad (8.40)$$

This follows from the condition that the total probability that the system is in *some* state does not change. If you start with a particle—an object or the world—then you've still got it as time goes on. The total probability of finding it somewhere is

$$\sum_i |C_i(t)|^2,$$

which must not vary with time. If this is to be true for any starting condition ϕ, then Eq. (8.40) must also be true.

As our first example, we take a situation in which the physical circumstances are not changing with time; we mean the *external* physical conditions, so that H is independent of time. Nobody is turning magnets on and off. We also pick a system for which only one base state is required for the description; it is an approximation we could make for a hydrogen atom at rest, or something similar. Equation (8.39) then says

$$i\hbar \frac{dC_1}{dt} = H_{11}C_1. \qquad (8.41)$$

Only one equation—that's all! And if H_{11} is constant, this differential equation is easily solved to give

$$C_1 = (\text{const})e^{-(i/\hbar)H_{11}t}. \qquad (8.42)$$

This is the time dependence of a state with a definite energy $E = H_{11}$. You see why H_{ij} ought to be called the energy matrix. It is the generalization of the energy for more complex situations.

Next, to understand a little more about what the equations mean, we look at a system which has two base states. Then Eq. (8.39) reads

$$i\hbar \frac{dC_1}{dt} = H_{11}C_1 + H_{12}C_2,$$

$$i\hbar \frac{dC_2}{dt} = H_{21}C_1 + H_{22}C_2. \qquad (8.43)$$

If the H's are again independent of time, you can easily solve these equations. We leave you to try for fun, and we'll come back and do them later. Yes, you can solve the quantum mechanics without knowing the H's, so long as they are independent of time.

8–6 The ammonia molecule

We want now to show you how the dynamical equation of quantum mechanics can be used to describe a particular physical circumstance. We have picked an interesting but simple example in which, by making some reasonable guesses about the Hamiltonian, we can work out some important—and even practical—results. We are going to take a situation describable by two states: the ammonia molecule.

The ammonia molecule has one nitrogen atom and three hydrogen atoms located in a plane below the nitrogen so that the molecule has the form of a pyramid, as drawn in Fig. 8–1(a). Now this molecule, like any other, has an infinite number of states. It can spin around any possible axis; it can be moving in any direction; it can be vibrating inside, and so on, and so on. It is, therefore, not a two-state system at all. But we want to make an approximation that all other states remain fixed, because they don't enter into what we are concerned with at the moment. We will consider only that the molecule is spinning around its axis of symmetry (as shown in the figure), that it has zero translational momentum, and that it is vibrating as little as possible. That specifies all conditions except one: *there are still the two possible positions for the nitrogen atom*—the nitrogen may be on one side of the plane of hydrogen atoms or on the other, as shown in Fig. 8–1(a) and (b). So we will discuss the molecule as though it were a two-state system. We mean that there are only two states we are going to really worry about, all other things being assumed to stay put. You see, even if we know that it is spinning with a certain angular momentum around the axis and that it is moving with a certain momentum and vibrating in a definite way, there are still two possible states. We will say that the molecule is in the state $| 1 \rangle$ when the nitrogen is "up," as in Fig. 8–1(a), and is in the state $| 2 \rangle$ when the nitrogen is "down," as in (b). The states $| 1 \rangle$ and $| 2 \rangle$ will be taken as the set of base states for our analysis of the behavior of the ammonia molecule. At any moment, the actual state $| \psi \rangle$ of the molecule can be represented by giving $C_1 = \langle 1 | \psi \rangle$, the amplitude to be in state $| 1 \rangle$, and $C_2 = \langle 2 | \psi \rangle$, the amplitude to be in state $| 2 \rangle$. Then, using Eq. (8.8) we can write the state vector $| \psi \rangle$ as

$$| \psi \rangle = | 1 \rangle\langle 1 | \psi \rangle + | 2 \rangle\langle 2 | \psi \rangle$$

or

$$| \psi \rangle = | 1 \rangle C_1 + | 2 \rangle C_2. \qquad (8.44)$$

Now the interesting thing is that if the molecule is known to be in some state at some instant, it will *not* be in the same state a little while later. The two C-coefficients will be changing with time according to the equations (8.43)—which hold for any two-state system. Suppose, for example, that you had made some observation—or had made some selection of the molecules—so that you *know* that the molecule is *initially* in the state $| 1 \rangle$. At some later time, there is some chance that it will be found in state $| 2 \rangle$. To find out what this chance is, we have to solve the differential equation which tells us how the amplitudes change with time.

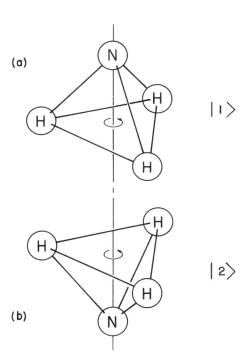

(a)

$| 1 \rangle$

(b)

$| 2 \rangle$

Fig. 8–1. Two equivalent geometric arrangements of the ammonia molecule.

The only trouble is that we don't know what to use for the coefficients H_{ij} in Eq. (8.43). There are some things we *can* say, however. Suppose that once the molecule was in the state $|1\rangle$ there was no chance that it could ever get into $|2\rangle$, and vice versa. Then H_{12} and H_{21} would both be zero, and Eq. (8.43) would read

$$i\hbar \frac{dC_1}{dt} = H_{11}C_1, \qquad i\hbar \frac{dC_2}{dt} = H_{22}C_2.$$

We can easily solve these two equations; we get

$$C_1 = (\text{const})e^{-(i/\hbar)H_{11}t}, \qquad C_2 = (\text{const})e^{-(i/\hbar)H_{22}t}. \tag{8.45}$$

These are just the amplitudes for *stationary* states with the energies $E_1 = H_{11}$ and $E_2 = H_{22}$. We note, however, that for the ammonia molecule the two states $|1\rangle$ and $|2\rangle$ have a definite symmetry. If nature is at all reasonable, the matrix elements H_{11} and H_{22} must be equal. We'll call them both E_0, because they correspond to the energy the states would have if H_{12} and H_{21} were zero. But Eqs. (8.45) do not tell us what ammonia really does. It turns out that it is possible for the nitrogen to push its way through the three hydrogens and flip to the other side. It is quite difficult; to get half-way through requires a lot of energy. How can it get through if it hasn't got enough energy? There is *some* amplitude that it *will* penetrate the energy barrier. It is possible in quantum mechanics to sneak quickly across a region which is illegal energetically. There is, therefore, some small amplitude that a molecule which starts in $|1\rangle$ will get to the state $|2\rangle$. The coefficients H_{12} and H_{21} are not really zero. Again, by symmetry, they should both be the same—at least in magnitude. In fact, we already know that, in general, H_{ij} must be equal to the complex conjugate of H_{ji}, so they can differ only by a phase. It turns out, as you will see, that there is no loss of generality if we take them equal to each other. For later convenience we set them equal to a negative number; we take $H_{12} = H_{21} = -A$. We then have the following pair of equations:

$$i\hbar \frac{dC_1}{dt} = E_0 C_1 - A C_2, \tag{8.46}$$

$$i\hbar \frac{dC_2}{dt} = E_0 C_2 - A C_1. \tag{8.47}$$

These equations are simple enough and can be solved in any number of ways. One convenient way is the following. Taking the sum of the two, we get

$$i\hbar \frac{d}{dt}(C_1 + C_2) = (E_0 - A)(C_1 + C_2),$$

whose solution is

$$C_1 + C_2 = ae^{-(i/\hbar)(E_0-A)t}. \tag{8.48}$$

Then, taking the difference of (8.46) and (8.47), we find that

$$i\hbar \frac{d}{dt}(C_1 - C_2) = (E_0 + A)(C_1 - C_2),$$

which gives

$$C_1 - C_2 = be^{-(i/\hbar)(E_0+A)t}. \tag{8.49}$$

We have called the two integration constants a and b; they are, of course, to be chosen to give the appropriate starting condition for any particular physical problem. Now, by adding and subtracting (8.48) and (8.49), we get C_1 and C_2:

$$C_1(t) = \frac{a}{2} e^{-(i/\hbar)(E_0-A)t} + \frac{b}{2} e^{-(i/\hbar)(E_0+A)t}, \tag{8.50}$$

$$C_2(t) = \frac{a}{2} e^{-(i/\hbar)(E_0-A)t} - \frac{b}{2} e^{-(i/\hbar)(E_0+A)t}. \tag{8.51}$$

They are the same except for the sign of the second term.

We have the solutions; now what do they mean? (The trouble with quantum mechanics is not only in solving the equations but in understanding what the solutions mean!) First, notice that if $b = 0$, both terms have the same frequency $\omega = (E_0 - A)/\hbar$. If everything changes at one frequency, it means that the system is in a state of definite energy—here, the energy $(E_0 - A)$. So there is a stationary state of this energy in which the two amplitudes C_1 and C_2 are equal. We get the result that *the ammonia molecule has a definite energy $(E_0 - A)$ if there are equal amplitudes for the nitrogen atom to be "up" and to be "down."*

There is another stationary state possible if $a = 0$; both amplitudes then have the frequency $(E_0 + A)/\hbar$. So there is another state with the definite energy $(E_0 + A)$ if the two amplitudes are equal but with the opposite sign; $C_2 = -C_1$. These are the only two states of definite energy. We will discuss the states of the ammonia molecule in more detail in the next chapter; we will mention here only a couple of things.

We conclude that *because* there is some chance that the nitrogen atom can flip from one position to the other, the energy of the molecule is not just E_0, as we would have expected, but that there are *two* energy levels $(E_0 + A)$ and $(E_0 - A)$. Every one of the possible states of the molecule, whatever energy it has, is "split" into two levels. We say *every* one of the states because, you remember, we picked out one particular state of rotation, and internal energy, and so on. For each possible condition of that kind there is a doublet of energy levels because of the flip-flop of the molecule.

Let's now ask the following question about an ammonia molecule. Suppose that at $t = 0$, we *know* that a molecule is in the state $|\,1\rangle$ or, in other words, that $C_1(0) = 1$ and $C_2(0) = 0$. What is the probability that the molecule will be found in the state $|\,2\rangle$ at the time t, or will still be found in state $|\,1\rangle$ at the time t? Our starting condition tells us what a and b are in Eqs. (8.50) and (8.51). Letting $t = 0$, we have that

$$C_1(0) = \frac{a + b}{2} = 1, \qquad C_2(0) = \frac{a - b}{2} = 0.$$

Clearly, $a = b = 1$. Putting these values into the formulas for $C_1(t)$ and $C_2(t)$ and rearranging some terms, we have

$$C_1(t) = e^{-(i/\hbar)E_0 t}\left(\frac{e^{(i/\hbar)At} + e^{-(i/\hbar)At}}{2}\right),$$

$$C_2(t) = e^{-(i/\hbar)E_0 t}\left(\frac{e^{(i/\hbar)At} - e^{-(i/\hbar)At}}{2}\right).$$

We can rewrite these as

$$C_1(t) = e^{-(i/\hbar)E_0 t}\cos\frac{At}{\hbar}, \tag{8.52}$$

$$C_2(t) = ie^{-(i/\hbar)E_0 t}\sin\frac{At}{\hbar}. \tag{8.53}$$

The two amplitudes have a magnitude that varies harmonically with time.

The probability that the molecule is found in state $|\,2\rangle$ at the time t is the absolute square of $C_2(t)$:

$$|C_2(t)|^2 = \sin^2\frac{At}{\hbar}. \tag{8.54}$$

The probability starts at zero (as it should), rises to one, and then oscillates back and forth between zero and one, as shown in the curve marked P_2 of Fig. 8–2. The probability of being in the $|\,1\rangle$ state does not, of course, stay at one. It "dumps" into the second state until the probability of finding the molecule in the first state is zero, as shown by the curve P_1 of Fig. 8–2. The probability sloshes back and forth between the two.

A long time ago we saw what happens when we have two equal pendulums with a slight coupling. (See Chapter 49, Vol. I.) When we lift one back and let go,

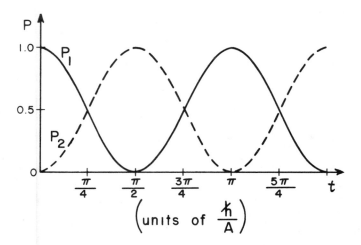

Fig. 8–2. The probability P_1 that an ammonia molecule in state $|1\rangle$ at $t = 0$ will be found in state $|1\rangle$ at t. The probability P_2 that it will be found in state $|2\rangle$.

it swings, but then gradually the other one starts to swing. Pretty soon the second pendulum has picked up all the energy. Then, the process reverses, and pendulum number one picks up the energy. It is exactly the same kind of a thing. The speed at which the energy is swapped back and forth depends on the coupling between the two pendulums—the rate at which the "oscillation" is able to leak across. Also, you remember, with the two pendulums there are two special motions—each with a definite frequency—which we call the fundamental modes. If we pull both pendulums out together, they swing together at one frequency. On the other hand, if we pull one out one way and the other out the other way, there is another stationary mode also at a definite frequency.

Well, here we have a similar situation—the ammonia molecule is mathematically like the pair of pendulums. These are the two frequencies—$(E_0 + A)/h$ and $(E_0 - A)/h$—for when they are oscillating together, or oscillating opposite.

The pendulum analogy is not much deeper than the principle that the same equations have the same solutions. The linear equations for the amplitudes (8.39) are very much like the linear equations of harmonic oscillators. (In fact, this is the reason behind the success of our classical theory of the index of refraction, in which we replaced the quantum mechanical atom by a harmonic oscillator, even though, classically, this is not a reasonable view of electrons circulating about a nucleus.) If you pull the nitrogen to one side, then you get a *superposition* of these two frequencies, and you get a kind of *beat* note, because the system is *not* in one or the other states of definite frequency. The splitting of the energy levels of the ammonia molecule is, however, strictly a quantum mechanical effect.

The splitting of the energy levels of the ammonia molecule has important practical applications which we will describe in the next chapter. At long last we have an example of a practical physical problem that you can understand with the quantum mechanics!

9

The Ammonia Maser

9–1 The states of an ammonia molecule

In this chapter we are going to discuss the application of quantum mechanics to a practical device, the ammonia maser. You may wonder why we stop our formal development of quantum mechanics to do a special problem, but you will find that many of the features of this special problem are quite common in the general theory of quantum mechanics, and you will learn a great deal by considering this one problem in detail. The ammonia maser is a device for generating electromagnetic waves, whose operation is based on the properties of the ammonia molecule which we discussed briefly in the last chapter. We begin by summarizing what we found there.

The ammonia molecule has many states, but we are considering it as a two-state system, thinking now only about what happens when the molecule is in any specific state of rotation or translation. A physical model for the two states can be visualized as follows. If the ammonia molecule is considered to be rotating about an axis passing through the nitrogen atom and perpendicular to the plane of the hydrogen atoms, as shown in Fig. 9–1, there are still two possible conditions —the nitrogen may be on one side of the plane of hydrogen atoms or on the other. We call these two states $|1\rangle$ and $|2\rangle$. They are taken as a set of base states for our analysis of the behavior of the ammonia molecule.

9–1 The states of an ammonia molecule

9–2 The molecule in a static electric field

9–3 Transitions in a time-dependent field

9–4 Transitions at resonance

9–5 Transitions off resonance

9–6 The absorption of light

MASER = Microwave Amplification by Stimulated Emission of Radiation

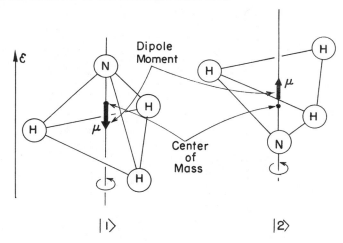

$|1\rangle$ $|2\rangle$

Fig. 9–1. A physical model of two base states for the ammonia molecule. These states have the electric dipole moments μ.

In a system with two base states, any state $|\psi\rangle$ of the system can always be described as a linear combination of the two base states; that is, there is a certain amplitude C_1 to be in one base state and an amplitude C_2 to be in the other. We can write its state vector as

$$|\psi\rangle = |1\rangle C_1 + |2\rangle C_2, \tag{9.1}$$

where

$$C_1 = \langle 1 | \psi \rangle \quad \text{and} \quad C_2 = \langle 2 | \psi \rangle.$$

These two amplitudes change with time according to the Hamiltonian equations, Eq. (8.43). Making use of the symmetry of the two states of the ammonia molecule, we set $H_{11} = H_{22} = E_0$, and $H_{12} = H_{21} = -A$, and get the solution [see Eqs. (8.50) and (8.51)]

$$C_1 = \frac{a}{2} e^{-(i/\hbar)(E_0-A)t} + \frac{b}{2} e^{-(i/\hbar)(E_0+A)t}, \tag{9.2}$$

$$C_2 = \frac{a}{2} e^{-(i/\hbar)(E_0-A)t} - \frac{b}{2} e^{-(i/\hbar)(E_0+A)t}. \tag{9.3}$$

We want now to take a closer look at these general solutions. Suppose that the molecule was initially put into a state $|\psi_{II}\rangle$ for which the coefficient b was equal to zero. Then at $t = 0$ the amplitudes to be in the states $|1\rangle$ and $|2\rangle$ are identical, *and they stay that way for all time*. Their phases both vary with time in the same way—with the frequency $(E_0 - A)/\hbar$. Similarly, if we were to put the molecule into a state $|\psi_I\rangle$ for which $a = 0$, the amplitude C_2 is the negative of C_1, and this relationship would stay that way forever. Both amplitudes would now vary with time with the frequency $(E_0 + A)/\hbar$. These are the only two possibilities of states for which the relation between C_1 and C_2 is independent of time.

We have found two special solutions in which the two amplitudes *do not vary in magnitude* and, furthermore, have phases which vary at the same frequencies. These are *stationary states* as we defined them in Section 7–1, which means that they are *states of definite energy*. The state $|\psi_{II}\rangle$ has the energy $E_{II} = E_0 - A$, and the state $|\psi_I\rangle$ has the energy $E_I = E_0 + A$. They are the only two stationary states that exist, so we find that the molecule has two energy levels, with the energy difference $2A$. (We mean, of course, two energy levels for the assumed state of rotation and vibration which we referred to in our initial assumptions.)[†]

If we hadn't allowed for the possibility of the nitrogen flipping back and forth, we would have taken A equal to zero and the two energy levels would be on top of each other at energy E_0. The actual levels are not this way; their *average* energy is E_0, but they are split apart by $\pm A$, giving a separation of $2A$ between the energies of the two states. Since A is, in fact, very small, the difference in energy is also very small.

In order to excite an *electron* inside an atom, the energies involved are relatively very high—requiring photons in the optical or ultraviolet range. To excite the *vibrations* of the molecules involves photons in the infrared. If you talk about exciting *rotations*, the energy differences of the states correspond to photons in the far infrared. But the energy difference $2A$ is lower than any of those and is, in fact, below the infrared and well into the microwave region. Experimentally, it has been found that there is a pair of energy levels with a separation of 10^{-4} electron volt—corresponding to a frequency 24,000 megacycles. Evidently this means that $2A = hf$, with $f = 24{,}000$ megacycles (corresponding to a wavelength of $1\frac{1}{4}$ cm). So here we have a molecule that has a transition which does not emit light in the ordinary sense, but emits microwaves.

For the work that follows we need to describe these two states of definite energy a little bit better. Suppose we were to construct an amplitude C_{II} by taking the sum of the two numbers C_1 and C_2:

$$C_{II} = C_1 + C_2 = \langle 1 | \Phi \rangle + \langle 2 | \Phi \rangle. \tag{9.4}$$

What would that mean? Well, this is just the amplitude to find the state $|\Phi\rangle$ in a new state $|II\rangle$ in which the amplitudes of the original base states are equal. That is, writing $C_{II} = \langle II | \Phi \rangle$, we can abstract the $|\Phi\rangle$ away from Eq. (9.4)—because it is true for any Φ—and get

$$\langle II | = \langle 1 | + \langle 2 |,$$

which means the same as

$$|II\rangle = |1\rangle + |2\rangle. \tag{9.5}$$

---

† In what follows it is helpful—in reading to yourself or in talking to someone else—to have a handy way of distinguishing between the Arabic 1 and 2 and the Roman I and II. We find it convenient to reserve the names "one" and "two" for the Arabic numbers, and to call I and II by the names "eins" and "zwei" (although "unus" and "duo" might be more logical!).

The amplitude for the state $|II\rangle$ to be in the state $|1\rangle$ is

$$\langle 1 \mid II\rangle = \langle 1 \mid 1\rangle + \langle 1 \mid 2\rangle,$$

which is, of course, just 1, since $|1\rangle$ and $|2\rangle$ are base states. The amplitude for the state $|II\rangle$ to be in the state $|2\rangle$ is also 1, so the state $|II\rangle$ is one which has equal amplitudes to be in the two base states $|1\rangle$ and $|2\rangle$.

We are, however, in a bit of trouble. The state $|II\rangle$ has a total probability greater than one of being in *some* base state *or other*. That simply means, however, that the state vector is not properly "normalized." We can take care of that by remembering that we should have $\langle II \mid II\rangle = 1$, which must be so for any state. Using the general relation that

$$\langle \chi \mid \Phi\rangle = \sum_i \langle \chi \mid i\rangle\langle i \mid \Phi\rangle,$$

letting both Φ and χ be the state II, and taking the sum over the base states $|1\rangle$ and $|2\rangle$, we get that

$$\langle II \mid II\rangle = \langle II \mid 1\rangle\langle 1 \mid II\rangle + \langle II \mid 2\rangle\langle 2 \mid II\rangle.$$

This will be equal to one as it should if we change our definition of C_{II}—in Eq. (9.4)—to read

$$C_{II} = \frac{1}{\sqrt{2}} [C_1 + C_2].$$

In the same way we can construct an amplitude

$$C_I = \frac{1}{\sqrt{2}} [C_1 - C_2],$$

or

$$C_I = \frac{1}{\sqrt{2}} [\langle 1 \mid \Phi\rangle - \langle 2 \mid \Phi\rangle]. \tag{9.6}$$

This amplitude is the projection of the state $|\Phi\rangle$ into a new state $|I\rangle$ which has opposite amplitudes to be in the states $|1\rangle$ and $|2\rangle$. Namely, Eq. (9.6) means the same as

$$\langle I| = \frac{1}{\sqrt{2}} [\langle 1| - \langle 2|],$$

or

$$|I\rangle = \frac{1}{\sqrt{2}} [|1\rangle - |2\rangle], \tag{9.7}$$

from which it follows that

$$\langle 1 \mid I\rangle = \frac{1}{\sqrt{2}} = -\langle 2 \mid I\rangle .$$

Now the reason we have done all this is that the states $|I\rangle$ and $|II\rangle$ *can be taken as a new set of base states* which are especially convenient for describing the stationary states of the ammonia molecule. You remember that the requirement for a set of base states is that

$$\langle i \mid j\rangle = \delta_{ij}.$$

We have already fixed things so that

$$\langle I \mid I\rangle = \langle II \mid II\rangle = 1.$$

You can easily show from Eqs. (9.5) and (9.7) that

$$\langle I \mid II\rangle = \langle II \mid I\rangle = 0.$$

The amplitudes $C_I = \langle I \mid \Phi\rangle$ and $C_{II} = \langle II \mid \Phi\rangle$ for any state Φ to be in our new base states $|I\rangle$ and $|II\rangle$ must also satisfy a Hamiltonian equation with the form of Eq. (8.39). In fact, if we just subtract the two equations (9.2) and (9.3) and differentiate with respect to t, we see that

$$i\hbar \frac{dC_I}{dt} = (E_0 + A)C_I = E_I C_I. \tag{9.8}$$

And taking the sum of Eqs. (9.2) and (9.3), we see that

$$i\hbar \frac{dC_{II}}{dt} = (E_0 - A)C_{II} = E_{II} C_{II}. \tag{9.9}$$

Using $| I \rangle$ and $| II \rangle$ for base states, the Hamiltonian matrix has the simple form

$$H_{I,I} = E_I, \qquad H_{I,II} = 0,$$

$$H_{I,II} = 0, \qquad H_{II,II} = E_{II}.$$

Note that each of the Eqs. (9.8) and (9.9) look just like what we had in Section 8–6 for the equation of a one-state system. They have a simple exponential time dependence corresponding to a single energy. As time goes on, the amplitudes to be in each state act independently.

The two stationary states $| \psi_I \rangle$ and $| \psi_{II} \rangle$ we found above are, of course, solutions of Eqs. (9.8) and (9.9). The state $| \psi_I \rangle$ (for which $C_1 = -C_2$) has

$$C_I = e^{-(i/\hbar)(E_0 + A)t}, \qquad C_{II} = 0. \tag{9.10}$$

And the state $| \psi_{II} \rangle$ (for which $C_1 = C_2$) has

$$C_I = 0, \qquad C_{II} = e^{-(i/\hbar)(E_0 - A)t}. \tag{9.11}$$

Remember that the amplitudes in Eq. (9.10) are

$$C_I = \langle I | \psi_I \rangle, \qquad \text{and} \qquad C_{II} = \langle II | \psi_I \rangle;$$

so Eq. (9.10) means the same thing as

$$| \psi_I \rangle = | I \rangle e^{-(i/\hbar)(E_0 + A)t}.$$

That is, the state vector of the stationary state $| \psi_I \rangle$ is the same as the state vector of the base state $| I \rangle$ except for the exponential factor appropriate to the energy of the state. In fact at $t = 0$

$$| \psi_I \rangle = | I \rangle;$$

the state $| I \rangle$ has the same physical configuration as the stationary state of energy $E_0 + A$. In the same way, we have for the second stationary state that

$$| \psi_{II} \rangle = | II \rangle e^{-(i/\hbar)(E_0 - A)t}.$$

The state $| II \rangle$ is just the stationary state of energy $E_0 - A$ at $t = 0$. Thus our two new base states $| I \rangle$ and $| II \rangle$ have physically the form of the states of definite energy, with the exponential time factor taken out so that they can be time-independent base states. (In what follows we will find it convenient not to have to distinguish always between the stationary states $| \psi_I \rangle$ and $| \psi_{II} \rangle$ and their base states $| I \rangle$ and $| II \rangle$, since they differ only by the obvious time factors.)

In summary, the state vectors $| I \rangle$ and $| II \rangle$ are a pair of base vectors which are appropriate for describing the definite energy states of the ammonia molecule. They are related to our original base vectors by

$$| I \rangle = \frac{1}{\sqrt{2}} [| 1 \rangle - | 2 \rangle], \qquad | II \rangle = \frac{1}{\sqrt{2}} [| 1 \rangle + | 2 \rangle]. \tag{9.12}$$

The amplitudes to be in $| I \rangle$ and $| II \rangle$ are related to C_1 and C_2 by

$$C_I = \frac{1}{\sqrt{2}} [C_1 - C_2], \qquad C_{II} = \frac{1}{\sqrt{2}} [C_1 + C_2]. \tag{9.13}$$

Any state at all can be represented by a linear combination of $|1\rangle$ and $|2\rangle$—with the coefficients C_1 and C_2—or by a linear combination of the definite energy base states $|I\rangle$ and $|II\rangle$—with the coefficients C_I and C_{II}. Thus,

$$|\Phi\rangle = |1\rangle C_1 + |2\rangle C_2$$

or

$$|\Phi\rangle = |I\rangle C_I + |II\rangle C_{II}.$$

The second form gives us the amplitudes for finding the state $|\Phi\rangle$ in a state with the energy $E_I = E_0 + A$ or in a state with the energy $E_{II} = E_0 - A$.

9–2 The molecule in a static electric field

If the ammonia molecule is in either of the two states of definite energy and we disturb it at a frequency ω such that $\hbar\omega = E_I - E_{II} = 2A$, the system may make a transition from one state to the other. Or, if it is in the upper state, it may change to the lower state and emit a photon. But in order to induce such transitions you must have a physical connection to the states—some way of disturbing the system. There must be some external machinery for affecting the states, such as magnetic or electric fields. In this particular case, these states are sensitive to an electric field. We will, therefore, look next at the problem of the behavior of the ammonia molecule in an external electric field.

To discuss the behavior in an electric field, we will go back to the original base system $|1\rangle$ and $|2\rangle$, rather than using $|I\rangle$ and $|II\rangle$. Suppose that there is an electric field in a direction perpendicular to the plane of the hydrogen atoms. Disregarding for the moment the possibility of flipping back and forth, would it be true that the energy of this molecule is the same for the two positions of the nitrogen atom? Generally, no. The electrons tend to lie closer to the nitrogen than to the hydrogen nuclei, so the hydrogens are slightly positive. The actual amount depends on the details of electron distribution. It is a complicated problem to figure out exactly what this distribution is, but in any case the net result is that the ammonia molecule has an electric dipole moment, as indicated in Fig. 9–1. We can continue our analysis without knowing in detail the direction or amount of displacement of the charge. However, to be consistent with the notation of others, let's suppose that the electric dipole moment is μ, with its direction point *from* the nitrogen atom and perpendicular to the plane of the hydrogen atoms.

Now, when the nitrogen flips from one side to the other, the center of mass will not move, but the electric dipole moment will flip over. As a result of this moment, the energy in an electric field $\mathcal{E}$ will depend on the molecular orientation.† With the assumption made above, the potential energy will be higher if the nitrogen atom points in the direction of the field, and lower if it is in the opposite direction; the separation in the two energies will be $2\mu\mathcal{E}$.

In the discussion up to this point, we have assumed values of E_0 and A without knowing how to calculate them. According to the correct physical theory, it should be possible to calculate these constants in terms of the positions and motions of all the nuclei and electrons. But nobody has ever done it. Such a system involves ten electrons and four nuclei and that's just too complicated a problem. As a matter of fact, there is no one who knows much more about this molecule than we do. All anyone can say is that when there is an electric field, the energy of the two states is different, the difference being proportional to the electric field. We have called the coefficient of proportionality 2μ, but its value must be determined experimentally. We can also say that the molecule has the amplitude A to flip over, but this will have to be measured experimentally. Nobody can give us accurate theoretical values of μ and A, because the calculations are too complicated to do in detail.

---

† We are sorry that we have to introduce a new notation. Since we have been using p and E for momentum and energy, we don't want to use them again for dipole moment and electric field. Remember, in this section μ is the *electric* dipole moment.

For the ammonia molecule in an electric field, our description must be changed. If we ignored the amplitude for the molecule to flip from one configuration to the other, we would expect the energies of the two states $|\,1\rangle$ and $|\,2\rangle$ to be $(E_0 \pm \mu\mathcal{E})$. Following the procedure of the last chapter, we take

$$H_{11} = E_0 + \mu\mathcal{E}, \qquad H_{22} = E_0 - \mu\mathcal{E}. \qquad (9.14)$$

Also we will assume that for the electric fields of interest the field does not affect appreciably the geometry of the molecule and, therefore, does not affect the amplitude that the nitrogen will jump from one position to the other. We can then take that H_{12} and H_{21} are not changed; so

$$H_{12} = H_{21} = -A. \qquad (9.15)$$

We must now solve the Hamiltonian equations, Eq. (8.43), with these new values of H_{ij}. We could solve them just as we did before, but since we are going to have several occasions to want the solutions for two-state systems, let's solve the equations once and for all in the general case of arbitrary H_{ij}—assuming only that they do not change with time.

We want the general solution of the pair of Hamiltonian equations

$$i\hbar \frac{dC_1}{dt} = H_{11}C_1 + H_{12}C_2, \qquad (9.16)$$

$$i\hbar \frac{dC_2}{dt} = H_{21}C_1 + H_{22}C_2. \qquad (9.17)$$

Since these are linear differential equations with constant coefficients, we can always find solutions which are exponential functions of the dependent variable t. We will first look for a solution in which C_1 and C_2 both have the same time dependence; we can use the trial functions

$$C_1 = a_1 e^{-i\omega t}, \qquad C_2 = a_2 e^{-i\omega t}.$$

Since such a solution corresponds to a state of energy $E = \hbar\omega$, we may as well write right away

$$C_1 = a_1 e^{-(i/\hbar)Et}, \qquad (9.18)$$

$$C_2 = a_2 e^{-(i/\hbar)Et}, \qquad (9.19)$$

where E is as yet unknown and to be determined so that the differential equations (9.16) and (9.17) are satisfied.

When we substitute C_1 and C_2 from (9.18) and (9.19) in the differential equations (9.16) and (9.17), the derivatives give us just $-iE/\hbar$ times C_1 or C_2, so the left sides become just EC_1 and EC_2. Cancelling the common exponential factors, we get

$$Ea_1 = H_{11}a_1 + H_{12}a_2, \qquad Ea_2 = H_{21}a_1 + H_{22}a_2.$$

Or, rearranging the terms, we have

$$(E - H_{11})a_1 - H_{12}a_2 = 0, \qquad (9.20)$$

$$-H_{21}a_1 + (E - H_{22})a_2 = 0. \qquad (9.21)$$

With such a set of homogeneous algebraic equations, there will be nonzero solutions for a_1 and a_2 only if the determinant of the coefficients of a_1 and a_2 is zero, that is, if

$$\mathrm{Det} \begin{pmatrix} E - H_{11} & -H_{12} \\ -H_{21} & E - H_{22} \end{pmatrix} = 0. \qquad (9.22)$$

However, when there are only two equations and two unknowns, we don't need such a sophisticated idea. The two equations (9.20) and (9.21) each give a ratio for the two coefficients a_1 and a_2, and these two ratios must be equal. From (9.20) we have that

$$\frac{a_1}{a_2} = \frac{H_{12}}{E - H_{11}}, \tag{9.23}$$

and from (9.21) that

$$\frac{a_1}{a_2} = \frac{E - H_{22}}{H_{21}}. \tag{9.24}$$

Equating these two ratios, we get that E must satisfy

$$(E - H_{11})(E - H_{22}) - H_{12}H_{21} = 0.$$

This is the same result we would get by solving Eq. (9.22). Either way, we have a quadratic equation for E which has two solutions:

$$E = \frac{H_{11} + H_{22}}{2} \pm \sqrt{\frac{(H_{11} - H_{22})^2}{4} + H_{12}H_{21}}. \tag{9.25}$$

There are two possible values for the energy E. Note that both solutions give *real numbers* for the energy, because H_{11} and H_{22} are real, and $H_{12}H_{21}$ is equal to $H_{12}H_{12}^* = |H_{12}|^2$, which is both real and positive.

Using the same convention we took before, we will call the upper energy E_I and the lower energy E_{II}. We have

$$E_I = \frac{H_{11} + H_{22}}{2} + \sqrt{\frac{(H_{11} - H_{22})^2}{4} + H_{12}H_{21}}, \tag{9.26}$$

$$E_{II} = \frac{H_{11} + H_{22}}{2} - \sqrt{\frac{(H_{11} - H_{22})^2}{4} + H_{12}H_{21}}. \tag{9.27}$$

Using each of these two energies separately in Eqs. (9.18) and (9.19), we have the amplitudes for the two stationary states (the states of definite energy). If there are no external disturbances, a system initially in one of these states will stay that way forever—only its phase changes.

We can check our results for two special cases. If $H_{12} = H_{21} = 0$, we have that $E_I = H_{11}$ and $E_{II} = H_{22}$. This is certainly correct, because then Eqs. (9.16) and (9.17) are uncoupled, and each represents a state of energy H_{11} and H_{22}. Next, if we set $H_{11} = H_{22} = E_0$ and $H_{21} = H_{12} = -A$, we get the solution we found before:

$$E_I = E_0 + A \quad \text{and} \quad E_{II} = E_0 - A.$$

For the general case, the two solutions E_I and E_{II} refer to two states—which we can again call the states

$$|\psi_I\rangle = |I\rangle e^{-(i/\hbar)E_I t} \quad \text{and} \quad |\psi_{II}\rangle = |II\rangle e^{-(i/\hbar)E_{II} t}.$$

These states will have C_1 and C_2 as given in Eqs. (9.18) and (9.19), where a_1 and a_2 are still to be determined. Their ratio is given by either Eq. (9.23) or Eq. (9.24). They must also satisfy one more condition. If the system is known to be in one of the stationary states, the sum of the probabilities that it will be found in $|1\rangle$ or $|2\rangle$ must equal one. We must have that

$$|C_1|^2 + |C_2|^2 = 1, \tag{9.28}$$

or, equivalently,

$$|a_1|^2 + |a_2|^2 = 1. \tag{9.29}$$

These conditions do not uniquely specify a_1 and a_2; they are still undetermined

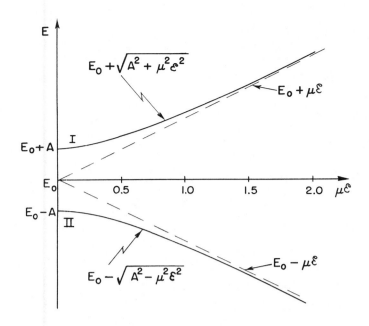

Fig. 9–2. Energy levels of the ammonia molecule in an electric field.

by an arbitrary phase—in other words, by a factor like $e^{i\delta}$. Although general solutions for the a's can be written down,† it is usually more convenient to work them out for each special case.

Let's go back now to our particular example of the ammonia molecule in an electric field. Using the values for H_{11}, H_{22}, and H_{12} given in (9.14) and (9.15), we get for the energies of the two stationary states

$$E_I = E_0 + \sqrt{A^2 + \mu^2\mathcal{E}^2}, \qquad E_{II} = E_0 - \sqrt{A^2 + \mu^2\mathcal{E}^2}. \qquad (9.30)$$

These two energies are plotted as a function of the electric field strength $\mathcal{E}$ in Fig. 9–2. When the electric field is zero, the two energies are, of course, just $E_0 \pm A$. When an electric field is applied, the splitting between the two levels increases. The splitting increases at first slowly with $\mathcal{E}$, but eventually becomes proportional to $\mathcal{E}$. (The curve is a hyperbola.) For enormously strong fields, the energies are just

$$E_I = E_0 + \mu\mathcal{E} = H_{11}, \qquad E_{II} = E_0 - \mu\mathcal{E} = H_{22}. \qquad (9.31)$$

The fact that there is an amplitude for the nitrogen to flip back and forth has little effect when the two positions have very different energies. This is an interesting point which we will come back to again later.

We are at last ready to understand the operation of the ammonia maser. The idea is the following. First, we find a way of separating molecules in the state $|\,I\rangle$ from those in the state $|\,II\rangle$.‡ Then the molecules in the higher energy state $|\,I\rangle$ are passed through a cavity which has a resonant frequency of 24,000 megacycles. The molecules can deliver energy to the cavity—in a way we will discuss later—and leave the cavity in the state $|\,II\rangle$. Each molecule that makes such a transition will deliver the energy $E = E_I - E_{II}$ to the cavity. The energy from the molecules will appear as electrical energy in the cavity.

How can we separate the two molecular states? One method is as follows. The ammonia gas is let out of a little jet and passed through a pair of slits to give a narrow beam, as shown in Fig. 9–3. The beam is then set through a

---

† For example, the following set is one acceptable solution, as you can easily verify:

$$a_1 = \frac{H_{12}}{[(E - H_{11})^2 + H_{12}H_{21}]^{1/2}}, \qquad a_2 = \frac{E - H_{11}}{[(E - H_{11})^2 + H_{12}H_{21}]^{1/2}}.$$

‡ From now on we will write $|\,I\rangle$ and $|\,II\rangle$ instead of $|\psi_I\rangle$ and $|\psi_{II}\rangle$. You must remember that the actual states $|\psi_I\rangle$ and $|\psi_{II}\rangle$ are the energy base states multiplied by the appropriate exponential factor.

region in which there is a large transverse electric field. The electrodes to produce the field are shaped so that the electric field varies rapidly across the beam. Then the square of the electric field $\varepsilon \cdot \varepsilon$ will have a large gradient perpendicular to the beam. Now a molecule in state $|I\rangle$ has an energy which increases with ε^2, and therefore this part of the beam will be deflected toward the region of lower ε^2. A molecule in state $|II\rangle$ will, on the other hand, be deflected toward the region of larger ε^2, since its energy decreases as ε^2 increases.

Incidentally, with the electric fields which can be generated in the laboratory, the energy $\mu\varepsilon$ is always much smaller than A. In such cases, the square root in Eqs. (9.30) can be approximated by

$$A\left(1 + \frac{1}{2}\frac{\mu^2\varepsilon^2}{A^2}\right). \tag{9.32}$$

So the energy levels are, for all practical purposes,

$$E_I = E_0 + A + \frac{\mu^2\varepsilon^2}{2A} \tag{9.33}$$

and

$$E_{II} = E_0 - A - \frac{\mu^2\varepsilon^2}{2A}. \tag{9.34}$$

And the energies vary approximately linearly with ε^2. The force on the molecules is then

$$F = \frac{\mu^2}{2A}\nabla\varepsilon^2. \tag{9.35}$$

Many molecules have an energy in an electric field which is proportional to ε^2. The coefficient is the polarizability of the molecule. Ammonia has an unusually high polarizability because of the small value of A in the denominator. Thus, ammonia molecules are unusually sensitive to an electric field. (What would you expect for the dielectric coefficient of NH_3 gas?)

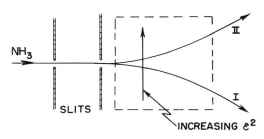

Fig. 9–3. The ammonia beam may be separated by an electric field in which ε^2 has a gradient perpendicular to the beam.

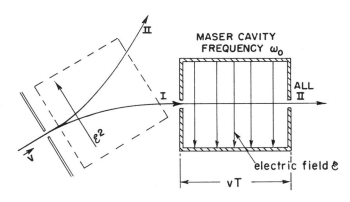

Fig. 9–4. Schematic diagram of the ammonia maser.

9–3 Transitions in a time-dependent field

In the ammonia maser, the beam with molecules in the state $|I\rangle$ and with the energy E_I is sent through a resonant cavity, as shown in Fig. 9–4. The other beam is discarded. Inside the cavity, there will be a time-varying electric field, so the next problem we must discuss is the behavior of a molecule in an electric field that varies with time. We have a completely different kind of a problem—one with a time-varying Hamiltonian. Since H_{ij} depends upon ε, the H_{ij} vary with time, and we must determine the behavior of the system in this circumstance.

To begin with, we write down the equations to be solved:

$$i\hbar\frac{dC_1}{dt} = (E_0 + \mu\varepsilon)C_1 - AC_2,$$

$$i\hbar\frac{dC_2}{dt} = -AC_1 + (E_0 - \mu\varepsilon)C_2. \tag{9.36}$$

To be definite, let's suppose that the electric field varies sinusoidally; then we can write

$$\mathcal{E} = 2\mathcal{E}_0 \cos \omega t = \mathcal{E}_0(e^{i\omega t} + e^{-i\omega t}). \tag{9.37}$$

In actual operation the frequency ω will be very nearly equal to the resonant frequency of the molecular transition $\omega_0 = 2A/\hbar$, but for the time being we want to keep things general, so we'll let it have any value at all. The best way to solve our equations is to form linear combinations of C_1 and C_2 as we did before. So we add the two equations, divide by the square root of 2, and use the definitions of C_I and C_{II} that we had in Eq. (9.13). We get

$$i\hbar \frac{dC_{II}}{dt} = (E_0 - A)C_{II} + \mu\mathcal{E}C_I. \tag{9.38}$$

You'll note that this is the same as Eq. (9.9) with an extra term due to the electric field. Similarly, if we subtract the two equations (9.36), we get

$$i\hbar \frac{dC_I}{dt} = (E_0 + A)C_I + \mu\mathcal{E}C_{II}. \tag{9.39}$$

Now the question is, how to solve these equations? They are more difficult than our earlier set, because $\mathcal{E}$ depends on t; and, in fact, for a general $\mathcal{E}(t)$ the solution is not expressible in elementary functions. However, we can get a good approximation so long as the electric field is small. First we will write

$$C_I = \gamma_I e^{-i(E_0+A)t/\hbar} = \gamma_I e^{-i(E_I)t/\hbar},$$
$$C_{II} = \gamma_{II} e^{-i(E_0-A)t/\hbar} = \gamma_{II} e^{-i(E_{II})t/\hbar}. \tag{9.40}$$

If there were no electric field, these solutions would be correct with γ_I and γ_{II} just chosen as two complex constants. In fact, since the probability of being in state $|I\rangle$ is the absolute square of C_I and the probability of being in state $|II\rangle$ is the absolute square of C_{II}, the probability of being in state $|I\rangle$ or in state $|II\rangle$ is just $|\gamma_I|^2$ or $|\gamma_{II}|^2$. For instance, if the system were to start originally in state $|II\rangle$ so that γ_I was zero and $|\gamma_{II}|^2$ was one, this condition would go on forever. There would be no chance, if the molecule were originally in state $|II\rangle$, ever to get into state $|I\rangle$.

Now the idea of writing our equations in the form of Eq. (9.40) is that if $\mu\mathcal{E}$ is small in comparison with A, the solutions can still be written in this way, but then γ_I and γ_{II} become slowly varying functions of time—where by "slowly varying" we mean slowly *in comparison* with the exponential functions. That is the trick. We use the fact that γ_I and γ_{II} vary slowly to get an approximate solution.

We want now to substitute C_I from (9.40) in the differential equation (9.39), but we must remember that γ_I is also a function of t. We have

$$i\hbar \frac{dC_I}{dt} = E_I \gamma_I e^{-iE_I t/\hbar} + i\hbar \frac{d\gamma_I}{dt} e^{-iE_I t/\hbar}.$$

The differential equation becomes

$$\left(E_I \gamma_I + i\hbar \frac{d\gamma_I}{dt}\right) e^{-(i/\hbar)E_I t} = E_I \gamma_I e^{-(i/\hbar)E_I t} + \mu\mathcal{E}\gamma_{II} e^{-(i/\hbar)E_{II} t}. \tag{9.41}$$

Similarly, the equation in dC_{II}/dt becomes

$$\left(E_{II} \gamma_{II} + i\hbar \frac{d\gamma_{II}}{dt}\right) e^{-(i/\hbar)E_{II} t} = E_{II} \gamma_{II} e^{-(i/\hbar)E_{II} t} + \mu\mathcal{E}\gamma_I e^{-(i/\hbar)E_I t}. \tag{9.42}$$

Now you will notice that we have equal terms on both sides of each equation. We cancel these terms, and we also multiply the first equation by $e^{+iE_I t/\hbar}$ and the second by $e^{+iE_{II}t/\hbar}$. Remembering that $(E_I - E_{II}) = 2A = \hbar\omega_0$, we have finally,

$$ i\hbar \frac{d\gamma_I}{dt} = \mu\mathcal{E}(t)e^{i\omega_0 t}\gamma_{II}, $$

$$ i\hbar \frac{d\gamma_{II}}{dt} = \mu\mathcal{E}(t)e^{-i\omega_0 t}\gamma_I. $$

(9.43)

Now we have an apparently simple pair of equations—and they are still exact, of course. The derivative of one variable is a function of time $\mu\mathcal{E}(t)e^{i\omega_0 t}$, multiplied by the second variable; the derivative of the second is a similar time function, multiplied by the first. Although these simple equations cannot be solved in general, we will solve them for some special cases.

We are, for the moment at least, interested only in the case of an oscillating electric field. Taking $\mathcal{E}(t)$ as given in Eq. (9.37), we find that the equations for γ_I and γ_{II} become

$$ i\hbar \frac{d\gamma_I}{dt} = \mu\mathcal{E}_0[e^{i(\omega+\omega_0)t} + e^{-i(\omega-\omega_0)t}]\gamma_{II}, $$

$$ i\hbar \frac{d\gamma_{II}}{dt} = \mu\mathcal{E}_0[e^{i(\omega-\omega_0)t} + e^{-i(\omega+\omega_0)t}]\gamma_I. $$

(9.44)

Now if $\mathcal{E}_0$ is sufficiently small, the rates of change of γ_I and γ_{II} are also small. The two γ's will not vary much with t, especially in comparison with the rapid variations due to the exponential terms. These exponential terms have real and imaginary parts that oscillate at the frequency $\omega + \omega_0$ or $\omega - \omega_0$. The terms with $\omega + \omega_0$ oscillate very rapidly about an average value of zero and, therefore, do not contribute very much on the average to the rate of change of γ. So we can make a reasonably good approximation by replacing these terms by their average value, namely, zero. We will just leave them out, and take as our approximation:

$$ i\hbar \frac{d\gamma_I}{dt} = \mu\mathcal{E}_0 e^{-i(\omega-\omega_0)t}\gamma_{II}, $$

$$ i\hbar \frac{d\gamma_{II}}{dt} = \mu\mathcal{E}_0 e^{i(\omega-\omega_0)t}\gamma_I. $$

(9.45)

Even the remaining terms, with exponents proportional to $(\omega - \omega_0)$, will also vary rapidly unless ω is near ω_0. Only then will the right-hand side vary slowly enough that any appreciable amount will accumulate when we integrate the equations with respect to t. In other words, with a *weak* electric field the only significant frequencies are those near ω_0.

With the approximation made in getting Eq. (9.45), the equations can be solved exactly, but the work is a little elaborate, so we won't do that until later when we take up another problem of the same type. Now we'll just solve them approximately—or rather, we'll find an exact solution for the case of perfect resonance, $\omega = \omega_0$, and an approximate solution for frequencies near resonance.

9–4 Transitions at resonance

Let's take the case of perfect resonance first. If we take $\omega = \omega_0$, the exponentials are equal to one in both equations of (9.45), and we have just

$$ \frac{d\gamma_I}{dt} = -\frac{i\mu\mathcal{E}_0}{\hbar}\gamma_{II}, \qquad \frac{d\gamma_{II}}{dt} = -\frac{i\mu\mathcal{E}_0}{\hbar}\gamma_I. $$

(9.46)

If we eliminate first γ_I and then γ_{II} from these equations, we find that each satisfies the differential equation of simple harmonic motion:

$$ \frac{d^2\gamma}{dt^2} = -\left(\frac{\mu\mathcal{E}_0}{\hbar}\right)^2 \gamma. $$

(9.47)

The general solutions for these equations can be made up of sines and cosines.

As you can easily verify, the following equations are a solution:

$$\gamma_I = a \cos\left(\frac{\mu\mathcal{E}_0}{\hbar}\right)t + b \sin\left(\frac{\mu\mathcal{E}_0}{\hbar}\right)t,$$

$$\gamma_{II} = ib \cos\left(\frac{\mu\mathcal{E}_0}{\hbar}\right)t - ia \sin\left(\frac{\mu\mathcal{E}_0}{\hbar}\right)t,$$

(9.48)

where a and b are constants to be determined to fit any particular physical situation.

For instance, suppose that at $t = 0$ our molecular system was in the upper energy state $|\,I\rangle$, which would require—from Eq. (9.40)—that $\gamma_I = 1$ and $\gamma_{II} = 0$ at $t = 0$. For this situation we would need $a = 1$ and $b = 0$. The probability that the molecule is in the state $|\,I\rangle$ at some later t is the absolute square of γ_I, or

$$P_I = |\gamma_I|^2 = \cos^2\left(\frac{\mu\mathcal{E}_0}{\hbar}\right)t.$$

(9.49)

Similarly, the probability that the molecule will be in the state $|\,II\rangle$ is given by the absolute square of γ_{II},

$$P_{II} = \gamma_{II}^2 = \sin^2\left(\frac{\mu\mathcal{E}_0}{\hbar}\right)t.$$

(9.50)

So long as $\mathcal{E}$ is small and we are on resonance, the probabilities are given by simple oscillating functions. The probability to be in state $|\,I\rangle$ falls from one to zero and back again, while the probability to be in the state $|\,II\rangle$ rises from zero to one and back. The time variation of the two probabilities is shown in Fig. 9–5. Needless to say, the sum of the two probabilities is always equal to one; the molecule is always in *some* state!

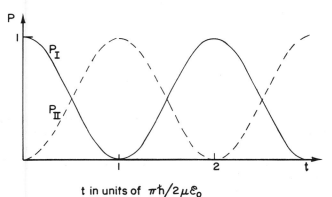

Fig. 9–5. Probabilities for the two states of the ammonia molecule in a sinusoidal electric field.

t in units of $\pi\hbar/2\mu\mathcal{E}_0$

Let's suppose that it takes the molecule the time T to go through the cavity. If we make the cavity just long enough so that $\mu\mathcal{E}_0 T/\hbar = \pi/2$, then a molecule which enters in state $|\,I\rangle$ will certainly leave it in state $|\,II\rangle$. If it enters the cavity in the upper state, it will leave the cavity in the lower state. In other words, its energy is decreased, and the loss of energy can't go anywhere else but into the machinery which generates the field. The details by which you can see how the energy of the molecule is fed into the oscillations of the cavity are not simple; however, we don't need to study these details, because we can use the principle of conservation of energy. (We could study them if we had to, but then we would have to deal with the quantum mechanics of the field in the cavity in addition to the quantum mechanics of the atom.)

In summary: the molecule enters the cavity, the cavity field—oscillating at exactly the right frequency—induces transitions from the upper to the lower state, and the energy released is fed into the oscillating field. In an operating maser the molecules deliver enough energy to maintain the cavity oscillations—not only providing enough power to make up for the cavity losses but even providing small amounts of excess power that can be drawn from the cavity. Thus, the molecular energy is converted into the energy of an external electromagnetic field.

Remember that before the beam enters the cavity, we have to use a filter which separates the beam so that only the upper state enters. It is easy to demonstrate that if you were to start with molecules in the lower state, the process will go the other way and take energy out of the cavity. If you put the unfiltered beam in, as many molecules are taking energy out as are putting energy in, so nothing much would happen. In actual operation it isn't necessary, of course, to make $(\mu\mathcal{E}_0 T/\hbar)$ exactly $\pi/2$. For any other value (except an exact integral multiple of π), there is some probability for transitions from state $|I\rangle$ to state $|II\rangle$. For other values, however, the device isn't 100 percent efficient; many of the molecules which leave the cavity could have delivered some energy to the cavity but didn't.

In actual use, the velocity of all the molecules is not the same; they have some kind of Maxwell distribution. This means that the ideal periods of time for different molecules will be different, and it is impossible to get 100 percent efficiency for all the molecules at once. In addition, there is another complication which is easy to take into account, but we don't want to bother with it at this stage. You remember that the electric field in a cavity usually varies from place to place across the cavity. Thus, as the molecules drift across the cavity, the electric field at the molecule varies in a way that is more complicated than the simple sinusoidal oscillation in time that we have assumed. Clearly, one would have to use a more complicated integration to do the problem exactly, but the general idea is still the same.

There are other ways of making masers. Instead of separating the atoms in state $|I\rangle$ from those in state $|II\rangle$ by a Stern-Gerlach apparatus, one can have the atoms already in the cavity (as a gas or a solid) and shift atoms from state $|II\rangle$ to state $|I\rangle$ by some means. One way is one used in the so-called three-state maser. For it, atomic systems are used which have three energy levels, as shown in Fig. 9–6, with the following special properties. The system will absorb radiation (say, light) of frequency $\hbar\omega_1$ and go from the lowest energy level E_{II} to some high-energy level E', and then will quickly emit photons of frequency $\hbar\omega_2$ and go to the state $|I\rangle$ with energy E_I. The state $|I\rangle$ has a long lifetime so its population can be raised, and the conditions are then appropriate for maser operation between states $|I\rangle$ and $|II\rangle$. Although such a device is called a "three-state" maser, the maser operation really works just as a two-state system such as we are describing.

A laser (*L*ight *A*mplification by *S*timulated *E*mission of *R*adiation) is just a maser working at optical frequencies. The "cavity" for a laser usually consists of just two plane mirrors between which standing waves are generated.

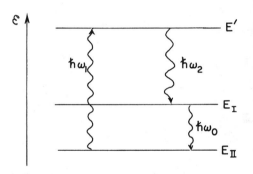

Fig. 9–6. The energy levels of a "three-state" maser.

9–5 Transitions off resonance

Finally, we would like to find out how the states vary in the circumstance that the cavity frequency is nearly, but not exactly, equal to ω_0. We could solve this problem exactly, but instead of trying to do that, we'll take the important case that the electric field is small and also the period of time T is small, so that $\mu\mathcal{E}_0 T/\hbar$ is much less than one. Then, even in the case of perfect resonance which we have just worked out, the probability of making a transition is small. Suppose that we start again with $\gamma_I = 1$ and $\gamma_{II} = 0$. During the time T we would expect γ_I to remain nearly equal to one, and γ_{II} to remain very small compared with unity. Then the problem is very easy. We can calculate γ_{II} from the second equation in (9.45), taking γ_I equal to one and integrating from $t = 0$ to $t = T$. We get

$$\gamma_{II} = \frac{\mu\mathcal{E}_0}{\hbar}\left[\frac{1 - e^{i(\omega - \omega_0)T}}{\omega - \omega_0}\right]. \qquad (9.51)$$

This γ_{11}, used with Eq. (9.40), gives the amplitude to have made a transition from the state $|I\rangle$ to the state $|II\rangle$ during the time interval T. The probability $P(I \rightarrow II)$ to make the transition is $|\gamma_{II}|^2$, or

$$P(I \rightarrow II) = |\gamma_{II}|^2 = \left[\frac{\mu\mathcal{E}_0 T}{\hbar}\right]^2 \frac{\sin^2[(\omega - \omega_0)T/2]}{[(\omega - \omega_0)T/2]^2}. \qquad (9.52)$$

It is interesting to plot this probability for a fixed length of time as a function of the frequency of the cavity in order to see how sensitive it is to frequencies near the resonant frequency ω_0. We show such a plot of $P(I \rightarrow II)$ in Fig. 9–7. (The vertical scale has been adjusted to be 1 at the peak by dividing by the value of the probability when $\omega = \omega_0$.) We have seen a curve like this in the diffraction theory, so you should already be familiar with it. The curve falls rather abruptly to zero for $(\omega - \omega_0) = 2\pi/T$ and never regains significant size for large frequency deviations. In fact, by far the greatest part of the area under the curve lies within the range $\pm \pi/T$. It is possible to show† that the area under the curve is just $2\pi/T$ and is equal to the area of the shaded rectangle drawn in the figure.

Let's examine the implication of our results for a real maser. Suppose that the ammonia molecule is in the cavity for a reasonable length of time, say for one millisecond. Then for $f_0 = 24,000$ megacycles, we can calculate that the probability for a transition falls to zero for a frequency deviation of $(f - f_0)/f_0 = 1/f_0 T$, which is five parts in 10^8. Evidently the frequency must be very close to ω_0 to get a significant transition probability. Such an effect is the basis of the great precision that can be obtained with "atomic" clocks, which work on the maser principle.

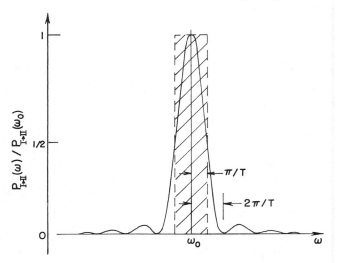

Fig. 9–7. Transition probability for the ammonia molecule as a function of frequency.

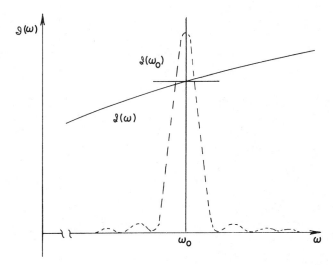

Fig. 9–8. The spectral intensity $\mathcal{I}(\omega)$ can be approximated by its value at ω_0.

9–6 The absorption of light

Our treatment above applies to a more general situation than the ammonia maser. We have treated the behavior of a molecule under the influence of an electric field, whether that field was confined in a cavity or not. So we could be simply shining a beam of "light"—at microwave frequencies—at the molecule and ask for the probability of emission or absorption. Our equations apply equally well to this case, but let's rewrite them in terms of the *intensity* of the radiation rather than the electric field. If we define the intensity $\mathcal{I}$ to be the average energy flow per unit area per second, then from Chapter 27 of Volume II, we can write

$$\mathcal{I} = \epsilon_0 c^2 |\mathcal{E} \times \boldsymbol{B}|_{\text{ave}} = \tfrac{1}{2}\epsilon_0 c^2 (\mathcal{E} \times \boldsymbol{B})_{\text{max}} = 2\epsilon_0 c \mathcal{E}_0^2.$$

(The maximum value of $\mathcal{E}$ is $2\mathcal{E}_0$.) The transition probability now becomes:

$$P(I \rightarrow II) = 2\pi \left[\frac{\mu^2}{4\pi\epsilon_0 \hbar^2 c}\right] \mathcal{I} T^2 \, \frac{\sin^2 [(\omega - \omega_0)T/2]}{[(\omega - \omega_0)T/2]^2}. \qquad (9.53)$$

———————

† Using the formula $\int_{-\infty}^{\infty} (\sin^2 x/x^2) \, dx = \pi$.

Ordinarily the light shining on such a system is not exactly monochromatic. It is, therefore, interesting to solve one more problem—that is, to calculate the transition probability when the light has intensity $\mathcal{I}(\omega)$ per unit frequency interval, covering a broad range which includes ω_0. Then, the probability of going from $|I\rangle$ to $|II\rangle$ will become an integral:

$$P(I \to II) = 2\pi \left[\frac{\mu^2}{4\pi\epsilon_0\hbar^2 c}\right] T^2 \int_0^\infty \mathcal{I}(\omega) \frac{\sin^2 [(\omega - \omega_0)T/2]}{[(\omega - \omega_0)T/2]^2} \, d\omega. \qquad (9.54)$$

In general, $\mathcal{I}(\omega)$ will vary much more slowly with ω than the sharp resonance term. The two functions might appear as shown in Fig. 9–8. In such cases, we can replace $\mathcal{I}(\omega)$ by its value $\mathcal{I}(\omega_0)$ at the center of the sharp resonance curve and take it outside of the integral. What remains is just the integral under the curve of Fig. 9–7, which is, as we have seen, just equal to $2\pi/T$. We get the result that

$$P(I \to II) = 4\pi^2 \left[\frac{\mu^2}{4\pi\epsilon_0\hbar^2 c}\right] \mathcal{I}(\omega_0)T. \qquad (9.55)$$

This is an important result, because it is the *general theory of the absorption of light by any molecular or atomic system*. Although we began by considering a case in which state $|I\rangle$ had a higher energy than state $|II\rangle$, none of our arguments depended on that fact. Equation (9.55) still holds if the state $|I\rangle$ has a *lower* energy than the state $|II\rangle$; then $P(I \to II)$ represents the probability for a transition with the *absorption* of energy from the incident electromagnetic wave. The absorption of light by any atomic system always involves the amplitude for a transition in an oscillating electric field between two states separated by an energy $E = \hbar\omega_0$. For any particular case, it is always worked out in just the way we have done here and gives an expression like Eq. (9.55). We, therefore, emphasize the following features of this result. First, the probability is proportional to T. In other words, there is a constant probability per unit time that transitions will occur. Second, this probability is proportional to the *intensity* of the light incident on the system. Finally, the transition probability is proportional to μ^2, where, you remember, $\mu\mathcal{E}$ defined the shift in energy due to the electric field $\mathcal{E}$. Because of this, $\mu\mathcal{E}$ also appeared in Eqs. (9.38) and (9.39) as the coupling term that is responsible for the transition between the otherwise stationary states $|I\rangle$ and $|II\rangle$. In other words, for the small $\mathcal{E}$ we have been considering, $\mu\mathcal{E}$ is the so-called "perturbation term" in the Hamiltonian matrix element which connects the states $|I\rangle$ and $|II\rangle$. In the general case, we would have that $\mu\mathcal{E}$ gets replaced by the matrix element $\langle II|H|I\rangle$ (see Section 5–6).

In Volume I (Section 42–5) we talked about the relations among light absorption, induced emission, and spontaneous emission in terms of the Einstein A- and B-coefficients. Here, we have at last the quantum mechanical procedure for computing these coefficients. What we have called $P(I \to II)$ for our two-state ammonia molecule corresponds precisely to the absorption coefficient B_{nm} of the Einstein radiation theory. For the complicated ammonia molecule—which is too difficult for anyone to calculate—we have taken the matrix element $\langle II |H| I\rangle$ as $\mu\mathcal{E}$, saying that μ is to be gotten from experiment. For simpler atomic systems, the μ_{mn} which belongs to any particular transition can be calculated from the *definition*

$$\mu_{mn}\mathcal{E} = \langle m |H| n \rangle = H_{mn}, \qquad (9.56)$$

where H_{mn} is the matrix element of the Hamiltonian which includes the effects of a weak electric field. The μ_{mn} calculated in this way is called the *electric dipole matrix element*. The quantum mechanical theory of the absorption and emission of light is, therefore, reduced to a calculation of these matrix elements for particular atomic systems.

Our study of a simple two-state system has thus led us to an understanding of the general problem of the absorption and emission of light.

10

Other Two-State Systems

10–1 The hydrogen molecular ion

In the last chapter we discussed some aspects of the ammonia molecule under the approximation that it can be considered as a two-state system. It is, of course, not really a two-state system—there are many states of rotation, vibration, translation, and so on—but each of these states of motion must be analyzed in terms of two internal states because of the flip-flop of the nitrogen atom. Here we are going to consider other examples of systems which, to some approximation or other, can be considered as two-state systems. Lots of things will be approximate because there are always many other states, and in a more accurate analysis they would have to be taken into account. But in each of our examples we will be able to understand a great deal by just thinking about two states.

Since we will only be dealing with two-state systems, the Hamiltonian we need will look just like the one we used in the last chapter. When the Hamiltonian is independent of time, we know that there are two stationary states with definite—and usually different—energies. Generally, however, we start our analysis with a set of base states which are *not* these stationary states, but states which may, perhaps, have some other simple physical meaning. Then, the stationary states of the system will be represented by a linear combination of these base states.

For convenience, we will summarize the important equations from Chapter 9. Let the original choice of base states be $|1\rangle$ and $|2\rangle$. Then any state $|\psi\rangle$ is represented by the linear combination

$$|\psi\rangle = |1\rangle\langle 1|\psi\rangle + |2\rangle\langle 2|\psi\rangle = |1\rangle C_1 + |2\rangle C_2. \qquad (10.1)$$

The amplitudes C_i (by which we mean either C_1 or C_2) satisfy the two linear differential equations

$$i\hbar \frac{dC_i}{dt} = \sum_j H_{ij} C_j, \qquad (10.2)$$

where both i and j take on the values 1 and 2.

When the terms of the Hamiltonian H_{ij} do not depend on t, the two states of definite energy (the stationary states), which we call

$$|\psi_I\rangle = |I\rangle e^{-(i/\hbar)E_I t} \qquad \text{and} \qquad |\psi_{II}\rangle = |II\rangle e^{-(i/\hbar)E_{II} t},$$

have the energies

$$E_I = \frac{H_{11} + H_{22}}{2} + \sqrt{\left(\frac{H_{11} - H_{22}}{2}\right)^2 + H_{12}H_{21}},$$

$$E_{II} = \frac{H_{11} + H_{22}}{2} - \sqrt{\left(\frac{H_{11} - H_{22}}{2}\right)^2 + H_{12}H_{21}}. \qquad (10.3)$$

The two C's for each of these states have the same time dependence. The state vectors $|I\rangle$ and $|II\rangle$ which go with the stationary states are related to our original base states $|1\rangle$ and $|2\rangle$ by

$$|I\rangle = |1\rangle a_1 + |2\rangle a_2,$$

$$|II\rangle = |1\rangle a_1' + |2\rangle a_2'. \qquad (10.4)$$

10–1 The hydrogen molecular ion

10–2 Nuclear forces

10–3 The hydrogen molecule

10–4 The benzene molecule

10–5 Dyes

10–6 The Hamiltonian of a spin one-half particle in a magnetic field

10–7 The spinning electron in a magnetic field

The a's are complex constants, which satisfy

$$|a_1|^2 + |a_2|^2 = 1,$$

$$\frac{a_1}{a_2} = \frac{H_{12}}{E_I - H_{11}}, \tag{10.5}$$

$$|a_1'|^2 + |a_2'|^2 = 1,$$

$$\frac{a_1'}{a_2'} = \frac{H_{12}}{E_{II} - H_{11}}. \tag{10.6}$$

If H_{11} and H_{22} are equal—say both are equal to E_0—and $H_{12} = H_{21} = -A$, then $E_I = E_0 + A$, $E_{II} = E_0 - A$, and the states $|I\rangle$ and $|II\rangle$ are particularly simple:

$$|I\rangle = \frac{1}{\sqrt{2}}\left[|1\rangle - |2\rangle\right], \qquad |II\rangle = \frac{1}{\sqrt{2}}\left[|1\rangle + |2\rangle\right]. \tag{10.7}$$

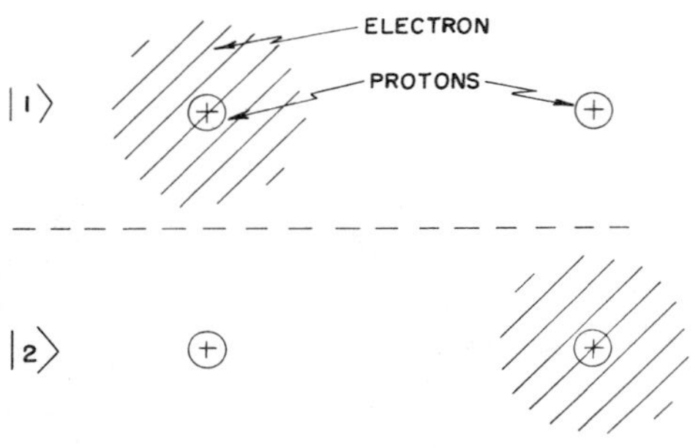

Fig. 10–1. A set of base states for two protons and an electron.

Now we will use these results to discuss a number of interesting examples taken from the fields of chemistry and physics. The first example is the hydrogen molecular ion. A positively ionized hydrogen molecule consists of two protons with one electron worming its way around them. If the two protons are very far apart, what states would we expect for this system? The answer is pretty clear: The electron will stay close to one proton and form a hydrogen atom in its lowest state, and the other proton will remain alone as a positive ion. So, if the two protons are far apart, we can visualize one physical state in which the electron is "attached" to one of the protons. There is, clearly, another state symmetric to that one in which the electron is near the other proton, and the first proton is the one that is an ion. We will take these two as our base states, and we'll call them $|1\rangle$ and $|2\rangle$. They are sketched in Fig. 10–1. Of course, there are really many states of an electron near a proton, because the combination can exist as any one of the excited states of the hydrogen atom. We are not interested in that variety of states now; we will consider only the situation in which the hydrogen atom is in the lowest state—its ground state—and we will, for the moment, disregard spin of the electron. We can just suppose that for all our states the electron has its spin "up" along the z-axis.†

Now to remove an electron from a hydrogen atom requires 13.6 electron volts of energy. So long as the two protons of the hydrogen molecular ion are far apart, it still requires about this much energy—which is for our present considerations a great deal of energy—to get the electron somewhere near the midpoint between the protons. So it is impossible, classically, for the electron to jump from one proton to the other. However, in quantum mechanics it is possible—though not very likely. There is some small amplitide for the electron to move from one proton to the other. As a first approximation, then, each of our base states $|1\rangle$ and $|2\rangle$ will have the energy E_0, which is just the energy of one hydrogen atom plus one proton. We can take that the Hamiltonian matrix elements H_{11} and H_{22} are both approximately equal to E_0. The other matrix elements H_{12} and H_{21}, which are the amplitudes for the electron to go back and forth, we will again write as $-A$.

You see that this is the same game we played in the last two chapters. If we disregard the fact that the electron can flip back and forth, we have two states of exactly the same energy. This energy will, however, be split into two energy levels by the possibility of the electron going back and forth—the greater the probability of the transition, the greater the split. So the two energy levels of the system are $E_0 + A$ and $E_0 - A$, and the states which have these definite energies are given by Eqs. (10.7).

---

† This is satisfactory so long as there are no important magnetic fields. We will discuss the effects of magnetic fields on the electron later in this chapter, and the very small effects of spin in the hydrogen atom in Chapter 12.

From our solution we see that if a proton and a hydrogen ion are put anywhere near together, the electron will not stay on one of the protons but will flip back and forth between the two protons. If it starts on one of the protons, it will oscillate back and forth between the states $|1\rangle$ and $|2\rangle$—giving a time-varying solution. In order to have the lowest energy solution (which does not vary with time), it is necessary to start the system with equal amplitudes for the electron to be around each proton. Remember, there are not two electrons—we are not saying that there is an electron around each proton. There is only *one* electron, and *it* has the same amplitude—$1/\sqrt{2}$ in magnitude—to be in either position.

Now the amplitude A for an electron which is near one proton to get to the other one depends on the separation between the protons. The closer the protons are together, the larger the amplitude. You remember that we talked in Chapter 7 about the amplitude for an electron to "penetrate a barrier," which it could not do classically. We have the same situation here. The amplitude for an electron to get across decreases roughly exponentially with the distance—for large distances. Since the transition probability, and therefore A, gets larger when the protons are closer together, the separation of the energy levels will also get larger. If the system is in the state $|I\rangle$, the energy $E_0 + A$ increases with decreasing distance, so these quantum mechanical effects make a *repulsive* force tending to keep the protons apart. On the other hand, if the system is in the state $|II\rangle$, the total energy *decreases* if the protons are brought closer together; there is an *attractive* force pulling the protons together. The variation of the two energies with the distance between the two protons should be roughly as shown in Fig. 10–2. We have, then, a quantum-mechanical explanation of the binding force that holds the H_2^+ ion together.

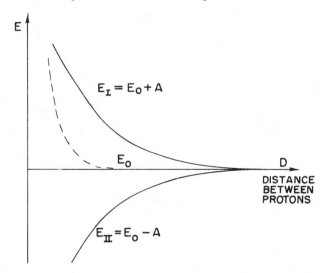

Fig. 10–2. The energies of the two stationary states of the H_2^+ ion as a function of the distance between the two protons.

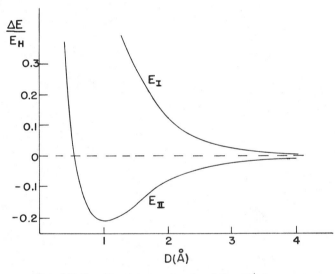

Fig. 10–3. The energy levels of the H_2^+ ion as a function of the interproton distance D. ($E_h = 13.6$ ev.)

We have, however, forgotten one thing. In addition to the force we have just described, there is also an electrostatic repulsive force between the two protons. When the two protons are far apart—as in Fig. 10–1—the "bare" proton sees only a neutral atom, so there is a negligible electrostatic force. At very close distances, however, the "bare" proton begins to get "inside" the electron distribution—that is, it is closer to the proton on the average than to the electron. So there begins to be some extra electrostatic energy which is, of course, positive. This energy—which also varies with the separation—should be included in E_0. So for E_0 we should take something like the broken-line curve in Fig. 10–2 which rises rapidly for distances less than the radius of a hydrogen atom. We should add and subtract the flip-flop energy A from this E_0. When we do that, the energies E_I and E_{II} will vary with the interproton distance D as shown in Fig. 10–3. [In this figure, we have plotted the results of a more detailed calculation. The interproton distance is given in units of 1 A(10^{-8} cm), and the excess energy over a proton plus a hydrogen atom is given in units of the binding energy of the hydrogen atom—the so-called "Rydberg" energy, 13.6 ev.] We see that the state $|II\rangle$ has a minimum-energy point. This will be the equilibrium configuration—the lowest energy condition—for the H_2^+ ion. The energy at this point is lower than the energy of a separated proton and hydrogen ion, so the system is bound. A single electron acts to hold the two protons together. A chemist would call it a "one-electron bond."

This kind of chemical binding is also often called "quantum mechanical resonance" (by analogy with the two coupled pendulums we have described before). But that really sounds more mysterious than it is, it's only a "resonance" if you start out by making a poor choice for your base states—as we did also! If you picked the state $|II\rangle$, you would have the lowest energy state—that's all.

We can see in another way why such a state should have a lower energy than a proton and a hydrogen atom. Let's think about an electron near two protons with some fixed, but not too large, separation. You remember that with a single proton the electron is "spread out" because of the uncertainty principle. It seeks a balance between having a low coulomb *potential* energy and not getting confined into too small a space, which would make a high *kinetic* energy (because of the uncertainty relation $\Delta p\, \Delta x \approx \hbar$). Now if there are two protons, there is more space where the electron can have a low potential energy. It can spread out—lowering its kinetic energy—without increasing its potential energy. The net result is a lower energy than a hydrogen atom. Then why does the other state $|I\rangle$ have a higher energy? Notice that this state is the *difference* of the states $|1\rangle$ and $|2\rangle$. Because of the symmetry of $|1\rangle$ and $|2\rangle$, the difference must have zero amplitude to find the electron half-way between the two protons. This means that the electron is somewhat more confined, which leads to a larger energy.

We should say that our approximate treatment of the H_2^+ ion as a two-state system breaks down pretty badly once the protons get as close together as they are at the minimum in the curve of Fig. 10–3, and so, will not give a good value for the actual binding energy. For small separations, the energies of the two "states" we imagined in Fig. 6–1 are not really equal to E_0; a more refined quantum mechanical treatment is needed.

Suppose we ask now what would happen if instead of two protons, we had two different objects—as, for example, one proton and one lithium positive ion (both particles still with a single positive charge). In such a case, the two terms H_{11} and H_{22} of the Hamiltonian would no longer be equal; they would, in fact, be quite different. If it should happen that the difference $(H_{11} - H_{22})$ is, in absolute value, much greater than $A = -H_{12}$, the attractive force gets very weak, as we can see in the following way.

If we put $H_{12}H_{21} = A^2$ into Eqs. (10.3) we get

$$E = \frac{H_{11} + H_{22}}{2} \pm \frac{H_{11} - H_{22}}{2} \sqrt{1 + \frac{4A^2}{(H_{11} - H_{22})^2}}.$$

When $H_{11} - H_{22}$ is much greater than A^2, the square root is very nearly equal to

$$1 + \frac{2A^2}{(H_{11} - H_{22})^2}.$$

The two energies are then

$$E_I = H_{11} + \frac{A^2}{(H_{11} - H_{22})},$$

$$E_{II} = H_{22} - \frac{A^2}{(H_{11} - H_{22})}. \tag{10.8}$$

They are now very nearly just the energies H_{11} and H_{22} of the isolated atoms, pushed apart only slightly by the flip-flop amplitude A.

The energy difference $E_I - E_{II}$ is

$$(H_{11} - H_{22}) + \frac{2A^2}{H_{11} - H_{22}}.$$

The *additional* separation from the flip-flop of the electron is no longer equal to $2A$; it is smaller by the factor $A/(H_{11} - H_{22})$, which we are now taking to be much less than one. Also, the dependence of $E_I - E_{II}$ on the separation of the two nuclei is much smaller than for the H_2^+ ion—it is also reduced by the factor $A/(H_{11} - H_{22})$. We can now see why the binding of unsymmetric diatomic molecules is generally very weak.

In our theory of the H_2^+ ion we have discovered an explanation for the mechanism by which an electron shared by two protons provides, in effect, an attractive force between the two protons which can be present even when the protons are at large distances. The attractive force comes from the reduced energy of the system due to the possibility of the electron jumping from one proton to the other. In such a jump the system changes from the configuration (hydrogen atom, proton) to the configuration (proton, hydrogen atom), or switches back. We can write the process symbolically as

$$(H, p) \rightleftharpoons (p, H).$$

The energy shift due to this process is proportional to the amplitude A that an electron whose energy is $-W_H$ (its binding energy in the hydrogen atom) can get from one proton to the other.

For large distances R between the two protons, the electrostatic potential energy of the electron is nearly zero over most of the space it must go when it makes its jump. In this space, then, the electron moves nearly like a free particle in empty space—but with a *negative* energy! We have seen in Chapter 3 [Eq. (3.7)] that the amplitude for a particle of definite energy to get from one place to another a distance r away is proportional to

$$\frac{e^{(i/\hbar)pr}}{r},$$

where p is the momentum corresponding to the definite energy. In the present case (using the nonrelativistic formula), p is given by

$$\frac{p^2}{2m} = -W_H. \tag{10.9}$$

This means that p is an imaginary number,

$$p = i\sqrt{2mW_H}$$

(the other sign for the radical gives nonsense here).

We should expect, then, that the amplitude A for the H_2^+ ion will vary as

$$A \sim \frac{e^{-(\sqrt{2mW_H}/\hbar)R}}{R} \tag{10.10}$$

for large separations R between the two protons. The energy shift due to the electron binding is proportional to A, so there is a force pulling the two protons together which is proportional—for large R—to the derivative of (10.10) with respect to R.

Finally, to be complete, we should remark that in the two-proton, one-electron system there is still one other effect which gives a dependence of the energy on R. We have neglected it until now because it is usually rather unimportant—the exception is just for those very large distances where the energy of the exchange term A has decreased exponentially to very small values. The new effect we are thinking of is the electrostatic attraction of the proton for the hydrogen atom, which comes about in the same way any charged object attracts a neutral object. The bare proton makes an electric field $\mathcal{E}$ (varying as $1/R^2$) at the neutral hydrogen atom. The atom becomes polarized, taking on an induced dipole moment μ proportional to $\mathcal{E}$. The energy of the dipole is $\mu\mathcal{E}$, which is proportional to $\mathcal{E}^2$—or to $1/R^4$. So there is a term in the energy of the system which decreases with the fourth power of the distance. (It is a correction to E_0.) This energy falls off with distance more slowly than the shift A given by (10.10); at some large separation R it becomes the only remaining important term giving a variation of energy with R—and, therefore, the only remaining force. Note that the electrostatic term has the same sign for both of the base states (the force is attractive, so the energy is negative) and so also for the two stationary states, whereas the electron exchange term A gives opposite signs for the two stationary states.

10–2 Nuclear forces

We have seen that the system of a hydrogen atom and a proton has an energy of interaction due to the exchange of the single electron which varies at large separations R as

$$\frac{e^{-\alpha R}}{R}, \tag{10.11}$$

with $\alpha = \sqrt{2mW_H}/\hbar$. (One usually says that there is an exchange of a "virtual" electron when—as here—the electron has to jump across a space where it would have a negative energy. More specifically, a "virtual exchange" means that the phenomenon involves a quantum mechanical interference between an exchanged state and a nonexchanged state.)

Now we might ask the following question: Could it be that forces between other kinds of particles have an analogous origin? What about, for example, the nuclear force between a neutron and a proton, or between two protons? In an attempt to explain the nature of nuclear forces, Yukawa proposed that the force between two nucleons is due to a similar exchange effect—only, in this case, due to the virtual exchange, not of an electron, but of a new particle, which he called a "meson." Today, we would identify Yukawa's meson with the π-meson (or "pion") produced in high-energy collisions of protons or other particles.

Let's see, as an example, what kind of a force we would expect from the exchange of a positive pion (π^+) of mass m_π between a proton and a neutron. Just as a hydrogen atom H^0 can go into a proton p^+ by giving up an electron e^-,

$$H^0 \rightarrow p^+ + e^-, \tag{10.12}$$

a proton p^+ can go into a neutron n^0 by giving up a π^+ meson:

$$p^+ \rightarrow n^0 + \pi^+. \tag{10.13}$$

So if we have a proton at a and a neutron at b separated by the distance R, the proton can become a neutron by emitting a π^+, which is then absorbed by the neutron at b, turning it into a proton. There is an energy of interaction of the two-nucleon (plus pion) system which depends on the amplitude A for the pion exchange—just as we found for the electron exchange in the H_2^+ ion.

In the process (10.12), the energy of the H^0 atom is less than that of the proton by W_H (calculating nonrelativistically, and omitting the rest energy mc^2 of the electron), so the electron has a negative *kinetic* energy—or imaginary momentum—as in Eq. (10.9). In the nuclear process (10.13), the proton and neutron have almost equal masses, so the π^+ will have zero *total* energy. The relation between the total energy E and the momentum p for a pion of mass m_π is

$$E^2 = p^2c^2 + m_\pi^2 c^4.$$

Since E is zero (or at least negligible in comparison with m_π), the momentum is again imaginary:

$$p = im_\pi c.$$

Using the same arguments we gave for the amplitude that a bound electron would penetrate the barrier in the space between two protons, we get for the nuclear case an exchange amplitude A which should—for large R—go as

$$\frac{e^{-(m_\pi c/\hbar)R}}{R}. \tag{10.14}$$

The interaction energy is proportional to A, and so varies in the same way. We get an energy variation in the form of the so-called *Yukawa potential* between two nucleons. Incidentally, we obtained this same formula earlier directly from the differential equation for the motion of a pion in free space [see Chapter 28, Vol. II, Eq. (28.18)].

We can, following the same line of argument, discuss the interaction between two protons (or between two neutrons) which results from the exchange of a *neutral* pion (π^0). The basic process is now

$$p^+ \to p^+ + \pi^0. \tag{10.15}$$

A proton can emit a virtual π^0, but then it remains still a proton. If we have two protons, proton No. 1 can emit a virtual π^0 which is absorbed by proton No. 2. At the end, we still have two protons. This is somewhat different from the H_2^+ ion. There the H^0 went into a different condition—the proton—after emitting the electron. Now we are assuming that a proton can emit a π^0 without changing its character. Such processes are, in fact, observed in high-energy collisions. The process is analogous to the way that an electron emits a photon and ends up still an electron:

$$e \to e + \text{photon}. \tag{10.16}$$

We do not "see" the photons inside the electrons before they are emitted or after they are absorbed, and their emission does not change the "nature" of the electron.

Going back to the two protons, there is an interaction energy which arises from the amplitude A that one proton emits a neutral pion which travels across (with imaginary momentum) to the other proton and is absorbed there. This amplitude is again proportional to (10.14), with m_π the mass of the neutral pion. All the same arguments give an equal interaction energy for two neutrons. Since the nuclear forces (disregarding electrical effects) between neutron and proton, between proton and proton, between neutron and neutron are the same, we conclude that the masses of the charged and neutral pions should be the same. Experimentally, the masses are indeed very nearly equal, and the small difference is about what one would expect from electric self-energy corrections (see Chapter 28, Vol. II).

There are other kinds of particles—like K-mesons—which can be exchanged between two nucleons. It is also possible for two pions to be exchanged at the same time. But all of these other exchanged "objects" have a rest mass m_x higher than the pion mass m_π, and lead to terms in the exchange amplitude which vary as

$$\frac{e^{-(m_x c/\hbar) R}}{R}.$$

These terms die out faster with increasing R than the one-meson term. No one knows, today, how to calculate these higher-mass terms, but for large enough values of R only the one-pion term survives. And, indeed, those experiments which involve nuclear interactions only at large distances do show that the interaction energy is as predicted from the one-pion exchange theory.

In the classical theory of electricity and magnetism, the coulomb electrostatic interaction and the radiation of light by an accelerating charge are closely related— both come out of the Maxwell equations. We have seen in the quantum theory that light can be represented as the quantum excitations of the harmonic oscillations of the classical electromagnetic fields in a box. Alternatively, the quantum theory can be set up by describing light in terms of particles—photons—which obey Bose statistics. We emphasized in Section 4–5 that the two alternative points of view always give identical predictions. Can the second point of view be carried through completely to include *all* electromagnetic effects? In particular, if we want to describe the electromagnetic field purely in terms of Bose particles—that is, in terms of photons—what is the coulomb force due to?

From the "particle" point of view the coulomb interaction between two electrons *comes from the exchange of a virtual photon.* One electron emits a photon —as in reaction (10.16)—which goes over to the second electron, where it is absorbed in the reverse of the same reaction. The interaction energy is again given by a formula like (10.14), but now with m_π replaced by the rest mass of the photon —which is zero. So the virtual exchange of a photon between two electrons gives an interaction energy that varies simply inversely as R, the distance between the two electrons—just the normal coulomb potential energy! In the "particle" theory of electromagnetism, the process of a virtual photon exchange gives rise to all the phenomena of electrostatics.

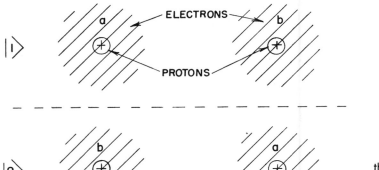

Fig. 10–4. A set of base states for the H_2 molecule.

10–3 The hydrogen molecule

As our next two-state system we will look at the neutral hydrogen molecule H_2. It is, naturally, more complicated to understand because it has two electrons. Again, we start by thinking of what happens when the two protons are well separated. Only now we have two electrons to add. To keep track of them, we'll call one of them "electron a" and the other "electron b." We can again imagine two possible states. One possibility is that "electron a" is around the first proton and "electron b" is around the second, as shown in Fig. 10–4(a). We have simply two hydrogen atoms. We will call this state $|\,1\,\rangle$. There is also another possibility: that "electron b" is around the first proton and that "electron a" is around the second. We call this state $|\,2\,\rangle$. From the symmetry of the situation, those two possibilities should be energetically equivalent, but, as we will see, the energy of the system is *not* just the energy of two hydrogen atoms. We should mention that there are many other possibilities. For instance, "electron a" might be near the first proton and "electron b" might be in another state around the *same* proton. We'll disregard such a case, since it will certainly have higher energy (because of the large coulomb repulsion between the two electrons). For greater accuracy, we would have to include such states, but we can get the essentials of the molecular binding by considering just the two states of Fig. 10.4. To this approximation we can describe any state by giving the amplitude $\langle 1\,|\,\phi\rangle$ to be in the state $|\,1\,\rangle$ and an amplitude $\langle 2\,|\,\phi\rangle$ to be in state $|\,2\,\rangle$. In other words, the state vector $|\,\phi\rangle$ can be written as the linear combination

$$|\,\phi\rangle = \sum_i |\,i\,\rangle\langle i\,|\,\phi\rangle.$$

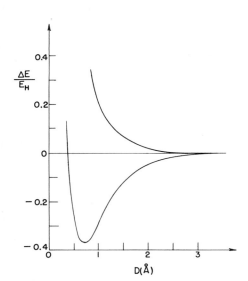

Fig. 10–5. The energy levels of the H_2 molecule for different interproton distances D. $(E_h = 13.6\ \text{ev.})$

To proceed, we assume—as usual—that there is some amplitude A that the electrons can move through the intervening space and exchange places. This possibility of exchange means that the energy of the system is split, as we have seen for other two-state systems. As for the hydrogen molecular ion, the splitting is very small when the distance between the protons is large. As the protons approach each other, the amplitude for the electrons to go back and forth increases, so the splitting increases. The decrease of the lower energy state means that there is an attractive force which pulls the atoms together. Again the energy levels rise when the protons get very close together because of the coulomb repulsion. The net final result is that the two stationary states have energies which vary with the separation as shown in Fig. 10–5. At a separation of about 0.74 A, the lower energy level reaches a minimum; this is the proton-proton distance of the true hydrogen molecule.

Now you have probably been thinking of an objection. What about the fact that the two electrons are identical particles? We have been calling them "electron *a*" and "electron *b*," but there really is no way to tell which is which. And we have said in Chapter 4 that for electrons—which are Fermi particles—if there are two ways something can happen by exchanging the electrons, the two amplitudes will interfere with a *negative* sign. This means that if we switch which electron is which, the sign of the amplitude must reverse. We have just concluded, however, that the bound state of the hydrogen molecule would be (at $t = 0$)

$$| II \rangle = \frac{1}{\sqrt{2}} (| 1 \rangle + | 2 \rangle).$$

However, according to our rules of Chapter 4, this state is not allowed. If we reverse which electron is which, we get the state

$$\frac{1}{\sqrt{2}} (| 2 \rangle + | 1 \rangle),$$

and we get the same sign instead of the opposite one.

These arguments are correct *if both electrons have the same spin*. It is true that if both electrons have spin up (or both have spin down), the only state that is permitted is

$$| I \rangle = \frac{1}{\sqrt{2}} (| 1 \rangle - | 2 \rangle).$$

For this state, an interchange of the two electrons gives

$$(| 2 \rangle - | 1 \rangle),$$

which is $- | I \rangle$, as required. So if we bring two hydrogen atoms near to each other with their electrons spinning in the same direction, they can go into the state $| I \rangle$ and not state $| II \rangle$. But notice that state $| I \rangle$ is the *upper* energy state. Its curve of energy versus separation has no minimum. The two hydrogens will always repel and will not form a molecule. So we conclude that the hydrogen molecule cannot exist with parallel electron spins. And that is right.

On the other hand, our state $| II \rangle$ is perfectly symmetric for the two electrons. In fact, if we interchange which electron we call *a* and which we call *b* we get back exactly the same state. We saw in Section 4–7 that if two Fermi particles are in the same state, they *must* have opposite spins. So, the bound hydrogen molecule must have one electron with spin up and one with spin down.

The whole story of the hydrogen molecule is really somewhat more complicated if we want to include the proton spins. It is then no longer right to think of the molecule as a *two*-state system. It should really be looked at as an *eight*-state system—there are four possible spin arrangements for each of our states $| 1 \rangle$ and $| 2 \rangle$—so we were cutting things a little short by neglecting the spins. Our final conclusions are, however, correct.

We find that the lowest energy state—the only bound state—of the H_2 molecule has the two electrons with spins opposite. The total spin angular momentum of the electrons is zero. On the other hand, two nearby hydrogen atoms with spins parallel—and so with a total angular momentum $\hbar$—must be in a higher (unbound) energy state; the atoms repel each other. There is an interesting correlation between the spins and the energies. It gives another illustration of something we mentioned before, which is that there appears to be an "interaction" energy between two spins because the case of parallel spins has a higher energy than the opposite case. In a certain sense you could say that the spins try to reach an antiparallel condition and, in doing so, have the potential to liberate energy—not because there is a large magnetic force, but because of the exclusion principle.

We saw in Section 10–1 that the binding of two *different* ions by a *single* electron is likely to be quite weak. This is *not* true for binding by *two* electrons. Suppose the two protons in Fig. 10–4 were replaced by any two ions (with closed inner electron shells and a single ionic charge), and that the binding energies of an electron at the two ions are different. The energies of states $|1\rangle$ and $|2\rangle$ would still be equal because in each of these states we have one electron bound to each ion. Therefore, we always have the splitting proportional to A. Two-electron binding is ubiquitous—it is the most common valence bond. Chemical binding usually involves this flip-flop game played by two electrons. Although two atoms can be bound together by only one electron, it is relatively rare—because it requires just the right conditions.

Finally, we want to mention that if the energy of attraction for an electron to one nucleus is much greater than to the other, then what we have said earlier about ignoring other possible states is no longer right. Suppose nucleus a (or it may be a positive ion) has a much stronger attraction for an electron than does nucleus b. It may then happen that the total energy is still fairly low even when both electrons are at nucleus a, and no electron is at nucleus b. The strong attraction may more than compensate for the mutual repulsion of the two electrons. If it does, the lowest energy state may have a large amplitude to find both electrons at a (making a negative ion) and a small amplitude to find any electron at b. The state looks like a negative ion with a positive ion. This is, in fact, what happens in an "ionic" molecule like NaCl. You can see that all the gradations between covalent binding and ionic binding are possible.

You can now begin to see how it is that many of the facts of chemistry can be most clearly understood in terms of a quantum mechanical description.

10–4 The benzene molecule

Chemists have invented nice diagrams to represent complicated organic molecules. Now we are going to discuss one of the most interesting of them—the benzene molecule shown in Fig. 10–6. It contains six carbon and six hydrogen atoms in a symmetrical array. Each bar of the diagram represents a *pair* of electrons, with spins opposite, doing the covalent bond dance. Each hydrogen atom contributes one electron and each carbon atom contributes four electrons to make up the total of 30 electrons involved. (There are two more electrons close to the nucleus of the carbon which form the first, or K, shell. These are not shown since they are so tightly bound that they are not appreciably involved in the covalent binding.) So each bar in the figure represents a *bond*, or pair of electrons, and the double bonds mean that there are *two pairs* of electrons between alternate pairs of carbon atoms.

There is a mystery about this benzene molecule. We can calculate what energy should be required to form this chemical compound, because the chemists have measured the energies of various compounds which involve pieces of the ring—for instance, they know the energy of a double bond by studying ethylene, and so on. We can, therefore, calculate the total energy we should expect for the benzene

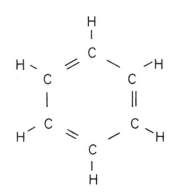

Fig. 10–6. The benzene molecule, C_6H_6.

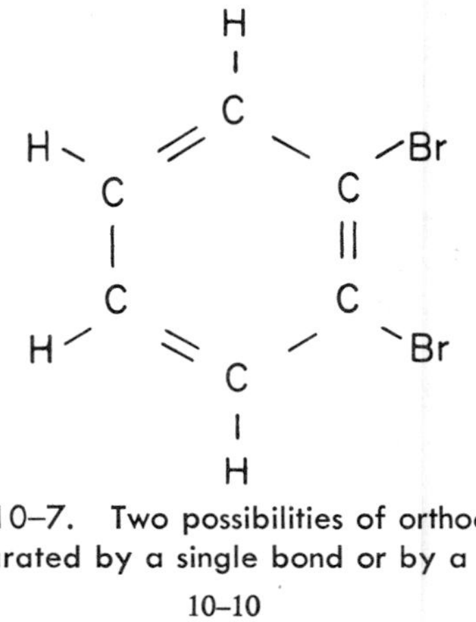

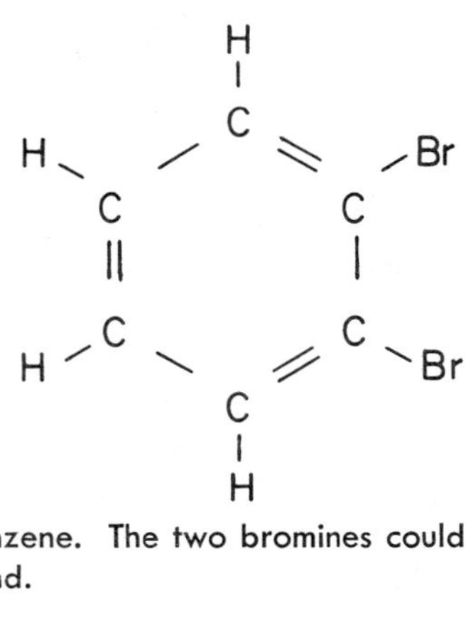

Fig. 10–7. Two possibilities of orthodibromobenzene. The two bromines could be separated by a single bond or by a double bond.

molecule. The actual energy of the benzene ring, however, is much lower than we get by such a calculation; it is more tightly bound than we would expect from what is called an "unsaturated double bond system." Usually a double bond system which is not in such a ring is easily attacked chemically because it has a relatively high energy—the double bonds can be easily broken by the addition of other hydrogens. But in benzene the ring is quite permanent and hard to break up. In other words, benzene has a much lower energy than you would calculate from the bond picture.

Then there is another mystery. Suppose we replace two adjacent hydrogens by bromine atoms to make ortho-dibromobenzene. There are two ways to do this, as shown in Fig. 10–7. The bromines could be on the opposite ends of a double bond as shown in part (a) of the figure, or could be on the opposite ends of a single bond as in (b). One would think that ortho-dibromobenzene should have two different forms, but it doesn't. There is only one such chemical.†

Now we want to resolve these mysteries—and perhaps you have already guessed how: by noticing, of course, that the "ground state" of the benzene ring is really a two-state system. We could imagine that the bonds in benzene could be in either of the two arrangements shown in Fig. 10–8. You say, "But they are really the same; they should have the same energy." Indeed, they should. And for that reason they must be analyzed as a two-state system. Each state represents a different configuration of the whole set of electrons, and there is some amplitude A that the whole bunch can switch from one arrangement to the other—there is a chance that the electrons can flip from one dance to the other.

As we have seen, this chance of flipping makes a mixed state whose energy is lower than you would calculate by looking separately at either of the two pictures in Fig. 10–8. Instead, there are two stationary states—one with an energy above and one with an energy below the expected value. So actually, the true normal state (lowest energy) of benzene is neither of the possibilities shown in Fig. 10–8, but it has the amplitude $1/\sqrt{2}$ to be in each of the states shown. It is the only state that is involved in the chemistry of benzene at normal temperatures. Incidentally, the upper state also exists; we can tell it is there because benzene has a strong absorption for ultraviolet light at the frequency $\omega = (E_I - E_{II})/\hbar$. You will remember that in ammonia, where the object flipping back and forth was three protons, the energy separation was in the microwave region. In benzene, the objects are electrons, and because they are much lighter, they find it easier to flip back and forth, which makes the coefficient A very much larger. The result is that the energy difference is much larger—about 1.5 ev, which is the energy of an ultraviolet photon.‡

What happens if we substitute bromine? Again the two "possibilities" (a) and (b) in Fig. 10–7 represent the two different electron configurations. The only difference is that the two base states we start with would have slightly different energies. The lowest energy stationary state will still involve a linear combination of the two states, but with unequal amplitudes. The amplitude for state $|1\rangle$ might have a value something like $\sqrt{2/3}$, say, whereas state $|2\rangle$ would have the magnitude

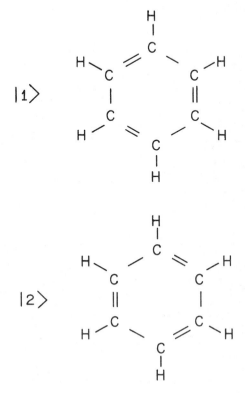

Fig. 10–8. A set of base states for the benzene molecule.

---

† We are oversimplifying a little. Originally, the chemists thought that there should be *four* forms of dibromobenzene: two forms with the bromines on adjacent carbon atoms (ortho-dibromobenzene), a third form with the bromines on next-nearest carbons (meta-dibromobenzene), and a fourth form with the bromines opposite to each other (para-dibromobenzene). However, they found only three forms—there is only *one* form of the ortho-molecule.

‡ What we have said is a little misleading. Absorption of ultraviolet light would be very weak in the two-state system we have taken for benzene, because the dipole moment matrix element between the two states is zero. [The two states are electrically symmetric, so in our formula Eq. (9.55) for the probability of a transition, the dipole moment μ is zero and no light is absorbed.] If these were the only states, the existence of the upper state would have to be shown in other ways. A more complete theory of benzene, however, which begins with more base states (such as those having adjacent double bonds) shows that the true stationary states of benzene are slightly distorted from the ones we have found. The resulting dipole moments permit the transition we mentioned in the text to occur by the absorption of ultraviolet light.

$\sqrt{1/3}$. We can't say for sure without more information, but once the two energies H_{11} and H_{22} are no longer equal, then the amplitudes C_1 and C_2 no longer have equal magnitudes. This means, of course, that one of the two possibilities in the figure is more likely than the other, but the electrons are mobile enough so that there is some amplitude for both. The other state has different amplitudes (like $\sqrt{1/3}$ and $-\sqrt{2/3}$) but lies at a higher energy. There is only one lowest state, not two as the naive theory of fixed chemical bonds would suggest.

10-5 Dyes

We will give you one more chemical example of the two-state phenomenon—this time on a larger molecular scale. It has to do with the theory of dyes. Many dyes—in fact, most artificial dyes—have an interesting characteristic; they have a kind of symmetry. Figure 10–9 shows an ion of a particular dye called magenta, which has a purplish red color. The molecule has three ring structures—two of which are benzene rings. The third is not exactly the same as a benzene ring because it has only two double bonds inside the ring. The figure shows two equally satisfactory pictures, and we would guess that they should have equal energies. But there is a certain amplitude that all the electrons can flip from one condition to the other, shifting the position of the "unfilled" position to the opposite end. With so many electrons involved, the flipping amplitude is somewhat lower than it is in the case of benzene, and the difference in energy between the two stationary states is smaller. There are, nevertheless, the usual two stationary states $|\,I\,\rangle$ and $|\,II\,\rangle$ which are the sum and difference combinations of the two base states shown in the figure. The energy separation of $|\,I\,\rangle$ and $|\,II\,\rangle$ comes out to be equal to the energy of a photon in the optical region. If one shines light on the molecule, there is a very strong absorption at one frequency, and it appears to be brightly colored. That's why it's a dye!

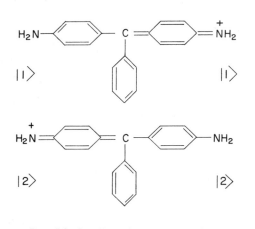

Fig. 10–9. Two base states for the molecule of the dye magenta.

Another interesting feature of such a dye molecule is that in the two base states shown, the center of electric charge is located at different places. As a result, the molecule should be strongly affected by an external electric field. We had a similar effect in the ammonia molecule. Evidently we can analyze it by using exactly the same mathematics, provided we know the numbers E_0 and A. Generally, these are obtained by gathering experimental data. If one makes measurements with many dyes, it is often possible to guess what will happen with some related dye molecule. Because of the large shift in the position of the center of electric charge the value of μ in formula (9.55) is large and the material has a high probability for absorbing light of the characteristic frequency $2A/\hbar$. Therefore, it is not only colored but very strongly so—a small amount of substance absorbs a lot of light.

The rate of flipping—and, therefore, A—is very sensitive to the complete structure of the molecule. By changing A, the energy splitting, and with it the color of the dye, can be changed. Also, the molecules do not have to be perfectly symmetrical. We have seen that the same basic phenomenon exists with slight modifications, even if there is some small asymmetry present. So, one can get some modification of the colors by introducing slight asymmetries in the molecules. For example, another important dye, malachite green, is very similar to magenta, but has two of the hydrogens replaced by CH_3. It's a different color because the A is shifted and the flip-flop rate is changed.

10-6 The Hamiltonian of a spin one-half particle in a magnetic field

Now we would like to discuss a two-state system involving an object of spin one-half. Some of what we will say has been covered in earlier chapters, but doing it again may help to make some of the puzzling points a little clearer. We can think of an electron at rest as a two-state system. Although we will be talking in this section about "an electron," what we find out will be true for *any* spin one-half particle. Suppose we choose for our base states $|\,1\,\rangle$ and $|\,2\,\rangle$ the states in which the z-component of the electron spin is $+\hbar/2$ and $-\hbar/2$.

These states are, of course, the same ones we have called (+) and (−) in earlier chapters. To keep the notation of this chapter consistent, though, we call the "plus" spin state $|1\rangle$ and the "minus" spin state $|2\rangle$—where "plus" and "minus" refer to the angular momentum in the z-direction.

Any possible state ψ for the electron can be described as in Eq. (10.1) by giving the amplitude C_1 that the electron is in state $|1\rangle$, and the amplitude C_2 that it is in state $|2\rangle$. To treat this problem, we will need to know the Hamiltonian for this two-state system—that is, for an electron in a magnetic field. We begin with the special case of a magnetic field in the z-direction.

Suppose that the vector $\boldsymbol{B}$ has only a z-component B_z. From the definition of the two base states (that is, spins parallel and antiparallel to $\boldsymbol{B}$) we know that they are already stationary states with a definite energy in the magnetic field. State $|1\rangle$ corresponds to an energy† equal to $-\mu B_z$ and state $|2\rangle$ to $+\mu B_z$. The Hamiltonian must be very simple in this case since C_1, the amplitude to be in state $|1\rangle$, is not affected by C_2, and vice versa:

$$i\hbar \frac{dC_1}{dt} = E_1 C_1 = -\mu B_z C_1,$$

$$i\hbar \frac{dC_2}{dt} = E_2 C = +\mu B_z C_2. \tag{10.17}$$

For this special case, the Hamiltonian is

$$H_{11} = -\mu B_z, \qquad H_{12} = 0,$$
$$H_{21} = 0, \qquad H_{22} = +\mu B_z. \tag{10.18}$$

So we know what the Hamiltonian is for the magnetic field in the z-direction, and we know the energies of the stationary states.

Now suppose the field is *not* in the z-direction. What is the Hamiltonian? How are the matrix elements changed if the field is not in the z-direction? We are going to make an assumption that there is a kind of superposition principle for the terms of the Hamiltonian. More specifically, we want to assume that if two magnetic fields are superposed, the terms in the Hamiltonian simply add—if we know the H_{ij} for a pure B_z and we know the H_{ij} for a pure B_x, then the H_{ij} for both B_z and B_x together is simply the sum. This is certainly true if we consider only fields in the z-direction—if we double B_z, then all the H_{ij} are doubled. So let's assume that H is linear in the field $\boldsymbol{B}$. That's all we need to be able to find the H_{ij} for any magnetic field.

Suppose we have a constant field $\boldsymbol{B}$. We *could* have chosen our z-axis in its direction, and we *would* have found two stationary states with the energies $\mp\mu B$. Just choosing our axes in a different direction won't change the *physics*. Our *description* of the stationary states will be different, but their energies *will still be* $\mp\mu B$—that is,

$$E_I = -\mu\sqrt{B_x^2 + B_y^2 + B_z^2}$$

and

$$E_{II} = +\mu\sqrt{B_x^2 + B_y^2 + B_z^2}. \tag{10.19}$$

The rest of the game is easy. We have here the formulas for the energies. We want a Hamiltonian which is linear in B_x, B_y, and B_z, and which will give these energies when used in our general formula of Eq. (10.3). The problem: find the Hamiltonian. First, notice that the energy splitting is symmetric, with an average value of zero. Looking at Eq. (10.3), we can see directly that that requires

$$H_{22} = -H_{11}.$$

(Note that this checks with what we already know when B_x and B_y are both zero;

---

† We are taking the rest energy $m_0 c^2$ as our "zero" of energy and treating the magnetic moment μ of the electron as a *negative* number, since it points opposite to the spin.

in that case $H_{11} = -\mu B_z$ and $H_{22} = \mu B_z$.) Now if we equate the energies of Eq. (10.3) with what we know from Eq. (10.19), we have

$$\left(\frac{H_{11} - H_{22}}{2}\right)^2 + |H_{12}|^2 = \mu^2(B_x^2 + B_y^2 + B_z^2). \qquad (10.20)$$

(We have also made use of the fact that $H_{21} = H_{12}^*$, so that $H_{12}H_{21}$ can also be written as $|H_{12}|^2$.) Again for the special case of a field in the z-direction, this gives

$$\mu^2 B_z^2 + |H_{12}|^2 = \mu^2 B_z^2.$$

Clearly, $|H_{12}|$ must be zero in this special case, which means that H_{12} cannot have any terms in B_z. (Remember, we have said that all terms must be linear in B_x, B_y, and B_z.)

So far, then, we have discovered that H_{11} and H_{22} have terms in B_z, while H_{12} and H_{21} do not. We can make a simple guess that will satisfy Eq. (10.20) if we say that

$$H_{11} = -\mu B_z,$$
$$H_{22} = \mu B_z, \qquad (10.21)$$

and
$$|H_{12}|^2 = \mu^2(B_x^2 + B_y^2).$$

And it turns out that that's the *only* way it can be done!

"Wait"—you say—"H_{12} is not linear in B; Eq. (10.21) gives $H_{12} = \mu\sqrt{B_x^2 + B_y^2}$." Not necessarily. There is another possibility which *is* linear, namely,

$$H_{12} = \mu(B_x + iB_y).$$

There are, in fact, several such possibilities—most generally, we could write

$$H_{12} = \mu(B_x \pm iB_y)e^{i\delta},$$

where δ is some arbitrary phase. Which sign and phase should we use? It turns out that you can choose either sign, and any phase you want, and the physical results will always be the same. So the choice is a matter of convention. People ahead of us have chosen to use the minus sign and to take $e^{i\delta} = -1$. We might as well follow suit and write

$$H_{12} = -\mu(B_x - iB_y), \qquad H_{21} = -\mu(B_x + iB_y).$$

(Incidentally, these conventions are related to, and consistent with, some of the arbitrary choices we made in Chapter 6.)

The complete Hamiltonian for an electron in an arbitrary magnetic field is, then

$$H_{11} = -\mu B_z, \qquad\qquad H_{12} = -\mu(B_x - iB_y),$$
$$H_{21} = -\mu(B_x + iB_y), \qquad H_{22} = +\mu B_z. \qquad (10.22)$$

And the equations for the amplitudes C_1 and C_2 are

$$i\hbar \frac{dC_1}{dt} = -\mu[B_z C_1 + (B_x - iB_y)C_2],$$

$$i\hbar \frac{dC_2}{dt} = -\mu[(B_x + iB_y)C_1 - B_z C_2]. \qquad (10.23)$$

So we have discovered the "equations of motion for the spin states" of an electron in a magnetic field. We guessed at them by making some physical argument, but the real test of any Hamiltonian is that it should give predictions in agreement with experiment. According to any tests that have been made, these equations are right. In fact, although we made our arguments only for constant fields, the Hamiltonian we have written is also right for magnetic fields which vary with time. So we can now use Eq. (10.23) to look at all kinds of interesting problems.

10–7 The spinning electron in a magnetic field

Example number one: We start with a constant field in the z-direction. There are just the two stationary states with energies $\mp\mu B_z$. Suppose we add a small field in the x-direction. Then the equations look like our old two-state problem. We get the flip-flop business once more, and the energy levels are split a little farther apart. Now let's let the x-component of the field vary with time—say, as $\cos\omega t$. The equations are then the same as we had when we put an oscillating electric field on the ammonia molecule in Chapter 9. You can work out the details in the same way. You will get the result that the oscillating field causes transitions from the $+z$-state to the $-z$-state—and vice versa—when the horizontal field oscillates near the resonant frequency $\omega_0 = 2\mu B_z/\hbar$. *This gives the quantum mechanical theory of the magnetic resonance phenomena we described in Chapter 35 of Volume II* (*see Appendix*).

It is also possible to make a maser which uses a spin one-half system. A Stern-Gerlach apparatus is used to produce a beam of particles polarized in, say, the $+z$-direction, which are sent into a cavity in a constant magnetic field. The oscillating fields in the cavity can couple with the magnetic moment and induce transitions which give energy to the cavity.

Now let's look at the following question. Suppose we have a magnetic field **B** which points in the direction whose polar angle is θ and azimuthal angle is ϕ, as in Fig. 10–10. Suppose, additionally, that there is an electron which has been prepared with its spin pointing along this field. What are the amplitudes C_1 and C_2 for such an electron? In other words, calling the state of the electron $|\psi\rangle$, we want to write

$$|\psi\rangle = |1\rangle C_1 + |2\rangle C_2,$$

where C_1 and C_2 are

$$C_1 = \langle 1|\psi\rangle, \qquad C_2 = \langle 2|\psi\rangle,$$

where by $|1\rangle$ and $|2\rangle$ we mean the same thing we used to call $|+\rangle$ and $|-\rangle$ (referred to our chosen z-axis).

The answer to this question is also in our general equations for two-state systems. First, we know that since the electron's spin is parallel to **B** it is in a stationary state with energy $E_I = -\mu B$. Therefore, both C_1 and C_2 must vary as $e^{-iE_I t/\hbar}$, as in (9.18); and their coefficients a_1 and a_2 are given by (10.5), namely,

$$\frac{a_1}{a_2} = \frac{H_{12}}{E_I - H_{11}}. \qquad (10.24)$$

An additional condition is that a_1 and a_2 should be normalized so that $|a_1|^2 + |a_2|^2 = 1$. We can take H_{11} and H_{12} from (10.22) using

$$B_z = B\cos\theta, \qquad B_x = B\sin\theta\cos\phi, \qquad B_y = B\sin\theta\sin\phi.$$

So we have

$$H_{11} = -\mu B\cos\theta,$$
$$H_{12} = -\mu B\sin\theta\,(\cos\phi - i\sin\phi). \qquad (10.25)$$

The last factor in the second equation is, incidentally, $e^{-i\phi}$, so it is simpler to write

$$H_{12} = -\mu B\sin\theta\, e^{-i\phi}. \qquad (10.26)$$

Using these matrix elements in Eq. (10.16)—and canceling $-\mu B$ from numerator and denominator—we find

$$\frac{a_1}{a_2} = \frac{\sin\theta\, e^{-i\phi}}{1 - \cos\theta}. \qquad (10.27)$$

With this ratio and the normalization condition, we can find both a_1 and a_2. That's not hard, but we can make a short cut with a little trick. Notice that

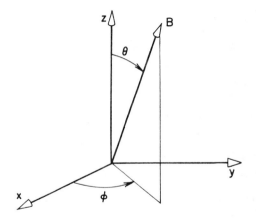

Fig. 10–10. The direction of B is defined by the polar angle θ and the azimuthal angle ϕ.

$1 - \cos \theta = 2 \sin^2 (\theta/2)$, and that $\sin \theta = 2 \sin (\theta/2) \cos (\theta/2)$. Then Eq. (10.27) is equivalent to

$$\frac{a_1}{a_2} = \frac{\cos \dfrac{\theta}{2} \, e^{-i\phi}}{\sin \dfrac{\theta}{2}} \, . \tag{10.28}$$

So one possible answer is

$$a_1 = \cos \frac{\theta}{2} e^{-i\phi}, \qquad a_2 = \sin \frac{\theta}{2}, \tag{10.29}$$

since it fits with (10.28) and also makes

$$|a_1|^2 + |a_2|^2 = 1.$$

As you know, multiplying both a_1 and a_2 by an arbitrary phase factor doesn't change anything. People generally prefer to make Eqs. (10.29) more symmetric by multiplying both by $e^{i\phi/2}$. So the form usually used is

$$a_1 = \cos \frac{\theta}{2} e^{-i\phi/2}, \qquad a_2 = \sin \frac{\theta}{2} e^{+i\phi/2}, \tag{10.30}$$

and this is the answer to our question. The numbers a_1 and a_2 are the amplitudes to find an electron with its spin up or down along the z-axis when we know that its spin is along the axis at θ and ϕ. (The amplitudes C_1 and C_2 are just a_1 and a_2 times $e^{-iE_I t/\hbar}$.)

Now we notice an interesting thing. The strength B of the magnetic field does not appear anywhere in (10.30). The result is clearly the same in the limit that B goes to zero. This means that we have answered *in general* the question of how to represent a particle whose spin is along an arbitrary axis. The amplitudes of (10.30) are the projection amplitudes for spin one-half particles corresponding to the projection amplitudes we gave in Chapter 5 [Eqs. (5.38)] for spin-one particles. We can now find the amplitudes for filtered beams of spin one-half particles to go through any particular Stern-Gerlach filter.

Let $|+z\rangle$ represent a state with spin up along the z-axis, and $|-z\rangle$ represent the spin down state. If $|+z'\rangle$ represents a state with spin up along a z'-axis which makes the polar angles θ and ϕ with the z-axis, then in the notation of Chapter 5, we have

$$\langle +z \mid +z'\rangle = \cos \frac{\theta}{2} e^{-i\phi/2}, \qquad \langle -z \mid +z'\rangle = \sin \frac{\theta}{2} e^{+i\phi/2}. \tag{10.31}$$

These results are equivalent to what we found in Chapter 6, Eq. (6.36), by purely geometrical arguments. (So if you decided to skip Chapter 6, you now have the essential results anyway.)

As our final example lets look again at one which we've already mentioned a number of times. Suppose that we consider the following problem. We start with an electron whose spin is in some given direction, then turn on a magnetic field in the z-direction for 25 minutes, and then turn it off. What is the final state? Again let's represent the state by the linear combination $|\psi\rangle = |1\rangle C_1 + |2\rangle C_2$. For this problem, however, the states of definite energy are also our base states $|1\rangle$ and $|2\rangle$. So C_1 and C_2 only vary in phase. We know that

$$C_1(t) = C_1(0)e^{-iE_I t/\hbar} = C_1(0)e^{+i\mu B t/\hbar},$$

and

$$C_2(t) = C_2(0)e^{-iE_{II} t/\hbar} = C_2(0)e^{-i\mu B t/\hbar}.$$

Now initially we said the electron spin was set in a given direction. That means that initially C_1 and C_2 are two numbers given by Eqs. (10.30). After we wait for a period of time T, the new C_1 and C_2 are the same two numbers multiplied respectively by $e^{i\mu B_z T/\hbar}$ and $e^{-i\mu B_z T/\hbar}$. What state is that? That's easy. It's exactly the same as if the angle ϕ had been changed by the subtraction of $2\mu B_z T/\hbar$ and the angle θ had been left unchanged. That means that at the end of the time

T, the state $|\psi\rangle$ represents an electron lined up in a direction which differs from the original direction only by a *rotation* about the z-axis through the angle $\Delta\phi = 2\mu B_z T/\hbar$. Since this angle is proportional to T, we can also say the direction of the spin *precesses* at the angular velocity $2\mu B_z/\hbar$ around the z-axis. This result we discussed several times previously in a less complete and rigorous manner. Now we have obtained a complete and accurate quantum mechanical description of the precession of atomic magnets.

It is interesting that the mathematical ideas we have just gone over for the spinning electron in a magnetic field can be applied to *any* two-state system. That means that by making a mathematical *analogy* to the spinning electron, *any problem* about two-state systems can be solved by pure geometry. It works like this. First you shift the zero of energy so that $(H_{11} + H_{22})$ is equal to zero so that $H_{11} = -H_{22}$. Then any two-state problem is *formally* the same as the electron in a magnetic field. All you have to do is *identify* $-\mu B_z$ with H_{11} and $-\mu(B_x - iB_y)$ with H_{12}. No matter what the physics is originally—an ammonia molecule, or whatever—you can translate it into a corresponding electron problem. So if we can solve the electron problem *in general*, we have solved *all* two-state problems.

And we have the general solution for the electron! Suppose you have some state to start with that has spin "up" in some direction, and you have a magnetic field B that points in some other direction. You just rotate the spin direction around the axis of B with the *vector* angular velocity $\omega(t)$ equal to a constant times the vector B (namely $\omega = 2\mu B/\hbar$). As B varies with time, you keep moving the axis of the rotation to keep it parallel with B, and keep changing the speed of rotation so that it is always proportional to the strength of B. See Fig. 10–11. If you keep doing this, you will end up with a certain final orientation of the spin axis, and the amplitudes C_1 and C_2 are just given by the projections—using (10.30)—into your coordinate frame. You see, it's just a geometric problem to keep track of where you end up after all the rotating. Although it's easy to see what's involved, this geometric problem (of finding the net result of a rotation with a varying angular velocity vector) is not easy to solve explicitly in the general case. Anyway, we see, *in principle*, the general solution to any two-state problem. In the next chapter we will look some more into the mathematical techniques for handling the important case of a spin one-half particle—and, therefore, for handling two-state systems in general.

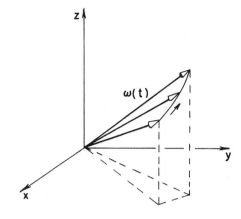

Fig. 10-11. The spin direction of an electron in a varying magnetic field $B(t)$ precesses at the frequency $\omega(t)$ about an axis parallel to B.

More Two-State Systems

11–1 The Pauli spin matrices

We continue our discussion of two-state systems. At the end of the last chapter we were talking about a spin one-half particle in a magnetic field. We described the spin state by giving the amplitude C_1 that the z-component of spin angular momentum is $+\hbar/2$ and the amplitude C_2 that it is $-\hbar/2$. In earlier chapters we have called these base states $|+\rangle$ and $|-\rangle$. We will now go back to that notation, although we may occasionally find it convenient to use $|+\rangle$ or $|1\rangle$, and $|-\rangle$ or $|2\rangle$, interchangeably.

We saw in the last chapter that when a spin one-half particle with a magnetic moment μ is in a magnetic field $\boldsymbol{B} = (B_x, B_y, B_z)$, the amplitudes $C_+(=C_1)$ and $C_-(=C_2)$ are connected by the following differential equations:

$$
i\hbar \frac{dC_+}{dt} = -\mu[B_z C_+ + (B_x - iB_y)C_-],
$$

$$
i\hbar \frac{dC_-}{dt} = -\mu[(B_x + iB_y)C_+ - B_z C_-].
$$

(11.1)

In other words, the Hamiltonian matrix H_{ij} is

$$
\begin{aligned}
H_{11} &= -\mu B_z, & H_{12} &= -\mu(B_x - iB_y), \\
H_{21} &= -\mu(B_x + iB_y), & H_{22} &= +\mu B_z.
\end{aligned}
$$

(11.2)

And Eqs. (11.1) are, of course, the same as

$$
i\hbar \frac{dC_i}{dt} = \sum_j H_{ij} C_j,
$$

(11.3)

where i and j take on the values $+$ and $-$ (or 1 and 2).

The two-state system of the electron spin is so important that it is very useful to have a neater way of writing things. We will now make a little mathematical digression to show you how people usually write the equations of a two-state system. It is done this way: First, note that each term in the Hamiltonian is proportional to μ and to some component of $\boldsymbol{B}$; we can then—*purely formally*—write that

$$
H_{ij} = -\mu[\sigma_{ij}^x B_x + \sigma_{ij}^y B_y + \sigma_{ij}^z B_z].
$$

(11.4)

There is no new physics here; this equation just means that the coefficients σ_{ij}^x, σ_{ij}^y, and σ_{ij}^z—there are $4 \times 3 = 12$ of them—can be figured out so that (11.4) is identical with (11.2).

Let's see what they have to be. We start with B_z. Since B_z appears only in H_{11} and H_{22}, everything will be O.K. if

$$
\begin{aligned}
\sigma_{11}^z &= 1, & \sigma_{12}^z &= 0, \\
\sigma_{21}^z &= 0, & \sigma_{22}^z &= -1.
\end{aligned}
$$

We often write the matrix H_{ij} as a little table like this:

$$
H_{ij} = {}^{i\downarrow} \overset{j\rightarrow}{\begin{pmatrix} H_{11} & H_{12} \\ H_{21} & H_{22} \end{pmatrix}}.
$$

11–1 The Pauli spin matrices

11–2 The spin matrices as operators

11–3 The solution of the two-state equations

11–4 The polarization states of the photon

11–5 The neutral K-meson†

11–6 Generalization to *N*-state systems

Review: Chapter 35, Vol. I, *Polarization*

---

† This section should be omitted on the first reading of this book. It is more advanced than is appropriate in a first course.

For the Hamiltonian of a spin one-half particle in the magnetic field B_z, this is the same as

$$H_{ij} = \begin{array}{l} \scriptstyle i \downarrow \end{array} \overset{\scriptstyle j \longrightarrow}{\begin{pmatrix} -\mu B_z & -\mu(B_x - iB_y) \\ -\mu(B_x + iB_y) & +\mu B_z \end{pmatrix}}.$$

In the same way, we can write the coefficients σ^z_{ij} as the matrix

$$\sigma^z_{ij} = \begin{array}{l} \scriptstyle i \downarrow \end{array} \overset{\scriptstyle j \longrightarrow}{\begin{pmatrix} 1 & 0 \\ 0 & -1 \end{pmatrix}}. \tag{11.5}$$

Working with the coefficients of B_x, we get that the terms of σ_x have to be

$$\sigma^x_{11} = 0, \qquad \sigma^x_{12} = 1,$$
$$\sigma^x_{21} = 1, \qquad \sigma^x_{22} = 0.$$

Or, in shorthand,

$$\sigma^x_{ij} = \begin{pmatrix} 0 & 1 \\ 1 & 0 \end{pmatrix}. \tag{11.6}$$

Finally, looking at B_y, we get

$$\sigma^y_{11} = 0, \qquad \sigma^y_{12} = -i,$$
$$\sigma^y_{21} = i, \qquad \sigma^y_{22} = 0;$$

or

$$\sigma^y_{ij} = \begin{pmatrix} 0 & -i \\ i & 0 \end{pmatrix}. \tag{11.7}$$

Table 11–1

The Pauli spin matrices

$$\sigma_z = \begin{pmatrix} 1 & 0 \\ 0 & -1 \end{pmatrix}$$

$$\sigma_x = \begin{pmatrix} 0 & 1 \\ 1 & 0 \end{pmatrix}$$

$$\sigma_y = \begin{pmatrix} 0 & -i \\ i & 0 \end{pmatrix}$$

$$1 = \begin{pmatrix} 1 & 0 \\ 0 & 1 \end{pmatrix}$$

With these three sigma matrices, Eqs. (11.2) and (11.4) are identical. To leave room for the subscripts i and j, we have shown which σ goes with which component of B by putting x, y, and z as superscripts. Usually, however, the i and j are omitted —it's easy to imagine they are there—and the x, y, z are written as subscripts. Then Eq. (11.4) is written

$$H = -\mu[\sigma_x B_x + \sigma_y B_y + \sigma_z B_z]. \tag{11.8}$$

Because the sigma matrices are so important—they are used all the time by the professionals—we have gathered them together in Table 11–1. (Anyone who is going to work in quantum physics really has to memorize them.) They are also called the *Pauli spin matrices* after the physicist who invented them.

In the table we have included one more two-by-two matrix which is needed if we want to be able to take care of a system which has two spin states of the same energy, or if we want to choose a different zero energy. For such situations we must add $E_0 C_+$ to the first equation in (11.1) and $E_0 C_-$ to the second equation. We can include this in the new notation if we define the *unit matrix* "1" as δ_{ij},

$$1 = \delta_{ij} = \begin{pmatrix} 1 & 0 \\ 0 & 1 \end{pmatrix}, \tag{11.9}$$

and rewrite Eq. (11.8) as

$$H = E_0 \delta_{ij} - \mu(\sigma_x B_x + \sigma_y B_y + \sigma_z B_z). \tag{11.10}$$

Usually, it is *understood* that any constant like E_0 is automatically to be multiplied by the unit matrix; then one writes simply

$$H = E_0 - \mu(\sigma_x B_x + \sigma_y B_y + \sigma_z B_z). \tag{11.11}$$

One reason the spin matrices are useful is that *any* two-by-two matrix at all can be written in terms of them. Any matrix you can write has four numbers in it, say,

$$M = \begin{pmatrix} a & b \\ c & d \end{pmatrix}.$$

It can always be written as a linear combination of four matrices. For example,

$$M = a \begin{pmatrix} 1 & 0 \\ 0 & 0 \end{pmatrix} + b \begin{pmatrix} 0 & 1 \\ 0 & 0 \end{pmatrix} + c \begin{pmatrix} 0 & 0 \\ 1 & 0 \end{pmatrix} + d \begin{pmatrix} 0 & 0 \\ 0 & 1 \end{pmatrix}.$$

There are many ways of doing it, but one special way is to say that M is a certain amount of σ_x, plus a certain amount of σ_y, and so on, like this:

$$M = \alpha 1 + \beta \sigma_x + \gamma \sigma_y + \delta \sigma_z,$$

where the "amounts" α, β, γ, and δ may, in general, be complex numbers.

Since any two-by-two matrix can be represented in terms of the unit matrix and the sigma matrices, we have all that we ever need for *any* two-state system. No matter what the two-state system—the ammonia molecule, the magenta dye, anything—the Hamiltonian equation can be written in terms of the sigmas. Although the sigmas seem to have a geometrical significance in the physical situation of an electron in a magnetic field, they can also be thought of as just useful matrices, which can be used for any two-state problem.

For instance, in one way of looking at things a proton and a neutron can be thought of as the same particle in either of two states. We say the *nucleon* (proton or neutron) is a two-state system—in this case, two states with respect to its charge. When looked at that way, the $| \, 1 \rangle$ state can represent the proton and the $| \, 2 \rangle$ state can represent the neutron. People say that the nucleon has two "isotopic-spin" states.

Since we will be using the sigma matrices as the "arithmetic" of the quantum mechanics of two-state systems, let's review quickly the conventions of matrix algebra. By the "sum" of any two or more matrices we mean just what was obvious in Eq. (11.4). In general, if we "add" two matrices A and B, the "sum" C means that each term C_{ij} is given by

$$C_{ij} = A_{ij} + B_{ij}.$$

Each term of C is the sum of the terms in the same slots of A and B.

In Section 5–6 we have already encountered the idea of a matrix "product." The same idea will be useful in dealing with the sigma matrices. In general, the "product" of two matrices A and B (in that order) is defined to be a matrix C whose elements are

$$C_{ij} = \sum_k A_{ik} B_{kj}. \tag{11.12}$$

It is the sum of products of terms taken in pairs from the ith row of A and the kth column of B. If the matrices are written out in tabular form as in Fig. 11–1, there is a good "system" for getting the terms of the product matrix. Suppose you are calculating C_{23}. You run your left index finger *along* the *second row* of A and your right index finger *down* the *third column* of B, multiplying each pair and adding as you go. We have tried to indicate how to do it in the figure.

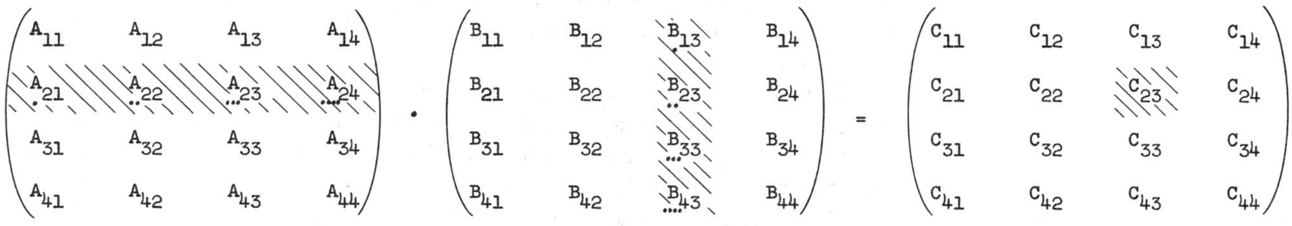

$$C_{ik} = \sum_j A_{ij} B_{jk}$$

Example: $C_{23} = A_{21} B_{13} + A_{22} B_{23} + A_{23} B_{33} + A_{24} B_{43}$

Fig. 11–1. Multiplying two matrices.

It is, of course, particularly simple for two-by-two matrices. For instance, if we multiply σ_x times σ_x, we get

$$\sigma_x^2 = \sigma_x \cdot \sigma_x = \begin{pmatrix} 0 & 1 \\ 1 & 0 \end{pmatrix} \cdot \begin{pmatrix} 0 & 1 \\ 1 & 0 \end{pmatrix} = \begin{pmatrix} 1 & 0 \\ 0 & 1 \end{pmatrix},$$

which is just the unit matrix 1. Or, for another example, let's work out $\sigma_x\sigma_y$:

$$\sigma_x\sigma_y = \begin{pmatrix} 0 & 1 \\ 1 & 0 \end{pmatrix} \cdot \begin{pmatrix} 0 & -i \\ i & 0 \end{pmatrix} = \begin{pmatrix} i & 0 \\ 0 & -i \end{pmatrix}.$$

Referring to Table 11–1, you see that the product is just i times the matrix σ_z. (Remember that a number times a matrix just multiplies each term of the matrix.) Since the products of the sigmas taken two at a time are important—as well as rather amusing—we have listed them all in Table 11–2. You can work them out as we have done for σ_x^2 and $\sigma_x\sigma_y$.

There's another very important and interesting point about these σ matrices. We can imagine, if we wish, that the three matrices σ_x, σ_y, and σ_z are analogous to the three components of a vector—it is sometimes called the "sigma vector" and is written $\boldsymbol{\sigma}$. It is really a "matrix vector" or a "vector matrix." It is three different matrices—one matrix associated with each axis, x, y, and z. With it, we can write the Hamiltonian of the system in a nice form which works in any coordinate system:

$$H = -\mu\boldsymbol{\sigma} \cdot \boldsymbol{B}. \tag{11.13}$$

Although we have written our three matrices in the representation in which "up" and "down" are in the z-direction—so that σ_z has a particular simplicity—we could figure out what the matrices would look like in some other representation. Although it takes a lot of algebra, you can show that they change among themselves like the components of a vector. (We won't, however, worry about proving it right now. You can check it if you want.) You can use $\boldsymbol{\sigma}$ in different coordinate systems as though it is a vector.

You remember that the H is related to energy in quantum mechanics. It is, in fact, just equal to the energy in the simple situation where there is only one state. Even for two-state systems of the electron spin, when we write the Hamiltonian as in Eq. (11.13), it looks very much like the *classical* formula for the energy of a little magnet with magnetic moment $\boldsymbol{\mu}$ in a magnetic field. $\boldsymbol{B}$ Classically, we would say

$$U = -\boldsymbol{\mu} \cdot \boldsymbol{B}, \tag{11.14}$$

where $\boldsymbol{\mu}$ is the property of the object and $\boldsymbol{B}$ is an external field. We can imagine that Eq. (11.14) can be converted to (11.13) if we replace the classical energy by the Hamiltonian and the classical $\boldsymbol{\mu}$ by the matrix $\mu\boldsymbol{\sigma}$. Then, after this purely formal substitution, we interpret the result as a matrix equation. It is sometimes said that to each quantity in classical physics there corresponds a matrix in quantum mechanics. It is really more correct to say that the Hamiltonian matrix corresponds to the energy, and any quantity that can be defined via energy has a corresponding matrix.

For example, the magnetic moment can be defined via energy by saying that the energy in an external field $\boldsymbol{B}$ is $-\boldsymbol{\mu} \cdot \boldsymbol{B}$. This *defines* the magnetic moment vector $\boldsymbol{\mu}$. Then we look at the formula for the Hamiltonian of a real (quantum) object in a magnetic field and try to identify whatever the matrices are that correspond to the various quantities in the classical formula. That's the trick by which *sometimes* classical quantities have their quantum counterparts.

You may try, if you want, to understand how a classical vector is equal to a matrix $\mu\boldsymbol{\sigma}$, and maybe you will discover something—but don't break your head on it. That's not the idea—they are *not equal*. Quantum mechanics is a different kind of a theory to represent the world. It just happens that there are certain correspondences which are hardly more than mnemonic devices—things to remember with. That is, you remember Eq. (11.14) when you learn classical physics;

Table 11–2

Products of the spin matrices

$$\sigma_x^2 = 1$$
$$\sigma_y^2 = 1$$
$$\sigma_z^2 = 1$$
$$\sigma_x\sigma_y = -\sigma_y\sigma_x = i\sigma_z$$
$$\sigma_y\sigma_z = -\sigma_z\sigma_y = i\sigma_x$$
$$\sigma_z\sigma_x = -\sigma_x\sigma_z = i\sigma_y$$

then if you remember the correspondence $\mu \rightarrow \mu\sigma$, you have a handle for remembering Eq. (11.13). Of course, nature knows the quantum mechanics, and the classical mechanics is only an approximation; so there is no mystery in the fact that in classical mechanics there is some shadow of quantum mechanical laws— which are truly the ones underneath. To reconstruct the original object from the shadow is not possible in any direct way, but the shadow does help you to remember what the object looks like. Equation (11.13) is the truth, and Eq. (11.14) is the shadow. Because we learn classical mechanics first, we would like to be able to get the quantum formula from it, but there is no sure-fire scheme for doing that. We must always go back to the real world and discover the correct quantum mechanical equations. When they come out looking like something in classical physics, we are in luck.

If the warnings above seem repetitious and appear to you to be belaboring self-evident truths about the relation of classical physics to quantum physics, please excuse the conditioned reflexes of a professor who has usually taught quantum mechanics to students who hadn't heard about Pauli spin matrices until they were in graduate school. Then they always seemed to be hoping that, somehow, quantum mechanics could be seen to follow as a logical consequence of classical mechanics which they had learned thoroughly years before. (Perhaps they wanted to avoid having to learn something new.) You have learned the classical formula, Eq. (11.14), only a few months ago—and then with warnings that it was inadequate—so maybe you will not be so unwilling to take the quantum formula, Eq. (11.13), as the basic truth.

11–2 The spin matrices as operators

While we are on the subject of mathematical notation, we would like to describe still *another* way of writing things—a way which is used very often because it is so compact. It follows directly from the notation introduced in Chapter 8. If we have a system in a state $|\psi(t)\rangle$, which varies with time, we can—as we did in Eq. (8.31)—write the amplitude that the system would be in the state $|i\rangle$ at $t + \Delta t$ as

$$\langle i \mid \psi(t + \Delta t)\rangle = \sum_j \langle i \mid U(t, t + \Delta t) \mid j\rangle\langle j \mid \psi(t)\rangle.$$

The matrix element $\langle i \mid U(t, t + \Delta t) \mid j\rangle$ is the amplitude that the base state $|j\rangle$ will be converted into the base state $|i\rangle$ in the time interval Δt. We then *defined* H_{ij} by writing

$$\langle i \mid U(t, t + \Delta t) \mid j\rangle = \delta_{ij} - \frac{i}{\hbar} H_{ij}(t)\,\Delta t,$$

and we showed that the amplitudes $C_i(t) = \langle i \mid \psi(t)\rangle$ were related by the differential equations

$$i\hbar \frac{dC_i}{dt} = \sum_j H_{ij}C_j. \tag{11.15}$$

If we write out the amplitudes C_i explicitly, the same equation appears as

$$i\hbar \frac{d}{dt} \langle i \mid \psi\rangle = \sum_j H_{ij}\langle j \mid \psi\rangle. \tag{11.16}$$

Now the matrix elements H_{ij} are also amplitudes which we can write as $\langle i \mid H \mid j\rangle$; our differential equation looks like this:

$$i\hbar \frac{d}{dt} \langle i \mid \psi\rangle = \sum_j \langle i \mid H \mid j\rangle\langle j \mid \psi\rangle. \tag{11.17}$$

We see that $-i/\hbar\,\langle i \mid H \mid j\rangle$ is the amplitude that—under the physical conditions described by H—a state $|j\rangle$ will, during the time dt, "generate" the state $|i\rangle$. (All of this is implicit in the discussion of Section 8–4.)

Now following the ideas of Section 8–2, we can drop out the common term $\langle i |$ in Eq. (11.17)—since it is true for any state $| i \rangle$—and write that equation simply as

$$i\hbar \frac{d}{dt} | \psi \rangle = \sum_j H | j \rangle\langle j | \psi \rangle. \qquad (11.18)$$

Or, going one step further, we can also remove the j and write

$$i\hbar \frac{d}{dt} | \psi \rangle = H | \psi \rangle. \qquad (11.19)$$

In Chapter 8 we pointed out that when things are written this way, the H in $H | j \rangle$ or $H | \psi \rangle$ is called an *operator*. From now on we will put the little hat (^) over an operator to remind you that it *is* an operator and not just a number. We will write $\hat{H} | \psi \rangle$. Although the two equations (11.18) and (11.19) *mean exactly the same thing* as Eq. (11.17) or Eq. (11.15), we can *think* about them in a different way. For instance, we would describe Eq. (11.18) in this way: "The time derivative of the *state vector* $| \psi \rangle$ is equal to what you get by operating with the Hamiltonian *operator* $\hat{H}$ on each base state, multiplying by the amplitude $\langle j | \psi \rangle$ that ψ is in the state j, and summing over all j." Or Eq. (11.19) is described this way. "The time derivative (times $i\hbar$) of a state $| \psi \rangle$ is equal to what you get if you operate with the Hamiltonian $\hat{H}$ on the state vector $| \psi \rangle$." It's just a short-hand way of saying what is in Eq. (11.17), but, as you will see, it can be a great convenience.

If we wish, we can carry the "abstraction" idea one more step. Equation (11.19) is true for *any state* $| \psi \rangle$. Also the left-hand side, $i\hbar d/dt$, is also an operator —it's the operation "differentiate by t and multiply by $i\hbar$." Therefore, Eq. (11.19) can also be thought of as an equation between operators—the operator equation

$$i\hbar \frac{d}{dt} = \hat{H}.$$

The Hamiltonian operator (within a constant) produces the same result as does d/dt when acting on any state. Remember that this equation—as well as Eq. (11.19)—is *not* a statement that the $\hat{H}$ operator is just the identical *operation* as d/dt. The equations are the dynamical law of nature—the law of motion—for a quantum system.

Just to get some practice with these ideas, we will show you another way we could get to Eq. (11.18). You know that we can write any state $| \psi \rangle$ in terms of its projections into some base set [see Eq. (8.8)],

$$| \psi \rangle = \sum_i | i \rangle\langle i | \psi \rangle. \qquad (11.20)$$

How does $| \psi \rangle$ change with time? Well, just take its derivative:

$$\frac{d}{dt} | \psi \rangle = \frac{d}{dt} \sum_i | i \rangle\langle i | \psi \rangle. \qquad (11.21)$$

Now the base states $| i \rangle$ do not change with time (at least *we* are always taking them as definite fixed states), but the amplitudes $\langle i | \psi \rangle$ are numbers which may vary. So Eq. (11.21) becomes

$$\frac{d}{dt} | \psi \rangle = \sum_i | i \rangle \frac{d}{dt} \langle i | \psi \rangle. \qquad (11.22)$$

Since we know $d\langle i | \psi \rangle/dt$ from Eq. (11.16), we get

$$\frac{d}{dt} | \psi \rangle = -\frac{i}{\hbar} \sum_i | i \rangle \sum_j H_{ij}\langle j | \psi \rangle$$

$$= -\frac{i}{\hbar} \sum_{ij} | i \rangle\langle i | H | j \rangle\langle j | \psi \rangle = -\frac{i}{\hbar} \sum_j H | j \rangle\langle j | \psi \rangle.$$

This is Eq. (11.18) all over again.

So we have many ways of looking at the Hamiltonian. We can think of the set of coefficients H_{ij} as just a bunch of numbers, or we can think of the "amplitudes" $\langle i \mid H \mid j \rangle$, or we can think of the "matrix" H_{ij}, or we can think of the operator" $\hat{H}$. It all means the same thing.

Now let's go back to our two-state systems. If we write the Hamiltonian in terms of the sigma matrices (with suitable numerical coefficients like B_x, etc.), we can clearly also think of σ_{ij}^x as an amplitude $\langle i \mid \sigma_x \mid j \rangle$ or, for short, as the operator $\hat{\sigma}_x$. If we use the operator idea, we can write the equation of motion of a state $\mid \psi \rangle$ in a magnetic field as

$$i\hbar \frac{d}{dt} \mid \psi \rangle = -\mu(B_x\hat{\sigma}_x + B_y\hat{\sigma}_y + B_z\hat{\sigma}_z) \mid \psi \rangle. \tag{11.23}$$

When we want to "use" such an equation we will normally have to express $\mid \psi \rangle$ in terms of base vectors (just as we have to find the components of space vectors when we want specific numbers). So we will usually want to put Eq. (11.23) in the somewhat expanded form:

$$i\hbar \frac{d}{dt} \mid \psi \rangle = -\mu \sum_i (B_z\hat{\sigma}_x + B_y\hat{\sigma}_y + B_z\hat{\sigma}_z) \mid i \rangle \langle i \mid \psi \rangle. \tag{11.24}$$

Now you will see why the operator idea is so neat. To use Eq. (11.24) we need to know what happens when the $\hat{\sigma}$ operators work on each of the base states. Let's find out. Suppose we have $\hat{\sigma}_z \mid + \rangle$; it is some vector $\mid ? \rangle$, but what? Well, let's multiply it on the left by $\langle + \mid$; we have

$$\langle + \mid \hat{\sigma}_z \mid + \rangle = \sigma_{11} = 1$$

(using Table 11-1). So we know that

$$\langle + \mid ? \rangle = 1. \tag{11.25}$$

Now let's multiply $\hat{\sigma}_z \mid + \rangle$ on the left by $\langle - \mid$. We get

$$\langle - \mid \hat{\sigma}_z \mid + \rangle = \sigma_{21} = 0;$$

so

$$\langle - \mid ? \rangle = 0. \tag{11.26}$$

There is only one state vector that satisfies both (11.25) and (11.26); it is $\mid + \rangle$. We discover then that

$$\hat{\sigma}_z \mid + \rangle = \mid + \rangle. \tag{11.27}$$

By this kind of argument you can easily show that all of the properties of the sigma matrices can be described in the operator notation by the set of rules given in Table 11-3.

If we have products of sigma matrices, they go over into products of operators. When two operators appear together as a product, you carry out first the operation with the operator which is farthest to the right. For instance, by $\hat{\sigma}_x\hat{\sigma}_y \mid + \rangle$ we are to understand $\hat{\sigma}_x(\hat{\sigma}_y \mid + \rangle)$. From Table 11-3, we get $\hat{\sigma}_y \mid + \rangle = i \mid - \rangle$, so

$$\hat{\sigma}_x\hat{\sigma}_y \mid + \rangle = \hat{\sigma}_x(i \mid - \rangle). \tag{11.28}$$

Now any number—like i—just moves through an operator (operators work only on state vectors); so Eq. (11.28) is the same as

$$\hat{\sigma}_x\hat{\sigma}_y \mid + \rangle = i\hat{\sigma}_x \mid - \rangle = i \mid + \rangle.$$

If you do the same thing for $\hat{\sigma}_x\hat{\sigma}_y \mid - \rangle$, you will find that

$$\hat{\sigma}_x\hat{\sigma}_y \mid - \rangle = -i \mid - \rangle.$$

Looking at Table 11-3, you see that $\hat{\sigma}_x\hat{\sigma}_y$ operating on $\mid + \rangle$ or $\mid - \rangle$ gives just what you get if you operate with $\hat{\sigma}_z$ and multiply by $-i$. We can, therefore, say

Table 11-3

Properties of the $\hat{\sigma}$-operator

$$\sigma_z \mid + \rangle = \mid + \rangle$$
$$\sigma_z \mid - \rangle = - \mid - \rangle$$
$$\sigma_x \mid + \rangle = \mid - \rangle$$
$$\sigma_x \mid - \rangle = \mid + \rangle$$
$$\sigma_y \mid + \rangle = i \mid - \rangle$$
$$\sigma_y \mid - \rangle = -i \mid + \rangle$$

that the operation $\hat{\sigma}_x\hat{\sigma}_y$ is identical with the operation $i\hat{\sigma}_z$, and write this statement as an operator equation:

$$\hat{\sigma}_x\hat{\sigma}_y = i\hat{\sigma}_z. \tag{11.29}$$

Notice that this equation is identical with one of our matrix equations of Table 11–2. So again we see the correspondence between the matrix and operator points of view. Each of the equations in Table 11–2 can, therefore, also be considered as equations about the sigma operators. You can check that they do indeed follow from Table 11–3. It is best, when working with these things, *not* to keep track of whether a quantity like σ or H is an operator or a matrix. All the equations are the same either way, so Table 11–2 is for sigma operators, or for sigma matrices, as you wish.

11–3 The solution of the two-state equations

We can now write our two-state equation in various forms, for example, either as

$$i\hbar\frac{dC_i}{dt} = \sum_j H_{ij}C_j$$

or $\tag{11.30}$

$$i\hbar\frac{d\,|\,\psi\rangle}{dt} = \hat{H}\,|\,\psi\rangle.$$

They both mean the same thing. For a spin one-half particle in a magnetic field, the Hamiltonian H is given by Eq. (11.8) or by Eq. (11.13).

If the field is in the z-direction, then—as we have seen several times by now—the solution is that the state $|\,\psi\rangle$, whatever it is, precesses around the z-axis (just as if you were to take the physical object and rotate it bodily around the z-axis) at an angular velocity equal to twice the magnetic field times $\mu/\hbar$. The same is true, of course, for a magnetic field along any other direction, because the physics is independent of the coordinate system. If we have a situation where the magnetic field varies from time to time in a complicated way, then we can analyze the situation in the following way. Suppose you start with the spin in the $+z$-direction and you have an x-magnetic field. The spin starts to turn. Then if the x-field is turned off, the spin stops turning. Now if a z-field is turned on, the spin precesses about z, and so on. So depending on how the fields vary in time, you can figure out what the final state is—along what axis it will point. Then you can refer that state back to the original $|+\rangle$ and $|-\rangle$ with respect to z by using the projection formulas we had in Chapter 10 (or Chapter 6). If the state ends up with its spin in the direction (θ, ϕ), it will have an up-amplitude $\cos(\theta/2)e^{-i\phi/2}$ and a down-amplitude $\sin(\theta/2)e^{+i\phi/2}$. That solves any problem. It is a word description of the solution of the differential equations.

The solution just described is sufficiently general to take care of *any two-state system*. Let's take our example of the ammonia molecule—including the effects of an electric field. If we describe the system in terms of the states $|\,I\rangle$ and $|\,II\rangle$, the equations look like this:

$$i\hbar\frac{dC_I}{dt} = +AC_I + \mu\mathcal{E}C_{II},$$

$$\tag{11.31}$$

$$i\hbar\frac{dC_{II}}{dt} = -AC_{II} + \mu\mathcal{E}C_I.$$

You say, "No, I remember there was an E_0 in there." Well, we have shifted the origin of energy to make the E_0 zero. (You can always do that by changing both amplitudes by the same factor—$e^{iE_0T/\hbar}$—and get rid of any constant energy.) Now if corresponding equations always have the same solutions, then we really don't have to do it twice. If we look at these equations and look at Eq. (11.1), then we can make the following identification. Let's call $|\,I\rangle$ the state $|+\rangle$ and $|\,II\rangle$ the state $|-\rangle$. That *does not mean* that we are lining-up the ammonia in space, or that $|+\rangle$ and $|-\rangle$ has anything to do with the z-axis. It is purely artificial.

We have an artificial space that we might "call the ammonia molecule representative space," or something—a three-dimensional "diagram" in which being "up" corresponds to having the molecule in the state $|I\rangle$ and being "down" along this false z-axis represents having a molecule in the state $|II\rangle$. Then, the equations will be identified as follows. First of all, you see that the Hamiltonian can be written in terms of the sigma matrices as

$$H = +A\sigma_z + \mu\mathcal{E}\sigma_x. \tag{11.32}$$

Or, putting it another way, μB_z in Eq. (11.1) corresponds to $-A$ in Eq. (11.32), and μB_x corresponds to $-\mu\mathcal{E}$. In our "model" space, then, we have a constant B field along the z-direction. If we have an electric field $\mathcal{E}$ which is changing with time, then we have a B field along the x-direction which varies in proportion. *So the behavior of an electron in a magnetic field with a constant component in the z-direction and an oscillating component in the x-direction is mathematically analogous and corresponds exactly to the behavior of an ammonia molecule in an oscillating electric field.* Unfortunately, we do not have the time to go any further into the details of this correspondence, or to work out any of the technical details. We only wished to make the point that *all* systems of two states can be made analogous to a spin one-half object precessing in a magnetic field.

11–4 The polarization states of the photon

There are a number of other two-state systems which are interesting to study, and the first new one we would like to talk about is the photon. To describe a photon we must first give its vector momentum. For a free photon, the frequency is determined by the momentum, so we don't have to say also what the frequency is. After that, though, we still have a property called the polarization. Imagine that there is a photon coming at you with a definite monochromatic frequency (which will be kept the same throughout all this discussion so that we don't have a variety of momentum states). Then there are two directions of polarization. In the classical theory, light can be described as having an electric field which oscillates horizontally or an electric field which oscillates vertically (for instance); these two kinds of light are called x-polarized and y-polarized light. The light can also be polarized in some other direction, which can be made up from the superposition of a field in the x-direction and one in the y-direction. Or if you take the x- and the y-components out of phase by 90°, you get an electric field that rotates—the light is elliptically polarized. (This is just a quick reminder of the classical theory of polarized light that we studied in Chapter 35, Vol. I.)

Now, however, suppose we have a *single* photon—just one. There is no electric field that we can discuss in the same way. All we have is *one photon*. But a photon has to have the analog of the classical phenomena of polarization. There must be at least two different kinds of photons. At first, you might think there should be an infinite variety—after all, the electric vector can point in all sorts of directions. We can, however, describe the polarization of a photon as a two-state system. A photon can be in the state $|x\rangle$ or in the state $|y\rangle$. By $|x\rangle$ we mean the polarization state of each one of the photons in a beam of light which *classically* is x-polarized light. On the other hand, by $|y\rangle$ we mean the polarization state of each of the photons in a y-polarized beam. And we can take $|x\rangle$ and $|y\rangle$ as our base states of a photon of given momentum pointing at you—in what we will call the z-direction. So there are two base states $|x\rangle$ and $|y\rangle$, and they are all that are needed to describe any photon at all.

For example, if we have a piece of polaroid set with its axis to pass light polarized in what we call the x-direction, and we send in a photon which we know is in the state $|y\rangle$, it will be absorbed by the polaroid. If we send in a photon which we know is in the state $|x\rangle$, it will come right through as $|x\rangle$. If we take a piece of calcite which takes a beam of polarized light and splits it into an $|x\rangle$ beam and a $|y\rangle$ beam, that piece of calcite is the complete analog of a Stern-Gerlach apparatus which splits a beam of silver atoms into the two states $|+\rangle$ and $|-\rangle$. So every-

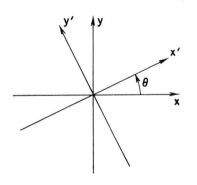

Fig. 11-2. Coordinates at right angles to the momentum vector of the photon.

thing we did before with particles and Stern-Gerlach apparatuses, we can do again with light and pieces of calcite. And what about light filtered through a piece of polaroid set at an angle θ? Is that another state? Yes, indeed, it *is* another state. Let's call the axis of the polaroid x' to distinguish it from the axes of our base states. See Fig. 11-2. A photon that comes out will be in the state $|x'\rangle$. However, any state can be represented as a linear combination of base states, and the formula for the combination is, here,

$$|x'\rangle = \cos\theta\,|x\rangle + \sin\theta\,|y\rangle. \tag{11.33}$$

That is, if a photon comes through a piece of polaroid set at the angle θ (with respect to x), it can still be resolved into $|x\rangle$ and $|y\rangle$ beams—by a piece of calcite, for example. Or you can, if you wish, just analyze it into x- and y-components in your imagination. Either way, you will find the amplitude $\cos\theta$ to be in the $|x\rangle$ state and the amplitude $\sin\theta$ to be in the $|y\rangle$ state.

Now we ask this question: Suppose a photon is polarized in the x'-direction by a piece of polaroid set at the angle θ and arrives at a polaroid at the angle zero—as in Fig. 11-3; what will happen? With what probability will it get through? The answer is the following. After it gets through the first polaroid, it is definitely in the state $|x'\rangle$. The second polaroid will let the photon through if it is in the state $|x\rangle$ (but absorb it if it is the state $|y\rangle$). So we are asking with what probability does the photon appear to be in the state $|x\rangle$? We get that probability from the absolute square of amplitude $\langle x\,|\,x'\rangle$ that a photon in the state $|x'\rangle$ is also in the state $|x\rangle$. What is $\langle x\,|\,x'\rangle$? Just multiply Eq. (11.33) by $\langle x|$ to get

$$\langle x\,|\,x'\rangle = \cos\theta\,\langle x\,|\,x\rangle + \sin\theta\,\langle x\,|\,y\rangle.$$

Now $\langle x\,|\,y\rangle = 0$, from the physics—as they *must* be if $|x\rangle$ and $|y\rangle$ are base states —and $\langle x\,|\,x\rangle = 1$. So we get

$$\langle x\,|\,x'\rangle = \cos\theta,$$

and the probability is $\cos^2\theta$. For example, if the first polaroid is set at $30°$, a photon will get through $3/4$ of the time, and $1/4$ of the time it will heat the polaroid by being absorbed therein.

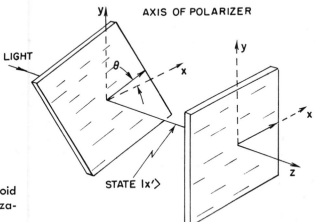

Fig. 11-3. Two sheets of polaroid with angle θ between planes of polarization.

Now let us see what happens classically in the same situation. We would have a beam of light with an electric field which is varying in some way or another—say "unpolarized." After it gets through the first polaroid, the electric field is oscillating in the x'-direction with a size $\mathcal{E}$; we would draw the field as an oscillating vector with a peak value $\mathcal{E}_0$ in a diagram like Fig. 11-4. Now when the light arrives at the second polaroid, only the x-component, $\mathcal{E}_0 \cos\theta$, of the electric field gets through. The *intensity* is proportional to the square of the field and, therefore, to $\mathcal{E}_0^2 \cos^2\theta$. So the energy coming through is $\cos^2\theta$ weaker than the energy which was entering the last polaroid.

11-10

The classical picture and the quantum picture give similar results. If you were to throw 10 billion photons at the second polaroid, and the average probability of each one going through is, say, 3/4, you would expect 3/4 of 10 billion would get through. Likewise, the energy that they would carry would be 3/4 of the energy that you attempted to put through. The classical theory says nothing about the statistics of the thing—it simply says that the energy that comes through will be precisely 3/4 of the energy which you were sending in. That is, of course, impossible if there is only one photon. There is no such thing as 3/4 of a photon. It is either *all* there, or it isn't there at all. Quantum mechanics tells us it is *all* there *3/4 of the time*. The relation of the two theories is clear.

What about the other kinds of polarization? For example, right-hand circular polarization? In the classical theory, right-hand circular polarization has equal components in x and y which are 90° out of phase. In the quantum theory, a right-hand circularly polarized (RHC) photon has equal amplitudes to be polarized $|x\rangle$ or $|y\rangle$, and the *amplitudes* are 90° out of phase. Calling a RHC photon a state $|R\rangle$ and a LHC photon a state $|L\rangle$, we can write (see Vol. I, Section 33-1)

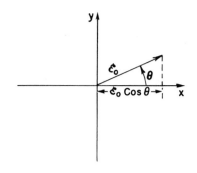

Fig. 11-4. The classical picture of the electric vector $\mathcal{E}$.

$$|R\rangle = \frac{1}{\sqrt{2}}\,(|x\rangle + i\,|y\rangle),$$

$$|L\rangle = -\frac{1}{\sqrt{2}}\,(|x\rangle - i\,|y\rangle). \qquad (11.34)$$

—the $1/\sqrt{2}$ is put in to get normalized states. With these states you can calculate any filtering or interference effects you want, using the laws of quantum theory. If you want, you can also choose $|R\rangle$ and $|L\rangle$ as base states and represent everything in terms of them. You only need to show first that $\langle R\,|\,L\rangle = 0$—which you can do by taking the conjugate form of the first equation above [see Eq. (8.13)] and multiplying it by the other. You can resolve light into x- and y-polarizations, or into x'- and y'-polarizations, or into right and left polarizations as a basis.

Just as an example, let's try to turn our formulas around. Can we represent the state $|x\rangle$ as a linear combination of right and left? Yes, here it is:

$$|x\rangle = \frac{1}{\sqrt{2}}\,(|R\rangle + |L\rangle),$$

$$|y\rangle = -\frac{i}{\sqrt{2}}\,(|R\rangle - |L\rangle). \qquad (11.35)$$

Proof: Add and subtract the two equations in (11.34). It is easy to go from one base to the other.

One curious point has to be made, though. If a photon is right circularly polarized, it shouldn't have anything to do with the x- and y-axes. If we were to look at the same thing from a coordinate system turned at some angle about the direction of flight, the light would still be right circularly polarized—and similarly for left. The right and left circularly polarized light are the same for any such rotation; the definition is independent of any choice of the x-direction (except that the photon direction is given). Isn't that nice—it doesn't take any axes to define it. Much better than x and y. On the other hand, isn't it rather a miracle that when you *add* the right and left together you can find out which direction x was? If "right" and "left" do not depend on x in any way, how is it that we can put them back together again and get x? We can answer that question in part by writing out the state $|R'\rangle$, which represents a photon RHC polarized in the frame x', y'. In that frame, you would write

$$|R'\rangle = \frac{1}{\sqrt{2}}\,(|x'\rangle + i\,|y'\rangle).$$

How does such a state look in the frame x, y? Just substitute x' from Eq. (11.33) and the corresponding $|y'\rangle$—we didn't write it down, but it is $(-\sin\theta)|x\rangle + (\cos\theta)|y\rangle$. Then

$$|R'\rangle = \frac{1}{\sqrt{2}}[\cos\theta\,|x\rangle + \sin\theta\,|y\rangle - i\sin\theta\,|x\rangle + i\cos\theta\,|y\rangle]$$

$$= \frac{1}{\sqrt{2}}[(\cos\theta - i\sin\theta)\,|x\rangle + i(\cos\theta - i\sin\theta)\,|y\rangle]$$

$$= \frac{1}{\sqrt{2}}(|x\rangle + i\,|y\rangle)(\cos\theta - i\sin\theta).$$

The first term is just $|R\rangle$, and the second is $e^{+i\theta}$; our result is that

$$|R'\rangle = e^{-i\theta}|R\rangle. \tag{11.36}$$

The states $|R'\rangle$ and $|R\rangle$ are the same except for the phase factor $e^{-i\theta}$. If you work out the same thing for $|L'\rangle$, you get that†

$$|L'\rangle = e^{+i\theta}|L\rangle. \tag{11.37}$$

Now you see what happens. If we add $|R\rangle$ and $|L\rangle$, we get something different from what we get when we add $|R'\rangle$ and $|L'\rangle$. For instance, an x-polarized photon is [Eq. (11.35)] the sum of $|R\rangle$ and $|L\rangle$, but a y-polarized photon is the sum with the phase of one shifted 90° backward and the other 90° forward. That is just what we would get from the sum of $|R'\rangle$ and $|L'\rangle$ for the special angle $\theta = 90°$, and that's right. An x-polarization in the *prime* frame is the same as a y-polarization in the original frame. So it is not exactly true that a circularly polarized photon looks the same for any set of axes. Its *phase* (the phase relation of the right and left circularly polarized states) keeps track of the x-direction.

11–5 The neutral K-meson‡

We will now describe a two-state system in the world of the strange particles—a system for which quantum mechanics gives a most remarkable prediction. To describe it completely would involve us in a lot of stuff about strange particles, so we will, unfortunately, have to cut some corners. We can only give an outline of how a certain discovery was made—to show you the kind of reasoning that was involved. It begins with the discovery by Gell-Mann and Nishijima of the concept of *strangeness* and of a new law of *conservation of strangeness*. It was when Gell-Mann and Pais were analyzing the consequences of these new ideas that they came across the prediction of a most remarkable phenomenon we are going to describe. First, though, we have to tell you a little about "strangeness."

We must begin with what are called the *strong interactions* of nuclear particles. These are the interactions which are responsible for the strong nuclear forces—as distinct, for instance, from the relatively weaker electromagnetic interactions. The interactions are "strong" in the sense that if two particles get close enough to interact at all, they interact in a big way and produce other particles very easily.

---

† It's similar to what we found (in Chapter 6) for a spin one-half particle when we rotated the coordinates about the z-axis—then we got the phase factors $e^{\pm i\phi/2}$. It is, in fact, exactly what we wrote down in Section 5–7 for the $|+\rangle$ and $|-\rangle$ states of a spin-one particle—which is no coincidence. The photon is a spin-one particle which has, however, no "zero" state.

‡ We now feel that the material of this section is longer and harder than is appropriate at this point in our development. We suggest that you skip it and continue with Section 11–6. If you are ambitious and have time you may wish to come back to it later. We leave it here, because it is a beautiful example—taken from recent work in high-energy physics—of what can be done with our formulation of the quantum mechanics of two-state systems.

The nuclear particles have also what is called a "weak interaction" by which certain things can happen, such as beta decay, but always very slowly on a nuclear time scale—the weak interactions are many, many orders of magnitude weaker than the strong interactions and even much weaker than electromagnetic interactions.

When the strong interactions were being studied with the big accelerators, people were surprised to find that certain things that "should" happen—that were expected to happen—did not occur. For instance, in some interactions a particle of a certain type did not appear when it was expected. Gell-Mann and Nishijima noticed that many of these peculiar happenings could be explained at once by inventing a new conservation law: the *conservation of strangeness*. They proposed that there was a new kind of attribute associated with each particle—which they called its "strangeness" number—and that in any strong interaction the "quantity of strangeness" is conserved.

Suppose, for instance, that a high-energy negative K-meson—with, say, an energy of many Bev—collides with a proton. Out of the interaction may come many other particles: π-mesons, K-mesons, lambda particles, sigma particles—any of the mesons or baryons listed in Table 2–2 of Vol. I. It is observed, however, that only *certain combinations* appear, and never others. Now certain conservation laws were already known to apply. First, energy and momentum are always conserved. The total energy and momentum after an event must be the same as before the event. Second, there is the conservation of electric charge which says that the total charge of the outgoing particles must be equal to the total charge carried by the original particles. In our example of a K-meson and a proton coming together, the following reactions *do* occur:

$$K^- + p \rightarrow p + K^- + \pi^+ + \pi^- + \pi^0$$

or   (11.38)

$$K^- + p \rightarrow \Sigma^- + \pi^+.$$

We would never get:

$$K^- + p \rightarrow p + K^- + \pi^+ \quad \text{or} \quad K^- + p \rightarrow \Lambda_0 + \pi^+, \quad (11.39)$$

because of the conservation of charge. It was also known that the *number* of *baryons* is conserved. The number of baryons *out* must be equal to the number of baryons *in*. For this law, an *antiparticle* of a baryon is counted as *minus* one baryon. This means that we can—and do—see

$$K^- + p \rightarrow \Lambda^0 + \pi^0$$

or   (11.40)

$$K^- + p \rightarrow p + K^- + p + \bar{p}$$

(where $\bar{p}$ is the antiproton, which carries a negative charge). But we *never* see

$$K^- + p \rightarrow K^- + \pi^+ + \pi^0$$

or   (11.41)

$$K^- + p \rightarrow p + K^- + n$$

(even when there is plenty of energy), because baryons would not be conserved.

These laws, however, do *not* explain the strange fact that the following reactions—which do not immediately appear to be especially different from some of those in (11.38) or (11.40)—are also never observed:

$$K^- + p \rightarrow p + K^- + K^0$$

or

$$K^- + p \rightarrow p + \pi^-$$ (11.42)

or

$$K^- + p \rightarrow \Lambda^0 + K^0.$$

The explanation is the conservation of strangeness. With each particle goes a number—its *strangeness S*—and there is a law that in any *strong* interaction, the

Table 11–4

The strangeness numbers of the strongly interacting particles

			S	
	-2	-1	0	$+1$
Baryons	Ξ^0 Ξ^-	Σ^+ Λ^0, Σ^0 Σ^-	p n	
Mesons		$\overline{K}^0$ K^-	π^+ π^0 π^-	K^+ K^0

Note: The π^- is the antiparticle of the π^+ (or *vice versa*).

total strangeness *out* must equal the total strangeness that went *in*. The proton and antiproton (p, $\overline{p}$), the neutron and antineutron (n, $\overline{n}$), and the π-mesons (π^+, π^0, π^-) all have the strangeness number *zero*; the K^+ and K^0 mesons have strangeness $+1$; the K^- and $\overline{K}^0$ (the anti-K^0),† the Λ^0 and the Σ-particles ($+$, 0, $-$) have strangeness -1. There is also a particle with strangeness -2—the Ξ-particle (capital "ksi")—and perhaps others as yet unknown. We have made a list of these strangenesses in Table 11–4.

Let's see how the strangeness conservation works in some of the reactions we have written down. If we start with a K^- and a proton, we have a total strangeness of $(-1 + 0) = -1$. The conservation of strangeness says that the strangeness of products *after* the reaction must also add up to -1. You see that that is so for the reactions of (11.38) and (11.40). But in the reactions of (11.42) the strangeness of the right-hand side is *zero* in each case. Such reactions do not conserve strangeness, and do not occur. Why? Nobody knows. Nobody knows any more than what we have just told you about this. Nature just works that way.

Now let's look at the following reaction: a π^- hits a proton. You might, for instance, get a Λ^0 particle plus a neutral K-particle—two neutral particles. Now which neutral K do you get? Since the Λ-particle has a strangeness -1 and the π and p^+ have a strangeness zero, and since this is a fast production reaction, the strangeness must not change. The K-particle must have strangeness $+1$—it must therefore be the K^0. The reaction is

$$\pi^- + p \rightarrow \Lambda^0 + K^0,$$

with

$$S = 0 + 0 = -1 + +1 \quad \text{(conserved)}.$$

If the $\overline{K}^0$ were there instead of the K^0, the strangeness on the right would be -2—which nature does not permit, since the strangeness on the left side is zero. On the other hand, a $\overline{K}^0$ can be produced in other reactions, such as

$$n + n \rightarrow n + p + \overline{K}^0 + K^+,$$

$$S = 0 + 0 = 0 + 0 + +1 + -1$$

or

$$K^- + p \rightarrow n + \overline{K}^0,$$

$$S = -1 + 0 = 0 + -1.$$

You may be thinking, "That's all a lot of stuff, because how do you *know* whether it is a $\overline{K}^0$ or a K^0? They look exactly the same. They are antiparticles of each other, so they have exactly the same mass, and both have zero electric charge.

---

† Read as: "K-naught-bar," or "K-zero-bar."

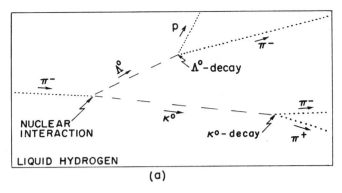
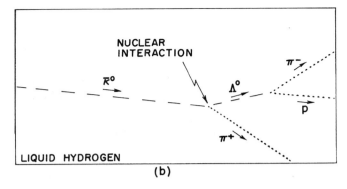

Fig. 11-5. High-energy events as seen in a hydrogen bubble chamber. (a) A π^- meson interacts with a hydrogen nucleus (proton) producing a Λ^0 particle and a K^0 meson. Both particles decay in the chamber. (b) A $\overline{K}^0$ meson interacts with a proton producing a π^+ meson and a Λ^0 particle which then decays. (The neutral particles leave no tracks. Their inferred trajectories are indicated here by light dashed lines.)

How do you distinguish them?" By the reactions *they* produce. For example, a $\overline{K}^0$ can interact with matter to produce a Λ-particle, like this:

$$\overline{K}^0 + p \to \Lambda^0 + \pi^+,$$

but a K^0 cannot. There is *no* way a K^0 can produce a Λ-particle when it interacts with ordinary matter (protons and neutrons).† So the experimental distinction between the K^0 and the $\overline{K}^0$ would be that one of them will and one of them will not produce Λ's.

One of the predictions of the strangeness theory is then this—if, in an experiment with high-energy pions, a Λ-particle is produced with a neutral K-meson, then *that* neutral K-meson going into other pieces of matter will never produce a Λ. The experiment might run something like this. You send a beam of π^--mesons into a large hydrogen bubble chamber. A π^- track disappears, but somewhere else a pair of tracks appear (a proton and a π^-) indicating that a Λ-particle has disintegrated‡—see Fig. 11-5. Then you know that there is a K^0 somewhere which you cannot see.

You can, however, figure out where it is going by using the conservation of momentum and energy. [It could reveal itself later by disintegrating into two charged particles, as shown in Fig. 11-5(a).] As the K^0 goes flying along, it may interact with one of the hydrogen nuclei (protons), producing perhaps some other particles. The prediction of the strangeness theory is that it will *never* produce a Λ-particle in a simple reaction like, say,

$$K^0 + p \to \Lambda^0 + \pi^0,$$

although a $\overline{K}^0$ can do just that. That is, in a bubble chamber a $\overline{K}^0$ might produce the event sketched in Fig. 11-5(b)—in which the Λ^0 is seen because it decays—but a K^0 will not. That's the first part of our story. That's the conservation of strangeness.

The conservation of strangeness is, however, *not perfect*. There are very slow disintegrations of the strange particles—decays taking a long¶ time like 10^{-10} second in which the strangeness is *not* conserved. These are called the "weak" decays. For example, the K^0 disintegrates into a pair of π-mesons (+ and −)

---

† Except, of course, if it *also* produces *two* K^+'s or other particles with a total strangeness of +2. We can think here of reactions in which there is insufficient energy to produce these additional strange particles.

‡ The free Λ-particle decays slowly via a *weak* interaction (so strangeness need not be conserved). The decay products are either a p and a π^-, or an n and a π^0. The lifetime is 2.2×10^{-10} sec.

¶ A typical time for strong interactions is more like 10^{-23} sec.

with a lifetime of 10^{-10} second. That was, in fact, the way K-particles were first seen. Notice that the decay reaction

$$K^0 \to \pi^+ + \pi^-$$

does not conserve strangeness, so it cannot go "fast" by the strong interaction; it can only go through the weak decay process.

Now the $\overline{K}^0$ also disintegrates *in the same way*—into a π^+ and a π^-— and also with the same lifetime

$$\overline{K}^0 \to \pi^- + \pi^+.$$

Again we have a weak decay because it does not conserve strangeness. There is a principle that for any reaction there is the corresponding reaction with "matter" replaced by "antimatter" and *vice versa*. Since the $\overline{K}^0$ is the antiparticle of the K^0, it should decay into the antiparticles of the π^+ and π^-, but the antiparticle of a π^+ is the π^-. (Or, if you prefer, *vice versa*. It turns out that for the π-mesons it doesn't matter which one you call "matter.") So as a consequence of the weak decays, the K^0 and $\overline{K}^0$ can go into the same final products. When "seen" through their decays—as in a bubble chamber—they look like the same particle. Only their strong interactions are different.

At last we are ready to describe the work of Gell-Mann and Pais. They first noticed that since the K^0 and the $\overline{K}^0$ can both turn into states of two π-mesons there must be some amplitude that a K^0 can turn into a $\overline{K}^0$, and also that a $\overline{K}^0$ can turn into a K^0. Writing the reactions as one does in chemistry, we would have

$$K^0 \leftrightarrows \pi^- + \pi^+ \leftrightarrows \overline{K}^0. \tag{11.43}$$

These reactions imply that there is some amplitude per unit time, say $-i/\hbar$ times $\langle \overline{K}^0 \mid W \mid K^0 \rangle$, that a K^0 will turn into a $\overline{K}^0$ through the weak interaction responsible for the decay into two π-mesons. And there is the corresponding amplitude $\langle K^0 \mid W \mid \overline{K}^0 \rangle$ for the reverse process. Because matter and antimatter behave in exactly the same way, these two amplitudes are numerically equal; we'll call them both A:

$$\langle \overline{K}^0 \mid W \mid K^0 \rangle = \langle K^0 \mid W \mid \overline{K}^0 \rangle = A. \tag{11.44}$$

Now—said Gell-Mann and Pais—here is an interesting situation. What people have been calling two distinct states of the world—the K^0 and the $\overline{K}^0$— should really be considered as *one* two-state *system*, because there is an amplitude to go from one state to the other. For a complete treatment, one would, of course, have to deal with more than two states, because there are also the states of 2π's, and so on; but since they were mainly interested in the relation of K^0 and $\overline{K}^0$, they did not have to complicate things and could make the approximation of a two-state system. The other states *were* taken into account to the extent that their effects appeared implicitly in the amplitudes of Eq. (11.44).

Accordingly, Gell-Mann and Pais analyzed the neutral particle as a two-state system. They began by choosing as their two base states the states $\mid K^0 \rangle$ and $\mid \overline{K}^0 \rangle$. (From here on, the story goes very much as it did for the ammonia molecule.) Any state $\mid \psi \rangle$ of the neutral K-particle could then be described by giving the amplitudes that it was in either base state. We'll call these amplitudes

$$C_+ = \langle K^0 \mid \psi \rangle, \qquad C_- = \langle \overline{K}^0 \mid \psi \rangle. \tag{11.45}$$

The next step was to write the Hamiltonian equations for this two-state system. If there were no coupling between the K^0 and the $\overline{K}^0$, the equations would be simply

$$i\hbar \frac{dC_+}{dt} = E_0 C_+,$$

$$i\hbar \frac{dC_-}{dt} = E_0 C_-. \tag{11.46}$$

But since there is the amplitude $\langle \overline{K}{}^0 \mid W \mid K^0 \rangle$ for the K^0 to turn into a $\overline{K}{}^0$ there should be the additional term

$$\langle \overline{K}{}^0 \mid W \mid K^0 \rangle C_- = AC_-$$

added to the right-hand side of the first equation. And similarly, the term AC_+ should be inserted in the equation for the rate of change of C_-.

But that's not all. When the two-pion effect is taken into account there is an *additional* amplitude for the K^0 to turn into *itself* through the process

$$K^0 \to \pi^- + \pi^+ \to K^0.$$

The additional amplitude, which we would write $\langle K^0 \mid W \mid K^0 \rangle$, is just equal to the amplitude $\langle \overline{K}{}^0 \mid W \mid K^0 \rangle$, since the amplitudes to go to and from a pair of π-mesons are identical for the K^0 and the $\overline{K}{}^0$. If you wish, the argument can be written out in detail like this. First write†

$$\langle \overline{K}{}^0 \mid W \mid K^0 \rangle = \langle \overline{K}{}^0 \mid W \mid 2\pi \rangle \langle 2\pi \mid W \mid K^0 \rangle$$

and

$$\langle K^0 \mid W \mid K^0 \rangle = \langle K^0 \mid W \mid 2\pi \rangle \langle 2\pi \mid W \mid K^0 \rangle.$$

Because of the symmetry of matter and antimatter

$$\langle 2\pi \mid W \mid K^0 \rangle = \langle 2\pi \mid W \mid \overline{K}{}^0 \rangle,$$

and also

$$\langle K^0 \mid W \mid 2\pi \rangle = \langle \overline{K}{}^0 \mid W \mid 2\pi \rangle.$$

It then follows that $\langle K^0 \mid W \mid K^0 \rangle = \langle \overline{K}{}^0 \mid W \mid K^0 \rangle$, and also that $\langle \overline{K}{}^0 \mid W \mid K^0 \rangle = \langle K^0 \mid W \mid \overline{K}{}^0 \rangle$, as we said earlier. Anyway, there are the two additional amplitudes $\langle K^0 \mid W \mid K^0 \rangle$ and $\langle \overline{K}{}^0 \mid W \mid \overline{K}{}^0 \rangle$, both equal to A, which should be included in the Hamiltonian equations. The first gives a term AC_+ on the right-hand side of the equation for dC_+/dt, and the second gives a new term AC_- in the equation for dC_-/dt. Reasoning this way, Gell-Mann and Pais concluded that the Hamiltonian equations for the $K^0 \overline{K}{}^0$ system should be

$$i\hbar \frac{dC_+}{dt} = E_0 C_+ + AC_- + AC_+,$$

$$i\hbar \frac{dC_-}{dt} = E_0 C_- + AC_+ + AC_-. \tag{11.47}$$

We must now correct something we have said in earlier chapters: that two amplitudes like $\langle K^0 \mid W \mid \overline{K}{}^0 \rangle$ and $\langle \overline{K}{}^0 \mid W \mid K^0 \rangle$ which are the reverse of each other, are always complex conjugates. That was true when we were talking about particles that did not decay. But if particles can decay—and can, therefore, become "lost"—the two amplitudes are not necessarily complex conjugates. So the equality of (11.44) does not mean that the amplitudes are real numbers; they are in fact complex numbers. The coefficient A is, therefore, complex; and we can't just incorporate it into the energy E_0.

Having played often with electron spins and such, our heroes knew that the Hamiltonian equations of (11.47) meant that there was *another* pair of base states which could also be used to represent the K-particle system and which would have especially simple behaviors. They said, "Let's take the sum and difference of these two equations. Also, let's measure all our energies from E_0, and use units for

---

† We are making a simplification here. The 2π-system can have many states corresponding to various momenta of the π-mesons, and we should make the right-hand side of this equation into a sum over the various base states of the π's. The complete treatment still leads to the same conclusions.

energy and time that make $\hbar = 1$." (That's what modern theoretical physicists always do. It doesn't change the physics but makes the equations take on a simple form.) Their result:

$$i\frac{d}{dt}(C_+ + C_-) = 2A(C_+ + C_-), \qquad i\frac{d}{dt}(C_+ - C_-) = 0. \qquad (11.48)$$

It is apparent that the combinations of amplitudes $(C_+ + C_-)$ and $(C_+ - C_-)$ act independently from each other (corresponding, of course, to the stationary states we have been studying earlier). So they concluded that it would be more convenient to use a different representation for the K-particle. They defined the two states

$$|K_1\rangle = \frac{1}{\sqrt{2}}(|K^0\rangle + |\bar{K}^0\rangle), \qquad |K_2\rangle = \frac{1}{\sqrt{2}}(|K^0\rangle - |\bar{K}^0\rangle). \qquad (11.49)$$

They said that instead of thinking of the K^0 and $\bar{K}^0$ mesons, we can equally well think in terms of the two "particles" (that is, "states") K_1 and K_2. (These correspond, of course, to the states we have usually called $|I\rangle$ and $|II\rangle$. We are not using our old notation because we want now to follow the notation of the original authors—and the one you will see in physics seminars.)

Now Gell-Mann and Pais didn't do all this just to get different names for the particles—there is also some strange new physics in it. Suppose that C_1 and C_2 are the amplitudes that some state $|\psi\rangle$ will be either a K_1 or a K_2 meson:

$$C_1 = \langle K_1 | \psi \rangle, \qquad C_2 = \langle K_2 | \psi \rangle.$$

From the equations of (11.49),

$$C_1 = \frac{1}{\sqrt{2}}(C_+ + C_-), \qquad C_2 = \frac{1}{\sqrt{2}}(C_+ - C_-). \qquad (11.50)$$

Then the Eqs. (11.48) become

$$i\frac{dC_1}{dt} = 2AC_1, \qquad i\frac{dC_2}{dt} = 0. \qquad (11.51)$$

The solutions are

$$C_1(t) = C_1(0)e^{-i2At}, \qquad C_2(t) = C_2(0), \qquad (11.52)$$

where, of course, $C_1(0)$ and $C_2(0)$ are the amplitudes at $t = 0$.

These equations say that if a neutral K-particle starts out in the state $|K_1\rangle$ at $t = 0$ [then $C_1(0) = 1$ and $C_2(0) = 0$], the amplitudes at the time t are

$$C_1(t) = e^{-i2At}, \qquad C_2(t) = 0.$$

Remembering that A is a complex number, it is convenient to take $A = \alpha - i\beta$. (Since the imaginary part of $2A$ turns out to be negative, we write it as minus $i\beta$.) With this substitution, $C_1(t)$ reads

$$C_1(t) = C_1(0)e^{-\beta t}e^{-i\alpha t}. \qquad (11.53)$$

The probability of finding a K_1 particle at t is the absolute square of this amplitude, which is $e^{-2\beta t}$. And, from Eqs. (11.52), the probability of finding the K_2 state at any time is zero. That means that if you make a K-particle in the state $|K_1\rangle$, the probability of finding it in the same state decreases exponentially with time—but you will never find it in state $|K_2\rangle$. Where does it go? It disintegrates into two π-mesons with the mean life $\tau = 1/2\beta$ which is, experimentally, 10^{-10} sec. We made provisions for that when we said that A was complex.

On the other hand, Eq. (11.52) says that if we make a K-particle completely in the K_2 state, it stays that way forever. Well, that's not really true. It is observed experimentally to disintegrate into *three* π-mesons, but 600 times slower than the two-pion decay we have described. So there are some other small terms we have left out in our approximation. But so long as we are considering only the two-pion decay, the K_2 lasts "forever."

Now to finish the story of Gell-Mann and Pais. They went on to consider what happens when a K-particle is produced *with a Λ^0 particle* in a *strong* interaction. Since it must then have a strangeness of $+1$, it must be produced in the K^0 state. So at $t = 0$ it is neither a K_1 nor a K_2 but a *mixture*. The initial conditions are

$$C_+(0) = 1, \qquad C_-(0) = 0.$$

But that means—from Eq. (11.50)—that

$$C_1(0) = \frac{1}{\sqrt{2}}, \qquad C_2(0) = \frac{1}{\sqrt{2}},$$

and—from Eq. (11.51)—that

$$C_1(t) = \frac{1}{\sqrt{2}} e^{-\beta t} e^{-i\alpha t}, \qquad C_2(t) = \frac{1}{\sqrt{2}}. \tag{11.54}$$

Now remember that K_1 and K_2 are each linear combinations of K^0 and $\overline{K}^0$. In Eqs. (11.54) the amplitudes have been chosen so that at $t = 0$ the $\overline{K}^0$ parts cancel each other out by interference, leaving only a K^0 state. But the $|K_1\rangle$ state *changes with time*, and the $|K_2\rangle$ state *does not*. After $t = 0$ the interference of C_1 and C_2 will give finite amplitudes for both K^0 and $\overline{K}^0$.

What does all this mean? Let's go back and think of the experiment we sketched in Fig. 11–5. A π^- meson has produced a Λ^0 particle and a K^0 meson which is tooting along through the hydrogen in the chamber. As it goes along, there is some small but uniform chance that it will collide with a hydrogen nucleus. At first, we thought that strangeness conservation would prevent the K-particle from making a Λ^0 in such an interaction. Now, however, we see that that is not right. For although our K-particle *starts out* as a K^0—which cannot make a Λ^0—it does not *stay* this way. After a while, there *is some amplitude* that it will have flipped to the $\overline{K}^0$ state. We can, therefore, sometimes expect to see a Λ^0 produced along the K-particle track. The chance of this happening is given by the amplitude C_-, which we can [by using Eq. (11.50) backwards] relate to C_1 and C_2. The relation is

$$C_- = \frac{1}{\sqrt{2}} (C_1 - C_2) = \tfrac{1}{2}(e^{-\beta t} e^{-i\alpha t} - 1). \tag{11.55}$$

As our K-particle goes along, the probability that it will "act like" a $\overline{K}^0$ is equal to $|C_-|^2$, which is

$$|C_-|^2 = \tfrac{1}{4}(1 + e^{-2\beta t} - 2e^{-\beta t} \cos \alpha t). \tag{11.56}$$

A complicated and strange result!

This, then, is the remarkable prediction of Gell-Mann and Pais: when a K^0 is produced, the chance that it will turn into a $\overline{K}^0$—as it can demonstrate by being able to produce a Λ^0—varies with time according to Eq. (11.56). This prediction came from using only sheer logic and the basic principles of the quantum mechanics—with no knowledge at all of the inner workings of the K-particle. Since nobody knows anything about the inner machinery, that is as far as Gell-Mann and Pais could go. They could not give any theoretical values for α and β. And nobody has been able to do so to this date. They were able to give a value of β obtained from the experimentally observed rate of decay into two π's ($2\beta = 10^{10}$ sec), but they could say nothing about α.

We have plotted the function of Eq. (11.56) for two values of α in Fig. 11–6. You can see that the form depends very much on the ratio of α to β. There is no $\overline{K}^0$ probability at first; then it builds up. If α is large, the probability would have large oscillations. If α is small, there will be little or no oscillation—the probability will just rise smoothly to 1/4.

Now, typically, the K-particle will be travelling at a constant speed near the speed of light. The curves of Fig. 11–6 then also represent the probability along the track of observing a $\overline{K}^0$—with typical distances of several centimeters. You can see why this prediction is so remarkably peculiar. You produce a single particle and instead of just disintegrating, it does something else. Sometimes it disintegrates, and other times it turns into a different kind of a particle. Its characteristic probability of producing an effect varies in a strange way as it goes along. There is nothing else quite like it in nature. And this most remarkable prediction was made solely by arguments about the interference of amplitudes.

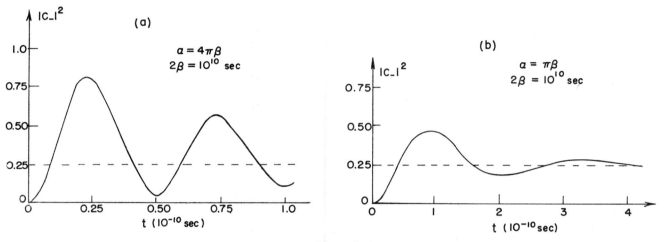

Fig. 11–6. The function of Eq. (11–56): (a) for $\alpha = \pi\beta$, (b) for $\alpha = 4\pi\beta$ (with $2\beta = 10^{10}$ sec).

If there is any place where we have a chance to test the main principles of quantum mechanics in the purest way—does the superposition of amplitudes work or doesn't it?—this is it. In spite of the fact that this effect has been predicted now for several years, there is no experimental determination that is very clear. There are some rough results which indicate that the α is not zero, and that the effect really occurs—they indicate that α is between 2β and 4β. That's all there is, experimentally. It would be very beautiful to check out the curve exactly to see if the principle of superposition really still works in such a mysterious world as that of the strange particles—with unknown reasons for the decays, and unknown reasons for the strangeness.

The analysis we have just described is very characteristic of the way quantum mechanics is being used today in the search for an understanding of the strange particles. All the complicated theories that you may hear about are no more and no less than this kind of elementary hocus-pocus using the principles of superposition and other principles of quantum mechanics of that level. Some people claim that they have theories by which it is possible to calculate the β and α, or at least the α given the β, but these theories are completely useless. For instance, the theory that predicts the value of α, given the β, tells us that the value of α should be infinite. The set of equations with which they originally start involves two π-mesons and then goes from the two π's back to a K^0, and so on. When it's all worked out, it does indeed produce a pair of equations like the ones we have here; but because there are an infinite number of states of two π's, depending on their momenta, integrating over all the possibilities gives an α which is infinite. But nature's α is *not* infinite. So the dynamical theories are wrong. It is really quite remarkable that the phenomena which can be predicted *at all* in the world of the strange particles come from the principles of quantum mechanics at the level at which you are learning them now.

11–20

11–6 Generalization to N-state systems

We have finished with all the two-state systems we wanted to talk about. In the following chapters we will go on to study systems with more states. The extension to N-state systems of the ideas we have worked out for two states is pretty straightforward. It goes like this.

If a system has N distinct states, we can represent any state $|\psi(t)\rangle$ as a linear combination of any set of base states $|i\rangle$, where $i = 1, 2, 3, \ldots, N$;

$$|\psi(t)\rangle = \sum_{\text{all } i} |i\rangle C_i(t). \tag{11.57}$$

The coefficients $C_i(t)$ are the amplitudes $\langle i \mid \psi(t)\rangle$. The behavior of the amplitudes C_i with time is governed by the equations

$$i\hbar \frac{dC_i(t)}{dt} = \sum_j H_{ij}C_j, \tag{11.58}$$

where the energy matrix H_{ij} describes the physics of the problem. It looks the same as for two states. Only now, both i and j must range over all N base states, and the energy matrix H_{ij}—or, if you prefer, the Hamiltonian—is an N by N matrix with N^2 numbers. As before, $H_{ij}^* = H_{ji}$—so long as particles are conserved —and the diagonal elements H_{ii} are real numbers.

We have found a general solution for the C's of a two-state system when the energy matrix is constant (doesn't depend on t). It is also not difficult to solve Eq. (11.58) for an N-state system when H is not time dependent. Again, we begin by looking for a possible solution in which the amplitudes all have the *same* time dependence. We try

$$C_i = a_i e^{-(i/\hbar) E t}. \tag{11.59}$$

When these C_i's are substituted into (11.58), the derivatives $dC_i(t)/dt$ become just $(-i/\hbar)EC_i$. Canceling the common exponential factor from all terms, we get

$$Ea_i = \sum_j H_{ij}a_j. \tag{11.60}$$

This is a set of N linear algebraic equations for the N unknowns $a_1, a_2, \ldots, a_n$, and there is a solution only if you are lucky—only if the determinant of the coefficients of all the a's is zero. But it's not necessary to be that sophisticated; you can just start to solve the equations any way you want, and you will find that they can be solved only for certain values of E. (Remember that E is the only adjustable thing we have in the equations.)

If you want to be formal, however, you can write Eq. (11.60) as

$$\sum_j (H_{ij} - \delta_{ij}E)a_j = 0. \tag{11.61}$$

Then you can use the rule—if you know it—that these equations will have a solution only for those values of E for which

$$\text{Det } (H_{ij} - \delta_{ij}E) = 0. \tag{11.62}$$

Each term of the determinant is just H_{ij}, except that E is subtracted from every diagonal element. That is, (11.62) means just

$$\text{Det} \begin{pmatrix} H_{11} - E & H_{12} & H_{13} & \cdots \\ H_{21} & H_{22} - E & H_{23} & \cdots \\ H_{31} & E_{32} & H_{33} - E & \cdots \\ \cdots & \cdots & \cdots & \cdots \end{pmatrix} = 0. \tag{11.63}$$

This is, of course, just a special way of writing an algebraic equation for E which is the sum of a bunch of products of all the terms taken a certain way. These products will give all the powers of E up to E^N.

So we have an Nth order polynomial equal to zero, and there are, in general, N roots. (We must remember, however, that some of them may be multiple roots—meaning that two or more roots are equal.) Let's call the N roots

$$E_I, E_{II}, E_{III}, \ldots, E_\mathbf{n}, \ldots, E_\mathbf{N}. \tag{11.64}$$

(We will use $\mathbf{n}$ to represent the nth Roman numeral, so that $\mathbf{n}$ takes on the values I, II, $\ldots$, N.) It may be that some of these energies are equal—say $E_{II} = E_{III}$—but we will still choose to call them by different names.

The equations (11.60)— or (11.61)—have one solution for each value of E. If you put any one of the E's—say $E_\mathbf{n}$—into (11.60) and solve for the a_i, you get a set which belongs to the energy $E_\mathbf{n}$. We will call this set $a_i(\mathbf{n})$.

Using these $a_i(\mathbf{n})$ in Eq. (11.59), we have the amplitudes $C_i(\mathbf{n})$ that the definite energy states are in the base state $|\,i\rangle$. Letting $|\,\mathbf{n}\rangle$ stand for the state vector of the definite energy state at $t = 0$, we can write

$$C_i(\mathbf{n}) = \langle i \,|\, \mathbf{n}\rangle e^{(i/\hbar)E_\mathbf{n}t},$$

with

$$\langle i \,|\, \mathbf{n}\rangle = a_i(\mathbf{n}). \tag{11.65}$$

The complete definite energy state $|\,\psi_\mathbf{n}(t)\rangle$ can then be written as

$$|\,\psi_\mathbf{n}(t)\rangle = \sum_i |\,i\rangle a_i(\mathbf{n})e^{-(i/\hbar)E_\mathbf{n}t},$$

or

$$|\,\psi_\mathbf{n}(t)\rangle = |\,\mathbf{n}\rangle e^{(i/\hbar)E_\mathbf{n}t}. \tag{11.66}$$

The state vectors $|\,\mathbf{n}\rangle$ describe the configuration of the definite energy states, but have the time dependence factored out. Then they are constant vectors which can be used as a new base set if we wish.

Each of the states $|\,\mathbf{n}\rangle$ has the property—as you can easily show—that when operated on by the Hamiltonian operator $\hat{H}$ it gives just $E_\mathbf{n}$ times the same state:

$$\hat{H}\,|\,n\rangle = E_\mathbf{n}\,|\,n\rangle. \tag{11.67}$$

The energy $E_\mathbf{n}$ is, then, a number which is a characteristic of the Hamiltonian operator $\hat{H}$. As we have seen, a Hamiltonian will, in general, have several characteristic energies. In the mathematician's world these would be called the "characteristic values" of the matrix H_{ij}. Physicists usually call them the "eigenvalues" of $\hat{H}$. ("Eigen" is the German word for "characteristic" or "proper.") With each eigenvalue of $\hat{H}$—in other words, for each energy—there is the state of definite energy, which we have called the "stationary state." Physicists usually call the states $|\,\mathbf{n}\rangle$ "the eigenstates of $\hat{H}$." Each eigenstate corresponds to a particular eigenvalue $E_\mathbf{n}$.

Now, generally, the states $|\,\mathbf{n}\rangle$—of which there are N—can also be used as a base set. For this to be true, all of the states must be orthogonal, meaning that for any two of them, say $|\,\mathbf{n}\rangle$ and $|\,\mathbf{m}\rangle$,

$$\langle \mathbf{n} \,|\, \mathbf{m}\rangle = 0. \tag{11.68}$$

This will be true automatically if all the energies are different. Also, we can multiply all the $a_i(\mathbf{n})$ by a suitable factor so that all the states are normalized—by which we mean that

$$\langle \mathbf{n} \,|\, \mathbf{n}\rangle = 1 \tag{11.69}$$

for all $\mathbf{n}$.

When it happens that Eq. (11.63) accidentally has two (or more) roots with the same energy, there are some minor complications. First, there are still two different sets of a_i's which go with the two equal energies, but the states they give may *not* be orthogonal. Suppose you go through the normal procedure and find two stationary states with equal energies—let's call them $|\mu\rangle$ and $|\nu\rangle$. Then it will not necessarily be so that they are orthogonal—if you are unlucky,

$$\langle\mu\,|\,\nu\rangle \neq 0.$$

It is, however, always true that you can cook up two new states, which we will call $|\mu'\rangle$ and $|\nu'\rangle$, that have the same energies and are also orthogonal, so that

$$\langle\mu'\,|\,\nu'\rangle = 0. \tag{11.70}$$

You can do this by making $|\mu'\rangle$ and $|\nu'\rangle$ a suitable linear combination of $|\mu\rangle$ and $|\nu\rangle$, with the coefficients chosen to make it come out so that Eq. (11.70) is true. It is always convenient to do this. We will generally assume that this has been done so that we can always assume that our proper energy states $|n\rangle$ are all orthogonal.

We would like, for fun, to prove that when two of the stationary states have different energies they are indeed orthogonal. For the state $|n\rangle$ with the energy E_n, we have that

$$\hat{H}\,|\,n\rangle = E_n\,|\,n\rangle. \tag{11.71}$$

This operator equation really means that there is an equation between numbers. Filling the missing parts, it means the same as

$$\sum_j \langle i\,|\,\hat{H}\,|\,j\rangle\langle j\,|\,n\rangle = E_n\langle i\,|\,n\rangle. \tag{11.72}$$

If we take the complex conjugate of this equation, we get

$$\sum_j \langle i\,|\,\hat{H}\,|\,j\rangle^*\langle j\,|\,n\rangle^* = E_n^*\langle i\,|\,n\rangle^*. \tag{11.73}$$

Remember now that the complex conjugate of an amplitude is the reverse amplitude, so (11.73) can be rewritten as

$$\sum_j \langle n\,|\,j\rangle\langle j\,|\,\hat{H}\,|\,i\rangle = E_n^*\langle n\,|\,i\rangle. \tag{11.74}$$

Since this equation is valid for *any* i, its "short form" is

$$\langle n\,|\,\hat{H} = E_n^*\langle n\,|, \tag{11.75}$$

which is called the *adjoint* to Eq. (11.71).

Now we can easily prove that E_n is a real number. We multiply Eq. (11.71) by $\langle n\,|$ to get

$$\langle n\,|\hat{H}|\,n\rangle = E_n, \tag{11.76}$$

since $\langle n\,|\,n\rangle = 1$. Then we multiply Eq. (11.75) on the left by $|\,n\rangle$ to get

$$\langle n\,|\hat{H}|\,n\rangle = E_n^*. \tag{11.77}$$

Comparing (11.76) with (11.77) it is clear that

$$E_n = E_n^*, \tag{11.78}$$

which means that E_n is real. We can erase the star on E_n in Eq. (11.75).

Finally we are ready to show that the different energy states are orthogonal. Let $|\,n\rangle$ and $|\,m\rangle$ be any two of the definite energy base states. Using Eq. (11.75) for the state **m**, and multiplying it by $|\,n\rangle$, we get that

$$\langle m\,|\hat{H}|\,n\rangle = E_m\langle m\,|\,n\rangle.$$

But if we multiply (11.71) by $\langle \mathbf{m} \,|$, we get

$$\langle \mathbf{m} \,|\hat{H}| \,\mathbf{n} \rangle \;=\; E_\mathbf{n}\langle \mathbf{m} \mid \mathbf{n} \rangle.$$

Since the left sides of these two equations are equal, the right sides are, also:

$$E_\mathbf{m}\langle \mathbf{m} \mid \mathbf{n} \rangle \;=\; E_\mathbf{n}\langle \mathbf{m} \mid \mathbf{n} \rangle. \tag{11.79}$$

If $E_\mathbf{m} = E_\mathbf{n}$ the equation does not tell us anything. But if the energies of the two states $|\,\mathbf{m} \rangle$ and $|\,\mathbf{n} \rangle$ *are different* $(E_\mathbf{m} \neq E_\mathbf{n})$, Eq. (11.79) says that $\langle \mathbf{m} \mid \mathbf{n} \rangle$ must be zero, as we wanted to prove. The two states are necessarily orthogonal so long as $E_\mathbf{n}$ and $E_\mathbf{m}$ are numerically different.

The Hyperfine Splitting in Hydrogen

12–1 Base states for a system with two spin one-half particles

In this chapter we take up the "hyperfine splitting" of hydrogen, because it is a physically interesting example of what we can already do with quantum mechanics. It's an example with more than two states, and it will be illustrative of the methods of quantum mechanics as applied to slightly more complicated problems. It is enough more complicated that once you see how this one is handled you can get immediately the generalization to all kinds of problems.

As you know, the hydrogen atom consists of an electron sitting in the neighborhood of the proton, where it can exist in any one of a number of discrete energy states in each one of which the pattern of motion of the electron is different. The first excited state, for example, lies 3/4 of a Rydberg, or about 10 electron volts, above the ground state. But even the so-called ground state of hydrogen is not really a single, definite-energy state, because of the spins of the electron and the proton. These spins are responsible for the "hyperfine structure" in the energy levels, which splits all the energy levels into several nearly equal levels.

The electron can have its spin either "up" or "down" and, the proton can also have *its* spin either "up" or "down." There are, therefore, *four* possible spin states for every dynamical condition of the atom. That is, when people say "the ground state" of hydrogen, they really mean the "four ground states," and not just the very lowest state. The four spin states do not all have exactly the same energy; there are slight shifts from the energies we would expect with no spins. The shifts are, however, much, much smaller than the 10 volts or so from the ground state to the next state above. As a consequence, each dynamical state has its energy split into a set of very close energy levels—the so-called *hyperfine splitting*.

The energy differences among the four spin states is what we want to calculate in this chapter. The hyperfine splitting is due to the interaction of the magnetic moments of the electron and proton, which gives a slightly different magnetic energy for each spin state. These energy shifts are only about ten-millionths of an electron volt—really very small compared with 10 volts! It is because of this large gap that we can think about the ground state of hydrogen as a "four-state" system, without worrying about the fact that there are really many more states at higher energies. We are going to limit ourselves here to a study of the hyperfine structure of the ground state of the hydrogen atom.

For our purposes we are not interested in any of the details about the *positions* of the electron and proton because that has all been worked out by the atom so to speak—it has worked itself out by getting into the ground state. We need know only that we have an electron and proton in the neighborhood of each other with some definite spatial relationship. In addition, they can have various different relative orientations of their spins. It is only the effect of the spins that we want to look into.

The first question we have to answer is: What are the *base states* for the system? Now the question has been put incorrectly. There is no such thing as *"the"* base states, because, of course, the set of base states you may choose is not unique. New sets can always be made out of linear combinations of the old. There are always many choices for the base states, and among them, any choice is equally legitimate. So the question is not what is *the* base set, but what *could* a base set be? We can choose any one we wish for our own convenience. It is usually best to start with a base set which is *physically* the clearest. It may not be the solution

12–1 Base states for a system with two spin one-half particles

12–2 The Hamiltonian for the ground state of hydrogen

12–3 The energy levels

12–4 The Zeeman splitting

12–5 The states in a magnetic field

12–6 The projection matrix for spin one

to any problem, or may not have any *direct* importance, but it will generally make it easier to understand what is going on.

We choose the following four base states:

State 1: The electron and proton are both spin "up."
State 2: The electron is "up" and the proton is "down."
State 3: The electron is "down" and the proton is "up."
State 4: The electron and proton are both "down."

We need a handy notation for these four states, so we'll represent them this way:

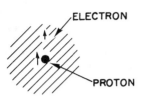

ELECTRON

PROTON

State 1: $|+\,+\rangle$; electron *up*, proton *up*.
State 2: $|+\,-\rangle$; electron *up*, proton *down*.
State 3: $|-\,+\rangle$; electron *down*, proton *up*.
State 4: $|-\,-\rangle$; electron *down*, proton *down*.

(12.1)

You will have to remember that the *first* plus or minus sign refers to the electron and the *second*, to the proton. For handy reference, we've also summarized the notation in Fig. 12–1. Sometimes it will also be convenient to call these states $|\,1\rangle$, $|\,2\rangle$, $|\,3\rangle$, and $|\,4\rangle$.

You may say, "But the particles interact, and maybe these aren't the right base states. It sounds as though you are considering the two particles independently." Yes, indeed! The interaction raises the problem: what is the *Hamiltonian* for the system, but the interaction is not involved in the question of how to *describe* the system. What we choose for the base states has nothing to do with what happens next. It may be that the atom cannot ever *stay* in one of these base states, even if it is started that way. That's another question. That's the question: How do the amplitudes change with time in a particular (fixed) base? In choosing the base states, we are just choosing the "unit vectors" for our description.

While we're on the subject, let's look at the general problem of finding a set of base states when there is more than one particle. You know the base states for a single particle. An electron, for example, is completely described in real life—not in our simplified cases, but in real life—by giving the amplitudes to be in each of the following states:

$$|\text{ electron "up" with momentum } \boldsymbol{p}\rangle$$

or

$$|\text{ electron "down" with momentum } \boldsymbol{p}\rangle.$$

Fig. 12–1. A set of base states for the ground state of the hydrogen atom.

There are really two infinite sets of states, one state for each value of $\boldsymbol{p}$. That is to say that an electron state $|\,\psi\rangle$ is completely described if you know all the amplitudes

$$\langle +, \boldsymbol{p}\,|\,\psi\rangle \quad \text{and} \quad \langle -, \boldsymbol{p}\,|\,\psi\rangle,$$

where the $+$ and $-$ represent the components of angular momentum along some axis—usually the z-axis—and $\boldsymbol{p}$ is the vector momentum. There must, therefore, be two amplitudes for every possible momentum (a multi-infinite set of base states). That is all there is to describing a single particle.

When there is more than one particle, the base states can be written in a similar way. For instance, if there were an electron and a proton in a more complicated situation than we are considering, the base states could be of the following kind:
$|$ an electron with spin "up," moving with momentum $\boldsymbol{p}_1$ and

a proton with spin "down," moving with momentum $\boldsymbol{p}_2\rangle.$

And so on for other spin combinations. If there are more than two particles—same idea. So you see that to write down the *possible* base states is really very easy. The only problem is, what is the Hamiltonian?

For our study of the ground state of hydrogen we don't need to use the full sets of base states for the various momenta. We are specifying particular momentum states for the proton and electron when we say "the ground state." The details of the configuration—the amplitudes for all the momentum base states—can be calculated, but that is another problem. Now we are concerned only with the effects of the spin, so we can take only the four base states of (12.1). Our next problem is: What is the Hamiltonian for this set of states?

12–2 The Hamiltonian for the ground state of hydrogen

We'll tell you in a moment what it is. But first, we should remind you of one thing: *any* state can always be written as a linear combination of the base states. For any state $|\psi\rangle$ we can write

$$|\psi\rangle = |+\,+\rangle\langle+\,+\,|\,\psi\rangle + |+\,-\rangle\langle+\,-\,|\,\psi\rangle + |-\,+\rangle\langle-\,+\,|\,\psi\rangle$$
$$+ |-\,-\rangle\langle-\,-\,|\,\psi\rangle. \qquad (12.2)$$

Remember that the complete brackets are just complex numbers, so we can also write them in the usual fashion as C_i, where $i = 1, 2, 3,$ or 4, and write Eq. (12.2) as

$$|\psi\rangle = |+\,+\rangle C_1 + |+\,-\rangle C_2 + |-\,+\rangle C_3 + |-\,-\rangle C_4. \qquad (12.3)$$

By giving the four amplitudes C_i we completely describe the spin state $|\psi\rangle$. If these four amplitudes change with time, as they will, the rate of change in time is given by the operator $\hat{H}$. The problem is to find the $\hat{H}$.

There is no general rule for writing down the Hamiltonian of an atomic system, and finding the right formula is much more of an art than finding a set of base states. We were able to tell you a general rule for writing a set of base states for any problem of a proton and an electron, but to describe the general Hamiltonian of such a combination is too hard at this level. Instead, we will lead you to a Hamiltonian by some heuristic argument—and you will have to accept it as the correct one because the results will agree with the test of experimental observation.

You will remember that in the last chapter we were able to describe the Hamiltonian of a single, spin one-half particle by using the sigma matrices—or the exactly equivalent sigma operators. The properties of the operators are summarized in Table 12–1. These operators—which are just a convenient, shorthand way of keeping track of the matrix elements of the type $\langle+\,|\,\sigma_z\,|\,+\rangle$—were useful for describing the behavior of a *single* particle of spin one-half. The question is: Can we find an analogous device to describe a system with two spins? The answer is yes, very simply, as follows. We invent a thing which we will call "sigma electron," which we represent by the vector operator $\boldsymbol{\sigma}^e$, and which has the x-, y-, and z-components, $\sigma_x^e, \sigma_y^e, \sigma_z^e$. We now make the *convention* that when one of these things operates on any one of our four base states of the hydrogen atom, it acts only on the *electron* spin, and in exactly the same way as if the electron were all by itself. Example: What is $\sigma_y^e\,|-\,+\rangle$? Since σ_y on an electron "down" is $-i$ times the corresponding state with the electron "up",

$$\sigma_y^e\,|-\,+\rangle = -i\,|+\,+\rangle.$$

(When σ_y^e acts on the combined state it flips over the electron, but does nothing to the proton and multiplies the result by $-i$.) Operating on the other states, σ_y^e would give

$$\sigma_y^e\,|+\,+\rangle = i\,|-\,+\rangle,$$
$$\sigma_y^e\,|+\,-\rangle = i\,|-\,-\rangle,$$
$$\sigma_y^e\,|-\,-\rangle = -i\,|+\,-\rangle.$$

Just remember that the operators $\boldsymbol{\sigma}^e$ work only on the *first* spin symbol—that is, on the *electron* spin.

Next we define the corresponding operator "sigma proton" for the proton spin. Its three components $\sigma_x^p, \sigma_y^p, \sigma_z^p$ act in the same way as $\boldsymbol{\sigma}^e$, only on the

Table 12–1

$$\sigma_z\,|+\rangle = +\,|+\rangle$$
$$\sigma_z\,|-\rangle = -\,|-\rangle$$
$$\sigma_x\,|+\rangle = +\,|-\rangle$$
$$\sigma_x\,|-\rangle = +\,|+\rangle$$
$$\sigma_y\,|+\rangle = +\,i\,|-\rangle$$
$$\sigma_y\,|-\rangle = -i\,|+\rangle$$

proton spin. For example, if we have σ_x^p acting on each of the four base states, we get—always using Table 12–1—

$$\sigma_x^p \,|++\rangle = |+-\rangle,$$
$$\sigma_x^p \,|+-\rangle = |++\rangle,$$
$$\sigma_x^p \,|-+\rangle = |--\rangle,$$
$$\sigma_x^p \,|--\rangle = |-+\rangle.$$

As you can see, it's not very hard.

Now in the most general case we could have more complex things. For instance, we could have products of the two operators like $\sigma_y^e \sigma_z^p$. When we have such a product we do first what the operator on the right says, and then do what the other one says.† For example, we would have that

$$\sigma_x^e \sigma_z^p \,|+-\rangle = \sigma_x^e(\sigma_z^p \,|+-\rangle) = \sigma_x^e(-\,|+-\rangle) = -\sigma_x^e \,|+-\rangle = -\,|--\rangle.$$

Note that these operators don't do anything on pure numbers—we have used this fact when we wrote $\sigma_x^e(-1) = (-1)\sigma_x^e$. We say that the operators "commute" with pure numbers, or that a number "can be moved through" the operator. You can practice by showing that the product $\sigma_x^e \sigma_z^p$ gives the following results for the four states:

$$\sigma_x^e \sigma_z^p \,|++\rangle = +\,|-+\rangle,$$
$$\sigma_x^e \sigma_z^p \,|+-\rangle = -\,|--\rangle,$$
$$\sigma_x^e \sigma_z^p \,|-+\rangle = +\,|++\rangle,$$
$$\sigma_x^e \sigma_z^p \,|--\rangle = -\,|+-\rangle.$$

If we take all the possible operators, using each kind of operator only once, there are sixteen possibilities. Yes, *sixteen*—provided we include also the "unit operator" $\hat{1}$. First, there are the three: $\sigma_x^e, \sigma_y^e, \sigma_z^e$. Then the three $\sigma_z^p, \sigma_y^p, \sigma_z^p$—that makes six. In addition, there are the nine possible products of the form $\sigma_x^e \sigma_y^p$, which makes a total of 15. And there's the unit operator which just leaves any state unchanged. Sixteen in all.

Now note that for a four-state system, the Hamiltonian matrix has to be a four-by-four matrix of coefficients—it will have sixteen entries. It is easily demonstrated that any four-by-four matrix—and, therefore, the Hamiltonian matrix in particular—can be written as a linear combination of the sixteen double-spin matrices corresponding to the set of operators we have just made up. Therefore, for the interaction between a proton and an electron that involves only their spins, we can expect that the Hamiltonian operator can be written as a linear combination of the same 16 operators. The only question is, how?

Well, first, we know that the interaction doesn't depend on our choice of axes for a coordinate system. If there is no external disturbance—like a magnetic field—to determine a unique direction in space, the Hamiltonian can't depend on our choice of the direction of the x-, y-, and z-axes. That means that the Hamiltonian can't have a term like σ_x^e all by itself. It would be ridiculous, because then somebody with a different coordinate system would get different results.

The only possibilities are a term with the unit matrix, say a constant a (times $\hat{1}$), and some combination of the sigmas that doesn't depend on the coordinates—some "invariant" combination. The only *scalar* invariant combination of two vectors is the dot product, which for our σ's is

$$\boldsymbol{\sigma}^e \cdot \boldsymbol{\sigma}^p = \sigma_x^e \sigma_x^p + \sigma_y^e \sigma_y^p + \sigma_z^e \sigma_z^p. \tag{12.4}$$

This operator is invariant with respect to any rotation of the coordinate system.

---

† For these *particular* operators, you will notice it turns out that the sequence of the operators doesn't matter.

12–4

So the only possibility for a Hamiltonian with the proper symmetry in space is a constant times the unit matrix plus a constant times this dot product, say,

$$\hat{H} = E_0 + A\,\boldsymbol{\sigma}^e \cdot \boldsymbol{\sigma}^p. \tag{12.5}$$

That's our Hamiltonian. It's the only thing that it can be, by the symmetry of space, *so long as there is no external field.* The constant term doesn't tell us much; it just depends on the level we choose to measure energies from. We may just as well take $E_0 = 0$. The second term tells us all we need to know to find the level splitting of the hydrogen.

If you want to, you can think of the Hamiltonian in a different way. If there are two magnets near each other with magnetic moments $\boldsymbol{\mu}_e$ and $\boldsymbol{\mu}_p$, the mutual energy will depend on $\boldsymbol{\mu}_e \cdot \boldsymbol{\mu}_p$—among other things. And, you remember, we found that the classical thing we call $\boldsymbol{\mu}_e$ appears in quantum mechanics as $\mu_e \boldsymbol{\sigma}_e$. Similarly, what appears classically as $\boldsymbol{\mu}_p$ will usually turn out in quantum mechanics to be $\mu_p \boldsymbol{\sigma}_p$ (where μ_p is the magnetic moment of the proton, which is about 1000 times smaller than μ_e, and has the opposite sign). So Eq. (12.5) says that the interaction energy is like the interaction between two magnets—only not quite, because the interaction of the two magnets depends on the radial distance between them. But Eq. (12.5) could be—and, in fact, *is*—some kind of an average interaction. The electron is moving all around inside the atom, and our Hamiltonian gives only the average interaction energy. All it says is that for a prescribed arrangement in space for the electron and proton there is an energy proportional to the cosine of the angle between the two magnetic moments, speaking classically. Such a classical qualitative picture may help you to understand where it comes from, but the important thing is that Eq. (12.5) is the correct quantum mechanical formula.

The order of magnitude of the classical interaction between two magnets would be the product of the two magnetic moments divided by the cube of the distance between them. The distance between the electron and the proton in the hydrogen atom is, speaking roughly, one half an atomic radius, or 0.5 angstrom. It is, therefore, possible to make a crude estimate that the constant A should be about equal to the product of the two magnetic moments μ_e and μ_p divided by the cube of 1/2 angstrom. Such an estimate gives a number in the right ball park. It turns out that A can be calculated accurately once you understand the complete quantum theory of the hydrogen atom—which we so far do not. It has, in fact, been calculated to an accuracy of about 30 parts in one million. So, unlike the flip-flop constant A of the ammonia molecule, which couldn't be calculated at all well by a theory, our constant A for the hydrogen *can* be calculated from a more detailed theory. But never mind, we will for our present purposes think of the A as a number which could be determined by experiment, and analyze the physics of the situation.

Taking the Hamiltonian of Eq. (12.5), we can use it with the equation

$$i\hbar \dot{C}_i = \sum_j H_{ij} C_j \tag{12.6}$$

to find out what the spin interactions do to the energy levels. To do that, we need to work out the sixteen matrix elements $H_{ij} = \langle i \,|\, H \,|\, j \rangle$ corresponding to each pair of the four base states in (12.1).

We begin by working out what $\hat{H}\,|\,j\rangle$ is for each of the four base states. For example,

$$\hat{H}\,|++\rangle = A\boldsymbol{\sigma}^e \cdot \boldsymbol{\sigma}^p \,|++\rangle = A\{\sigma_x^e \sigma_x^p + \sigma_y^e \sigma_y^p + \sigma_z^e \sigma_z^p\}\,|++\rangle. \tag{12.7}$$

Using the method we described a little earlier—it's easy if you have memorized Table 12–1—we find what each pair of σ's does on $|++\rangle$. The answer is

$$
\begin{aligned}
\sigma_x^e \sigma_x^p \,|++\rangle &= +\,|--\rangle, \\
\sigma_y^e \sigma_y^p \,|++\rangle &= -\,|--\rangle, \\
\sigma_z^e \sigma_z^p \,|++\rangle &= +\,|++\rangle.
\end{aligned}
\tag{12.8}
$$

So (12.7) becomes

$$\hat{H}\,|++\rangle = A\{|--\rangle - |--\rangle + |++\rangle\} = A\,|++\rangle. \qquad (12.9)$$

Since our four base states are all orthogonal, that gives us immediately that

$$\begin{aligned}
\langle ++\,|\,H\,|++\rangle &= A\langle ++\,|++\rangle = A,\\
\langle +-\,|\,H\,|++\rangle &= A\langle +-\,|++\rangle = 0,\\
\langle -+\,|\,H\,|++\rangle &= A\langle -+\,|++\rangle = 0,\\
\langle --\,|\,H\,|++\rangle &= A\langle --\,|++\rangle = 0.
\end{aligned} \qquad (12.10)$$

Remembering that $\langle j\,|\,H\,|\,i\rangle = \langle i\,|\,H\,|\,j\rangle^*$, we can already write down the differential equation for the amplitudes C_1:

$$i\hbar\dot{C}_1 = H_{11}C_1 + H_{12}C_2 + H_{13}C_3 + H_{14}C_4$$

or

$$i\hbar\dot{C}_1 = AC_1. \qquad (12.11)$$

That's all! We get only the one term.

Now to get the rest of the Hamiltonian equations we have to crank through the same procedure for $\hat{H}$ operating on the other states. First, we will let you practice by checking out all of the sigma products we have written down in Table 12–2. Then we can use them to get:

$$\begin{aligned}
\hat{H}\,|+-\rangle &= A\{2\,|-+\rangle - |+-\rangle\},\\
\hat{H}\,|-+\rangle &= A\{2\,|+-\rangle - |-+\rangle\},\\
\hat{H}\,|--\rangle &= A\,|--\rangle.
\end{aligned} \qquad (12.12)$$

Then, multiplying each one in turn on the left by all the other state vectors, we get the following Hamiltonian matrix, H_{ij}:

$$H_{ij} = \overset{i\downarrow}{\overset{\overrightarrow{j}}{\begin{pmatrix} A & 0 & 0 & 0 \\ 0 & -A & 2A & 0 \\ 0 & 2A & -A & 0 \\ 0 & 0 & 0 & A \end{pmatrix}}}. \qquad (12.13)$$

It means, of course, nothing more than that our differential equations for the four amplitudes C_i are

$$\begin{aligned}
i\hbar\dot{C}_1 &= AC_1,\\
i\hbar\dot{C}_2 &= -AC_2 + 2AC_3,\\
i\hbar\dot{C}_3 &= 2AC_2 - AC_3,\\
i\hbar\dot{C}_4 &= AC_4.
\end{aligned} \qquad (12.14)$$

Before solving these equations we can't resist telling you about a clever rule due to Dirac—it will make you feel that you are really advanced—although we don't need it for our work. We have—from the equations (12.9) and (12.12)—that

$$\begin{aligned}
\boldsymbol{\sigma}^e \cdot \boldsymbol{\sigma}^p\,|++\rangle &= |++\rangle,\\
\boldsymbol{\sigma}^e \cdot \boldsymbol{\sigma}^p\,|+-\rangle &= 2\,|-+\rangle - |+-\rangle,\\
\boldsymbol{\sigma}^e \cdot \boldsymbol{\sigma}^p\,|-+\rangle &= 2\,|+-\rangle - |-+\rangle,\\
\boldsymbol{\sigma}^e \cdot \boldsymbol{\sigma}^p\,|--\rangle &= |--\rangle.
\end{aligned} \qquad (12.15)$$

Table 12–2

Spin operators for the hydrogen atom

$$\begin{aligned}
\sigma_x^e\sigma_x^p\,|++\rangle &= +\,|--\rangle\\
\sigma_x^e\sigma_x^p\,|+-\rangle &= +\,|-+\rangle\\
\sigma_x^e\sigma_x^p\,|-+\rangle &= +\,|+-\rangle\\
\sigma_x^e\sigma_x^p\,|--\rangle &= +\,|++\rangle\\[6pt]
\sigma_y^e\sigma_y^p\,|++\rangle &= -\,|--\rangle\\
\sigma_y^e\sigma_y^p\,|+-\rangle &= +\,|-+\rangle\\
\sigma_y^e\sigma_y^p\,|-+\rangle &= +\,|+-\rangle\\
\sigma_y^e\sigma_y^p\,|--\rangle &= -\,|++\rangle\\[6pt]
\sigma_z^e\sigma_z^p\,|++\rangle &= +\,|++\rangle\\
\sigma_z^e\sigma_z^p\,|+-\rangle &= -\,|+-\rangle\\
\sigma_z^e\sigma_z^p\,|-+\rangle &= -\,|-+\rangle\\
\sigma_z^e\sigma_z^p\,|--\rangle &= +\,|--\rangle
\end{aligned}$$

Look, said Dirac, I can also write the first and last equations as

$$\boldsymbol{\sigma}^e \cdot \boldsymbol{\sigma}^p \, | + + \rangle = 2 \, | + + \rangle - | + + \rangle,$$

$$\boldsymbol{\sigma}^e \cdot \boldsymbol{\sigma}^p \, | - - \rangle = 2 \, | - - \rangle - | - - \rangle;$$

then they are all quite similar. Now I invent a new operator, which I will call $P_{\text{spin exch}}$ and which I *define* to have the following properties:†

$$P_{\text{spin exch}} \, | + + \rangle = | + + \rangle,$$

$$P_{\text{spin exch}} \, | + - \rangle = | - + \rangle,$$

$$P_{\text{spin exch}} \, | - + \rangle = | + - \rangle,$$

$$P_{\text{spin exch}} \, | - - \rangle = | - - \rangle.$$

All the operator does is interchange the spin directions of the two particles. Then I can write the whole set of equations in (12.15) as a simple operator equation:

$$\boldsymbol{\sigma}^e \cdot \boldsymbol{\sigma}^p = 2 P_{\text{spin exch}} - 1. \tag{12.16}$$

That's the formula of Dirac. His "spin exchange operator" gives a handy rule for figuring out $\boldsymbol{\sigma}^e \cdot \boldsymbol{\sigma}^p$. (You see, you can do everything now. The gates are open.)

12–3 The energy levels

Now we are ready to work out the energy levels of the ground state of hydrogen by solving the Hamiltonian equations (12.14). We want to find the energies of the stationary states. This means that we want to find those special states $| \psi \rangle$ for which each amplitude $C_i = \langle i \, | \, \psi \rangle$ in the set belonging to $| \psi \rangle$ has the same time dependence—namely, $e^{-i\omega t}$. Then the state will have the energy $E = \hbar\omega$. So we want a set for which

$$C_i = a_i e^{(-i/\hbar)Et}, \tag{12.17}$$

where the four coefficients a_i are independent of time. To see whether we can get such amplitudes, we substitute (12.17) into Eq. (12.14) and see what happens. Each $i\hbar \, dC/dt$ in Eq. (12.14) turns into EC, and—after cancelling out the common exponential factor—each C becomes an a; we get

$$
\begin{aligned}
E a_1 &= A a_1, \\
E a_2 &= -A a_2 + 2A a_3, \\
E a_3 &= 2A a_2 - A a_3, \\
E a_4 &= A a_4,
\end{aligned}
\tag{12.18}
$$

which we have to solve for a_1, a_2, a_3, and a_4. Isn't it nice that the first equation is independent of the rest—that means we can see one solution right away. If we choose $E = A$,

$$a_1 = 1, \qquad a_2 = a_3 = a_4 = 0,$$

gives a solution. (Of course, taking all the a's equal to zero also gives a solution, but that's no state at all!) Let's call our first solution the state $| I \rangle$:‡

$$| I \rangle = | 1 \rangle = | + + \rangle. \tag{12.19}$$

Its energy is

$$E_I = A.$$

---

† This operator is now called the "Pauli spin exchange operator."
‡ The state is really $| I \rangle e^{-(i/\hbar)E_I t}$; but, as usual we will identify the states by the constant vectors which are equal to the complete vectors at $t = 0$.

With that clue you can immediately see another solution from the last equation in (12.18):

$$a_1 = a_2 = a_3 = 0, \qquad a_4 = 1,$$

$$E = A.$$

We'll call that solution state $| II \rangle$:

$$| II \rangle = | 4 \rangle = | - - \rangle, \tag{12.20}$$

$$E_{II} = A.$$

Now it gets a little harder; the two equations left in (12.18) are mixed up. But we've done it all before. Adding the two, we get

$$E(a_2 + a_3) = A(a_2 + a_3). \tag{12.21}$$

Subtracting, we have

$$E(a_2 - a_3) = -3A(a_2 - a_3). \tag{12.22}$$

By inspection—and remembering ammonia—we see that there are two solutions:

$$a_2 = a_3, \qquad E = A$$

and

$$a_2 = -a_3, \qquad E = -3A. \tag{12.23}$$

They are mixtures of $| 2 \rangle$ and $| 3 \rangle$. Calling these states $| III \rangle$ and $| IV \rangle$, and putting in a factor $1/\sqrt{2}$ to make the states properly normalized, we have

$$| III \rangle = \frac{1}{\sqrt{2}} (| 2 \rangle + | 3 \rangle) = \frac{1}{\sqrt{2}} (| + - \rangle + | - + \rangle),$$

$$E_{III} = A \tag{12.24}$$

and

$$| IV \rangle = \frac{1}{\sqrt{2}} (| 2 \rangle - | 3 \rangle) = \frac{1}{\sqrt{2}} (| + - \rangle - | - + \rangle),$$

$$E_{IV} = -3A. \tag{12.25}$$

We have found four stationary states and their energies. Notice, incidentally, that our four states are orthogonal, so they also can be used for base states if desired. Our problem is completely solved.

Three of the states have the energy A, and the last has the energy $-3A$. The average is zero—which means that when we took $E_0 = 0$ in Eq. (12.5), we were choosing to measure all the energies from the average energy. We can draw the energy-level diagram for the ground state of hydrogen as shown in Fig. 12–2.

Now the difference in energy between state $| IV \rangle$ and any one of the others is $4A$. An atom which happens to have gotten into state $| I \rangle$ could fall from there to state $| IV \rangle$ and emit light. Not optical light, because the energy is so tiny—it would emit a microwave quantum. Or, if we shine microwaves on hydrogen gas, we will find an absorption of energy as the atoms in state $| IV \rangle$ pick up energy and go into one of the upper states—but only at the frequency $\omega = 4A/\hbar$. This frequency has been measured experimentally; the best result, obtained very recently,† is

$$f = \omega/2\pi = (1{,}420{,}405{,}751.800 \pm 0.028) \text{ cycles per second.} \tag{12.26}$$

The error is only two parts in 100 billion! Probably no basic physical quantity is measured better than that—it's one of the most remarkably accurate measurements in physics. The theorists were very happy that they could compute the energy to an accuracy of 3 parts in 10^5, but in the meantime it has been measured to 2 parts in 10^{11}—a million times more accurate than the theory. So the experimenters are

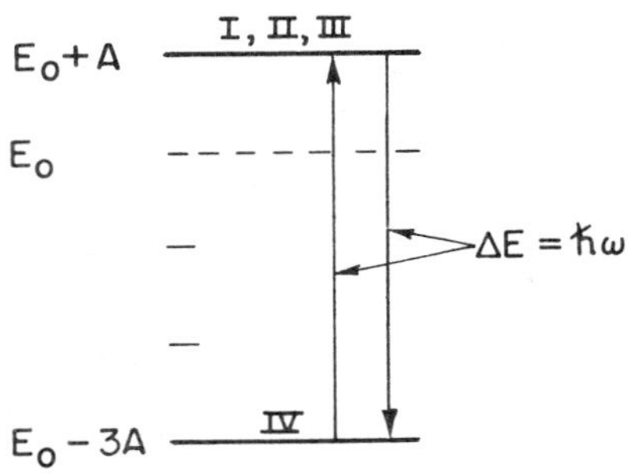

Fig. 12–2. Energy-level diagram for the ground state of atomic hydrogen.

---

† Crampton, Kleppner, and Ramsey; *Physical Review Letters*, Vol. **11**, page 338 (1963).

way ahead of the theorists. In the theory of the ground state of the hydrogen atom *you* are as good as anybody. You, too, can just take your value of A from experiment—that's what everybody has to do in the end.

You have probably heard before about the "21-centimeter line" of hydrogen. That's the wavelength of the 1420 megacycle spectral line between the hyperfine states. Radiation of this wavelength is emitted or absorbed by the atomic hydrogen gas in the galaxies. So with radio telescopes tuned in to 21-cm waves (or 1420 megacycles approximately) we can observe the velocities and the location of concentrations of atomic hydrogen gas. By measuring the intensity, we can estimate the amount of hydrogen. By measuring the frequency shift due to the Doppler effect, we can find out about the motion of the gas in the galaxy. That is one of the big programs of radio astronomy. So now we are talking about something that's very real—it is not an artificial problem.

12–4 The Zeeman splitting

Although we have finished the problem of finding the energy levels of the hydrogen ground state, we would like to study this interesting system some more. In order to say anything more about it—for instance, in order to calculate the rate at which the hydrogen atom absorbs or emits radio waves at 21 centimeters—we have to know what happens when the atom is disturbed. We have to do as we did for the ammonia molecule—after we found the energy levels we went on and studied what happened when the molecule was in an electric field. We were then able to figure out the effects from the electric field in a radio wave. For the hydrogen atom, the electric field does nothing to the levels, except to move them all by some constant amount proportional to the square of the field—which is not of any interest because that won't change the energy *differences*. It is now the *magnetic* field which is important. So the next step is to write the Hamiltonian for a more complicated situation in which the atom sits in an external magnetic field.

What, then, is the Hamiltonian? We'll just tell you the answer, because we can't give you any "proof" except to say that this is the way the atom works.

The Hamiltonian is

$$\hat{H} = A(\boldsymbol{\sigma}^{e} \cdot \boldsymbol{\sigma}^{p}) - \mu_{e}\boldsymbol{\sigma}^{e} \cdot \boldsymbol{B} - \mu_{p}\boldsymbol{\sigma}^{p} \cdot \boldsymbol{B}. \qquad (12.27)$$

It now consists of three parts. The first term $A\boldsymbol{\sigma}^{e} \cdot \boldsymbol{\sigma}^{p}$ represents the magnetic interaction between the electron and the proton—it is the same one that would be there if there were no magnetic field. This is the term we have already had; and the influence of the magnetic field on the constant A is negligible. The effect of the external magnetic field shows up in the last two terms. The second term, $-\mu_{e}\boldsymbol{\sigma}^{e} \cdot \boldsymbol{B}$, is the energy the electron would have in the magnetic field if it were there alone.† In the same way, the last term $-\mu_{p}\boldsymbol{\sigma}^{p} \cdot \boldsymbol{B}$, would have been the energy of a proton alone. Classically, the energy of the two of them together would be the sum of the two, and that works also quantum mechanically. In a magnetic field, the energy of interaction due to the magnetic field is just the sum of the energy of interaction of the electron with the external field, and of the proton with the field—both expressed in terms of the sigma operators. In quantum mechanics these terms are not really the energies, but thinking of the classical formulas for the energy is a way of remembering the rules for writing down the Hamiltonian. Anyway, the correct Hamiltonian is Eq. (12.27).

Now we have to go back to the beginning and do the problem all over again. Much of the work is, however, done—we need only to add the effects of the new terms. Let's take a constant magnetic field $\boldsymbol{B}$ in the z-direction. Then we have to

---

† Remember that classically $U = -\boldsymbol{\mu} \cdot \boldsymbol{B}$, so the energy is lowest when the moment is along the field. For positive particles, the magnetic moment is parallel to the spin and for negative particles it is opposite. So in Eq. (12.27), μ_{p} is a *positive* number, but μ_{e} is a *negative* number.

add to our Hamiltonian operator $\hat{H}$ the two new pieces—which we can call $\hat{H}'$:

$$\hat{H}' = -(\mu_{\mathrm{e}}\sigma_z^{\mathrm{e}} + \mu_{\mathrm{p}}\sigma_z^{\mathrm{p}})B.$$

Using Table 12–1, we get right away that

$$\hat{H}'\,|++\rangle = -(\mu_{\mathrm{e}} + \mu_{\mathrm{p}})B\,|++\rangle,$$
$$\hat{H}'\,|+-\rangle = -(\mu_{\mathrm{e}} - \mu_{\mathrm{p}})B\,|+-\rangle,$$
$$\hat{H}'\,|-+\rangle = -(-\mu_{\mathrm{e}} + \mu_{\mathrm{p}})B\,|-+\rangle, \tag{12.28}$$
$$\hat{H}'\,|--\rangle = (\mu_{\mathrm{e}} + \mu_{\mathrm{p}})B\,|--\rangle.$$

How very convenient! The $\hat{H}'$ operating on each state just gives a number times that state. The matrix $\langle i \mid H' \mid j\rangle$ has, therefore, only *diagonal* elements—we can just add the coefficients in (12.28) to the corresponding diagonal terms of (12.13), and the Hamiltonian equations of (12.14) become

$$i\hbar dC_1/dt = \{A - (\mu_{\mathrm{e}} + \mu_{\mathrm{p}})B\}C_1,$$
$$i\hbar dC_2/dt = -\{A + (\mu_{\mathrm{e}} - \mu_{\mathrm{p}})B\}C_2 + 2AC_3,$$
$$i\hbar dC_3/dt = 2AC_2 - \{A - (\mu_{\mathrm{e}} - \mu_{\mathrm{p}})B\}C_3, \tag{12.29}$$
$$i\hbar dC_4/dt = \{A + (\mu_{\mathrm{e}} + \mu_{\mathrm{p}})B\}C_4.$$

The form of the equations is not different—only the coefficients. So long as B doesn't vary with time, we can continue as we did before. Substituting $C_i = a_i e^{-(i/\hbar)Et}$, we get—as a modification of (12.18)—

$$Ea_1 = A\,\{-(\mu_{\mathrm{e}} + \mu_{\mathrm{p}})B\}a_1,$$
$$Ea_2 = -\{A + (\mu_{\mathrm{e}} - \mu_{\mathrm{p}})B\}a_2 + 2Aa_3,$$
$$Ea_3 = 2Aa_2 - \{A - (\mu_{\mathrm{e}} - \mu_{\mathrm{p}})B\}a_3, \tag{12.30}$$
$$Ea_4 = \{A + (\mu_{\mathrm{e}} + \mu_{\mathrm{p}})B\}a_4.$$

Fortunately, the first and fourth equations are still independent of the rest, so the same technique works again.

One solution is the state $\mid I\rangle$ for which $a_1 = 1$, $a_2 = a_3 = a_4 = 0$, or

$$\mid I\rangle = \mid 1\rangle = \mid ++\rangle,$$

with
$$E_I = A - (\mu_{\mathrm{e}} + \mu_{\mathrm{p}})B. \tag{12.31}$$

Another is
$$\mid II\rangle = \mid 4\rangle = \mid --\rangle,$$

with
$$E_{II} = A + (\mu_{\mathrm{e}} + \mu_{\mathrm{p}})B. \tag{12.32}$$

A little more work is involved for the remaining two equations, because the coefficients of a_2 and a_3 are no longer equal. But they are just like the pair we had for the ammonia molecule. Looking back at Eq. (9.20), we can make the following analogy (remembering that the labels 1 and 2 there correspond to 2 and 3 here):

$$H_{11} \to -A - (\mu_{\mathrm{e}} - \mu_{\mathrm{p}})B,$$
$$H_{12} \to 2A,$$
$$H_{21} \to 2A, \tag{12.33}$$
$$H_{22} \to -A + (\mu_{\mathrm{e}} - \mu_{\mathrm{p}})B.$$

The energies are then given by (9.25), which was

$$E = \frac{H_{11} + H_{22}}{2} \pm \sqrt{\frac{(H_{11} - H_{22})^2}{4} + H_{12}H_{21}}. \tag{12.34}$$

Making the substitutions from (12.33), the energy formula becomes

$$E = -A \pm \sqrt{(\mu_{\mathrm e} - \mu_{\mathrm p})^2 B^2 + 4A^2}.$$

Although in Chapter 9 we used to call these energies E_I and E_{II}, and we are in this problem calling them E_{III} and E_{IV},

$$E_{III} = A\{-1 + 2\sqrt{1 + (\mu_{\mathrm e} - \mu_{\mathrm p})^2 B^2/4A^2}\},$$
$$E_{IV} = -A\{1 + 2\sqrt{1 + (\mu_{\mathrm e} - \mu_{\mathrm p})^2 B^2/4A^2}\}. \tag{12.35}$$

So we have found the energies of the four stationary states of a hydrogen atom in a constant magnetic field. Let's check our results by letting B go to zero and seeing whether we get the same energies we had in the preceding section. You see that we do. For $B = 0$, the energies E_I, E_{II}, and E_{III} go to $+A$, and E_{IV} goes to $-3A$. Even our labeling of the states agrees with what we called them before. When we turn on the magnetic field though, all of the energies change in a different way. Let's see how they go.

First, we have to remember that for the electron, $\mu_{\mathrm e}$ is negative, and about 1000 times larger than $\mu_{\mathrm p}$—which is positive. So $\mu_{\mathrm e} + \mu_{\mathrm p}$ and $\mu_{\mathrm e} - \mu_{\mathrm p}$ are both negative numbers, and nearly equal. Let's call them $-\mu$ and $-\mu'$:

$$\mu = -(\mu_{\mathrm e} + \mu_{\mathrm p}), \qquad \mu' = -(\mu_{\mathrm e} - \mu_{\mathrm p}). \tag{12.36}$$

(Both μ and μ' are positive numbers, nearly equal to magnitude of $\mu_{\mathrm e}$—which is about one Bohr magneton.) Then our four energies are

$$E_I = A + \mu B,$$
$$E_{II} = A - \mu B,$$
$$E_{III} = A\{-1 + 2\sqrt{1 + \mu'^2 B^2/4A^2}\}, \tag{12.37}$$
$$E_{IV} = -A\{1 + 2\sqrt{1 + \mu'^2 B^2/4A^2}\}.$$

The energy E_I starts at A and increases linearly with B—with the slope μ. The

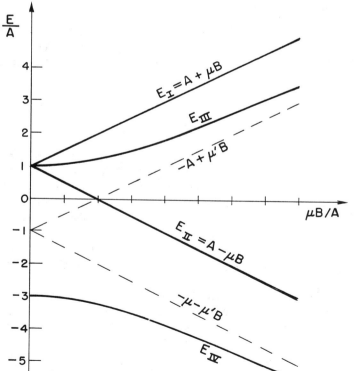

Fig. 12–3. The energy levels of the ground state of hydrogen in a magnetic field **B**.

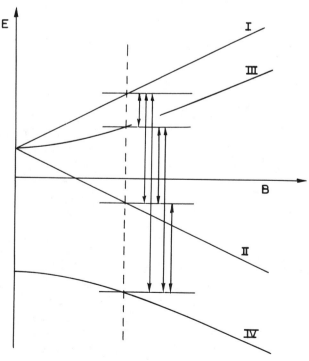

Fig. 12–4. Transitions between the levels of ground state energy levels of hydrogen in some particular magnetic field **B**.

energy E_{II} also starts at A but *decreases* linearly with increasing B—its slope is $-\mu$. These two levels vary with B as shown in Fig. 12–3. We show also in the figure the energies E_{III} and E_{IV}. They have a different B-dependence. For small B, they depend quadratically on B, so they start out with horizontal slopes. Then they begin to curve, and for *large B* they approach straight lines with slopes $\pm\mu'$, which are nearly the same as the slopes of E_I and E_{II}.

The shift of the energy levels of an atom due to a magnetic field is called the *Zeeman effect*. We say that the curves in Fig. 12–3 show the *Zeeman splitting* of the ground state of hydrogen. When there is no magnetic field, we get just one spectral line from the hyperfine structure of hydrogen. The transitions between state $| IV \rangle$ and any one of the others occurs with the absorption or emission of a photon whose frequency 1420 megacycles is $1/h$ times the energy difference $4A$. When the atom is in a magnetic field B, however, there are many more lines. There can be transitions between any two of the four states. So if we have atoms in all four states, energy can be absorbed—or emitted—in any one of the six transitions shown by the vertical arrows in Fig. 12–4. Many of these transitions can be observed by the Rabi molecular beam technique we described in Volume II, Section 35–3 (see Appendix).

What makes the transitions go? The transitions will occur if you apply a small disturbing magnetic field that varies with time (in addition to the steady strong field B). It's just as we saw for a varying electric field on the ammonia molecule. Only here, it is the magnetic field which couples with the magnetic moments and does the trick. But the theory follows through in the same way that we worked it out for the ammonia. The theory is the simplest if you take a perturbing magnetic field that rotates in the xy-plane—although any horizontal oscillating field will do. When you put in this perturbing field as an additional term in the Hamiltonian, you get solutions in which the amplitudes vary with time—as we found for the ammonia molecule. So you can calculate easily and accurately the probability of a transition from one state to another. And you find that it all agrees with experiment.

12–5 The states in a magnetic field

We would like now to discuss the shapes of the curves in Fig. 12–3. In the first place, the energies for large fields are easy to understand, and rather interesting. For B large enough (namely for $\mu B/A \gg 1$) we can neglect the 1 in the formulas of (12.37). The four energies become

$$
\begin{aligned}
E_I &= A + \mu B, & E_{II} &= A - \mu B, \\
E_{III} &= -A + \mu' B, & E_{IV} &= -A - \mu' B.
\end{aligned}
\tag{12.38}
$$

These are the equations of the four straight lines in Fig. 12–3. We can understand these energies physically in the following way. The nature of the stationary states in a *zero* field is determined completely by the interaction of the two magnetic moments. The mixtures of the base states $| + - \rangle$ and $| - + \rangle$ in the stationary states $| III \rangle$ and $| IV \rangle$ are due to this interaction. In *large external* fields, however, the proton and electron will be influenced hardly at all by the field of the other; each will act as if it were alone in the external field. Then—as we have seen many times—the electron spin will be either parallel to or opposite to the external magnetic field.

Suppose the electron spin is "up"—that is, along the field; its energy will be $-\mu_e B$. The proton can still be either way. If the proton spin is also "up," its energy is $-\mu_p B$. The sum of the two is $-(\mu_e + \mu_p)B = \mu B$. That is just what we find for E_I—which is fine, because we are describing the state $| + + \rangle = | I \rangle$. There is still the small additional term A (now $\mu B \gg A$) which represents the interaction energy of the proton and electron when their spins are parallel. (We originally took A as positive because the theory we spoke of says it should be, and experimentally it is indeed so.) On the other hand, the proton can have its spin down. Then its energy in the external field goes to $-\mu_p B$, so it and the electron have the energy $-(\mu_e - \mu_p)B = \mu' B$. And the interaction energy becomes $-A$.

The sum is just the energy E_{III} in (12.38). So the state $|III\rangle$ must for large fields become the state $|+-\rangle$.

Suppose now the electron spin is "down." Its energy in the external field is $\mu_e B$. If the proton is also "down," the two together have the energy $(\mu_e + \mu_p)B = \mu B$, *plus* the interaction energy A—since their spins are parallel. That makes just the energy E_{II} in (12.38) and corresponds to the state $|--\rangle = |II\rangle$—which is nice. Finally if the electron is "down" and the proton is "up," we get the energy $(\mu_e - \mu_p)B - A$ (*minus* A for the interaction because the spins are opposite) which is just E_{IV}. And the state corresponds to $|-+\rangle$.

"But, wait a moment!", you are probably saying, "The states $|III\rangle$ and $|IV\rangle$ *are not* the states $|+-\rangle$ and $|-+\rangle$; they are *mixtures* of the two." Well, only slightly. They are indeed mixtures for $B = 0$, but we have not yet figured out what they are for large B. When we used the analogies of (12.33) in our formulas of Chapter 9 to get the energies of the stationary states, we could also have taken the amplitudes that go with them. They come from Eq. (9.23), which is

$$\frac{a_2}{a_3} = \frac{E - H_{22}}{H_{11}}.$$

The ratio a_2/a_3 is, of course, just C_2/C_3. Plugging in the analogous quantities from (12.33), we get

or

$$\frac{C_2}{C_3} = \frac{E + A - (\mu_e - \mu_p)B}{2A}$$

$$\frac{C_2}{C_3} = \frac{E + A + \mu'B}{2A}, \tag{12.39}$$

where for E we are to use the appropriate energy—either E_{III} or E_{IV}. For instance, for state $|III\rangle$ we have

$$\left(\frac{C_2}{C_3}\right)_{III} \approx \frac{\mu'B}{A}. \tag{12.40}$$

So for large B the state $|III\rangle$ has $C_2 \gg C_3$; the state becomes almost completely the state $|2\rangle = |+-\rangle$. Similarly, if we put E_{IV} into (12.39) we get $(C_2/C_3)_{IV} \ll 1$; for high fields state $|IV\rangle$ becomes just the state $|3\rangle = |-+\rangle$. You see that the coefficients in the linear combinations of our base states which make up the stationary states depend on B. The state we call $|III\rangle$ is a 50–50 mixture of $|+-\rangle$ and $|-+\rangle$ at very low fields, but shifts completely over to $|+-\rangle$ at high fields. Similarly, the state $|IV\rangle$, which at low fields is also a 50–50 mixture (with opposite signs) of $|+-\rangle$ and $|-+\rangle$, goes over into the state $|-+\rangle$ when the spins are uncoupled by a strong external field.

We would also like to call your attention particularly to what happens at *very low* magnetic fields. There is one energy—at $-3A$—which *does not change* when you turn on a small magnetic field. And there is another energy—at $+A$—which splits into three different energy levels when you turn on a small magnetic field. For weak fields the energies vary with B as shown in Fig. 12–5. Suppose that we have somehow selected a bunch of hydrogen atoms which all have the energy $-3A$. If we put them through a Stern-Gerlach experiment—with fields that are not too strong—we would find that they just go straight through. (Since their energy doesn't depend on B, there is—according to the principle of virtual work—no force on them in a magnetic field gradient.) Suppose, on the other hand, we were to select a bunch of atoms with the energy $+A$, and put them through a Stern-Gerlach apparatus, say an S apparatus. (Again the fields in the apparatus should not be so great that they disrupt the insides of the atom, by which we mean a field small enough that the energies vary linearly with B.) We would find *three* beams. The states $|I\rangle$ and $|II\rangle$ get opposite forces—their energies vary linearly with B with the slopes $\pm\mu$ so the *forces* are like those on a dipole with $\mu_z = \mp\mu$; but the state $|III\rangle$ goes straight through. So we are right back in Chapter 5. *A hydrogen atom with the energy $+A$ is a spin-one particle.* This energy state is a "particle" for which $j = 1$, and it can be described—with respect to some set of

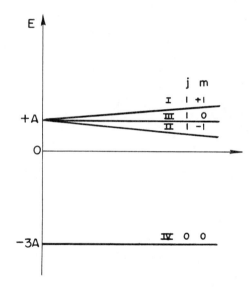

Fig. 12–5. The states of the hydrogen atom for small magnetic fields.

12–13

axes in space—in terms of the base states $|+S\rangle$, $|0\,S\rangle$, and $|-S\rangle$ we used in Chapter 5. On the other hand, when a hydrogen atom has the energy $-3A$, it is a spin-zero particle. (Remember, what we are saying is only strictly true for infinitesimal magnetic fields.) So we can group the states of hydrogen in zero magnetic field this way:

$$\left.\begin{array}{l} |\,I\rangle = |+\,+\rangle \\[2mm] |\,III\rangle = \dfrac{|+\,-\rangle + |-\,+\rangle}{\sqrt{2}} \\[2mm] |\,II\rangle = |-\,-\rangle \end{array}\right\} \text{spin 1} \left\{\begin{array}{l} |+S\rangle \\[2mm] |0\,S\rangle \\[2mm] |-S\rangle \end{array}\right. \tag{12.41}$$

$$|\,IV\rangle = \frac{|+\,-\rangle - |-\,+\rangle}{\sqrt{2}} \quad \text{spin 0.} \tag{12.42}$$

We have said in Chapter 35 of Volume II (Appendix) that for any particle its component of angular momentum along any axis can have only certain values always $\hbar$ apart. The z-component of angular momentum J_z can be $j\hbar$, $(j-1)\hbar$, $(j-2)\hbar, \ldots, (-j)\hbar$, where j is the spin of the particle (which can be an integer or half-integer). Although we neglected to say so at the time, people usually write

$$J_z = m\hbar, \tag{12.43}$$

where m stands for one of the numbers $j, j-1, j-2, \ldots, -j$. You will, therefore, see people in books label the four ground states of hydrogen by the so-called *quantum numbers* j and m [often called the "total angular momentum quantum number" (j), and "magnetic quantum number" (m)]. Then, instead of our state symbols $|\,I\rangle$, $|\,II\rangle$, and so on, they will write a state as $|\,j, m\rangle$. So they would write our little table of states for zero field in (12.41) and (12.42) as shown in Table 12–3. It's not new physics, it's all just a matter of notation.

Table 12–3

Zero field states of the hydrogen atom

State $|\,j, m\rangle$	j	m	Our notation
$|\,1, +1\rangle$	1	+1	$|\,I\rangle = |+S\rangle$
$|\,1, 0\rangle$	1	0	$|\,III\rangle = |0\,S\rangle$
$|\,1, -1\rangle$	1	-1	$|\,II\rangle = |-S\rangle$
$|\,0, 0\rangle$	0	0	$|\,IV\rangle$

12–6 The projection matrix for spin one †

We would like now to use our knowledge of the hydrogen atom to do something special. We discussed in Chapter 5 that a particle of *spin one* which was in one of the base states ($+$, 0, or $-$) with respect to a Stern-Gerlach apparatus of a particular orientation—say an S apparatus—would have a certain amplitude to be in each of the three states with respect to a T apparatus with a different orientation in space. There are nine such amplitudes $\langle jT\,|\,iS\rangle$ which make up the projection matrix. In Section 5–7 we gave without proof the terms of this matrix for various orientations of T with respect to S. Now we will show you one way they can be derived.

In the hydrogen atom we have found a spin-one system which is made up of two spin one-half particles. We have already worked out in Chapter 6 how to transform the spin one-half amplitudes. We can use this information to calculate the transformation for spin one. This is the way it works: We have a system—a hydrogen atom with the energy $+A$—which has spin one. Suppose we run it through a Stern-Gerlach filter S, so that we know it is in one of the base states with respect to S, say $|+S\rangle$. What is the amplitude that it will be in one of the base states, say $|+T\rangle$, with respect to the T apparatus? If we call the coordinate system of the S apparatus the x, y, z system, the $|+S\rangle$ state is what we have been calling the state $|+\,+\rangle$. But suppose another guy took his z-axis along the axis of T. He will be referring his states to what we will call the x', y', z' frame. His "up" and "down" states for the electron and proton would be different from ours. *His* "plus-plus" state—which we can write $|+'\,+'\rangle$, referring to the "prime" frame—is the $|+T\rangle$ state of the spin-one particle. What we want is $\langle +T\,|\,+S\rangle$ which is just another way of writing the amplitude $\langle +'\,+'\,|\,+\,+\rangle$.

---

† Those who chose to jump over Chapter 6 should skip this section also.

We can find the amplitude $\langle +' +' \mid + + \rangle$ in the following way. In *our* frame the *electron* in the $\mid + + \rangle$ state has its spin "up". That means that it has some amplitude $\langle +' \mid + \rangle_e$ of being "up" in *his* frame, and some amplitude $\langle -' \mid + \rangle_e$ of being "down" in that frame. Similarly, the *proton* in the $\mid + + \rangle$ state has spin "up" in our frame and the amplitudes $\langle +' \mid + \rangle_p$ and $\langle -' \mid + \rangle_p$ of having spin "up" or spin "down" in the "prime" frame. Since we are talking about two distinct particles, the amplitude that *both* particles will be "up" *together* in *his* frame is the product of the two amplitudes,

$$\langle +' +' \mid + + \rangle = \langle +' \mid + \rangle_e \langle +' \mid + \rangle_p. \tag{12.44}$$

We have put the subscripts e and p on the amplitudes $\langle +' \mid + \rangle$ to make it clear what we were doing. But they are both just the transformation amplitudes for a spin one-half particle, so they are really identical numbers. They are, in fact, just the amplitude we have called $\langle +T \mid +S \rangle$ in Chapter 6, and which we listed in the tables at the end of that chapter.

Now, however, we are about to get into trouble with notation. We have to be able to distinguish the amplitude $\langle +T \mid +S \rangle$ for a *spin one-half* particle from what we have *also* called $\langle +T \mid +S \rangle$ for a *spin-one* particle—yet they are completely different! We hope it won't be too confusing, but *for the moment* at least, we will have to use some different symbols for the spin one-half amplitudes. To help you keep things straight, we summarize the new notation in Table 12-4. We will continue to use the notation $\mid +S \rangle$, $\mid 0\,S \rangle$, and $\mid -S \rangle$ for the states of a spin-one particle.

With our new notation, Eq. (12.44) becomes simply

$$\langle +' +' \mid + + \rangle = a^2,$$

and this is just the spin-*one* amplitude $\langle +T \mid +S \rangle$. Now, let's suppose, for instance, that the other guy's coordinate frame—that is, the T, or "primed," apparatus—is just rotated with respect to *our* z-axis by the angle ϕ; then from Table 6-2,

$$a = \langle +' \mid + \rangle = e^{i\phi/2}.$$

So from (12.44) we have that the spin-one amplitude is

$$\langle +T \mid +S \rangle = \langle +' +' \mid + + \rangle = (e^{i\phi/2})^2 = e^{i\phi}. \tag{12.45}$$

You can see how it goes.

Now we will work through the general case for all the states. If the proton and electron are both "up" in *our* frame—the S-frame—the amplitudes that it will be in *any one of the four* possible states in the other guy's frame—the T-frame—are

$$\begin{aligned}
\langle +' +' \mid + + \rangle &= \langle +' \mid + \rangle_e \langle +' \mid + \rangle_p = a^2, \\
\langle +' -' \mid + + \rangle &= \langle +' \mid + \rangle_e \langle -' \mid + \rangle_p = ab, \\
\langle -' +' \mid + + \rangle &= \langle -' \mid + \rangle_e \langle +' \mid + \rangle_p = ba, \\
\langle -' -' \mid + + \rangle &= \langle -' \mid + \rangle_e \langle -' \mid + \rangle_p = b^2.
\end{aligned} \tag{12.46}$$

We can, then, write the state $\mid + + \rangle$ as the following linear combination:

$$\mid + + \rangle = a^2 \mid +' +' \rangle + ab\{\mid +' -' \rangle + \mid -' +' \rangle\} + b^2 \mid -' -' \rangle. \tag{12.47}$$

Now we notice that $\mid +' +' \rangle$ is the state $\mid +T \rangle$, that $\{\mid +' -' \rangle + \mid -' +' \rangle\}$ is just $\sqrt{2}$ *times* the state $\mid 0\,T \rangle$—see (12.41)—and that $\mid -' -' \rangle = \mid -T \rangle$. In other words, Eq. (12.47) can be rewritten as

$$\mid +S \rangle = a^2 \mid +T \rangle + \sqrt{2}\,ab \mid 0\,T \rangle + b^2 \mid -T \rangle. \tag{12.48}$$

In a similar way you can easily show that

$$\mid -S \rangle = c^2 \mid +T \rangle + \sqrt{2}\,cd \mid 0\,T \rangle + d^2 \mid -T \rangle. \tag{12.49}$$

Table 12-4

Spin one-half amplitudes

This chapter	Chapter 6
$a = \langle +' \mid + \rangle$	$\langle +T \mid +S \rangle$
$b = \langle -' \mid + \rangle$	$\langle -T \mid +S \rangle$
$c = \langle +' \mid - \rangle$	$\langle +T \mid -S \rangle$
$d = \langle -' \mid - \rangle$	$\langle -T \mid -S \rangle$

For $|0\,S\rangle$ it's a little more complicated, because

$$|0\,S\rangle = \frac{1}{\sqrt{2}}\{|+\,-\rangle + |-\,+\rangle\}.$$

But we can express each of the states $|+\,-\rangle$ and $|-\,+\rangle$ in terms of the "prime" states and take the sum. That is,

$$|+\,-\rangle = ac\,|+'\,+'\rangle + ad\,|+'\,-'\rangle + bc\,|-'\,+'\rangle + bd\,|-'\,-'\rangle \quad (12.50)$$

and

$$|-\,+\rangle = ac\,|+'\,+'\rangle + bc\,|+'\,-'\rangle + ad\,|-'\,+'\rangle + bd\,|-'\,-'\rangle. \quad (12.51)$$

Taking $1/\sqrt{2}$ times the sum, we get

$$|0\,S\rangle = \frac{2}{\sqrt{2}}\,ac\,|+'\,+'\rangle + \frac{ad+bc}{\sqrt{2}}\{|+'\,-'\rangle + |-'\,+'\rangle\} + \frac{2}{\sqrt{2}}\,bd\,|-'\,-'\rangle.$$

It follows that

$$|0\,S\rangle = \sqrt{2}\,ac\,|+T\rangle + (ad+bc)\,|0\,T\rangle + \sqrt{2}\,bd\,|-T\rangle. \quad (12.52)$$

We have now all of the amplitudes we wanted. The coefficients of Eqs. (12.48), (12.49), and (12.52) are the matrix elements $\langle jT\,|\,iS\rangle$. Let's pull them all together:

$$\langle jT\,|\,iS\rangle = \quad {}_{jT}\downarrow \overset{\overrightarrow{iS}}{\begin{pmatrix} a^2 & \sqrt{2}\,ac & c^2 \\ \sqrt{2}\,ab & ad+bc & \sqrt{2}\,cd \\ b^2 & \sqrt{2}\,bd & d^2 \end{pmatrix}}. \quad (12.53)$$

We have expressed the spin-one transformation in terms of the spin one-half amplitudes a, b, c, and d.

For instance, if the T-frame is rotated with respect to S by the angle α about the y-axis—as in Fig. 5–6—the amplitudes in Table 12–4 are just the matrix elements of $R_y(\alpha)$ in Table 6–2.

$$a = \cos\frac{\alpha}{2}, \qquad b = -\sin\frac{\alpha}{2},$$

$$c = \sin\frac{\alpha}{2}, \qquad d = \cos\frac{\alpha}{2}. \quad (12.54)$$

Using these in (12.53), we get the formulas of (5.38), which we gave there without proof.

What ever happened to the state $|IV\rangle$?! Well, it is a spin-zero system, so it has only one state—it is the *same in all coordinate systems*. We can check that everything works out by taking the difference of Eq. (12.50) and (12.51); we get that

$$|+\,-\rangle - |-\,+\rangle = (ad-bc)\{|+'\,-'\rangle - |-'\,+'\rangle\}.$$

But $(ad-bc)$ is the determinant of the spin one-half matrix, and so is equal to 1. We get that

$$|IV'\rangle = |IV\rangle$$

for any relative orientation of the two coordinate frames.

Propagation in a Crystal Lattice

13–1 States for an electron in a one-dimensional lattice

You would, at first sight, think that a low-energy electron would have great difficulty passing through a solid crystal. The atoms are packed together with their centers only a few angstroms apart, and the effective diameter of the atom for electron scattering is roughly an angstrom or so. That is, the atoms are large, relative to their spacing, so that you would expect the mean free path between collisions to be of the order of a few angstroms—which is practically nothing. You would expect the electron to bump into one atom or another almost immediately. Nevertheless, it is a ubiquitous phenomenon of nature that if the lattice is perfect, the electrons are able to travel through the crystal smoothly and easily—almost as if they were in a vacuum. This strange fact is what lets metals conduct electricity so easily; it has also permitted the development of many practical devices. It is, for instance, what makes it possible for a transistor to imitate the radio tube. In a radio tube electrons move freely through a vacuum, while in the transistor they move freely through a crystal lattice. The machinery behind the behavior of a transistor will be described in this chapter; the next one will describe the application of these principles in various practical devices.

The conduction of electrons in a crystal is one example of a very common phenomenon. Not only can electrons travel through crystals, but other "things" like atomic excitations can also travel in a similar manner. So the phenomenon which we want to discuss appears in many ways in the study of the physics of the solid state.

You will remember that we have discussed many examples of two-state systems. Let's now think of an electron which can be in either one of two positions, in each of which it is in the same kind of environment. Let's also suppose that there is a certain amplitude to go from one position to the other, and, of course, the same amplitude to go back, just as we have discussed for the hydrogen molecular ion in Section 10–1. The laws of quantum mechanics then give the following results. There are two possible states of definite energy for the electron. Each state can be described by the amplitude for the electron to be in each of the two basic positions. In either of the definite-energy states, the magnitudes of these two amplitudes are constant in time, and the phases vary in time with the same frequency. On the other hand, if we start the electron in one position, it will later have moved to the other, and still later will swing back again to the first position. The amplitude is analogous to the motions of two coupled pendulums.

Now consider a perfect crystal lattice in which we imagine that an electron can be situated in a kind of "pit" at one particular atom and with some particular energy. Suppose also that the electron has some amplitude to move into a different pit at one of the nearby atoms. It is something like the two-state system—but with an additional complication. When the electron arrives at the neighboring atom, it can afterward move on to still another position as well as return to its starting point. Now we have a situation analogous not to *two* coupled pendulums, but to an *infinite number* of pendulums all coupled together. It is something like what you see in one of those machines—made with a long row of bars mounted on a torsion wire—that is used in first-year physics to demonstrate wave propagation.

If you have a harmonic oscillator which is coupled to another harmonic oscillator, and that one to another, and so on . . . , and if you start an irregularity in one place, the irregularity will propagate as a wave along the line. The same situation exists if you place an electron at one atom of a long chain of atoms.

13–1 States for an electron in a one-dimensional lattice

13–2 States of definite energy

13–3 Time-dependent states

13–4 An electron in a three-dimensional lattice

13–5 Other states in a lattice

13–6 Scattering by imperfections in the lattice

13–7 Trapping by a lattice imperfection

13–8 Scattering amplitudes and bound states

Usually, the simplest way of analyzing the mechanical problem is not to think in terms of what happens if a pulse is started at a definite place, but rather in terms of steady-wave solutions. There exist certain patterns of displacements which propagate through the crystal as a wave of a single, fixed frequency. Now the same thing happens with the electron—and for the same reason, because it's described in quantum mechanics by similar equations.

You must appreciate one thing, however; the amplitude for the electron to be at a place is an *amplitude*, not a probability. If the electron were simply leaking from one place to another, like water going through a hole, the behavior would be completely different. For example, if we had two tanks of water connected by a tube to permit some leakage from one to the other, then the levels would approach each other exponentially. But for the electron, what happens is amplitude leakage and not just a plain probability leakage. And it's a characteristic of the imaginary term—the i in the differential equations of quantum mechanics—which changes the exponential solution to an oscillatory solution. What happens then is quite different from the leakage between interconnected tanks.

We want now to analyze quantitatively the quantum mechanical situation. Imagine a one-dimensional system made of a long line of atoms as shown in Fig. 13–1(a). (A crystal is, of course, three-dimensional but the physics is very much the same; once you understand the one-dimensional case you will be able to understand what happens in three dimensions.) Next, we want to see what happens if we put a single electron on this line of atoms. Of course, in a real crystal there are already millions of electrons. But most of them (nearly all for an insulating crystal) take up positions in some pattern of motion each around its own atom—and everything is quite stationary. However, we now want to think about what happens if we put an *extra* electron in. We will not consider what the other ones are doing because we suppose that to change their motion involves a lot of excitation energy. We are going to add an electron as if to produce one slightly bound negative ion. In watching what the *one* extra electron does we are making an approximation which disregards the mechanics of the inside workings of the atoms.

Of course the electron could then move to another atom, transferring the negative ion to another place. We will suppose that just as in the case of an electron jumping between two protons, the electron can jump from one atom to the neighbor on either side with a certain amplitude.

Now how do we describe such a system? What will be reasonable base states? If you remember what we did when we had only two possible positions, you can guess how it will go. Suppose that in our line of atoms the spacings are all equal; and that we number the atoms in sequence, as shown in Fig. 13–1(a). One of the base states is that the electron is at atom number 6, another base state is that the electron is at atom number 7, or at atom number 8, and so on. We can describe the nth base state by saying that the electron is at atom number n. Let's say that this is the base state $|n\rangle$. Figure 13–1 shows what we mean by the three base states

$$|n-1\rangle, \quad |n\rangle, \quad \text{and} \quad |n+1\rangle.$$

Using these base states, any state $|\phi\rangle$ of our one-dimensional crystal can be described by giving all the amplitudes $\langle n|\phi\rangle$ that the state $|\phi\rangle$ is in one of the base states—which means the amplitude that it is located at one particular atom. Then we can write the state $|\phi\rangle$ as a superposition of the base states

$$|\phi\rangle = \sum_n |n\rangle\langle n|\phi\rangle. \tag{13.1}$$

Next, we are going to suppose that when the electron is at one atom, there is a certain amplitude that it will leak to the atom on either side. And we'll take the simplest case for which it can only leak to the nearest neighbors—to get to the next-nearest neighbor, it has to go in two steps. We'll take that the amplitudes for the electron jump from one atom to the next is $iA/\hbar$ (per unit time).

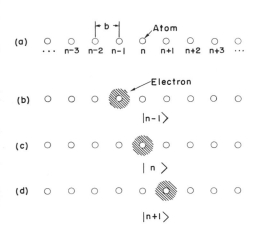

Fig. 13–1. The base states of an electron in a one-dimensional crystal.

For the moment we would like to write the amplitude $\langle n \mid \phi \rangle$ to be on the nth atom as C_n. Then Eq. (13.1) will be written

$$| \phi \rangle = \sum_n | n \rangle C_n. \tag{13.2}$$

If we knew each of the amplitudes C_n at a given moment, we could take their absolute squares and get the probability that you would find the electron if you looked at atom n at that time.

What will the situation be at some later time? By analogy with the two-state systems we have studied, we would propose that the Hamiltonian equations for this system should be made up of equations like this:

$$i\hbar \frac{dC_n(t)}{dt} = E_0 C_n(t) - A C_{n+1}(t) - A C_{n-1}(t). \tag{13.3}$$

The first coefficient on the right, E_0, is, physically, the energy the electron would have if it couldn't leak away from one of the atoms. (It doesn't matter what we call E_0; as we have seen many times, it represents really nothing but our choice of the zero of energy.) The next term represents the amplitude per unit time that the electron is leaking into the nth pit from the $(n + 1)$st pit; and the last term is the amplitude for leakage from the $(n - 1)$st pit. As usual, we'll assume that A is a constant (independent of t).

For a full description of the behavior of any state $| \phi \rangle$, we would have one equation like (13.3) for every one of the amplitudes C_n. Since we want to consider a crystal with a very large number of atoms, we'll assume that there are an indefinitely large number of states—that the atoms go on forever in both directions. (To do the finite case, we will have to pay special attention to what happens at the ends.) If the number N of our base states is indefinitely large, then also our full Hamiltonian equations are infinite in number! We'll write down just a sample:

$$\vdots \qquad\qquad \vdots$$
$$i\hbar \frac{dC_{n-1}}{dt} = E_0 C_{n-1} - A C_{n-2} - A C_n,$$
$$i\hbar \frac{dC_n}{dt} = E_0 C_n - A C_{n-1} - A C_{n+1}, \tag{13.4}$$
$$i\hbar \frac{dC_{n+1}}{dt} = E_0 C_{n+1} - A C_n - A C_{n+2},$$
$$\vdots \qquad\qquad \vdots$$

13–2 States of definite energy

We could study many things about an electron in a lattice, but first let's try to find the states of definite energy. As we have seen in earlier chapters this means that we have to find a situation in which the amplitudes all change at the same frequency if they change with time at all. We look for solutions of the form

$$C_n = a_n e^{-iEt/\hbar}. \tag{13.5}$$

The complex number a_n tell us about the non-time-varying part of the amplitude to find the electron at the nth atom. If we put this trial solution into the equations of (13.4) to test them out, we get the result

$$E a_n = E_0 a_n - A a_{n+1} - A a_{n-1}. \tag{13.6}$$

We have an infinite number of such equations for the infinite number of unknowns a_n—which is rather petrifying.

All we have to do is take the determinant ... but wait! Determinants are fine when there are 2, 3, or 4 equations. But if there are a large number—or an infinite number—of equations, the determinants are not very convenient. We'd better just try to solve the equations directly. First, let's label the atoms by their *positions;* we'll say that the atom n is at x_n and the atom $(n + 1)$ is at x_{n+1}. If the atomic spacing is b—as in Fig. 13–1—we will have that $x_{n+1} = x_n + b$. By choosing our origin at atom zero, we can even have it that $x_n = nb$. We can rewrite Eq. (13.5) as

$$C_n = a(x_n)e^{-iEt/\hbar}, \tag{13.7}$$

and Eq. (13.6) would become

$$Ea(x_n) = E_0 a(x_{n+1}) - Aa(x_{n+1}) - Aa(x_{n-1}). \tag{13.8}$$

Or, using the fact that $x_{n+1} = x_n + b$, we could also write

$$Ea(x_n) = E_0 a(x_n) - Aa(x_n + b) - Aa(x_n - b). \tag{13.9}$$

This equation is somewhat similar to a differential equation. It tells us that a quantity, $a(x)$, at one point, (x_n), is related to the same physical quantity at some neighboring points, $(x_n \pm b)$. (A differential equation relates the value of a function at a point to the values at infinitesimally nearby points.) Perhaps the methods we usually use for solving differential equations will also work here; let's try.

Linear differential equations with constant coefficients can always be solved in terms of exponential functions. We can try the same thing here; let's take as a trial solution

$$a(x_n) = e^{ikx_n}. \tag{13.10}$$

Then Eq. (13.9) becomes

$$Ee^{ikx_n} = E_0 e^{ikx_n} - Ae^{ik(x_n+b)} - Ae^{ik(x_n-b)}. \tag{13.11}$$

We can now divide out the common factor e^{ikx_n}; we get

$$E = E_0 - Ae^{ikb} - Ae^{-ikb}. \tag{13.12}$$

The last two terms are just equal to $(2A \cos kb)$, so

$$E = E_0 - 2A \cos kb. \tag{13.13}$$

We have found that for *any* choice at all for the constant k there is a solution whose energy is given by this equation. There are various possible energies depending on k, and each k corresponds to a different solution. There are an infinite number of solutions—which is not surprising, since we started out with an infinite number of base states.

Let's see what these solutions mean. For each k, the a's are given by Eq. (13.10). The amplitudes C_n are then given by

$$C_n = e^{ikx_n}e^{-(i/\hbar)Et}, \tag{13.14}$$

where you should remember that the energy E also depends on k as given in Eq. (13.13). The *space dependence* of the amplitudes is e^{ikx_n}. The amplitudes oscillate as we go along from one atom to the next.

We mean that, in space, the amplitude goes as a *complex* oscillation—the *magnitude* is the same at every atom, but the phase at a given time advances by the amount (ikb) from one atom to the next. We can visualize what is going on by plotting a vertical line to show just the real part at each atom as we have done in Fig. 13–2. The envelope of these vertical lines (as shown by the broken-line curve)

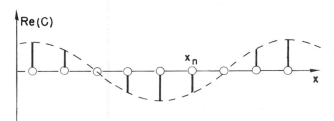

Fig. 13–2. Variation of the real part of C_n with x_n.

is, of course, a cosine curve. The imaginary part of C_n is also an oscillating function, but is shifted 90° in phase so that the absolute square (which is the sum of the squares of the real and imaginary parts) is the same for all the C's.

Thus if we pick a k, we get a stationary state of a particular energy E. And for any such state, the electron is equally likely to be found at every atom—there is no preference for one atom or the other. Only the phase is different for different atoms. Also, as time goes on the phases vary. From Eq. (13.14) the real and imaginary parts propagate along the crystal as waves—namely as the real or imaginary parts of

$$e^{i[kx_n-(E/\hbar)t]}. \tag{13.15}$$

The wave can travel toward positive or negative x depending on the sign we have picked for k.

Notice that we have been assuming that the number k that we put in our trial solution, Eq. (13.10), was a real number. We can see now why that must be so if we have an infinite line of atoms. Suppose that k were an imaginary number, say ik'. Then the amplitudes a_n would go as $e^{k'x_n}$, which means that the amplitude would get larger and larger as we go toward large x's—or toward large negative x's if k' is a negative number. This kind of solution would be O.K. if we were dealing with line of atoms that ended, but cannot be a physical solution for an infinite chain of atoms. It would give infinite amplitudes—and, therefore, infinite probabilities—which can't represent a real situation. Later on we will see an example in which an imaginary k does make sense.

The relation between the energy E and the wave number k as given in Eq. (13.13) is plotted in Fig. 13–3. As you can see from the figure, the energy can go from $(E_0 - 2A)$ at $k = 0$ to $(E_0 + 2A)$ at $k = \pm\pi/b$. The graph is plotted for positive A; if A were negative, the curve would simply be inverted, but the range would be the same. The significant result is that any energy is possible within a certain range or "band" of energies, but no others. According to our assumptions, if an electron in a crystal is in a stationary state, it can have no energy other than values in this band.

According to Eq. (13.10), the smallest k's correspond to low-energy states— $E \approx (E_0 - 2A)$. As k increases in magnitude (toward either positive or negative values) the energy at first increases, but then reaches a maximum at $k = \pm\pi/b$, as shown in Fig. 13–3. For k's larger than π/b, the energy would start to decrease again. But we do not really need to consider such values of k, because they do not give new states—they just repeat states we already have for smaller k. We can see that in the following way. Consider the lowest energy state for which $k = 0$. The coefficient $a(x_n)$ is the same for all x_n. Now we would get the same energy for $k = 2\pi/b$. But then, using Eq. (13.10), we have that

$$a(x_n) = e^{i(2\pi/b)x_n}.$$

However, taking x_0 to be at the origin, we can set $x_n = nb$; then $a(x_n)$ becomes

$$a(x_n) = e^{i2\pi n} = 1.$$

The state described by these $a(x_n)$ is physically the same state we got for $k = 0$. It does not represent a different solution.

As another example, suppose that k were $\pi/4b$. The real part of $a(x_n)$ would vary as shown by curve 1 in Fig. 13–4. If k were seven times larger ($k = 7\pi/4$), the real part of $a(x_n)$ would vary as shown by curve 2 in the figure. (The complete

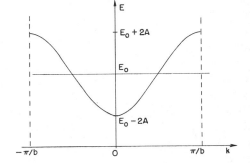

Fig. 13–3. The energy of the stationary states as a function of the parameter k.

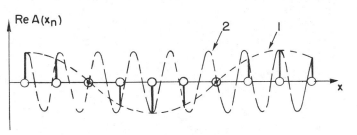

Fig. 13–4. Two values of k which represent the same physical situation; curve 1 is for $k = \pi/4$, curve 2 is for $k = 7\pi/4$.

cosine curves don't mean anything, of course; all that matters is their values *at the points* x_n. The curves are just to help you see how things are going.) You see that both values of k give the same amplitudes at all of the x_n's.

The upshot is that we have all the possible solutions of our problem if we take only k's in a certain limited range. We'll pick the range between $-\pi/b$ and $+\pi/b$—the one shown in Fig. 13–3. In this range, the energy of the stationary states increases uniformly with an increase in the magnitude of k.

One side remark about something you can play with. Suppose that the electron cannot only jump to the nearest neighbor with amplitude iA/h, but also has the possibility to jump in one direct leap to the *next nearest* neighbor with some other amplitude iB/h. You will find that the solution can again be written in the form $a_n = e^{ikx_n}$—this type of solution is universal. You will also find that the stationary states with wave number k have an energy equal to $(E_0 - 2A\cos kb - 2B\cos 2kb)$. This shows that the shape of the curve of E against k is not universal, but depends upon the particular assumptions of the problem. It is not always a cosine wave—it's not even necessarily symmetrical about some horizontal line. It is true, however, that the curve always repeats itself outside of the interval from $-\pi/b$ to π/b, so you never need to worry about other values of k.

Let's look a little more closely at what happens for small k—that is, when the variations of the amplitudes from one x_n to the next are quite slow. Suppose we choose our zero of energy by defining $E_0 = 2A$; then the minimum of the curve in Fig. 13–3 is at the zero of energy. For small enough k, we can write that

$$\cos kb \approx 1 - k^2b^2/2,$$

and the energy of Eq. (13.13) becomes

$$E = Ak^2b^2. \tag{13.16}$$

We have that the energy of the state is proportional to the square of the wave number which describes the spatial variations of the amplitudes C_n.

13–3 Time-dependent states

In this section we would like to discuss the behavior of states in the one-dimensional lattice in more detail. If the amplitude for an electron to be at x_n is C_n, the probability of finding it there is $|C_n|^2$. For the *stationary* states described by Eq. (13.12), this probability is the same for all x_n and does not change with time. How can we represent a situation which we would describe roughly by saying an electron of a certain energy is localized in a certain region—so that it is more likely to be found at one place than at some other place? We can do that by making a superposition of several solutions like Eq. (13.12) with slightly different values of k—and, therefore, slightly different energies. Then at $t = 0$, at least, the amplitude C_n will vary with position because of the interference between the various terms, just as one gets beats when there is a mixture of waves of different wavelengths (as we discussed in Chapter 48, Vol. I). So we can make up a "wave packet" with a predominant wave number k_0, but with various other wave numbers near k_0.†

In our superposition of stationary states, the amplitudes with different k's will represent states of slightly different energies, and, therefore, of slightly different frequencies; the interference pattern of the total C_n will, therefore, also vary with time—there will be a pattern of "beats." As we have seen in Chapter 48 of Volume I, the peaks of the beats [the place where $|C(x_n)|^2$ is large] will move along in x as time goes on; they move with the speed we have called the "group velocity." We found that this group velocity was related to the variation of k with frequency by

$$v_{\text{group}} = \frac{d\omega}{dk}; \tag{13.17}$$

---

† Provided we do not try to make the packet too narrow.

the same derivation would apply equally well here. An electron state which is a "clump"—namely one for which the C_n vary in space like the wave packet of Fig. 13–5—will move along our one-dimensional "crystal" with the speed v equal to $d\omega/dk$, where $\omega = E/\hbar$. Using (13.16) for E, we get that

$$v = \frac{2Ab^2}{\hbar}\, k. \tag{13.18}$$

In other words, the electrons move along with a speed proportional to the typical k. Equation (13.16) then says that the energy of such an electron is proportional to the square of its velocity—*it acts like a classical particle.* So long as we look at things on a scale gross enough that we don't see the fine structure, our quantum mechanical picture begins to give results like classical physics. In fact, if we solve Eq. (13.18) for k and substitute into (13.16), we can write

$$E = \tfrac{1}{2}m_{\text{eff}}\, v^2, \tag{13.19}$$

where m_{eff} is a constant. The extra "energy of motion" of the electron in a packet depends on the velocity just as for a classical particle. The constant m_{eff}—called the "effective mass"—is given by

$$m_{\text{eff}} = \frac{\hbar^2}{2Ab^2}. \tag{13.20}$$

Also notice that we can write

$$m_{\text{eff}}\, v = \hbar k. \tag{13.21}$$

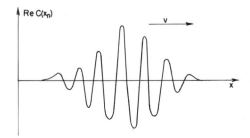

Fig. 13–5. The real part of $C(x_n)$ as a function of x for a superposition of several states of similar energy. (The spacing b is very small on the scale of x shown.)

If we choose to call $m_{\text{eff}}\, v$ the "momentum," it is related to the wave number k in the way we have described earlier for a free particle.

Don't forget that m_{eff} has nothing to do with the real mass of an electron. It may be quite different—although in real crystals it often happens to turn out to be the same general order of magnitude, about 2 to 20 times the free-space mass of the electron.

We have now explained a remarkable mystery—how an electron in a crystal (like an extra electron put into germanium) can ride right through the crystal and flow perfectly freely even though it has to hit all the atoms. It does so by having its amplitudes going pip-pip-pip from one atom to the next, working its way through the crystal. That is how a solid can conduct electricity.

13–4 An electron in a three-dimensional lattice

Let's look for a moment at how we could apply the same ideas to see what happens to an electron in three dimensions. The results turn out to be very similar. Suppose we have a rectangular lattice of atoms with lattice spacings of a, b, c in the three directions. (If you want a cubic lattice, take the three spacings all equal.) Also suppose that the amplitude to leap in the x-direction to a neighbor is $(iA_x/\hbar)$, to leap in the y-direction is $(iA_y/\hbar)$, and to leap in the z-direction is $(iA_z/\hbar)$. Now how should we describe the base states? As in the one-dimensional case, one base state is that the electron is at the atom whose locations are x, y, z, where (x, y, z) is one of the lattice points. Choosing our origin at one atom, these points are all at

$$x = n_x a, \qquad y = n_y b, \qquad \text{and} \qquad z = n_z c,$$

where n_x, n_y, n_z are any three integers. Instead of using subscripts to indicate such points, we will now just use x, y, and z, understanding that they take on only their values at the lattice points. Thus the base state is represented by the symbol $|\text{electron at } x, y, z\rangle$, and the amplitude for an electron in some state $|\psi\rangle$ to be in this base state is $C(x, y, z) = \langle\text{electron at } x, y, z \mid \psi\rangle$.

As before, the amplitudes $C(x, y, z)$ may vary with time. With our assumptions, the Hamiltonian equations should be like this:

$$i\hbar \frac{dC(x, y, z)}{dt} = E_0 C(x, y, z) - A_x C(x + a, y, z) - A_x C(x - a, y, z)$$
$$- A_y C(x, y + b, z) - A_y C(x, y - b, z)$$
$$- A_z C(x, y, z + c) - A_z C(x, y, z - c). \quad (13.22)$$

It looks rather long, but you can see where each term comes from.

Again we can try to find a stationary state in which all the C's vary with time in the same way. Again the solution is an exponential:

$$C(x, y, z) = e^{-iEt/\hbar} e^{i(k_x x + k_y y + k_z z)}. \quad (13.23)$$

If you substitute this into (13.22) you see that it works, provided that the energy E is related to k_x, k_y, and k_z in the following way:

$$E = E_0 - 2A_x \cos k_x a - 2A_y \cos k_y b - 2A_z \cos k_z c. \quad (13.24)$$

The energy now depends on the *three* wave numbers k_x, k_y, k_z, which, incidentally, are the components of a three-dimensional vector k. In fact, we can write Eq. (13.23) in vector notation as

$$C(x, y, z) = e^{-iEt/\hbar} e^{-i\mathbf{k} \cdot \mathbf{r}} \quad (13.25)$$

The amplitude varies as a complex *plane wave* in three dimensions, moving in the direction of k, and with the wave number $k = (k_x^2 + k_y^2 + k_z^2)^{1/2}$.

The energy associated with these stationary states depends on the three components of k in the complicated way given in Eq. (13.24). The nature, of the variation of E with k depends on relative signs and magnitudes of A_x, A_y, and A_z. If these three numbers are all positive, and if we are interested in small values of k, the dependence is relatively simple.

Expanding the cosines as we did before to get Eq. (13.16), we can now get that

$$E = E_{\min} + A_x a^2 k_x^2 + A_y b k_y^2 + A_z c k_z^2. \quad (13.26)$$

For a simple cubic lattice with lattice spacing a we expect that A_x and A_y and A_z would be equal—say all are just A—and we would have just

$$E = E_{\min} + Aa^2(k_x^2 + k_y^2 + k_z^2),$$

or

$$E = E_{\min} + Aa^2 k^2. \quad (13.27)$$

This is just like Eq. (13.16). Following the arguments used there, we would conclude that an electron packet in *three* dimensions (made up by superposing many states with nearly equal energies) also moves like a classical particle with some effective mass.

In a crystal with a lower symmetry than cubic (or even in a cubic crystal in which the state of the electron at each atom is not symmetrical) the three coefficients A_x, A_y, and A_z are different. Then the "effective mass" of an electron localized in a small region *depends on its direction of motion*. It could, for instance, have a different inertia for motion in the x-direction than for motion in the y-direction. (The details of such a situation are sometimes described in terms of an "effective mass tensor.")

13–5 Other states in a lattice

According to Eq. (13.24) the electron states we have been talking about can have energies only in a certain "band" of energies which covers the energy range from the minimum energy

$$E_0 - 2(A_x + A_y + A_z)$$

to the maximum energy

$$E_0 + 2(A_x + A_y + A_z).$$

Other energies are possible, but they belong to a different class of electron states. For the states we have described, we imagined base states in which an electron is placed on an atom of the crystal in some particular state, say the lowest energy state.

If you have an atom in empty space, and add an electron to make an ion, the ion can be formed in many ways. The electron can go on in such a way as to make the state of lowest energy, or it can go on to make one or another of many possible "excited states" of the ion each with a definite energy above the lowest energy. The same thing can happen in a crystal. Let's suppose that the energy E_0 we picked above corresponds to base states which are ions of the lowest possible energy. We could also imagine a new set of base states in which the electron sits near the nth atom in a different way—in one of the excited states of the ion—so that the energy E_0 is now quite a bit higher. As before there is some amplitude A (different from before) that the electron will jump from its excited state at one atom to the same excited state at a neighboring atom. The whole analysis goes as before; we find a band of possible energies centered at a higher energy. There can, in general, be many such bands each corresponding to a different level of excitation.

There are also other possibilities. There may be some amplitude that the electron jumps from an excited condition at one atom to an unexcited condition at the next atom. (This is called an interaction between bands.) The mathematical theory gets more and more complicated as you take into account more and more bands and add more and more coefficients for leakage between the possible states. No new ideas are involved, however; the equations are set up much as we have done in our simple example.

We should remark also that there is not much more to be said about the various coefficients, such as the amplitude A, which appear in the theory. Generally they are very hard to calculate, so in practical cases very little is known theoretically about these parameters and for any particular real situation we can only take values determined experimentally.

There are other situations where the physics and mathematics are almost exactly like what we have found for an electron moving in a crystal, but in which the "object" that moves is quite different. For instance, suppose that our original crystal—or rather linear lattice—was a line of neutral atoms, each with a loosely bound outer electron. Then imagine that we were to remove one electron. Which atom has lost its electron? Let C_n now represent the amplitude that the electron *is missing* from the atom at x_n. There will, in general, be some amplitude $iA/\hbar$ that the electron at a neighboring atom—say the $(n-1)$st atom—will jump to the nth leaving the $(n-1)$st atom without its electron. This is the same as saying that there is an amplitude A for the "missing electron" to jump from the nth atom to the $(n-1)$st atom. You can see that the equations will be exactly the same—of course, the value of A need not be the same as we had before. Again we will get the same formulas for the energy levels, for the "waves" of probability which move through the crystal with the group velocity of Eq. (13.18), for the effective mass, and so on. Only now the waves describe the behavior of the *missing electron*—or "hole" as it is called. So a "hole" acts just like a particle with a certain mass m_{eff}. You can see that this particle will appear to have a positive charge. We'll have some more to say about such holes in the next chapter.

As another example, we can think of a line of identical *neutral* atoms one of which has been put into an excited state—that is, with more than its normal ground state energy. Let C_n be the amplitude that the nth atom has the excitation. It can interact with a neighboring atom by handing over to it the extra energy and returning to the ground state. Call the amplitude for this process $iA/\hbar$. You can see that it's the same mathematics all over again. Now the object which moves is called an *exciton*. It behaves like a neutral "particle" moving through the crystal, carrying the excitation energy. Such motion may be involved in certain biological processes such as vision, or photosynthesis. It has been guessed that the absorption of light in the retina produces an "exciton" which moves through some periodic structure (such as the layers in the rods we described in Chapter 36, Vol. 1; see Fig. 36–5) to be accumulated at some special station where the energy is used to induce a chemical reaction.

13–6 Scattering from imperfections in the lattice

We want now to consider the case of a single electron in a crystal which is not perfect. Our earlier analysis says that perfect crystals have perfect conductivity —that electrons can go slipping through the crystal, as in a vacuum, without friction. One of the most important things that can stop an electron from going on forever is an imperfection or irregularity in the crystal. As an example, suppose that somewhere in the crystal there is a missing atom; or suppose that someone put one wrong atom at one of the atomic sites so that things there are different than at the other atomic sites. Say the energy, E_0 or the amplitude A could be different. How would we describe what happens then?

To be specific, we will return to the one-dimensional case and we will assume that atom number "zero" is an "impurity" atom and has a different value of E_0 than any of the other atoms. Let's call this energy $(E_0 + F)$. What happens? When an electron arrives at atom "zero" there is some probability that the electron is scattered backwards. If a wave packet is moving along and it reaches a place where things are a little bit different, some of it will continue onward and some of it will bounce back. It's quite difficult to analyze such a situation using a wave packet, because everything varies in time. It is much easier to work with steady-state solutions. So we will work with stationary states, which we will find can be made up of continuous waves which have transmitted and reflected parts. In three dimensions we would call the reflected part the scattered wave, since it would spread out in various directions.

We start out with a set of equations which are just like the ones in Eq. (13.6) except that the equation for $n = 0$ is different from all the rest. The five equations for $n = -2, -1, 0, +1,$ and $+2$ look like this:

$$\vdots \qquad \qquad \vdots$$

$$
\begin{aligned}
Ea_{-2} &= E_0 a_{-2} - Aa_{-1} - Aa_{-3}, \\
Ea_{-1} &= E_0 a_{1-} - Aa_0 - Aa_{-2}, \\
Ea_0 &= (E_0 + F)a_0 - Aa_1 - Aa_{-1}, \qquad (13.28) \\
Ea_1 &= E_0 a_1 - Aa_2 - Aa_0, \\
Ea_2 &= E_0 a_2 - Aa_3 - Aa_1,
\end{aligned}
$$

$$\vdots \qquad \qquad \vdots$$

There are, of course, all the other equations for $|n|$ is greater than 2. They will look just like Eq. (13.16).

For the general case, we really ought to use a different A for the amplitude that the electron jumps to or from atom "zero," but the main features of what goes on will come out of a simplified example in which all the A's are equal.

Equation (13.10) would still work as a solution for all of the equations except the one for atom "zero"—it isn't right for that one equation. We need a different solution which we can cook up in the following way. Equation (13.10) represents a wave going in the positive x-direction. A wave going in the negative x-direction would have been an equally good solution. It would be written

$$a(x_n) = e^{-ikx_n}.$$

The most general solution we could have taken for Eq. (13.6) would be a combination of a forward and a backward wave, namely

$$a_n = \alpha e^{ikx_n} + \beta e^{-ikx_n}. \tag{13.29}$$

This solution represents a complex wave of amplitude α moving in the $+x$-direction and a wave of amplitude β moving in the $-x$-direction.

Now take a look at the set of equations for our new problem—the ones in (13.28) together with those for all the other atoms. The equations involving a_n's with $n \leq 1$ are all satisfied by Eq. (13.29), with the condition that k is related to E and the lattice spacing b by

$$E = E_0 - 2A \cos kb. \tag{13.30}$$

The physical meaning is an "incident" wave of amplitude α approaching atom "zero" (the "scatterer") from the left, and a "scattered" or "reflected" wave of amplitude β going back toward the left. We do not loose any generality if we set the amplitude α of the incident wave equal to 1. Then the amplitude β is, in general, a complex number.

We can say all the same things about the solutions of a_n for $n \geq 1$. The coefficients could be different, so we would have for them

$$a_n = \gamma e^{ikx_n} + \delta e^{-ikx_n}, \qquad \text{for} \qquad n \geq 1. \tag{13.31}$$

Here, γ is the amplitude of a wave going to the right and δ a wave coming from the right. We want to consider the *physical* situation in which a wave is originally started only from the left, and there is only a "transmitted" wave that comes out beyond the scatterer—or impurity atom. We will try for a solution in which $\delta = 0$. We can, certainly, satisfy all of the equations for the a_n except for the middle three in Eq. (13.28) by the following trial solutions.

$$a_n \text{ (for } n < 0) = e^{ikx_n} + \beta e^{-ikx_n},$$
$$a_n \text{ (for } n > 0) = \gamma e^{ikx_n}. \tag{13.32}$$

The situation we are talking about is illustrated in Fig. 13–6.

By using the formulas in Eq. (13.32) for a_{-1} and a_{+1}, the three middle equations of Eq. (13.28) will allow us to solve for a_0 and also for the two coefficients β and γ. So we have found a complete solution. Setting $x_n = nb$, we have to solve the three equations

$$(E - E_0)\{e^{ik(-b)} + \beta e^{-ik(-b)}\} = -A\{a_0 + e^{ik(-2b)} + \beta e^{-ik(-2b)}\},$$
$$(E - E_0 - F)a_0 = -A\{\gamma e^{ikb} + e^{ik(-b)} + \beta e^{-ik(-b)}\}, \tag{13.33}$$
$$(E - E_0)\gamma e^{ikb} = -A\{\gamma e^{ik(2b)} + a_0\}.$$

Remember that E is given in terms of k by Eq. (13.30). If you substitute this value for E into the equations, and remember that $\cos x = \frac{1}{2}(e^{ix} + e^{-ix})$, you get from the first equation that

$$a_0 = 1 + \beta; \tag{13.34}$$

and from the third equation that

$$a_0 = \gamma. \tag{13.35}$$

These are consistent only if

$$\gamma = 1 + \beta \tag{13.36}$$

This equation says that the transmitted wave (γ) is just the original incident wave (1) with an added wave (β) equal to the reflected wave. This is not always true, but happens to be so for a scattering at one atom only. If there were a clump of impurity atoms, the amount added to the forward wave would not necessarily be the same as the reflected wave.

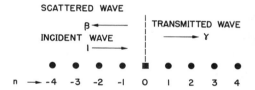

Fig. 13–6. Waves in a one-dimensional lattice with one "impurity" atom at $n = 0$.

We can get the amplitude β of the reflected wave from the middle equation of Eq. (13.33); we find that

$$\beta = \frac{-F}{F - 2iA \sin kb}. \tag{13.37}$$

We have the complete solution for the lattice with one unusual atom.

You may be wondering how the transmitted wave can be "more" than the incident wave as it appears in Eq. (13.34). Remember, though, that β and γ are complex numbers and that the number of particles (or rather, the probability of finding a particle) in a wave is proportional to the absolute square of the amplitude. In fact, there will be "conservation of electrons" only if

$$|\beta|^2 + |\gamma|^2 = 1. \tag{13.38}$$

You can show that this is true for our solution.

13–7 Trapping by a lattice imperfection

There is another interesting situation that can arise if F is a negative number. If the energy of the electron is lower at the impurity atom (at $n = 0$) than it is anywhere else, then the electron can get caught on this atom. That is, if $(E_0 + F)$ is below the bottom of the band at $(E_0 - 2A)$, then the electron can get "trapped" in a state with $E < E_0 - 2A$. Such a solution cannot come out of what we have done so far. We can get this solution, however, if we permit the trial solution we took in Eq. (13.15) to have an imaginary number for k. Let's set $k = i\kappa$. Again, we can have different solutions for $n < 0$ and for $n > 0$. A possible solution for $n < 0$ might be

$$a_n \text{ (for } n < 0) = ce^{+\kappa x_n}. \tag{13.39}$$

We have to take a plus sign in the exponent; otherwise the amplitude would get indefinitely large for large negative values of n. Similarly, a possible solution for $n > 0$ would be

$$a_n \text{ (for } n > 0) = c'e^{-\kappa x_n}. \tag{13.40}$$

If we put these trial solutions into Eq. (13.28) all but the middle three are satisfied provided that

$$E = E_0 - A(e^{\kappa b} + e^{-\kappa b}). \tag{13.41}$$

Since the sum of the two exponential terms is always greater than 2, this energy is below the regular band, and is what we are looking for. The remaining three equations in Eq. (13.28) are satisfied if $c = c'$ and if κ is chosen so that

$$A(e^{\kappa b} - e^{-\kappa b}) = -F. \tag{13.42}$$

Combining this equation with Eq. (13.41) we can find the energy of the trapped electron; we get

$$E = E_0 - \sqrt{4A^2 + F^2}. \tag{13.43}$$

The trapped electron has a unique energy—located somewhat below the conduction band.

Notice that the amplitudes we have in Eq. (13.39) and (13.40) do *not* say that the trapped electron sits right on the impurity atom. The probability of finding the electron at nearby atoms is given by the square of these amplitudes. For one particular choice of the parameters it might vary as shown in the bar graph of Fig. 13–7. The probability is greatest for finding the electron on the impurity atom. For nearby atoms the probability drops off exponentially with the distance from the impurity atom. This is another example of "barrier penetration." From the point-of-view of classical physics the electron doesn't have enough energy to get away from the energy "hole" at the trapping center. But quantum mechanically it can leak out a little way.

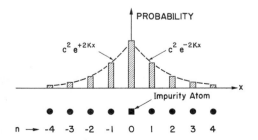

Fig. 13–7. The relative probabilities of finding a trapped electron at atomic sites near the trapping impurity atom.

13–8 Scattering amplitudes and bound states

Finally, our example can be used to illustrate a point which is very useful these days in the physics of high-energy particles. It has to do with a relationship between scattering amplitudes and bound states. Suppose we have discovered—through experiment and theoretical analysis—the way that pions scatter from protons. Then a new particle is discovered and someone wonders whether maybe it is just a combination of a pion and a proton held together in some bound state (in an analogy to the way an electron is bound to a proton to make a hydrogen atom). By a bound state we mean a combination which has a lower energy than the two free-particles.

There is a general theory which says that a bound state will exist at that energy at which the scattering amplitude becomes infinite if extrapolated algebraically (the mathematical term is "analytically continued") to energy regions outside of the permitted band.

The physical reason for this is as follows. A bound state is a situation in which there are only waves tied on to a point and there's no wave coming in to get it started, it just exists there by itself. The relative proportion between the so-called "scattered" or created wave and the wave being "sent in" is infinite. We can test this idea in our example. Let's write our expression Eq. (13.37) for the scattered amplitude directly in terms of the energy E of the particle being scattered (instead of in terms of k). Since Equation (13.30) can be rewritten as

$$2A \sin kb = \sqrt{4A^2 - (E - E_0)^2},$$

the scattered amplitude is

$$\beta = \frac{-F}{F - i\sqrt{4A^2 - (E - E_0)^2}}. \tag{13.44}$$

From our derivation, this equation should be used only for real states—those with energies in the energy band, $E = E_0 \pm 2A$. But suppose we forget that fact and extend the formula into the "unphysical" energy regions where $|E - E_0| > 2A$. For these unphysical regions we can write†

$$\sqrt{4A^2 - (E - E_0)^2} = i\sqrt{(E - E_0)^2 - 4A^2}.$$

Then the "scattering amplitude," whatever it may mean, is

$$\beta = \frac{-F}{F + \sqrt{(E - E_0)^2 - 4A^2}}. \tag{13.45}$$

Now we ask: Is there any energy E for which β becomes infinite (i.e., for which the expression for β has a "pole")? Yes, so long as F is negative, the denominator of Eq. (13.45) will be zero when

$$(E - E_0)^2 - 4A^2 = F^2,$$

or when

$$E = E_0 \pm \sqrt{4A^2 + F^2}.$$

The minus sign gives just the energy we found in Eq. (13.43) for the trapped energy.

What about the plus sign? This gives an energy *above* the allowed energy band. And indeed there is another bound state there which we missed when we solved the equations of Eq. (13.28). We leave it as a puzzle for you to find the energy and amplitudes a_n for this bound state.

The relation between scattering and bound states provides one of the most useful clues in the current search for an understanding of the experimental observations about the new strange particles.

---

† The sign of the root to be chosen here is a technical point related to the allowed signs of κ in Eqs. (13.39) and (13.40). We won't go into it here.

14

Semiconductors

14-1 Electrons and holes in semiconductors

One of the remarkable and dramatic developments in recent years has been the application of solid state science to technical developments in electrical devices such as transistors. The study of semiconductors led to the discovery of their useful properties and to a large number of practical applications. The field is changing so rapidly that what we tell you today may be incorrect next year. It will certainly be incomplete. And it is perfectly clear that with the continuing study of these materials many new and more wonderful things will be possible as time goes on. You will not need to understand this chapter for what comes later in this volume, but you may find it interesting to see that at least something of what you are learning has some relation to the practical world.

There are large numbers of semiconductors known, but we'll concentrate on those which now have the greatest technical application. They are also the ones that are best understood, and in understanding them we will obtain a degree of understanding of many of the others. The semiconductor substances in most common use today are silicon and germanium. These elements crystallize in the diamond lattice, a kind of cubic structure in which the atoms have tetrahedral bonding with their four nearest neighbors. They are insulators at very low temperatures—near absolute zero—although they do conduct electricity somewhat at room temperature. They are not metals; they are called *semiconductors*.

If we somehow put an extra electron into a crystal of silicon or germanium which is at a low temperature, we will have just the situation we described in the last chapter. The electron will be able to wander around in the crystal jumping from one atomic site to the next. Actually, we have looked only at the behavior of electrons in a rectangular lattice, and the equations would be somewhat different for the real lattice of silicon or germanium. All of the essential points are, however, illustrated by the results for the rectangular lattice.

As we saw in Chapter 13, these electrons can have energies only in a certain energy band—called the *conduction band*. Within this band the energy is related to the wave-number k of the probability amplitude C (see Eq. 13.24) by

$$E = E_0 - 2A_x \cos k_x a - 2A_y \cos k_y b - 2A \cos k_z c. \quad (14.1)$$

The A's are the amplitudes for jumping in the x-, y-, and z-directions, and a, b, and c are the lattice spacings in these directions.

For energies near the bottom of the band, we can approximate Eq. (14.1) by

$$E = E_{\min} + A_x a^2 k_x^2 + A_y b^2 k_y^2 + A_z c^2 k_z^2 \quad (14.2)$$

(see Section 13-4).

If we think of electron motion in some particular direction, so that the components of k are always in the same ratio, the energy is a quadratic function of the wave number—and as we have seen of the momentum of the electron. We can write

$$E = E_{\min} + \alpha k^2, \quad (14.3)$$

where α is some constant, and we can make a graph of E versus k as in Fig. 14-1. We'll call such a graph an "energy diagram." An electron in a particular state of energy and momentum can be indicated by a point such as S in the figure.

14-1 Electrons and holes in semiconductors

14-2 Impure semiconductors

14-3 The Hall effect

14-4 Semiconductor junctions

14-5 Rectification at a semiconductor junction

14-6 The transistor

Reference: C. Kittel, *Introduction to Solid State Physics*, Chapters 13, 14, and 18.

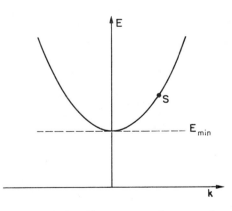

Fig. 14-1. The energy diagram for an electron in an insulating crystal.

As we also mentioned in Chapter 13, we can have a similar situation if we *remove* an electron from a neutral insulator. Then, an electron can jump over from a nearby atom and fill the "hole," but leaving another "hole" at the atom it started from. We can describe this behavior by writing an amplitude to find the *hole* at any particular atom, and by saying that the *hole* can jump from one atom to the next. (Clearly, the amplitudes A that the hole jumps from atom a to atom b is just the same as the amplitude that an electron on atom b jumps into the hole at atom a.) The mathematics is just the same for the *hole* as it was for the extra electron, and we get again that the energy of the hole is related to its wave number by an equation just like Eq. (14.1) or (14.2), except, of course, with different numerical values for the amplitudes A_x, A_y, and A_z. The hole has an energy related to the wave number of its probability amplitudes. Its energy lies in a restricted band, and near the bottom of the band its energy varies quadratically with the wave number—or momentum—just as in Fig. 14–1. Following the arguments of Section 13–3, we would find that *the hole also behaves like a classical particle* with a certain effective mass—except that in noncubic crystals the mass depends on the direction of motion. So the hole behaves like a *positive particle* moving through the crystal. The charge of the hole-particle is positive, because it is located at the site of a missing electron; and when it moves in one direction there are actually electrons moving in the opposite direction.

If we put several electrons into a neutral crystal, they will move around much like the atoms of a low-pressure gas. If there are not too many, their interactions will not be very important. If we then put an electric field across the crystal, the electrons will start to move and an electric current will flow. Eventually they would all be drawn to one edge of the crystal, and, if there is a metal electrode there, they would be collected, leaving the crystal neutral.

Similarly we could put many holes into a crystal. They would roam around at random unless there is an electric field. With a field they would flow toward the negative terminal, and would be "collected"—what actually happens is that they are neutralized by electrons from the metal terminal.

One can also have both holes and electrons together. If there are not too many, they will all go their way independently. With an electric field, they will all contribute to the current. For obvious reasons, electrons are called the *negative carriers* and the holes are called the *positive carriers*.

We have so far considered that electrons are put into the crystal from the outside, or are removed to make a hole. It is also possible to "create" an electron-hole pair by taking a bound electron away from one neutral atom and putting it some distance away in the same crystal. We then have a free electron and a free hole, and the two can move about as we have described.

The energy required to put an electron *into* a state S—we say to "create" the state S—is the energy E^- shown in Fig. 14–2. It is some energy above E^-_{min}. The energy required to "create" a hole in some state S' is the energy E^+ of Fig. 14–3, which is some energy greater than E^+_{min}. Now if we create a pair in the states S and S', the energy required is just $E^- + E^+$.

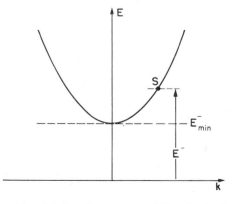

Fig. 14–2. The energy E^- is required to "create" a free electron.

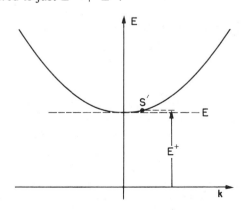

Fig. 14–3. The energy E^+ is required to "create" a hole in the state S'.

The creation of pairs is a common process (as we will see later), so many people like to put Fig. 14–2 and Fig. 14–3 together on the same graph—with the *hole* energy plotted *downward*, although it is, of course a *positive* energy. We have combined our two graphs in this way in Fig. 14–4. The advantage of such a graph is that the energy $E_{pair} = E^- + E^+$ required to create a pair with the electron in S and the hole in S' is just the vertical distance between S and S' as shown in Fig. 14–4. The minimum energy required to create a pair is called the "gap" energy and is equal to $E_{min}^- + E_{min}^+$.

Sometimes you will see a simpler diagram called an energy level diagram which is drawn when people are not interested in the k variable. Such a diagram—shown in Fig. 14–5—just shows the possible energies for the electrons and holes.†

How can electron-hole pairs be created? There are several ways. For example, photons of light (or x-rays) can be absorbed and create a pair if the photon energy is above the energy of the gap. The rate at which pairs are produced is proportional to the light intensity. If two electrodes are plated on a wafer of the crystal and a "bias" voltage is applied, the electrons and holes will be drawn to the electrodes. The circuit current will be proportional to the intensity of the light. This mechanism is responsible for the phenomenon of photoconductivity and the operation of photoconductive cells.

Electron hole pairs can also be produced by high-energy particles. When a fast-moving charged particle—for instance, a proton or a pion with an energy of tens or hundreds of Mev—goes through a crystal, its electric field will knock electrons out of their bound states creating electron-hole pairs. Such events occur hundreds of thousands of times per millimeter of track. After the passage of the particle, the carriers can be collected and in doing so will give an electrical pulse. This is the mechanism at play in the semiconductor counters recently put to use for experiments in nuclear physics. Such counters do not require semiconductors, they can also be made with crystalline insulators. In fact, the first of such counters was made using a diamond crystal which is an insulator at room temperature. Very pure crystals are required if the holes and electrons are to be able to move freely to the electrodes without being trapped. The semiconductors silicon and germanium are used because they can be produced with high purity in reasonable large sizes (centimeter dimensions).

So far we have been concerned with semiconductor crystals at temperatures near absolute zero. At any finite temperature there is still another mechanism by which electron-hole pairs can be created. The pair energy can be provided from the thermal energy of the crystal. The thermal vibrations of the crystal can transfer their energy to a pair—giving rise to "spontaneous" creation.

The probability per unit time that the energy as large as the gap energy E_{gap} will be concentrated at one atomic site is proportional to $e^{-E_{gap}/\kappa T}$, where T is the temperature and κ is Boltzmann's constant (see Chapter 40, Vol. I). Near absolute zero there is no appreciable probability, but as the temperature rises there is an increasing probability of producing such pairs. At any finite temperature the production should continue forever at a constant rate giving more and more negative and positive carriers. Of course that does not happen because after awhile the electrons and holes accidentally find each other—the electron drops into the hole and the excess energy is given to the lattice. We say that the electron and hole "annihilate." There is a certain probability per second that a hole meets an electron and the two things annihilate each other.

If the number of electrons per unit volume is N_n (n for negative carriers) and the density of positive carriers is N_p, the chance per unit time that an electron and a hole will find each other and annihilate is proportional to the product $N_n N_p$. In equilibrium this rate must equal the rate that pairs are created. You see that in

† In many books this same energy diagram is interpreted in a different way. The energy scale refers only to *electrons*. Instead of thinking of the energy of the hole, they think of the energy an electron *would* have if it filled the hole. This energy is *lower* than the free-electron energy—in fact, just the amount lower that you see in Fig. 14–5. With this interpretation of the energy scale, the gap energy is the minimum energy which must be given *to an electron* to move it from its bound state to the conduction band.

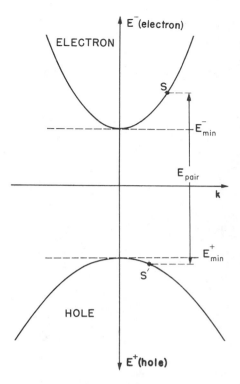

Fig. 14–4. Energy diagrams for an electron and a hole drawn together.

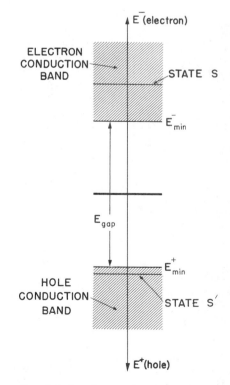

Fig. 14–5. Energy level diagram for electrons and holes.

equilibrium the product of N_n and N_p should be given by some constant times the Boltzmann factor:

$$N_n N_p = \text{const } e^{-E_{\text{gap}}/\kappa T}. \tag{14.4}$$

When we say constant, we mean nearly constant. A more complete theory—which includes more details about how holes and electrons "find" each other—shows that the "constant" is slightly dependent upon temperature, but the major dependence on temperature is in the exponential.

Let's consider, as an example, a pure material which is originally neutral. At a finite temperature you would expect the number of positive and negative carriers to be equal, $N_n = N_p$. Then each of them should vary with temperature as $e^{-E_{\text{gap}}/2\kappa T}$. The variation of many of the properties of a superconductor—the conductivity for example—is mainly determined by the exponential factor because all the other factors vary much more slowly with temperature. The gap energy for germanium is about 0.72 ev and for silicon 1.1 ev.

At room temperature κT is about 1/40 of an electron volt. At these temperatures there are enough holes and electrons to give a significant conductivity, while at, say, 30°K—one-tenth of room temperature—the conductivity is imperceptible. The gap energy of diamond is 6 or 7 ev and diamond is a good insulator at room temperature.

14–2 Impure semiconductors

So far we have talked about two ways that extra electrons can be put into an otherwise ideally perfect crystal lattice. One way was to inject the electron from an outside source; the other way, was to knock a bound electron off a neutral atom creating simultaneously an electron and a hole. It is possible to put electrons into the conduction band of a crystal in still another way. Suppose we imagine a crystal of germanium in which one of the germanium atoms is replaced by an arsenic atom. The germanium atoms have a valence of 4 and the crystal structure is controlled by the four valence electrons. Arsenic, on the other hand, has a valence of 5. It turns out that a single arsenic atom can sit in the germanium lattice (because it has approximately the correct size), but in doing so it must act as a valence 4 atom—using four of its valence electrons to form the crystal bonds and having one electron left over. This extra electron is very loosely attached—the binding energy is less than 1/10 of a volt. At room temperature the electron easily picks up that much energy from the thermal energy of the crystal, and then takes off on its own—moving about in the lattice as a free electron. An impurity atom such as the arsenic is called a *donor site* because it can give up a negative carrier to the crystal. If a crystal of germanium is grown from a melt to which a very small amount of arsenic has been added, the arsenic donor sites will be distributed throughout the crystal and the crystal will have a certain density of negative carriers built in.

You might think that these carriers would get swept away as soon as any small electric field was put across the crystal. This will not happen, however, because the arsenic atoms in the body of the crystal each have a positive charge. If the body of the crystal is to remain neutral, the average density of negative carrier electrons must be equal to the density of donor sites. If you put two electrodes on the edges of such a crystal and connect them to a battery, a current will flow; but as the carrier electrons are swept out at one end, new conduction electrons must be introduced from the electrode on the other end so that the average density of conduction electrons is left very nearly equal to the density of donor sites.

Since the donor sites are positively charged, there will be some tendency for them to capture some of the conduction electrons as they diffuse around inside the crystal. A donor site can, therefore, act as a trap such as those we discussed in the last section. But if the trapping energy is sufficiently small—as it is for arsenic—the number of carriers which are trapped at any one time is a small fraction of the total. For a complete understanding of the behavior of semiconductors one must take into account this trapping. For the rest of our discussion, however, we will assume that the trapping energy is sufficiently low and the temperature is sufficiently high, that all of the donor sites have given up their electrons. This is, of course, just an approximation.

It is also possible to build into a germanium crystal some impurity atom whose valence is 3, such as aluminum. The aluminum atom tries to act as a valence 4 object by stealing an extra electron. It can steal an electron from some nearby germanium atom and end up as a negatively charged atom with an effective valence of 4. Of course, when it steals the electron from a germanium atom, it leaves a hole there; and this hole can wander around in the crystal as a positive carrier. An impurity atom which can produce a hole in this way is called an *acceptor* because it "accepts" an electron. If a germanium or a silicon crystal is grown from a melt to which a small amount of aluminum impurity has been added, the crystal will have built-in a certain density of holes which can act as positive carriers.

When a donor or an acceptor impurity is added to a semiconductor, we say that the material has been "doped."

When a germanium crystal with some built-in donor impurities is at room temperature, some conduction electrons are contributed by the thermally induced electron-hole pair creation as well as by the donor sites. The electrons from both sources are, naturally, equivalent, and it is the total number N_n which comes into play in the statistical processes that lead to equilibrium. If the temperature is not too low, the number of negative carriers contributed by the donor impurity atoms is roughly equal to the number of impurity atoms present. In equilibrium Eq. (14.4) must still be valid; at a given temperature the product $N_n N_p$ is determined. This means that if we add some donor impurity which increases N_n, the number N_p of positive carriers will have to decrease by such an amount that $N_n N_p$ is unchanged. If the impurity concentration is high enough, the number N_n of negative carriers is determined by the number of donor sites and is nearly independent of temperature—all of the variation in the exponential factor is supplied by N_p, even though it is much less than N_n. An otherwise pure crystal with a small concentration of donor impurity will have a majority of negative carriers; such a material is called an "*n*-type" semiconductor.

If an acceptor-type impurity is added to the crystal lattice, some of the new holes will drift around and annihilate some of the free electrons produced by thermal fluctuation. This process will go on until Eq. (14.4) is satisfied. Under equilibrium conditions the number of positive carriers will be increased and the number of negative carriers will be decreased, leaving the product a constant. A material with an excess of positive carriers is called a "*p*-type" semiconductor.

If we put two electrodes on a piece of semiconductor crystal and connect them to a source of potential difference, there will be an electric field inside the crystal. The electric field will cause the positive and the negative carriers to move, and an electric current will flow. Let's consider first what will happen in an *n*-type material in which there is a large majority of negative carriers. For such material we can disregard the holes; they will contribute very little to the current because there are so few of them. In an ideal crystal the carriers would move across without any impediment. In a real crystal at a finite temperature, however,—especially in a crystal with some impurities—the electrons do not move completely freely. They are continually making collisions which knock them out of their original trajectories, that is, changing their momentum. These collisions are just exactly the scatterings we talked about in the last chapter and occur at any irregularity in the crystal lattice. In an *n*-type material the main causes of scattering are the very donor sites that are producing the carriers. Since the conduction electrons have a very slightly different energy at the donor sites, the probability waves are scattered from that point. Even in a perfectly pure crystal, however, there are (at any finite temperature) irregularities in the lattice due to thermal vibrations. From the classical point of view we can say that the atoms aren't lined up exactly on a regular lattice, but are, at any instant, slightly out of place due to their thermal vibrations. The energy E_0 associated with each lattice point in the theory we described in Chapter 13 varies a little bit from place to place so that the waves of probability amplitude are not transmitted perfectly but are scattered in an irregular fashion. At very high temperatures or for very pure materials this scattering may become important, but in most doped materials used in practical devices the impurity atoms contribute most of the scattering. We would like now to make an estimate of the electrical conductivity of such a material.

When an electric field is applied to an n-type semiconductor, each negative carrier will be accelerated in this field, picking up velocity until it is scattered from one of the donor sites. This means that the carriers which are ordinarily moving about in a random fashion with their thermal energies will pick up an average drift velocity along the lines of the electric field and give rise to a current through the crystal. The drift velocity is in general rather small compared with the typical thermal velocities so that we can estimate the current by assuming that the average time that the carrier travels between scatterings is a constant. Let's say that the negative carrier has an effective electric charge q_n. In an electric field $\mathcal{E}$, the force on the carrier will be $q_n\mathcal{E}$. In Section 43–3 of Volume I we calculated the average drift velocity under such circumstances and found that it is given by $F\tau/m$, where F is the force on the charge, τ is the mean free time between collisions, and m is the mass. We should use the effective mass we calculated in the last chapter but since we want to make a rough calculation we will suppose that this effective mass is the same in all directions. Here we will call it m_n. With this approximation the average drift velocity will be

$$v_{\text{drift}} = \frac{q_n \mathcal{E} \tau_n}{m_n} . \tag{14.5}$$

Knowing the drift velocity we can find the current. Electric current density j is just the number of carriers per unit volume, N_n, multiplied by the average drift velocity, and by the charge on each carrier. The current density is therefore

$$j = N_n v_{\text{drift}} q_n \mathcal{E} = \frac{N_n q_n^2 \tau_n}{m_n} \mathcal{E} . \tag{14.6}$$

We see that the current density is proportional to the electric field; such a semiconductor material obeys Ohm's law. The coefficient of proportionality between j and $\mathcal{E}$, the conductivity σ, is

$$\sigma = \frac{N_n q_n^2 \tau_n}{m_n} . \tag{14.7}$$

For an n-type material the conductivity is relatively independent of temperature. First, the number of majority carriers N_n is determined primarily by the density of donors in the crystal (so long as the temperature is not so low that too many of the carriers are trapped). Second, the mean time between collisions τ_n is mainly controlled by the density of impurity atoms, which is, of course, independent of the temperature.

We can apply all the same arguments to a p-type material, changing only the values of the parameters which appear in Eq. (14.7). If there are comparable numbers of both negative and positive carriers present at the same time, we must add the contributions from each kind of carrier. The total conductivity will be given by

$$\sigma = \frac{N_n q_n^2 \tau_n}{m_n} + \frac{N_p q_p^2 \tau_p}{m_p} . \tag{14.8}$$

For very pure materials, N_p and N_n will be nearly equal. They will be smaller than in a doped material, so the conductivity will be less. Also they will vary rapidly with temperature (like $e^{-E_{\text{gap}}/\kappa T}$, as we have seen), so the conductivity may change extremely fast with temperature.

14–3 The Hall effect

It is certainly a peculiar thing that in a substance where the only relatively free objects are electrons, there should be an electrical current carried by holes that behave like positive particles. We would like, therefore, to describe an experiment that shows in a rather clear way that the sign of the carrier of electric current is quite definitely positive. Suppose we have a block made of semiconductor material—it could also be a metal—and we put an electric field on it so as to draw a current in some direction, say the horizontal direction as drawn in Fig. 14–6. Now suppose we put a magnetic field on the block pointing at a right angle to the current, say *into* the plane of the figure. The moving carriers will feel a magnetic force $q(v \times B)$. And since the average drift velocity is either right or left—depending on the sign of the charge on the carrier—the average magnetic force on the carriers will be either up or down. No, that is not right! For the directions we have assumed for the current and the magnetic field the magnetic force on the moving charges will always be *up*. Positive charges moving in the direction of j (to the right) will feel an upward force. If the current is carried by negative charges, they will be moving left (for the same sign of the conduction current) and they will also feel an upward force. Under steady conditions, however, there is no upward motion of the carriers because the current can flow only from left to right. What happens is that a few of the charges initially flow upward, producing a surface charge density along the upper surface of semiconductor—leaving an equal and opposite surface charge density along the bottom surface of the crystal. The charges pile up on the top and bottom surfaces until the electric forces they produce on the moving charges just exactly cancel the magnetic force (on the average) so that the steady current flows horizontally. The charges on the top and bottom surfaces will produce a potential difference vertically across the crystal which can be measured with a high-resistance voltmeter, as shown in Fig. 14–7. The sign of the potential difference registered by the voltmeter will depend on the sign of the carrier charges responsible for the current.

When such experiments were first done it was expected that the sign of the potential difference would be negative as one would expect for negative conduction electrons. People were, therefore, quite surprised to find that for some materials the sign of the potential difference was in the opposite direction. It appeared that the current carrier was a particle with a positive charge. From our discussion of doped semiconductors it is understandable that an *n*-type semiconductor should produce the sign of potential difference appropriate to negative carriers, and that a *p*-type semiconductor should give an opposite potential difference, since the current is carried by the positively charged holes.

The original discovery of the anomalous sign of the potential difference in the Hall effect was made in a metal rather than a semiconductor. It had been assumed that in metals the conduction was always by electron; however, it was found out that for berylium the potential difference had the wrong sign. It is now understood that in metals as well as in semiconductors it is possible, in certain circumstances, that the "objects" responsible for the conduction are holes. Although it is ultimately the electrons in the crystal which do the moving, nevertheless, the relationship of the momentum and the energy, and the response to external fields is exactly what one would expect for an electric current carried by positive particles.

Let's see if we can make a quantitative estimate of the magnitude of the voltage difference expected from the Hall effect. If the voltmeter in Fig. 14–7 draws a negligible current, then the charges inside the semiconductor must be moving from left to right and the vertical magnetic force must be precisely cancelled by a vertical electric field which we will call $\mathcal{E}_{tr}$ (the "tr" is for "transverse"). If this electric field is to cancel the magnetic forces, we must have

$$\mathcal{E}_{tr} = -v_{drift} \times B. \tag{14.9}$$

Using the relation between the drift velocity and the electric current density given

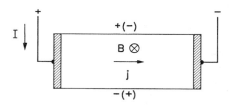

Fig. 14–6. The Hall effect comes from the magnetic forces on the carriers.

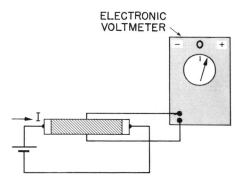

Fig. 14–7. Measuring the Hall effect.

in Eq. (14.6), we get

$$\mathcal{E}_{tr} = -\frac{1}{qN}\, jB.$$

The potential difference between the top and the bottom of the crystal is, of course, this electric field strength multiplied by the height of the crystal. The electric field strength $\mathcal{E}_{tr}$ in the crystal is proportional to the current density and to the magnetic field strength. The constant of proportionality $1/qN$ is called the Hall coefficient and is usually represented by the symbol R_H. The Hall coefficient depends just on the density of carriers—provided that carriers of one sign are in a large majority. Measurement of the Hall effect is, therefore, one convenient way of determining experimentally the density of carriers in a semiconductor.

14–4 Semiconductor junctions

We would like to discuss now what happens if we take two pieces of germanium or silicon with different internal characteristics—say different kinds or amounts of doping—and put them together to make a "junction." Let's start out with what is called a *p-n* junction in which we have *p*-type germanium on one side of the boundary and *n*-type germanium on the other side of the boundary—as sketched in Fig. 14-8. Actually, it is not practical to put together two separate pieces of crystal and have them in uniform contact on an atomic scale. Instead, junctions are made out of a single crystal which has been modified in the two separate regions. One way is to add some suitable doping impurity to the "melt" after only half of the crystal has grown. Another way is to paint a little of the impurity element on the surface and then heat the crystal causing some impurity atoms to diffuse into the body of the crystal. Junctions made in these ways do not have a sharp boundary, although the boundaries can be made as thin as 10^{-4} centimeters or so. For our discussions we will imagine an ideal situation in which these two regions of the crystal with different properties meeting at a sharp boundary.

On the *n*-type side of *p-n* junction there are free electrons which can move about, as well as the fixed donor sites which balance the overall electric charge. On the *p*-type side there are free holes moving about and an equal number of negative acceptor sites keeping the charge balanced. Actually, that describes the situation before we put the two materials in contact. Once they are connected together the situation will change near the boundary. When the electrons in the *n*-type material arrive at the boundary they will not be reflected back as they would at a free surface, but are able to go right on into the *p*-type material. Some of the electrons of the *n*-type material will, therefore, tend to diffuse over into the *p*-type material where there are fewer electrons. This cannot go on forever because as we lose electrons from the *n*-side the net positive charge there increases until finally an electric voltage is built up which retards the diffusion of electrons into the *p*-side. In a similar way, the positive carriers of the *p*-type material can diffuse across the junction into the *n*-type material. When they do this they leave behind an excess of negative charge. Under equilibrium conditions the net diffusion current must be zero. This brought about by the electric fields which are established in such a way as to draw the positive carriers back toward the *p*-type material.

The two diffusion processes we have been describing go on simultaneously and, you will notice, both act in the direction which will charge up the *n*-type material in a positive sense and the *p*-type material in a negative sense. Because of the finite conductivity of the semiconductor material, the change in potential from the *p*-side to the *n*-side will occur in a relatively narrow region near the boundary; the main body of each block of material will have a uniform potential. Let's imagine an *x*-axis in a direction perpendicular to the boundary surface. Then the electric potential will vary with x, as shown in Fig. 14–9(b). We have also shown in part (c) of the figure the expected variation of the density N_n of *n*-carriers and the density N_p of *p*-carriers. Far away from the junction the carrier densities N_p and N_n should be just the equilibrium density we would expect for individual blocks of materials at the same temperature. (We have drawn the figure for a

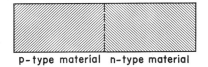

p-type material   n-type material

Fig. 14–8. A p-n junction.

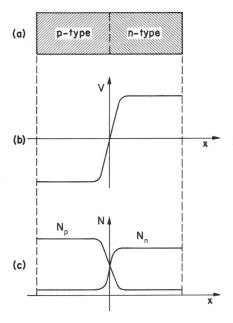

Fig. 14–9. The electric potential and the carrier densities in an unbiased semiconductor junction.

junction in which the *p*-type material is more heavily doped than the *n*-type material.) Because of the potential gradient at the junction, the positive carriers have to climb up a potential hill to get to the *p*-type side. This means that under equilibrium conditions there can be fewer positive carriers in the *n*-type material than there are in the *p*-type material. Remembering the laws of statistical mechanics, we expect that the ratio of *p*-type carriers on the two sides to be given by the following equation:

$$\frac{N_p(p\text{-side})}{N_p(n\text{-side})} = e^{-q_pV/\kappa T}. \tag{14.10}$$

The product q_pV in the numerator of the exponential is just the energy required to carry a charge of q_p through a potential difference V.

We have a precisely similar equation for the densities of the *n*-type carriers:

$$\frac{N_n(n\text{-side})}{N_n(p\text{-side})} = e^{-q_nV/\kappa T}. \tag{14.11}$$

If we know the equilibrium densities in each of the two materials, we can use either of the two equations above to determine the potential difference across the junction.

Notice that if Eqs. (14.10) and (14.11) are to give the same value for the potential difference V, the product N_pN_n must be the same for the *p*-side as for the *n*-side. (Remember that $q_n = -q_p$.) We have seen earlier, however, that this product depends only on the temperature and the gap energy of the crystal. Provided both sides of the crystal are at the same temperature, the two equations are consistent with the same value of the potential difference.

Since there is a potential difference from one side of the junction to the other, it looks something like a battery. Perhaps if we connect a wire from the *n*-type side to the *p*-type side we will get an electrical current. That would be nice because then the current would flow forever without using up any material and we would have an infinite source of energy in violation of the second law of thermodynamics! There is, however, no current if you connect a wire from the *p*-side to the *n*-side. And the reason is easy to see. Suppose we imagine first a wire made out of a piece of undoped material. When we connect this wire to the *n*-type side, we have a junction. There will be a potential difference across this junction. Let's say that it is just one-half the potential difference from the *p*-type material to the *n*-type material. When we connect our undoped wire to the *p*-type side of the junction, there is also a potential difference at this junction—again, one-half the potential drop across the *p-n* junction. At all the junctions the potential differences adjust themselves so that there is no net current flow in the circuit. Whatever kind of wire you use to connect together the two sides of the *n-p* junction, you are producing two new junctions, and so long as all the junctions are at the same temperature, the potential jumps at the junctions all compensate each other and no current will flow in the circuit. It does turn out, however—if you work out the details—that if some of the junctions are at a different temperature than the other junctions, currents will flow. Some of the junctions will be heated and others will be cooled by this current and thermal energy will be converted into electrical energy. This effect is responsible for the operation of thermocouples which are used for measuring temperatures, and of thermoelectric generators. The same effect is also used to make small refrigerators.

If we cannot measure the potential difference between the two sides of an *n-p* junction, how can we really be sure that the potential gradient shown in Fig. 14–9 really exists? One way is to shine light on the junction. When the light photons are absorbed they can produce an electron-hole pair. In the strong electric field that exists at the junction (equal to the slope of the potential curve of Fig. 14–9) the hole will be driven into the *p*-type region and the electron will be driven into the *n*-type region. If the two sides of the junction are now connected to an external circuit, these extra charges will provide a current. The energy of the light will be converted into electrical energy in the junction. The solar cells which generate electrical power for the operation of some of our satellites operate on this principle.

In our discussion of the operation of a semiconductor junction we have been assuming that the holes and the electrons act more-or-less independently—except that they somehow get into proper statistical equilibrium. When we were describing the current produced by light shining on the junction, we were assuming that an electron or a hole produced in the junction region would get into the main body of the crystal before being annihilated by a carrier of the opposite polarity. In the immediate vicinity of the junction, where the density of carriers of both signs is approximately equal, the effect of electron-hole annihilation (or as it is often called, "recombination") is an important effect, and in a detailed analysis of a semiconductor junction must be properly taken into account. We have been assuming that a hole or an electron produced in a junction region has a good chance of getting into the main body of the crystal before recombining. The typical time for an electron or a hole to find an opposite partner and annihilate it is for typical semiconductor materials in the range between 10^{-3} and 10^{-7} seconds. This time is, incidentally, much longer than the mean free time τ between collisions with scattering sites in the crystal which we used in the analysis of conductivity. In a typical n-p junction, the time for an electron or hole formed in the junction region to be swept away into the body of the crystal is generally much shorter than the recombination time. Most of the pairs will, therefore, contribute to an external current.

14–5 Rectification at a semiconductor junction

We would like to show next how it is that a p-n junction can act like a rectifier. If we put a voltage across the junction, a large current will flow if the polarity is in one direction, but a very small current will flow if the same voltage is applied in the opposite direction. If an alternating voltage is applied across the junction, a net current will flow in one direction—the current is "rectified." Let's look again at what is going on in the equilibrium condition described by the graphs of Fig. 14–9. In the p-type material there is a large concentration N_p of positive carriers. These carriers are diffusing around and a certain number of them each second approach the junction. This current of positive carriers which approaches the junction is proportional to N_p. Most of them, however, are turned back by the high potential hill at the junction and only the fraction $e^{-qV/\kappa T}$ gets through. There is also a current of positive carriers approaching the junction from the other side. This current is also proportional to the density of positive carriers in the n-type region, but the carrier density here is much smaller than the density on the p-type side. When the positive carriers approach the junction from the n-type side, they find a hill with a negative slope and immediately slide downhill to the p-type side of the junction. Let's call this current I_0. Under equilibrium the currents from the two directions are equal. We expect then the following relation:

$$I_0 \sim N_p(n\text{-side}) = N_p(p\text{-side})e^{-qV/\kappa T}. \qquad (14.12)$$

You will notice that this equation is really just the same as Eq. (14–10). We have just derived it in a different way.

Suppose, however, that we lower the voltage on the n-side of the junction by an amount ΔV—which we can do by applying an external potential difference to the junction. Now the difference in potential across the potential hill is no longer V but $V - \Delta V$. The current of positive carriers from the p-side to the n-side will now have this potential difference in its exponential factor. Calling this current I_1, we have

$$I_1 \sim N_p(p\text{-side})e^{-q(V-\Delta V)/\kappa T}.$$

This current is larger than I_0 by just the factor $e^{q\Delta V/\kappa T}$. So we have the following relation between I_1 and I_0:

$$I_1 = I_0 e^{+q\Delta V/\kappa T}. \qquad (14.13)$$

The current from the p-side increases exponentially with the externally applied voltage ΔV. The current of positive carriers from the n-side, however, remains constant so long as ΔV is not too large. When they approach the barrier, these carriers will still find a downhill potential and will all fall down to the p-side. (If ΔV is larger than the natural potential difference V, the situation would change, but we will not consider what happens at such high voltages.) The net current I of positive carriers which flows across the junction is then the difference between the currents from the two sides:

$$I = I_0(e^{+q\Delta V/\kappa T} - 1). \qquad (14.14)$$

The net current I of holes flows into the n-type region. There the holes diffuse into the body of the n-region, where they are eventually annihilated by the majority n-type carriers—the electrons. The electrons which are lost in this annihilation will be made up by a current of electrons from the external terminal of the n-type material.

When ΔV is zero, the net current in Eq. (14.14) is zero. For positive ΔV the current increases rapidly with the applied voltage. For negative ΔV the current reverses in sign, but the exponential term soon becomes negligible and the negative current never exceeds I_0—which under our assumptions is rather small. This back current I_0 is limited by the small density of the minority carriers on the n-side of the junction.

If you go through exactly the same analysis for the current of negative carriers which flows across the junction, first with no potential difference and then with a small externally applied potential difference ΔV, you get again an equation just like (14.14) for the net electron current. Since the total current is the sum of the currents contributed by the two carriers, Eq. (14.14) still applies for the total current provided we identify I_0 as the maximum current which can flow for a reversed voltage.

The voltage-current characteristic of Eq. (14.14) is shown in Fig. 14–10. It shows the typical behavior of solid state diodes—such as those used in modern computers. We should remark that Eq. (14.14) is true only for small voltages. For voltages comparable to or larger than the natural internal voltage difference V, other effects come into play and the current no longer obeys the simple equation.

You may remember, incidentally, that we got exactly the same equation we have found here in Eq. (14.14) when we discussed the "mechanical rectifier"—the ratchet and pawl—in Chapter 46 of Volume I. We get the same equations in the two situations because the basic physical processes are quite similar.

14–6 The transistor

Perhaps the most important application of semiconductors is in the transistor. The transistor consists of two semiconductor junctions very close together. Its operation is based in part on the same principles that we just described for the semiconductor diode—the rectifying junction. Suppose we make a little bar of germanium with three distinct regions, a p-type region, an n-type region, and another p-type region, as shown in Fig. 14–11(a). This combination is called a p-n-p transistor. Each of the two junctions in the transistor will behave much in the way we have described in the last section. In particular, there will be a potential gradient at each junction having a certain potential drop from the n-type region to each p-type region. If the two p-type regions have the same internal properties, the variation in potential as we go across the crystal will be as shown in the graph of Fig. 14–11(b).

Now let's imagine that we connect each of the three regions to external voltage sources as shown in part (a) of Fig. 14–12. We will refer all voltages to the terminal connected to the left-hand p-region so it will be, by definition, at zero potential. We will call this terminal the *emitter*. The n-type region is called the *base* and it is connected to a slightly negative potential. The right-hand p-type region is called the *collector*, and is connected to a somewhat larger negative potential. Under these circumstances the variation of potential across the crystal will be as shown in the graph of Fig. 14–12(b).

Let's first see what happens to the positive carriers, since it is primarily their behavior which controls the operation of the p-n-p transistor. Since the emitter is

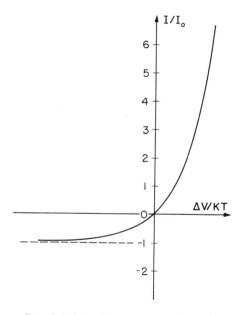

Fig. 14–10. The current through a junction as a function of the voltage across it.

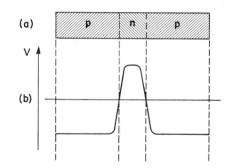

Fig. 14–11. The potential distribution in a transistor with no applied voltages.

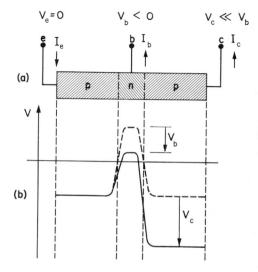

Fig. 14–12. The potential distribution in an operating transistor.

at a relatively more positive potential than the base, a current of positive carriers will flow from the emitter region into the base region. A relatively large current flows, since we have a junction operating with a "forward voltage"—corresponding to the right-hand half of the graph in Fig. 14–10. With these conditions, positive carriers or holes are being "emitted" from the p-type region into the n-type region. You might think that this current would flow out of the n-type region through the base terminal b. Now, however, comes the secret of the transistor. The n-type region is made very thin—typically 10^{-3} cm or less, much narrower than its transverse dimensions. This means that as the holes enter the n-type region they have a very good chance of diffusing across to the other junction before they are annihilated by the electrons in the n-type region. When they get to the right-hand boundary of the n-type region they find a steep downward potential hill and immediately fall into the right-hand p-type region. This side of the crystal is called the collector because it "collects" the holes after they have diffused across the n-type region. In a typical transistor, all but a fraction of a percent of the hole current which leaves the emitter and enters the base is collected in the collector region, and only the small remainder contributes to the net base current. The sum of the base and collector currents is, of course, equal to the emitter current.

Now imagine what happens if we vary slightly the potential V_b on the base terminal. Since we are on a relatively steep part of the curve of Fig. 14–10, a small variation of the potential V_b will cause a rather large change in the emitter current I_e. Since the collector voltage V_c is much more negative than the base voltage, these slight variations in potential will not effect appreciably the steep potential hill between the base and the collector. Most of the positive carriers emitted into the n-region will still be caught by the collector. Thus as we vary the potential of the base electrode, there will be a corresponding variation in the collector current I_c. The essential point, however, is that the base current I_b always remains a small fraction of the collector current. The transistor is an amplifier; a small current I_b introduced into the base electrode gives a large current —100 or so times higher—at the collector electrode.

What about the electrons—the negative carriers that we have been neglecting so far? First, note that we do not expect any significant electron current to flow between the base and the collector. With a large negative voltage on the collector, the electrons in the base would have to climb a very high potential energy hill and the probability of doing that is very small. There is a very small current of electrons to the collector.

On the other hand, the electrons in the base *can* go into the emitter region. In fact, you might expect the electron current in this direction to be comparable to the hole current from the emitter into the base. Such an electron current isn't useful, and, on the contrary, is bad because it increases the total base current required for a given current of holes to the collector. The transistor is, therefore, designed to minimize the electron current to the emitter. The electron current is proportional to N_n(base), the density of negative carriers in the base material while the hole current from the emitter depends on N_p(emitter), the density of positive carriers in the emitter region. By using relatively little doping in the n-type material N_n(base) can be made much smaller than N_p(emitter). (The very thin base region also helps a great deal because the sweeping out of the holes in this region by the collector increases significantly the average hole current from the emitter into the base, while leaving the electron current unchanged.) The net result is that the electron current across the emitter-base junction can be made much less than the hole current, so that the electrons do not play any significant role in operation of the p-n-p transistor. The currents are dominated by motion of the holes, and the transistor performs as an amplifier as we have described above.

It is also possible to make a transistor by interchanging the p-type and n-type materials in Fig. 14–11. Then we have what is called an n-p-n transistor. In the n-p-n transistor the main currents are carried by the electrons which flow from the emitter into the base and from there to the collector. Obviously, all the arguments we have made for the p-n-p transistor also apply to the n-p-n transistor if the potentials of the electrodes are chosen with the opposite signs.

15

The Independent Particle Approximation

15–1 Spin waves

In Chapter 13 we worked out the theory for the propagation of an electron or of some other "particle," such as an atomic excitation, through a crystal lattice. In the last chapter we applied the theory to semiconductors. But when we talked about situations in which there are many electrons we disregarded any interactions between them. To do this is of course only an approximation. In this chapter we will discuss further the idea that you can disregard the interaction between the electrons. We will also use the opportunity to show you some more applications of the theory of the propagation of particles. Since we will generally continue to disregard the interactions between particles, there is very little really new in this chapter except for the new applications. The first example to be considered is, however, one in which it is possible to write down quite exactly the correct equations when there is more than one "particle" present. From them we will be able to see how the approximation of disregarding the interactions is made. We will not, though, analyze the problem very carefully.

As our first example we will consider a "spin wave" in a ferromagnetic crystal. We have discussed the theory of ferromagnetism in Chapter 36 of Volume II. At zero temperature all the electron spins that contribute to the magnetism in the body of a ferromagnetic crystal are parallel. There is an interaction energy between the spins, which is lowest when all the spins are down. At any nonzero temperature, however, there is some chance that some of the spins are turned over. We calculated the probability in an approximate manner in Chapter 36. This time we will describe the quantum mechanical theory—so you will see what you would have to do if you wanted to solve the problem more exactly. (We will still make some idealizations by assuming that the electrons are localized at the atoms and that the spins interact only with neighboring spins.)

We consider a model in which the electrons at each atom are all paired except one, so that all of the magnetic effects come from one spin-$\frac{1}{2}$ electron per atom. Further, we imagine that these electrons are localized at the atomic sites in the lattice. The model corresponds roughly to metallic nickel.

We also assume that there is an interaction between any two adjacent spinning electrons which gives a term in the energy of the system

$$E = -\sum_{i,\,j} K\boldsymbol{\sigma}_i \cdot \boldsymbol{\sigma}_j, \tag{15.1}$$

where $\boldsymbol{\sigma}$'s represent the spins and the summation is over all adjacent pairs of electrons. We have already discussed this kind of interaction energy when we considered the hyperfine splitting of hydrogen due to the interaction of the magnetic moments of the electron and proton in a hydrogen atom. We expressed it then as $A\boldsymbol{\sigma}_e \cdot \boldsymbol{\sigma}_p$. Now, for a given pair, say the electrons at atom 4 and at atom 5, the Hamiltonian would be $-K\boldsymbol{\sigma}_4 \cdot \boldsymbol{\sigma}_5$. We have a term for each such pair, and the Hamiltonian is (as you would expect for classical energies) the sum of these terms for each interacting pair. The energy is written with the factor $-K$ so that a positive K will correspond to ferromagnetism—that is, the lowest energy results when adjacent spins are parallel. In a real crystal, there may be other terms which are the interactions of *next* nearest neighbors, and so on, but we don't need to consider such complications at this stage.

With the Hamiltonian of Eq. (15.1) we have a complete description of the ferromagnet—within our approximation—and the properties of the magnetization

15–1 Spin waves

15–2 Two spin waves

15–3 Independent particles

15–4 The benzene molecule

15–5 More organic chemistry

15–6 Other uses of the approximation

should come out. We should also be able to calculate the thermodynamic properties due to the magnetization. If we can find all the energy levels, the properties of the crystal at a temperature T can be found from the principle that the probability that a system will be found in a given state of energy E is proportional to $e^{-E/\kappa T}$. This problem has never been completely solved.

We will show some of the problems by taking a simple example in which all the atoms are in a line—a one-dimensional lattice. You can easily extend the ideas to three dimensions. At each atomic location there is an electron which has two possible states, either spin up or spin down, and the whole system is described by telling how all of the spins are arranged. We take the Hamiltonian of the system to be the operator of the interaction energy. Interpreting the spin vectors of Eq. (15.1) as the sigma-operators—or the sigma-matrices—we write for the linear lattice

$$\hat{H} = \sum_n -\frac{A}{2}\,\hat{\boldsymbol{\sigma}}_n \cdot \hat{\boldsymbol{\sigma}}_{n+1}. \tag{15.2}$$

In this equation we have written the constant as $A/2$ for convenience (so that some of the later equations will be exactly the same as the ones in Chapter 13).

Now what is the lowest state of this system? The state of lowest energy is the one in which all the spins are parallel—let's say, all up.† We can write this state as $|\cdots + + + + \cdots\rangle$, or $|\,\text{gnd}\rangle$ for the "ground," or lowest, state. It's easy to figure out the energy for this state. One way is to write out all the vector sigmas in terms of $\hat{\sigma}_x$, $\hat{\sigma}_y$, and $\hat{\sigma}_z$, and work through carefully what each term of the Hamiltonian does to the ground state, and then add the results. We can, however, also use a good short cut. We saw in Section 12–2, that $\hat{\boldsymbol{\sigma}}_i \cdot \hat{\boldsymbol{\sigma}}_j$ could be written in terms of the Pauli spin exchange operator like this:

$$\hat{\boldsymbol{\sigma}}_i \cdot \hat{\boldsymbol{\sigma}}_j = (2\hat{P}_{ij}^{\text{spin ex}} - 1), \tag{15.3}$$

where the operator $\hat{P}_{ij}^{\text{spin ex}}$ interchanges the spins of the ith and jth electrons. With this substitution the Hamiltonian becomes

$$\hat{H} = -A \sum_n (\hat{P}_{n,\,n+1}^{\text{spin ex}} - \tfrac{1}{2}). \tag{15.4}$$

It is now easy to work out what happens to different states. For instance if i and j are both up, then exchanging the spins leaves everything unchanged, so $\hat{P}_{ij}$ acting on the state just gives the same state back, and is equivalent to multiplying by $+1$. The expression $(\hat{P}_{ij} - \tfrac{1}{2})$ is just equal to one-half. (From now on we will leave off the descriptive superscript on the P.)

For the ground state all spins are up; so if you exchange a particular pair of spins, you get back the original state. The ground state is a stationary state. If you operate on it with the Hamiltonian you get the same state again multiplied by a sum of terms, $-(A/2)$ for each pair of spins. That is, the energy of the system in the ground state is $-A/2$ per atom

Next we would like to look at the energies of some of the excited states. It will be convenient to measure the energies with respect to the ground state—that is, to choose the ground state as our zero of energy. We can do that by adding the energy $A/2$ to each term in the Hamiltonian. That just changes the "$\tfrac{1}{2}$" in Eq. (15.4) to "1." Our new Hamiltonian is

$$\hat{H} = -A \sum_n (\hat{P}_{n,n+1} - 1). \tag{15.5}$$

With this Hamiltonian the energy of the lowest state is zero; the spin exchange operator is equivalent to multiplying by unity (for the ground state) which is cancelled by the "1" in each term.

---

† The ground state here is really "degenerate"; there are other states with the same energy—for example, all spins down, or all in any other direction. The slightest external field in the z-direction will give a different energy to all these states, and the one we have chosen will be the true ground state.

For describing states other than the ground state we will need a suitable set of base states. One convenient approach is to group the states according to whether one electron has spin down, or two electrons have spin down, and so on. There are, of course, many states with one spin down. The down spin could be at atom "4," or at atom "5," or at atom "6," . . . We can, in fact, choose just such states for our base states. We could write them this way: $|4\rangle$, $|5\rangle$, $|6\rangle$, . . . It will, however, be more convenient later if we label the "odd atom"—the one with the down-spinning electron—by its coordinate x. That is, we'll define the state $|x_5\rangle$ to be one with all the electrons spinning up except for the one on the atom at x_5, which has a down-spinning electron (see Fig. 15–1). In general, $|x_n\rangle$ is the state with one down spin that is located at the coordinate x_n of the nth atom.

What is the action of the Hamiltonian (15.5) on the state $|x_5\rangle$? One term of the Hamiltonian is say $-A(\hat{P}_{7,8} - 1)$. The operator $\hat{P}_{7,8}$ exchanges the two spins of the adjacent atoms 7, 8. But in the state $|x_5\rangle$ these are both up, and nothing happens; $\hat{P}_{7,8}$ is equivalent to multiplying by 1:

$$\hat{P}_{7,8} |x_5\rangle = |x_5\rangle.$$

It follows that

$$(\hat{P}_{7,8} - 1)|x_5\rangle = 0.$$

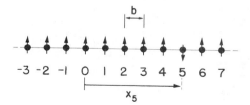

Fig. 15–1. The base state $|x_5\rangle$ of a linear array of spins. All the spins are up except the one at x_5, which is down.

Thus all the terms of the Hamiltonian give zero—except those involving atom 5, of course. On the state $|5\rangle$, the operation $\hat{P}_{4,5}$ exchanges the spin of atom 4 (up) and atom 5 (down). The result is the state with all spins up except the atom at 4. That is

$$\hat{P}_{4,5} |x_5\rangle = |x_4\rangle.$$

In the same way

$$\hat{P}_{5,6} |x_5\rangle = |x_6\rangle.$$

Hence, the only terms of the Hamiltonian which survive are $-A(\hat{P}_{4,5} - 1)$ and $-A(\hat{P}_{5,6} - 1)$. Acting on $|x_5\rangle$ they produce $-A|x_4\rangle + A|x_5\rangle$ and $-A|x_6\rangle + A|x_5\rangle$, respectively. The result is

$$\hat{H}|x_5\rangle = -A \sum_n (\hat{P}_{n,n+1} - 1)|x_5\rangle = -A\{|x_6\rangle + |x_4\rangle - 2|x_5\rangle\}. \quad (15.6)$$

When the Hamiltonian acts on state $|x_5\rangle$ it gives rise to some amplitude to be in states $|x_4\rangle$ and $|x_6\rangle$. That just means that there is a certain amplitude to have the down spin jump over to the next atom. So because of the interaction between the spins, if we begin with one spin down, then there is some probability that at a later time another one will be down instead. Operating on the general state $|x_n\rangle$, the Hamiltonian gives

$$\hat{H}|x_n\rangle = -A\{|x_{n+1}\rangle + |x_{n-1}\rangle - 2|x_n\rangle\}. \quad (15.7)$$

Notice particularly that if we take a complete set of states with only one spin down, they will only be mixed among themselves. The Hamiltonian will never mix these states with others that have more spins down. So long as you only exchange spins you never change the total number of down spins.

It will be convenient to use the matrix notation for the Hamiltonian, say $H_{n,m} \equiv \langle x_n | \hat{H} | x_m \rangle$; Eq. (15.7) is equivalent to

$$H_{n,n} = A;$$

$$H_{n,n+1} = H_{n,n-1} = -A; \quad (15.8)$$

$$H_{n,m} = 0, \quad \text{for} \quad |n - m| > 1.$$

Now what are the energy levels for states with one spin down? As usual we let C_n be the amplitude that some state $|\psi\rangle$ is in the state $|x_n\rangle$. If $|\psi\rangle$ is to be a definite energy state, all the C_n's must vary with time in the same way, namely,

$$C_n = a_n e^{-iEt/\hbar}. \quad (15.9)$$

We can put this trial solution into our usual Hamiltonian equation

$$ i\hbar \frac{dC_n}{dt} = \sum_m H_{nm} C_m, \tag{15.10} $$

using Eq. (15.8) for the matrix elements. Of course we get an infinite number of equations, but they can all be written as

$$ E a_n = 2A a_n - A a_{n-1} - A a_{n+1}. \tag{15.11} $$

We have again exactly the same problem we worked out in Chapter 13, except that where we had E_0 we now have $2A$. The solutions correspond to amplitudes C_n (the down-spin amplitude) which propagate along the lattice with a propagation constant k and an energy

$$ E = 2A(1 - \cos kb), \tag{15.12} $$

where b is the lattice constant.

The definite energy solutions correspond to "waves" of down spin—called "spin waves." And for each wavelength there is a corresponding energy. For large wavelengths (small k) this energy varies as

$$ E = A b^2 k^2. \tag{15.13} $$

Just as before, we can consider a localized wave packet (containing, however, only long wavelengths) which corresponds to a spin-down electron in one part of the lattice. This down spin will behave like a "particle." Because its energy is related to k by (15.13) the "particle" will have an effective mass:

$$ m_{\text{eff}} = \frac{\hbar^2}{2Ab^2}. \tag{15.14} $$

These "particles" are sometimes called "magnons."

15–2 Two spin waves

Now we would like to discuss what happens if there are *two* down spins. Again we pick a set of base states. We'll choose states in which there are down spins at two atomic locations, such as the state shown in Fig. 15-2. We can label such a state by the x-coordinates of the two sites with down spins. The one shown can be called $|x_2, x_5\rangle$. In general the base states are $|x_n, x_m\rangle$—a doubly infinite set! In this system of description, the state $|x_4, x_9\rangle$ and the state $|x_9, x_4\rangle$ are exactly the same state, because each simply says that there is a down spin at 4 and one at 9; there is no meaning to the order. Furthermore, the state $|x_4, x_4\rangle$ has no meaning, there isn't such a thing. We can describe any state $|\psi\rangle$ by giving the amplitudes to be in each of the base states. Thus $C_{m,n} = \langle x_m, x_n | \psi \rangle$ now means the amplitude for a system in the state $|\psi\rangle$ to be in a state in which both the mth and nth atoms have a down spin. The complications which now arise are not complications of ideas—they are merely complexities in bookkeeping. (One of the complexities of quantum mechanics is just the bookkeeping. With more and more down spins, the notation becomes more and more elaborate with lots of indices and the equations always look very horrifying; but the ideas are not necessarily more complicated than in the simplest case.)

The equations of motion of the spin system are the differential equations for the $C_{n,m}$. They are

$$ i\hbar \frac{dC_{n,m}}{dt} = \sum_{i,j} (H_{nm,ij}) C_{ij}. \tag{15.15} $$

Suppose we want to find the stationary states. As usual, the derivatives with respect to time become E times the amplitudes and the $C_{m,n}$ can be replaced by the

-3 -2 -1 0 1 2 3 4 5 6 7

Fig. 15–2. A state with two down spins.

coefficients $a_{m,n}$. Next we have to work out carefully the effect of H on a state with spins m and n down. It is not hard to figure out. Suppose for a moment that m and n are far enough apart so that we don't have to worry about the obvious trouble. The operation of exchange at the location x_n will move the down spin either to the $(n + 1)$ or $(n - 1)$ atom, and so there's an amplitude that the present state has come from the state $| x_m, x_{n+1}\rangle$ and also an amplitude that it has come from the state $| x_m, x_{n-1}\rangle$. Or it may have been the other spin that moved; so there's a certain amplitude that $C_{m,n}$ is fed from $C_{m+1,n}$ or from $C_{m-1,n}$. These effects should all be equal. The final result for the Hamiltonian equation on $C_{m,n}$ is

$$Ea_{m,n} = -A(a_{m+1,n} + a_{m-1,n} + a_{m,n+1} + a_{m,n-1}) + 4Aa_{m,n}. \qquad (15.16)$$

This equation is correct except in two situations. If $m = n$ there is no equation at all, and if $m = n \pm 1$, then two of the terms in Eq. (15.16) should be missing. *We are going to disregard these exceptions.* We simply ignore the fact that some few of these equations are slightly altered. After all, the crystal is supposed to be infinite, and we have an infinite number of terms; neglecting a few might not matter much. So for a first rough approximation let's forget about the altered equations. In other words, we assume that Eq. (15.16) is true for all m and n, even for m and n next to each other. *This is the essential part of our approximation.*

Then the solution is not hard to find. We get immediately

$$C_{m,n} = a_{m,n}e^{-iEt/\hbar}, \qquad (15.17)$$

with

$$a_{m,n} = (\text{const.})\, e^{ik_1 x_m}e^{ik_2 x_n}, \qquad (15.18)$$

where

$$E = 4A - 2A\cos k_1 b - 2A\cos k_2 b. \qquad (15.19)$$

Think for a moment what would happen if we had two *independent, single* spin waves (as in the previous section) corresponding to $k = k_1$ and $k = k_2$; they would have energies, from Eq. (15.12), of

$$\epsilon_1 = (2A - 2A\cos k_1 b)$$

and

$$\epsilon_2 = (2A - 2A\cos k_2 b).$$

Notice that the energy E in Eq. (15.19) is just their sum,

$$E = \epsilon(k_1) + \epsilon(k_2). \qquad (15.20)$$

In other words we can think of our solution in this way. There are two particles—that is, two spin waves. One of them has a momentum described by k_1, the other by k_2, and the energy of the system is the sum of the energies of the two objects. The two particles act completely independently. That's all there is to it.

Of course we have made some approximations, but we do not wish to discuss the precision of our answer at this point. However, you might guess that in a reasonable size crystal with billions of atoms—and, therefore, billions of terms in the Hamiltonian—leaving out a few terms wouldn't make much of an error. If we had so many down spins that there was an appreciable density, then we would certainly have to worry about the corrections.

[Interestingly enough, an exact solution can be written down if there are just the *two* down spins. The result is not particularly important. But it is interesting that the equations can be solved exactly for this case. The solution is:

$$a_{m,n} = \exp^{[ik_c(x_m+x_n)]} \sin k\, | x_m - x_n |, \qquad (15.21)$$

with the energy

$$E = 4A - 2A\cos k_1 b - 2A\cos k_2 b,$$

and with the wave numbers k_c and k related to k_1 and k_2 by

$$k_1 = k_c - k, \qquad k_2 = k_c + k. \tag{15.22}$$

This solution includes the "interaction" of the two spins. It describes the fact that when the spins come together there is a certain chance of scattering. The spins act very much like particles with an interaction. But the detailed theory of their scattering goes beyond what we want to talk about here.]

15–3 Independent particles

In the last section we wrote down a Hamiltonian, Eq. (15.15), for a two-particle system. Then using an approximation which is equivalent to neglecting any "interaction" of the two particles, we found the stationary states described by Eqs. (15.17) and (15.18). This state is just the product of two single-particle states. The solution we have given for $a_{m,n}$ in Eq. (15.18) is, however, really not satisfactory. We have very carefully pointed out earlier that the state $\mid x_9, x_4 \rangle$ is *not* a different state from $\mid x_4, x_9 \rangle$—the *order* of x_m and x_n has no significance. In general, the algebraic expression for the amplitude $C_{m,n}$ must be unchanged if we interchange the values of x_m and x_n, since that doesn't change the state. Either way, it should represent the amplitude to find a down spin at x_m and a down spin at x_n. But notice that (15.18) is *not* symmetric in x_m and x_n—since k_1 and k_2 can in general be different.

The trouble is that we have not forced our solution of Eq. (15.15) to satisfy this additional condition. Fortunately it is easy to fix things up. Notice first that a solution of the Hamiltonian equation just as good as (15.18) is

$$a_{m,n} = K e^{ik_2 x_m} e^{ik_1 x_n}. \tag{15.23}$$

It even has the same energy we got for (15.18). Any linear combination of (15.15) and (15.23) is also a good solution, and has an energy still given by Eq. (15.19). The solution we should have chosen—because of our symmetry requirement—is just the sum of (15.15) and (15.23):

$$a_{m,n} = K[e^{ik_1 x_m} e^{ik_2 x_n} + e^{ik_2 x_m} e^{ik_1 x_n}]. \tag{15.24}$$

Now, given any k_1 and k_2 the amplitude $C_{m,n}$ is independent of which way we put x_m and x_n—if we should happen to define x_m and x_n reversed we get the same amplitude. Our interpretation of Eq. (15.24) in terms of "magnons" must also be different. We can no longer say that the equation represents *one* particle with wave number k_1 and a *second* particle with wave number k_2. The amplitude (15.24) represents *one* state with two particles (magnons). The *state* is characterized by the two wave numbers k_1 and k_2. Our solution looks like a compound state of one particle with the momentum $p_1 = \hbar/k_1$ and another particle with the momentum $p_2 = \hbar/k_2$, but in our state we can't say which particle is which.

By now, this discussion should remind you of Chapter 4 and our story of identical particles. We have just been showing that the particles of the spin waves—the magnons—behave like identical Bose particles. All amplitudes must be symmetric in the coordinates of the two particles—which is the same as saying that if we "interchange the two particles," we get back the same amplitude and with the same sign. But, you may be thinking, why did we choose to *add* the two terms in making Eq. (15.24). Why not subtract? With a minus sign, interchanging x_m and x_n would just change the sign of $a_{m,n}$ which doesn't matter. But interchanging x_m and x_n *doesn't change anything*—all the electrons of the crystal are exactly where they were before, so there is no reason for even the sign of the amplitude to change. The magnons will behave like Bose particles.†

---

† In general, the quasi particles of the kind we are discussing may act like either Bose particles or Fermi particles, and as for free particles, the particles with integral spin are bosons and those with half-integral spins are fermions. The "magnon" stands for a spin-up electron turned over. The *change* in spin is *one*. The magnon has an integral spin, and is a boson.

The main points of this discussion have been twofold: First, to show you something about spin waves, and, second, to demonstrate a state whose amplitude is a *product* of two amplitudes, and whose energy is the *sum* of the energies corresponding to the two amplitudes. For *independent particles* the amplitude is the product and the energy is the sum. You can easily see why the energy is the sum. The energy is the coefficient of t in an imaginary exponential—it is proportional to the frequency. If two objects are doing something, one of them with the amplitude $e^{-iE_1 t/\hbar}$ and the other with the amplitude $e^{-iE_2 t/\hbar}$, and if the amplitude for the two things to happen together is the product of the amplitudes for each, then there is a single frequency in the product which is the sum of the two frequencies. The energy corresponding to the amplitude product is the sum of the two energies.

We have gone through a rather long-winded argument to tell you a simple thing. When you don't take into account any interaction between particles, you can think of each particle independently. They can individually exist in the various different states they would have alone, and they will each contribute the energy they would have had if they were alone. However, you must remember that if they are identical particles, they may behave either as Bose or as Fermi particles depending upon the problem. Two extra electrons added to a crystal, for instance, would have to behave like Fermi particles. When the positions of two electrons are interchanged, the amplitude must reverse sign. In the equation corresponding to Eq. (15.24) there would have to be a minus sign between the two terms on the right. As a consequence, two Fermi particles cannot be in exactly the same condition—with equal spins and equal k's. The amplitude for this state is zero.

15–4 The benzene molecule

Although quantum mechanics provides the basic laws that determine the structures of molecules, these laws can be applied exactly only to the most simple compounds. The chemists have, therefore, worked out various approximate methods for calculating some of the properties of complicated molecules. We would now like to show you how the independent particle approximation is used by the organic chemists. We begin with the benzene molecule.

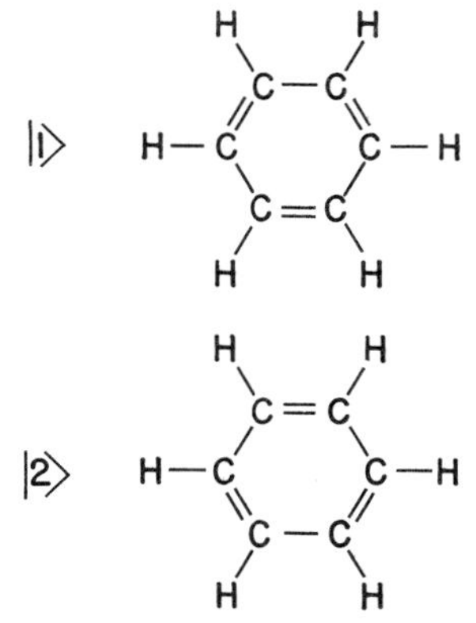

We discussed the benzene molecule from another point of view in Chapter 10. There we took an approximate picture of the molecule as a two-state system, with the two base states shown in Fig.15–3. There is a ring of six carbons with a hydrogen bonded to the carbon at each location. With the conventional picture of valence bonds it is necessary to assume double bonds between half of the carbon atoms, and in the lowest energy condition there are the two possibilities shown in the figure. There are also other, higher-energy states. When we discussed benzene in Chapter 10, we just took the two states and forgot all the rest. We found that the ground-state energy of the molecule was not the energy of one of the states in the figure, but was lower than that by an amount proportional to the amplitude to flip from one of these states to the other.

Now we're going to look at the same molecule from a completely different point of view—using a different kind of approximation. The two points of view will give us different answers, but if we improve either approximation it should lead to the truth, a valid description of benzene. However, if we don't bother to improve them, which is of course the usual situation, then you should not be surprised if the two descriptions do not agree exactly. We shall at least show that also with the new point-of-view the lowest energy of the benzene molecule is lower than either of the three-bond structures of Fig. 15–3.

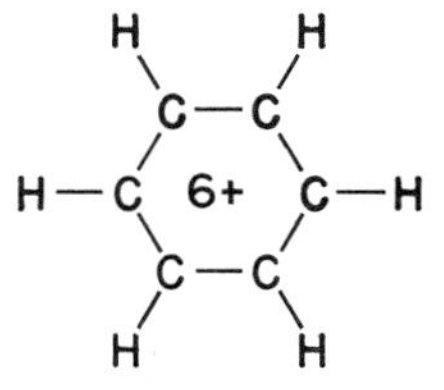

Now we want to use the following picture. Suppose we imagine the six carbon atoms of a benzene molecule connected only by single bonds as in Fig. 15–4. We have removed six electrons—since a bond stands for a pair of electrons—so we have a six-times ionized benzene molecule. Now we will consider what happens when we put back the six electrons one at a time, imagining that each one can run freely around the ring. We assume also that all the bonds shown in Fig. 15–4 are satisfied, and don't need to be considered further.

Fig. 15–3. The two base states for the benzene molecule used in Chapter 10.

Fig. 15–4. A benzene ring with six electrons removed.

Fig. 15–5. The ethylene molecule.

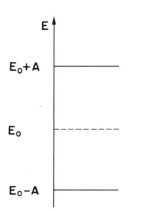

Fig. 15–6. The possible energy levels for the "extra" electrons in the ethylene molecule.

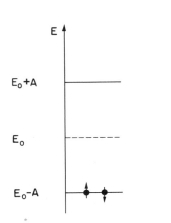

Fig. 15–7. In the extra bond of the ethylene molecule two electrons (one spin up, one spin down) can occupy the lowest energy level.

What happens when we put one electron back into the molecular ion? It might, of course, be located in any one of the six positions around the ring—corresponding to six base states. It would also have a certain amplitude, say A, to go from one position to the next. If we analyze the stationary states, there would be certain possible energy levels. That's only for one electron.

Next put a second electron in. And now we make the most ridiculous approximation that you can think of—*that what one electron does is not affected by what the other is doing.* Of course they really will interact; they repel each other through the Coulomb force, and furthermore when they are both at the same site, they must have considerably different energy than twice the energy for one being there. Certainly the approximation of independent particles is not legitimate when there are only six sites—particularly when we want to put in *six* electrons. Nevertheless the organic chemists have been able to learn a lot by making this kind of an approximation.

Before we work out the benzene molecule in detail, let's consider a simpler example—the ethylene molecule which contains just two carbon atoms with two hydrogen atoms on either side as shown in Fig. 15–5. This molecule has one "extra" bond involving two electrons between the two carbon atoms. Now remove one of these electrons; what do we have? We can look at it as a two-state system—the remaining electron can be at one carbon or the other. We can analyze it as a two-state system. The possible energies for the single electron are either $(E_0 - A)$ or $(E_0 + A)$, as shown in Fig. 15–6.

Now add the second electron. Good, if we have two electrons, we can put the first one in the lower state and the second one in the upper. Not quite; we forgot something. Each one of the states is really double. When we say there's a possible state with the energy $(E_0 - A)$, there are really two. Two electrons can go into the same state if one has its spin up and the other, its spin down. (No more can be put in because of the exclusion principle.) So there really are two possible states of energy $(E_0 - A)$. We can draw a diagram, as in Fig. 15–7, which indicates both the energy levels and their occupancy. In the condition of lowest energy both electrons will be in the lowest state with their spins opposite. The energy of the extra bond in the ethylene molecule therefore is $2(E_0 - A)$ if we neglect the interaction between the two electrons.

Let's go back to the benzene. Each of the two states of Fig. 15–3 has three double bonds. Each of these is just like the bond in ethylene, and contributes $2(E_0 - A)$ to the energy if E_0 is now the energy to put an electron on a site in benzene and A is the amplitude to flip to the next site. So the energy should be roughly $6(E_0 - A)$. But when we studied benzene before, we got that the energy was lower than the energy of the structure with three extra bonds. Let's see if the energy for benzene comes out lower than three bonds from our new point of view.

We start with the six-times ionized benzene ring and add one electron. Now we have a six-state system. We haven't solved such a system yet, but we know what to do. We can write six equations in the six amplitudes, and so on. But let's save some work—by noticing that we've already solved the problem, when we worked out the problem of an electron on an infinite line of atoms. Of course, the benzene is not an infinite line, it has 6 atomic sites in a circle. But imagine that we open out the circle to a line, and number the atoms along the line from 1 to 6. In an infinite line the next location would be 7, but if we insist that this location be identical with number 1 and so on, the situation will be just like the benzene ring. In other words we can take the solution for an infinite line *with an added requirement* that the solution must be periodic with a cycle six atoms long. From Chapter 13 the electron on a line has states of definite energy when the amplitude at each site is $e^{ikx_n} = e^{ikbn}$. For each k the energy is

$$E = E_0 - 2A \cos kb. \qquad (15.25)$$

We want to use now only those solutions which repeat every 6 atoms. Let's do first the general case for a ring of N atoms. If the solution is to have a period of N atomic spacing, e^{ikbN} must be unity; or kbN must be a multiple of 2π. Taking s to represent any integer, our condition is that

$$kbN = 2\pi s. \qquad (15.26)$$

We have seen before that there is no meaning to taking k's outside the range $\pm \pi/b$. This means that we get all possible states by taking values of s in the range $\pm N/2$.

We find then that for an N-atom ring there are N definite energy states† and they have wave numbers k_s given by

$$k_s = \frac{2\pi}{Nb} s. \qquad (15.27)$$

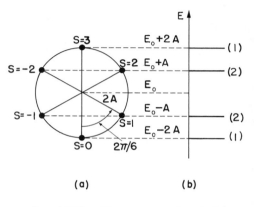

(a) (b)

Fig. 15–8. The energy levels in a ring with six electron locations (for example, a benzene ring).

Each state has the energy (15.25). We have a line spectrum of possible energy levels. The spectrum for benzene ($N = 6$) is shown in Fig. 15–8(b). (The numbers in parentheses indicate the number of *different* states with the same energy.)

There's a nice way to visualize the six energy levels, as we have shown in part (a) of the figure. Imagine a circle centered on a level with E_0, and with a radius of $2A$. If we start at the bottom and mark off six equal arcs (at angles from the bottom point of $k_s b = 2\pi s/N$, or $2\pi s/6$ for benzene), then the vertical heights of the points on the circle are the solutions of Eq. (15.25). The six points represent the six possible states. The lowest energy level is at ($E_0 - 2A$); there are two states with the same energy ($E_0 - A$), and so on.‡ These are possible states for one electron. If we have more than one electron, two—with opposite spins—can go into each state.

For the benzene molecule we have to put in six electrons. For the ground state they will go into the lowest possible energy states—two at $s = 0$, two at $s = +1$, and two at $s = -1$. According to the independent particle approximation the energy of the ground state is

$$\begin{aligned} E_{\text{ground}} &= 2(E_0 - 2A) + 4(E_0 - A) \\ &= 6E_0 - 8A. \qquad (15.28) \end{aligned}$$

The energy is indeed less than that of three separate double bonds—by the amount $2A$.

By comparing the energy of benzene to the energy of ethylene it is possible to determine A. It comes out to be 0.8 electron volt, or, in the units the chemists like, 18 kilocalories per mole.

We can use this description to calculate or understand other properties of benzene. For example, using Fig. 15–8 we can discuss the excitation of benzene by light. What would happen if we tried to excite one of the electrons? It could move up to one of the empty higher states. The lowest energy of excitation would be a transition from the highest filled level to the lowest empty level. That takes the energy $2A$. Benzene will absorb light of frequency ν when $h\nu = 2A$. There will also be absorption of photons with the energies $3A$ and $4A$. Needless to say, the absorption spectrum of benzene has been measured and the pattern of spectral lines is more or less correct except that the lowest transition occurs in the ultraviolet; and to fit the data one would have to choose a value of A between 1.4 and 2.4 electron volts. That is, the numerical value of A is two or three times larger than is predicted from the chemical binding energy.

What the chemist does in situations like this is to analyze many molecules of a similar kind and get some empirical rules. He learns, for example: For calculating binding energy use such and such a value of A, but for getting the absorption spectrum approximately right use another value of A. You may feel

---

† You might think that for N an even number there are $N + 1$ states. That is not so because $s = \pm N/2$ give the same state.

‡ When there are two states (which will have different amplitude distributions) with the same energy, we say that the two states are "degenerate." Notice that *four* electrons can have the energy $E_0 - A$.

that this sounds a little absurd. It is not very satisfactory from the point of view of a physicist who is trying to understand nature from first principles. But the problem of the chemist is different. He must try to guess ahead of time what is going to happen with molecules that haven't been made yet, or which aren't understood completely. What he needs is a series of empirical rules; it doesn't make much difference where they come from. So he uses the theory in quite a different way than the physicist. He takes equations that have some shadow of the truth in them, but then he must alter the constants in them—making empirical corrections.

In the case of benzene, the principal reason for the inconsistency is our assumption that the electrons are independent—the theory we started with is really not legitimate. Nevertheless, it has some shadow of the truth because its results seem to be going in the right direction. With such equations plus some empirical rules—including various exceptions—the organic chemist makes his way through the morass of complicated things he chooses to study. (Don't forget that the reason a physicist can really calculate from first principles is that he chooses only simple problems. He never solves a problem with 42 or even 6 electrons in it. So far, he has been able to calculate reasonably accurately only the hydrogen atom and the helium atom.)

15–5 More organic chemistry

Let's see how the same ideas can be used to study other molecules. Consider a molecule like butadiene (1, 3)—it is drawn in Fig. 15–9 according to the usual valence bond picture.

We can play the same game with the extra four electrons corresponding to the two double bonds. If we remove four electrons, we have four carbon atoms in a line. You already know how to solve a line. You say, "Oh no, I only know how to solve an *infinite* line." But the solutions for the infinite line also include the ones for a finite line. Watch. Let N be the number of atoms on the line and number them from 1 to N as shown in Fig. 15–10. In writing the equations for the amplitude at position 1 you would not have a term feeding from position 0. Similarly, the equation for position N would differ from the one that we used for an infinite line because there would be nothing feeding from position $N + 1$. But suppose that we can obtain a solution for the infinite line which has the following property: the amplitude to be at atom 0 is zero and the amplitude to be at atom $(N + 1)$ is also zero. Then the set of equations for all the locations from 1 to N on the finite line are also satisfied. You might think no such solution exists for the infinite line because our solutions all looked like e^{ikx_n} which has the same absolute value of the amplitude everywhere. But you will remember that the energy depends only on the absolute value of k, so that another solution, which is equally legitimate for the same energy, would be e^{-ikx_n}. And the same is true of any superposition of these two solutions. By subtracting them we can get the solution $\sin kx_n$, which satisfies the requirement that the amplitude be zero at $x = 0$. It still corresponds to the energy $(E_0 - 2A \cos kb)$. Now by a suitable choice for the value of k we can also make the amplitude zero at x_{N+1}. This requires that $(N + 1)kb$ be a multiple of π, or that

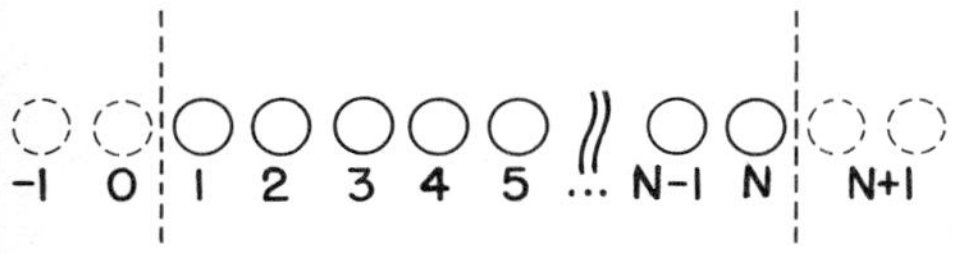

$$kb = \frac{\pi}{(N + 1)} s, \tag{15.29}$$

where s is an integer from 1 to N. (We take only positive k's because each solution contains $+k$ and $-k$; changing the sign of k gives the same state all over again.) For the butadiene molecule, $N = 4$, so there are four states with

$$kb = \pi/5, \quad 2\pi/5, \quad 3\pi/5, \quad \text{and} \quad 4\pi/5. \tag{15.30}$$

We can represent the energy levels using a circle diagram similar to the one for benzene. This time we use a semicircle divided into five equal parts as shown in Fig. 15–11. The point at the bottom corresponds to $s = 0$, which gives no

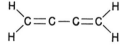

Fig. 15–9. The valence bond representation of the molecule butadiene (1, 3).

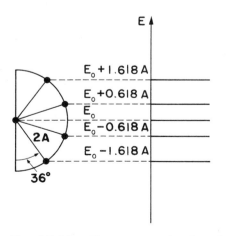

Fig. 15–10. A line of N molecules.

Fig. 15–11. The energy levels of butadiene.

15-10

state at all. The same is true of the point at the top, which corresponds to $s = N + 1$. The remaining 4 points give us four allowed energies. There are four stationary states, which is what we expect having started with four base states. In the circle diagram, the angular intervals are $\pi/5$ or 36 degrees. The lowest energy comes out $(E_0 - 1.618A)$. (Ah, what wonders mathematics holds; the golden mean of the Greeks† gives us the lowest energy state of the butadiene molecule according to this theory!)

Now we can calculate the energy of the butadiene molecule when we put in four electrons. With four electrons, we fill up the lowest two levels, each with two electrons of opposite spin. The total energy is

$$E = 2(E_0 - 1.618A) + 2(E_0 - 0.618A) = 4(E_0 - A) - 0.477A.$$

$$(15.31)$$

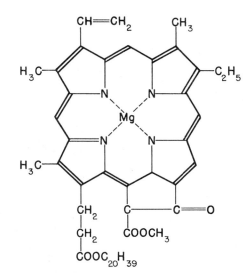

Fig. 15-12. A chlorophyll molecule.

This result seems reasonable. The energy is a little lower than for two simple double bonds, but the binding is not so strong as in benzene. Anyway this is the way the chemist analyzes some organic molecules.

The chemist can use not only the energies but the probability amplitudes as well. Knowing the amplitudes for each state, and which states are occupied, he can tell the probability of finding an electron anywhere in the molecule. Those places where the electrons are more likely to be are apt to be reactive in chemical substitutions which require that an electron be shared with some other group of atoms. The other sites are more likely to be reactive in those substitutions which have a tendency to yield an extra electron to the system.

The same ideas we have been using can give us some understanding of a molecule even as complicated as chlorophyll, one version of which is shown in Fig. 15-12. Notice that the double and single bonds we have drawn with heavy lines form a long closed ring with twenty intervals. The extra electrons of the double bonds can run around this ring. Using the independent particle method we can get a whole set of energy levels. There are strong absorption lines from transitions between these levels which lie in the visible part of the spectrum, and give this molecule its strong color. Similar complicated molecules such as the xanthophylls, which make leaves turn red, can be studied in the same way.

There is one more idea which emerges from the application of this kind of theory in organic chemistry. It is probably the most successful or, at least in a certain sense, the most accurate. This idea has to do with the question: In what situations does one get a particularly strong chemical binding? The answer is very interesting. Take the example, first, of benzene, and imagine the sequence of events that occurs as we start with the six-times ionized molecule and add more and more electrons. We would then be thinking of various benzene ions—negative or positive. Suppose we plot the energy of the ion (or neutral molecule) as a function of the number of electrons. If we take $E_0 = 0$ (since we don't know what it is), we get the curve shown in Fig. 15-13. For the first two electrons the slope of the function is a straight line. For each successive group the slope increases, and there is a discontinuity in slope between the groups of electrons. The slope changes when one has just finished filling a set of levels which all have the same energy and must move up to the next higher set of levels for the next electron.

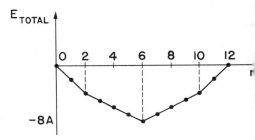

Fig. 15-13. The sum of all the electron energies when the lowest states in Fig. 15-8 are occupied by n electrons if we take that $E_0 = 0$.

The actual energy of the benzene ion is really quite different from the curve of Fig. 15-13 because of the interactions of the electrons and because of electrostatic energies we have been neglecting. These corrections will, however, vary with n in a rather smooth way. Even if we were to make all these corrections, the resulting energy curve would still have kinks at those values of n which just fill up a particular energy level.

Now consider a very smooth curve that fits the points on the average like the one drawn in Fig. 15-14. We can say that the points *above* this curve have "higher-than-normal" energies, and the points *below* the curve have "lower-than-normal"

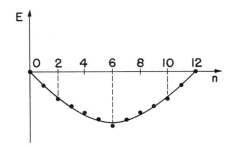

Fig. 15-14. The points of Fig. 15-12 with a smooth curve. Molecules with $n = 2, 6, 10$ are more stable than the others.

_____

† The ratio of the sides of a rectangle which can be divided into a square and a similar rectangle.

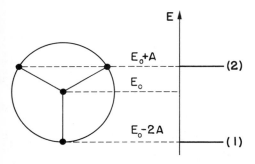

E

$E_0 + A$ — — — (2)

E_0 — — —

$E_0 - 2A$ — — — (1)

Fig. 15-15. Energy diagram for a ring of three.

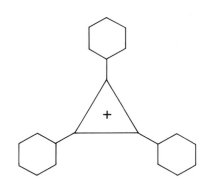

Fig. 15-16. The triphenyl cyclo-propanyl cation.

energies. We would, in general, expect that those configurations with a lower-than-normal energy would have an above average stability—chemically speaking. Notice that the configurations farther below the curve always occur at the end of one of the straight line segments—namely when there are enough electrons to fill up an "energy shell," as it is called. This is the very accurate prediction of the theory. Molecules—or ions—are particularly stable (in comparison with other similar configurations) when the available electrons just fill up an energy shell.

This theory has explained and predicted some very peculiar chemical facts. To take a very simple example, consider a ring of three. It's almost unbelievable that the chemist can make a ring of three and have it stable, but it has been done. The energy circle for three electrons is shown in Fig. 15-15. Now if you put two electrons in the lower state, you have only two of the three electrons that you require. The third electron must be put in at a much higher level. By our argument this molecule should not be particularly stable, whereas the two-electron structure should be stable. It does turn out, in fact, that the neutral molecule of triphenyl cyclopropenyl is very hard to make, but that the positive ion shown in Fig. 15-16 is relatively easy to make. The ring of three is never really easy because there is always a large stress when the bonds in an organic molecule make an equilateral triangle. To make a stable compound at all, the structure must be stabilized in some way. Anyway if you add three benzene rings on the corners, the positive ion can be made. (The reason for this requirement of added benzene rings is not really understood.)

In a similar way the five-sided ring can also be analyzed. If you draw the energy diagram, you can see in a qualitative way that the six-electron structure should be an especially stable structure, so that such a molecule should be most stable as a negative ion. Now the five-ring is well known and easy to make and always acts as a negative ion. Similarly, you can easily verify that a ring of 4 or 8 is not very interesting, but that a ring of 14 or 10—like a ring of 6—should be especially stable as a neutral object.

15-6 Other uses of the approximation

There are two other similar situations which we will describe only briefly. In considering the structure of an atom, we can consider that the electrons fill successive shells. The Schrödinger theory of electron motion can be worked out easily only for a *single* electron moving in a "central" field—one which varies only with the distance from a point. How can we then understand what goes on in an atom which has 22 electrons?! One way is to use a kind of independent particle approximation. First you calculate what happens with one electron. You get a number of energy levels. You put an electron into the lowest energy state. You can, for a rough model, continue to ignore the electron interactions and go on filling successive shells, but there is a way to get better answers by taking into account—in an approximate way at least—the effect of the electric charge carried by the electron. Each time you add an electron you compute its amplitude to be at various places, and then use this amplitude to estimate a kind of spherically symmetric charge distribution. You use the field of this distribution—together with the field of the positive nucleus and all the previous electrons—to calculate the states available for the next electron. In this way you can get reasonably correct estimates for the energies for the neutral atom and for various ionized states. You find that there are energy shells, just as we saw for the electrons in a ring molecule. With a partially filled shell, the atom will show a preference for taking on one or more extra electrons, or for losing some electrons so as to get into the most stable state of a filled shell.

This theory explains the machinery behind the fundamental chemical properties which show up in the periodic table of the elements. The inert gases are those elements in which a shell has just been completed, and it is especially difficult to make them react. (Some of them do react of course—with fluorine and oxygen, for example; but such compounds are very weakly bound; the so-called inert gases are nearly inert.) An atom which has one electron more or one electron less than an inert gas will easily lose or gain an electron to get into the especially stable (low-energy) condition which comes from having a completely filled shell—they are the very active chemical elements of valence $+1$ or -1.

The other situation is found in nuclear physics. In atomic nuclei the protons and neutrons interact with each other quite strongly. Even so, the independent particle model can again be used to analyze nuclear structure. It was first discovered experimentally that nuclei were especially stable if they contained certain particular numbers of neutrons—namely 2, 8, 20, 28, 50, 82. Nuclei containg protons in these numbers are also especially stable. Since there was initially no explanation for these numbers they were called the "magic numbers" of nuclear physics. It is well known that neutrons and protons interact strongly with each other; people were, therefore, quite surprised when it was discovered that an independent particle model predicted a shell structure which came out with the first few magic numbers. The model assumed that each nucleon (proton or neutron) moved in a central potential which was created by the average effects of all the other nucleons. This model failed, however, to give the correct values for the higher magic numbers. Then it was discovered by Maria Mayer, and independently by Jensen and his collaborators, that by taking the independent particle model and adding only a correction for what is called the "spin-orbit interaction," one could make an improved model which gave all of the magic numbers. (The spin-orbit interaction causes the energy of a nucleon to be lower if its spin has the same direction as its orbital angular momentum from motion in the nucleus.) The theory gives even more—its picture of the so-called "shell structure" of the nuclei enables us to predict certain characteristics of nuclei and of nuclear reactions.

The independent particle approximation has been found useful in a wide range of subjects—from solid-state physics, to chemistry, to biology, to nuclear physics. It is often only a crude approximation, but is able to give an understanding of why there are especially stable conditions—in shells. Since it omits all of the complexity of the interactions between the individual particles, we should not be surprised that it often fails completely to give correctly many important details.

16

The Dependence of Amplitudes on Position

16–1 Amplitudes on a line

We are now going to discuss how the probability amplitudes of quantum mechanics vary in space. In some of the earlier chapters you may have had a rather uncomfortable feeling that some things were being left out. For example, when we were talking about the ammonia molecule, we chose to describe it in terms of two base states. For one base state we picked the situation in which the nitrogen atom was "above" the plane of the three hydrogen atoms, and for the other base state we picked the condition in which the nitrogen atom was "below" the plane of the three hydrogen atoms. Why did we pick just these two states? Why is it not possible that the nitrogen atom could be at 2 angstroms above the plane of the three hydrogen atoms, or at 3 angstroms, or at 4 angstroms above the plane? Certainly, there are many positions that the nitrogen atom could occupy. Again when we talked about the hydrogen molecular ion, in which there is one electron shared by two protons, we imagined two base states: one for the electron in the neighborhood of proton number one, and the other for the electron in the neighborhood of proton number two. Clearly we were leaving out many details. The electron is not exactly at proton number two but is only in the neighborhood. It could be somewhere above the proton, somewhere below the proton, somewhere to the left of the proton, or somewhere to the right of the proton.

We intentionally avoided discussing these details. We said that we were interested in only certain features of the problem, so we were imagining that when the electron was in the vicinity of proton number one, it would take up a certain rather definite condition. In that condition the probability to find the electron would have some rather definite distribution around the proton, but we were not interested in the details.

We can also put it another way. In our discussion of a hydrogen molecular ion we chose an approximate description when we described the situation in terms of two base states. In reality there are lots and lots of these states. An electron can take up a condition around a proton in its lowest, or ground, state, but there are also many excited states. For each excited state the distribution of the electron around the proton is different. We ignored these excited states, saying that we were interested in only the conditions of low energy. But it is just these other excited states which give the possibility of various distributions of the electron around the proton. If we want to describe in detail the hydrogen molecular ion, we have to take into account also these other possible base states. We could do this in several ways, and one way is to consider in greater detail states in which the location of the electron in space is more carefully described.

We are now ready to consider a more elaborate procedure which will allow us to talk in detail about the position of the electron, by giving a probability amplitude to find the electron anywhere and everywhere in a given situation. This more complete theory provides the underpinning for the approximations we have been making in our earlier discussions. In a sense, our early equations can be derived as a kind of approximation to the more complete theory.

You may be wondering why we did not begin with the more complete theory and make the approximations as we went along. We have felt that it would be much easier for you to gain an understanding of the basic machinery of quantum mechanics by beginning with the two-state approximations and working gradually up to the more complete theory than to approach the subject the other way around. For this reason our approach to the subject appears to be in the reverse order to the one you will find in many books.

16–1 Amplitudes on a line

16–2 The wave function

16–3 States of definite momentum

16–4 Normalization of the states in x

16–5 The Schrödinger equation

16–6 Quantized energy levels

As we go into the subject of this chapter you will notice that we are breaking a rule we have always followed in the past. Whenever we have taken up any subject we have always tried to give a more or less complete description of the physics—showing you as much as we could about where the ideas led to. We have tried to describe the general consequences of a theory as well as describing some specific detail so that you could see where the theory would lead. We are now going to break that rule; we are going to describe how one can talk about probability amplitudes in space and show you the differential equations which they satisfy. We will not have time to go on and discuss many of the obvious implications which come out of the theory. Indeed we will not even be able to get far enough to relate this theory to some of the approximate formulations we have used earlier—for example, to the hydrogen molecule or to the ammonia molecule. For once, we must leave our business unfinished and open-ended. We are approaching the end of our course, and we must satisfy ourselves with trying to give you an introduction to the general ideas and with indicating the connections between what we have been describing and some of the other ways of approaching the subject of quantum mechanics. We hope to give you enough of an idea that you can go off by yourself and by reading books learn about many of the implications of the equations we are going to describe. We must, after all, leave something for the future.

Let's review once more what we have found out about how an electron can move along a line of atoms. When an electron has an amplitude to jump from one atom to the next, there are definite energy states in which the probability amplitude for finding the electron is distributed along the lattice in the form of a traveling wave. For long wavelengths—for small values of the wave number k—the energy of the state is proportional to the square of the wave number. For a crystal lattice with the spacing b, in which the amplitude per unit time for the electron to jump from one atom to the next is $iA/\hbar$, the energy of the state is related to k (for small kb) by

$$E = Ak^2b^2 \tag{16.1}$$

(see Section 13–3). We also saw that groups of such waves with similar energies would make up a wave packet which would behave like a classical particle with a mass m_{eff} given by:

$$m_{\text{eff}} = \frac{\hbar^2}{2Ab^2}. \tag{16.2}$$

Since waves of probability amplitude in a crystal behave like a particle, one might well expect that the general quantum mechanical description of a particle would show the same kind of wave behavior we observed for the lattice. Suppose we were to think of a lattice on a line and imagine that the lattice spacing b were to be made smaller and smaller. In the limit we would be thinking of a case in which the electron could be anywhere along the line. We would have gone over to a continuous distribution of probability amplitudes. We would have the amplitude to find an electron anywhere along the line. This would be one way to describe the motion of an electron in a vacuum. In other words, if we imagine that space can be labeled by an infinity of points all very close together and we can work out the equations that relate the amplitudes at one point to the amplitudes at neighboring points, we will have the quantum mechanical laws of motion of an electron in space.

Let's begin by recalling some of the general principles of quantum mechanics. Suppose we have a particle which can exist in various conditions in a quantum mechanical system. Any particular condition an electron can be found in, we call a "state," which we label with a state vector, say $|\phi\rangle$. Some other condition would be labeled with another state vector, say $|\psi\rangle$. We then introduce the idea of base states. We say that there is a set of states $|1\rangle$, $|2\rangle$, $|3\rangle$, $|4\rangle$, and so on, which have the following properties. First, all of these states are quite distinct—we say they are orthogonal. By this we mean that for any two of the base states $|i\rangle$ and $|j\rangle$, the amplitude $\langle i\,|\,j\rangle$ that an electron known to be in the state $|i\rangle$ is also in the state $|j\rangle$ is equal to zero—unless, of course, $|i\rangle$ and $|j\rangle$ stand for the same state. We represent this symbolically by

$$\langle i \,|\, j \rangle = \delta_{ij}. \tag{16.3}$$

You will remember that $\delta_{ij} = 0$ if i and j are different, and $\delta_{ij} = 1$ if i and j are the same number.

Second, the base states $|i\rangle$ must be a complete set, so that any state at all can be described in terms of them. That is, any state $|\phi\rangle$ at all can be described completely by giving all of the amplitudes $\langle i \,|\, \phi \rangle$ that a particle in the state $|\phi\rangle$ will also be found in the state $|i\rangle$. In fact, the state vector $|\phi\rangle$ is equal to the sum of the base states each multiplied by a coefficient which is the amplitude of the state $|\phi\rangle$ is also in the state $|i\rangle$:

$$|\phi\rangle = \sum_i |i\rangle\langle i \,|\, \phi \rangle. \tag{16.4}$$

Finally, if we consider any two states $|\phi\rangle$ and $|\psi\rangle$, the amplitude that the state $|\psi\rangle$ will also be in the state $|\phi\rangle$ can be found by first projecting the state $|\psi\rangle$ into the base states and then projecting from each base state into the state $|\phi\rangle$. We write that in the following way:

$$\langle \phi \,|\, \psi \rangle = \sum_i \langle \phi \,|\, i \rangle\langle i \,|\, \psi \rangle. \tag{16.5}$$

The summation is, of course, to be carried out over the whole set of base state $|i\rangle$.

In Chapter 13 when we were working out what happens with an electron placed on a linear array of atoms, we chose a set of base states in which the electron was localized at one or other of the atoms in the line. The base state $|n\rangle$ represented the condition in which the electron was localized at atom number "n." (There is, of course, no significance to the fact that we called our base states $|n\rangle$ instead of $|i\rangle$.) A little later, we found it convenient to label the base states by the coordinate x_n of the atom rather than by the number of the atom in the array. The state $|x_n\rangle$ is just another way of writing the state $|n\rangle$. Then, following the general rules, any state at all, say $|\psi\rangle$ is described by giving the amplitudes and that an electron in the state $|\psi\rangle$ is also in one of the states $|x_n\rangle$. For convenience we have chosen to let the symbol C_n stand for these amplitudes,

$$C_n =\text{'} \langle x_n \,|\, \psi \rangle. \tag{16.6}$$

Since the base states are associated with a location along the line, we can think of the amplitude C_n as a function of the coordinate x and write it as $C(x_n)$. The amplitudes $C(x_n)$ will, in general, vary with time and are, therefore, also functions of t. We will not generally bother to show explicitly this dependence.

In Chapter 13 we then proposed that the amplitudes $C(x_n)$ should vary with time in a way described by the Hamiltonian equation (Eq. 13.3). In our new notation this equation is

$$i\hbar \frac{\partial C(x_n)}{\partial t} = E_0 C(x_n) - AC(x_n + b) - AC(x_n - b). \tag{16.7}$$

The last two terms on the right-hand side represent the process in which an electron at atom $(n + 1)$ or at atom $(n - 1)$ can feed into atom n.

We found that Eq. (16.7) has solutions corresponding to definite energy states, which we wrote as

$$C(x_n) = e^{iEt/\hbar}e^{ikx_n}. \tag{16.8}$$

For the low-energy states the wavelengths are large (k is small), and the energy is related to k by

$$E = (E_0 - 2A) + Ak^2b^2, \tag{16.9}$$

or, choosing our zero of energy so that $(E_0 - 2A) = 0$, the energy is given by Eq. (16.1).

Let's see what might happen if we were to let the lattice spacing b go to zero, keeping the wave number k fixed. If that is all that were to happen the last term in Eq. (16.9) would just go to zero and there would be no physics. But suppose A and b are varied together so that as b goes to zero the product Ab^2 is kept constant†—using Eq. (16.2) we will write Ab^2 as the constant $\hbar^2/2m_{\text{eff}}$. Under these circumstances, Eq. (16.9) would be unchanged, but what would happen to the differential equation (16.7)?

First we will rewrite Eq. (16.7) as

$$ i\hbar \frac{\partial C(x_n)}{\partial t} = (E_0 - 2A)C(x_n) + A[2C(x_n) - C(x_n + b) - C(x_n - b)]. $$

$$ (16.10) $$

For our choice of E_0, the first term drops out. Next, we can think of a continuous function $C(x)$ that goes smoothly through the proper values $C(x_n)$ at each x_n. As the spacing b goes to zero, the points x_n get closer and closer together, and (if we keep the variation of $C(x)$ fairly smooth) the quantity in the brackets is just proportional to the second derivative of $C(x)$. We can write—as you can see by making a Taylor expansion of each term—the equality

$$ 2C(x) - C(x + b) - C(x - b) \approx -b^2 \frac{\partial^2 C(x)}{\partial x^2}. \qquad (16.11) $$

In the limit, then, as b goes to zero, keeping b^2A equal to K, Eq. (16.7) goes over into

$$ i\hbar \frac{\partial C(x)}{\partial t} = -\frac{\hbar^2}{2m_{\text{eff}}} \frac{\partial^2 C(x)}{\partial x^2}. \qquad (16.12) $$

We have an equation which says that the time rate of change of $C(x)$—the amplitude to find the electron at x—depends on the amplitude to find the electron at nearby points in a way which is proportional to the second derivative of the amplitude with respect to position.

The correct quantum mechanical equation for the motion of an electron in free space was first discovered by Schrödinger. For motion along a line it has exactly the form of Eq. (16.12) if we replace m_{eff} by m, the free-space mass of the electron. For motion along a line in free space the Schrödinger equation is

$$ i\hbar \frac{\partial C(x)}{\partial t} = -\frac{\hbar^2}{2m} \frac{\partial^2 C(x)}{\partial x^2}. \qquad (16.13) $$

We do not intend to have you think we have derived the Schrödinger equation but only wish to show you one way of thinking about it. When Schrödinger first wrote it down, he gave a kind of derivation based on some heuristic arguments and some brilliant intuitive guesses. Some of the arguments he used were even false, but that does not matter; the only important thing is that the ultimate equation gives a correct description of nature. The purpose of our discussion is then simply to show you that the correct fundamental quantum mechanical equation (16.13) has the same form you get for the limiting case of an electron moving along a line of atoms. This means that we can think of the differential equation in (16.13) as describing the diffusion of a probability amplitude from one point to the next along the line. That is, if an electron has a certain amplitude to be at one point, it will, a little time later, have some amplitude to be at neighboring points. In fact, the equation looks something like the diffusion equations which we have used in Volume I. But there is one main difference: the imaginary coefficient in front of the time derivative makes the behavior completely different from the ordinary diffusion such as you would have for a gas spreading out along a thin tube. Ordinary diffusion gives rise to real exponential solutions, whereas the solutions of Eq. (16.13) are complex waves.

---

† You can imagine that as the points x_n get closer together, the amplitude A to jump from $x_{n \pm 1}$ to x_n will increase.

16–2 The wave function

Now that you have some idea about how things are going to look, we want to go back to the beginning and study the problem of describing the motion of an electron along a line without having to consider states connected with atoms on a lattice. We want to go back to the beginning and see what ideas we have to use if we want to describe the motion of a free particle in space. Since we are interested in the behavior of a particle along a continuum, we will be dealing with an infinite number of possible states and, as you will see, the ideas we have developed for dealing with a finite number of states will need some technical modifications.

We begin by letting the state vector $|x\rangle$ stand for a state in which a particle is located precisely at the coordinate x. For every value x along the line—for instance 1.73, or 9.67, or 10.00—there is the corresponding state. We will take these states $|x\rangle$ as our base states and, if we include all the points on the line, we will have a complete set for motion in one dimension. Now suppose we have a different kind of a state, say $|\psi\rangle$, in which an electron is distributed in some way along the line. One way of describing this state is to give all the amplitudes that the electron will be also found in each of the base states $|x\rangle$. We must give an infinite set of amplitudes, one for each value of x. We will write these amplitudes as $\langle x | \psi \rangle$. Each of these amplitudes is a complex number and since there is one such complex number for each value of x, the amplitude $\langle x | \psi \rangle$ is indeed just a function of x, We will also write it as $C(x)$,

$$C(x) \equiv \langle x | \psi \rangle. \tag{16.14}$$

We have already considered such amplitudes which vary in a continuous way with the coordinates when we talked about the variations of amplitude with time in Chapter 7. We showed there, for example, that a particle with a definite momentum should be expected to have a particular variation of its amplitude in space. If a particle has a definite momentum p and a corresponding definite energy E, the amplitude to be found at any position x would look like

$$\langle x | \psi \rangle = C(x) \propto e^{+ipx/\hbar}. \tag{16.15}$$

This equation expresses an important general principle of quantum mechanics which connects the base states corresponding to different positions in space to another system of base states—all the states of definite momentum. The definite momentum states are often more convenient than the states in x for certain kinds of problems. Either set of base states is, of course, equally acceptable for a description of a quantum mechanical situation. We will come back later to the matter of the connection between them. For the moment we want to stick to our discussion of a description in terms of the states $|x\rangle$.

Before proceeding, we want to make one small change in notation which we hope will not be too confusing. The function $C(x)$, defined in Eq. (16.14), will of course have a form which depends on the particular state $|\psi\rangle$ under consideration. We should indicate that in some way. We could, for example, specify which function $C(x)$ we are talking about by a subscript say, $C_\psi(x)$. Although this would be a perfectly satisfactory notation, it is a little bit cumbersome and is not the one you will find in most books. Most people simply omit the letter C and use the symbol ψ to define the function

$$\psi(x) \equiv C_\psi(x) = \langle x | \psi \rangle. \tag{16.16}$$

Since this is the notation used by everybody else in the world, you might as well get used to it so that you will not be frightened when you come across it somewhere else. Remember though, that we will now be using ψ in two different ways. In Eq. (16.14), ψ stands for a label we have given to a particular physical state of the electron. On the left-hand side of Eq. (16.16), on the other hand, the symbol ψ is used to define a mathematical function of x which is equal to the amplitude to be associated with each point x along the line. We hope it will not be too confusing once you get accustomed to the idea. Incidentally, the function $\psi(x)$ is usually called "the wave function"—because it more often than not has the form of a complex wave in its variables.

Since we have defined $\psi(x)$ to be the amplitude that an electron in the state ψ will be found at the location x, we would like to interpret the absolute square of ψ to be the probability of finding an electron at the position x. Unfortunately, the probability of finding a particle exactly at any particular point is zero. The electron will, in general, be smeared out in a certain region of the line, and since, in any small piece of the line, there are an infinite number of points, the probability that it will be at any one of them cannot be a finite number. We can only describe the probability of finding an electron in terms of a *probability distribution*† which gives the *relative* probability of finding the electron at various approximate locations along the line. Let's let prob $(x, \Delta x)$ stand for the chance of finding the electron in a small interval Δx located near x. If we go to a small enough scale in any physical situation, the probability will be varying smoothly from place to place, and the probability of finding the electron in any small finite line segment Δx will be proportional to Δx. We can modify our definitions to take this into account.

We can think of the amplitude $\langle x \mid \psi \rangle$ as representing a kind of "amplitude density" for all the base states $\mid x \rangle$ in a small region. Since the probability of finding an electron in a small interval Δx at x should be proportional to the interval Δx, we choose our definition of $\langle x \mid \psi \rangle$ so that the following relation holds:

$$\text{prob } (x, \Delta x) = |\langle x \mid \psi \rangle|^2 \, \Delta x.$$

The amplitude $\langle x \mid \psi \rangle$ is therefore proportional to the amplitude that an electron in the state ψ will be found in the base state x and the constant of proportionality is chosen so that the absolute square of the amplitude $\langle x \mid \psi \rangle$ gives the *probability density* of finding an electron in any small region. We can write, equivalently,

$$\text{prob } (x, \Delta x) = |\psi(x)|^2 \, \Delta x. \tag{16.17}$$

We will now have to modify some of our earlier equations to make them compatible with this new definition of a probability amplitude. Suppose we have an electron in the state $\mid \psi \rangle$ and we want to know the amplitude for finding it in a different state $\mid \phi \rangle$ which may correspond to a different spread-out condition of the electron. When we were talking about a finite set of discrete states, we would have used Eq. (16.5). Before modifying our definition of the amplitudes we would have written

$$\langle \phi \mid \psi \rangle = \sum_{\text{all } x} \langle \phi \mid x \rangle \langle x \mid \psi \rangle. \tag{16.18}$$

Now if both of these amplitudes are normalized in the same way as we have described above, then a sum of all the states in a small region of x would be equivalent to multiplying by Δx, and the sum over all values of x simply becomes an integral. With our modified definitions, the correct form becomes

$$\langle \phi \mid \psi \rangle = \int_{\text{all } x} \langle \phi \mid x \rangle \langle x \mid \psi \rangle \, dx. \tag{16.19}$$

The amplitude $\langle x \mid \psi \rangle$ is what we are now calling $\psi(x)$ and, in a similar way, we will choose to let the amplitude $\langle x \mid \phi \rangle$ be represented by $\phi(x)$. Remembering that $\langle \phi \mid x \rangle$ is the complex conjugate of $\langle x \mid \phi \rangle$, we can write Eq. (16.18) as

$$\langle \phi \mid \psi \rangle = \int \phi^*(x)\psi(x) \, dx. \tag{16.20}$$

With our new definitions everything follows with the same formulas as before if you always replace a summation sign by an integral over x.

We should mention one qualification to what we have been saying. Any suitable set of base states must be complete if it is to be used for an adequate

---

† For a discussion of probability distributions see Vol. I, Section 6-4.

description of what is going on. For an electron in one dimension it is not really sufficient to specify only the base states $|x\rangle$, because for each of these states the electron may have a spin which is either up or down. One way of getting a complete set is to take two sets of states in x, one for up spin and the other for down spin. We will, however, not worry about such complications for the time being.

16–3 States of definite momentum

Suppose we have an electron in a state $|\psi\rangle$ which is described by the probability amplitude $\langle x | \psi \rangle = \psi(x)$. We know that this represents a state in which the electron is spread out along the line in a certain distribution so that the probability of finding the electron in a small interval dx at the location x is just

$$\text{prob } (x, dx) = |\psi(x)|^2 \, dx.$$

What can we say about the momentum of this electron? We might ask what is the probability that this electron has the momentum p? Let's start out by calculating the amplitude that the state $|\psi\rangle$ is in another state $|\text{mom } p\rangle$ which we define to be a state with the definite momentum p. We can find this amplitude by using our basic equation for the resolution of amplitudes, Eq. (16.20). In terms of the state $|\text{mom } p\rangle$

$$\langle \text{mom } p \,|\, \psi \rangle = \int_{x=-\infty}^{+\infty} \langle \text{mom } p \,|\, x \rangle \langle x \,|\, \psi \rangle \, dx. \tag{16.21}$$

And the probability that the electron will be found with the momentum p should be given in terms of the absolute square of this amplitude. We have again, however, a small problem about the normalizations. In general we can only ask about the probability of finding an electron with a momentum in a small range dp at the momentum p. The probability that the momentum is exactly some value p must be zero (unless the state $|\psi\rangle$ happens to be a state of definite momentum). Only if we ask for the probability of finding the momentum in a small range dp at the momentum p will we get a finite probability. There are several ways the normalizations can be adjusted. We will choose one of them which we think to be the most convenient, although that may not be apparent to you just now.

We take our normalizations so that the probability is related to the amplitude by

$$\text{prob } (p, dp) = |\langle \text{mom } p \,|\, \psi \rangle|^2 \, \frac{dp}{2\pi\hbar}. \tag{16.22}$$

With this definition the normalization of the amplitude $\langle \text{mom } p \,|\, x \rangle$ is determined. The amplitude $\langle \text{mom } p \,|\, x \rangle$ is, of course, just the complex conjugate of the amplitude $\langle x \,|\, \text{mom } p \rangle$, which is just the one we have written down in Eq. (16.15). With the normalization we have chosen, it turns out that the proper constant of proportionality in front of the exponential is just 1. Namely,

$$\langle \text{mom } p \,|\, x \rangle = \langle x \,|\, \text{mom } p \rangle^* = e^{-ipx/\hbar}. \tag{16.23}$$

Equation (16.21) then becomes

$$\langle \text{mom } p \,|\, \psi \rangle = \int_{-\infty}^{+\infty} e^{-ipx/\hbar} \langle x \,|\, \psi \rangle \, dx. \tag{16.24}$$

This equation together with Eq. (16.22) allows us to find the momentum distribution for any state $|\psi\rangle$.

Let's look at a particular example—for instance one in which an electron is localized in a certain region around $x = 0$. Suppose we take a wave function which has the following form:

$$\psi(x) = Ke^{-x^2/4\sigma^2}. \tag{16.25}$$

The probability distribution in x for this wave function is the absolute square, or

$$\text{prob } (x, dx) = P(x) \, dx = K^2 e^{-x^2/2\sigma^2} \, dx. \tag{16.26}$$

The probability density function $P(x)$ is the Gaussian curve shown in Fig. 16–1. Most of the probability is concentrated between $x = +\sigma$ and $x = -\sigma$. We say that the "half-width" of the curve is σ. (More precisely, σ is equal to the root-mean-square of the coordinate x for something spread out according to this distribution.) We would normally choose the constant K so that the probability density $P(x)$ is not merely *proportional* to the probability per unit length in x of finding the electron, but has a scale such that $P(x)\,\Delta x$ is *equal* to the probability of finding the electron in Δx near x. The constant K which does this can be found by requiring that $\int_{-\infty}^{+\infty} P(x)\,dx = 1$, since there must be unit probability that the electron is found somewhere. Here, we get that $K = (2\pi\sigma^2)^{-1/4}$. [We have used the fact that $\int_{-\infty}^{+\infty} e^{-t^2}\,dt = \sqrt{\pi}$; see Vol. I, page 40–6.]

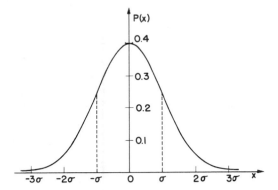

Fig. 16–1. The probability density for the wave function of Eq. (16.24).

Now let's find the distribution in momentum. Let's let $\phi(p)$ stand for the amplitude to find the electron with the momentum p,

$$\phi(p) \equiv \langle \mathrm{mom}\ p \mid \psi \rangle. \tag{16.27}$$

Substituting Eq. (16.25) into Eq. (16.24) we get

$$\phi(p) = \int_{-\infty}^{+\infty} e^{-ipx/\hbar} \cdot K e^{-x^2/4\sigma^2}\,dx. \tag{16.28}$$

the intregral can also be rewritten as

$$K e^{-p^2\sigma^2/\hbar^2} \int_{-\infty}^{+\infty} e^{-(1/4\sigma^2)(x+2ip\sigma^2/\hbar)^2}dx. \tag{16.29}$$

We can now make the substitution $u = x + 2ip\sigma^2/\hbar$, and the integral is

$$\int_{-\infty}^{+\infty} e^{-u^2/4\sigma^2}\,du = 2\sigma\sqrt{\pi}. \tag{16.30}$$

(The mathematicians would probably object to the way we got there, but the result is, nevertheless, correct.)

$$\phi(p) = (8\pi\sigma^2)^{1/4} e^{-p^2\sigma^2/\hbar^2}. \tag{16.31}$$

We have the interesting result that the amplitude function in p has precisely the same mathematical form as the amplitude function in x; only the width of the Gaussian is different. We can write this as

$$\phi(p) = (2\pi\eta^2)^{-1/4} e^{-p^2/4\eta^2}, \tag{16.32}$$

where the half-width η of the p-distribution function is related to the half-width σ of the x-distribution by

$$\eta = \frac{\hbar}{2\sigma}. \tag{16.33}$$

Our result says: if we make the width of the distribution in x very small by making σ small, η becomes large and the distribution in p is very much spread out. Or, conversely: if we have a narrow distribution in p, it must correspond to a spread-out distribution in x. We can, if we like, consider η and σ to be some measure of the uncertainty in the localization of the momentum and of the position of the electron in the state we are studying. If we call them Δp and Δx respectively Eq. (16.33) becomes

$$\Delta p \, \Delta x = \frac{\hbar}{2}. \tag{16.34}$$

Interestingly enough, it is possible to prove that for any other form of a a distribution in x or in p, the product $\Delta p \, \Delta x$ cannot be smaller than the one we have found here. The Gaussian distribution gives the smallest possible value for the product of the root-mean-square widths. In general, we can say

$$\Delta p \, \Delta x \geq \frac{\hbar}{2}. \tag{16.35}$$

This is a quantatative statement of the Heisenberg uncertainty principle, which we have discussed qualitatively many times before. We have usually made the approximate statement that the minimum value of the product $\Delta p \, \Delta x$ is of the same order as $\hbar$.

16–4 Normalization of the states in x

We return now to the discussion of the modifications of our basic equations which are required when we are dealing with a continuum of base states. When we have a finite number of discrete states, a fundamental condition which must be satisfied by the set of base states is

$$\langle i \, | \, j \rangle = \delta_{ij}. \tag{16.36}$$

If a particle is in one base state, the amplitude to be in another base state is 0. By choosing a suitable normalization, we have defined the amplitude $\langle i \, | \, i \rangle$ to be 1. These two conditions are described by Eq. (16.36). We want now to see how this relation must be modified when we use the base states $|x\rangle$ of a particle on a line. If the particle is known to be in one of the base states $|x\rangle$, what is the amplitude that it will be in another base state $|x'\rangle$? If x and x' are two different locations along the line, then the amplitude $\langle x \, | \, x' \rangle$ is certainly 0, so that is consistent with Eq. (16.36). But if x and x' are equal, the amplitude $\langle x \, | \, x' \rangle$ will not be 1, because of the same old normalization problem. To see how we have to patch things up, we go back to Eq. (16.19), and apply this equation to the special case in which the state $|\phi\rangle$ is just the base state $|x'\rangle$. We would have then

$$\langle x' \, | \, \psi \rangle = \int \langle x' \, | \, x \rangle \, \psi(x) \, dx. \tag{16.37}$$

Now the amplitude $\langle x \, | \, \psi \rangle$ is just what we have been calling the function $\psi(x)$. Similarly the amplitude $\langle x' \, | \, \psi \rangle$, since it refers to the same state ψ, is the same function of the variable x', namely $\psi(x')$. We can, therefore, rewrite Eq. (16.37) as

$$\psi(x') = \int \langle x' \, | \, x \rangle \, \psi(x) \, dx. \tag{16.38}$$

This equation must be true for any state ψ and, therefore, for any arbitrary function $\psi(x)$. This requirement should completely determine the nature of the amplitude $\langle x \, | \, x' \rangle$—which is, of course, just a function that depends on x and x'.

Our problem now is to find a function $f(x, x')$ which when multiplied into $\psi(x)$, and integrated over all x gives just the quantity $\psi(x')$. It turns out that there is no mathematical function which will do this! At least nothing like what we ordinarily mean by a "function."

Suppose we pick x' to be the special number 0 and define the amplitude $\langle 0 \mid x \rangle$ to be some function of x, let's say $f(x)$. Then Eq. (16.36) would read as follows:

$$\psi(0) = \int f(x)\psi(x)\,dx. \qquad (16.39)$$

What kind of function $f(x)$ could possibly satisfy this equation? Since the integral must not depend on what values $\psi(x)$ takes for values of x other than 0, $f(x)$ must clearly be 0 for all values of x except 0. But if $f(x)$ is 0 everywhere, the integral will be 0, too, and Eq. (16.39) will not be satisfied. So we have an impossible situation: we wish a function to be 0 everywhere but at a point, and still to give a finite integral. Since we can't find a function that does this, the easiest way out is just to *say* that the function $f(x)$ is *defined* by Eq. (16.37). Namely, $f(x)$ is that function which makes (16.39) correct. The function which does this was first invented by Dirac and carries his name. We write it $\delta(x)$. All we are saying is that the function $\delta(x)$ has the strange property that if it is substituted for $f(x)$ in the Eq. (16.39), the integral picks out the value that $\psi(x)$, takes on when x is equal 0; and, since the integral must be independent of $\psi(x)$ for all values of x other than 0, the function $\delta(x)$ must be 0 everywhere except at $x = 0$. Summarizing, we write

$$\langle 0 \mid x \rangle = \delta(x), \qquad (16.40)$$

where $\delta(x)$ is defined by

$$\psi(0) = \int \delta(x)\psi(x)\,dx. \qquad (16.41)$$

Notice what happens if we use the special function "1" for the function ψ in Eq. (16.41). Then we have the result

$$1 = \int \delta(x)\,dx. \qquad (16.42)$$

That is, the function $\delta(x)$ has the property that it is 0 everywhere except at $x = 0$ but has a finite integral equal to unity. We must imagine that the function $\delta(x)$ has such a fantastic infinity at one point that the total area comes out equal to one.

One way of imagining what the Dirac δ-function is like is to think of a sequence of rectangles—or any other peaked function you care to—which gets narrower and narrower and higher and higher, always keeping a unit area, as sketched in Fig. 16–2. The integral of this function from $-\infty$ to $+\infty$ is always 1. If you multiply it by any function $\psi(x)$ and integrate the product, you get something which is approximately the value of the function at $x = 0$, the approximation getting better and better as you use the narrower and narrower rectangles. You can if you wish, imagine the δ-function in terms of this kind of limiting process. The only important thing, however, is that the δ-function is defined so that Eq. (16.41) is true for every possible function $\psi(x)$. That uniquely defines the δ-function. Its properties are then as we have described.

If we change the argument of the δ-function from x to $x - x'$, the corresponding relations are

$$\delta(x - x') = 0, \qquad x' \neq x,$$

$$\int \delta(x - x')\psi(x)\,dx = \psi(x'). \qquad (16.43)$$

If we use $\delta(x - x')$ for the amplitude $\langle x \mid x' \rangle$ in Eq. (16.38), that equation is satisfied. Our result then is that for our base states in x, the condition corresponding to (16.36) is

$$\langle x' \mid x \rangle = \delta(x - x'). \qquad (16.44)$$

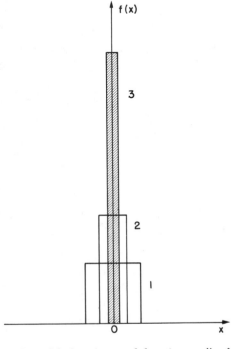

Fig. 16–2. A set of functions, all of unit area, which look more and more like $\delta(x)$.

We have now completed the necessary modifications of our basic equations which are necessary for dealing with the continuum of base states corresponding to the points along a line. The extension to three dimensions is fairly obvious; first we replace the coordinate x by the vector r. Then integrals over x become replaced by integrals over x, y, and z. In other words, they become volume integrals. Finally, the one-dimensional δ-function must be replaced by just the product of three δ-functions, one in x, one in y, and the other in z, $\delta(x - x')\,\delta(y - y')\,\delta(z - z')$. Putting everything together we get the following set of equations for the amplitudes for particle in three dimensions:

$$\langle \phi \mid \psi \rangle = \int \langle \phi \mid r \rangle \langle r \mid \psi \rangle \, d\,\mathrm{Vol}, \tag{16.45}$$

$$\langle r \mid \psi \rangle = \psi(r),$$
$$\langle r \mid \phi \rangle = \phi(r), \tag{16.46}$$

$$\langle \phi \mid \psi \rangle = \int \phi^*(r)\psi(r) \, d\,\mathrm{Vol}, \tag{16.47}$$

$$\langle r' \mid r \rangle = \delta(x - x')\,\delta(y - y')\,\delta(z - z'), \tag{16.48}$$

What happens when there is more than one particle? We will tell you about how to handle two particles and you will easily see what you must do if you want to deal with a larger number. Suppose there are two particles, which we can call particle No. 1 and particle No. 2. What shall we use for the base states? One perfectly good set can be described by saying that particle 1 is at x_1 and particle 2 is at x_2, which we can write as $\mid x_1 x_2 \rangle$. Notice that describing the position of *only one particle does not define* a base state. Each base state must define the condition of the entire system. You must not think that each particle moves independently as a wave in three dimensions. Any physical state $\mid \psi \rangle$ can be defined by giving all of the amplitudes $\langle x_1, x_2 \mid \psi \rangle$ to find the two particles at x_1 and x_2. This generalized amplitude is therefore a function of the *two* sets of coordinates x_1 and x_2. You see that such a function is not a wave in the sense of an oscillation that moves along in three dimensions. Neither is it generally simply a product of two individual waves, one for each particle. It is, in general, some kind of a wave in the six dimensions defined by x_1 and x_2. If there are two particles in nature which are interacting, there is no way of describing what happens to one of the particles by trying to write down a wave function for it alone. The famous paradoxes that we considered in earlier chapters—where the measurements made on one particle were claimed to be able to tell what was going to happen to another particle, or were able to destroy an interference—have caused people all sorts of trouble because they have tried to think of the wave function of one particle alone, rather than the correct wave function in the coordinates of both particles. The complete description can be given correctly only in terms of functions of the coordinates of both particles.

16–5 The Schrödinger equation

So far we have just been worrying about how we can describe states which may involve an electron being anywhere at all in space. Now we have to worry about putting into our description the physics of what can happen in various circumstances. As before, we have to worry about how states can change with time. If we have a state $\mid \psi \rangle$ which goes over into another state $\mid \psi' \rangle$ sometime later, we can describe the situation for all times by making the wave function—which is just the amplitude $\langle r \mid \psi \rangle$—a function of time as well as a function of the coordinate. A particle in a given situation can then be described by giving a time-varying wave function $\psi(r, t) = \psi(x, y, z, t)$. This time-varying wave function describes the evolution of successive states that occur as time develops. This so-called "coordinate representation"—which gives the projections of the state $\mid \psi \rangle$ into the base states $\mid r \rangle$ may not always be the most convenient one to use—but we will consider it first.

In Chapter 8 we described how states varied in time in terms of the Hamiltonian H_{ij}. We saw that the time variation of the various amplitudes was given in terms of the matrix equation

$$ i\hbar \frac{dC_i}{dt} = \sum_j H_{ij} C_j. \qquad (16.49) $$

This equation says that the time variation of each amplitude C_i is proportional to all of the other amplitudes C_j, with the coefficients H_{ij}.

How would we expect Eq. (16.49) to look when we are using the continuum of base states $|x\rangle$? Let's first remember that Eq. (16.49) can also be written as

$$ i\hbar \frac{d}{dt} \langle i \mid \psi \rangle = \sum_j \langle i \mid \hat{H} \mid j \rangle \langle j \mid \psi \rangle. $$

Now it is clear what we should do. For the x-representation we would expect

$$ i\hbar \frac{d}{dt} \langle x \mid \psi \rangle = \int \langle x \mid \hat{H} \mid x' \rangle \langle x' \mid \psi \rangle \, dx'. \qquad (16.50) $$

The sum over the base states $|j\rangle$, gets replaced by an integral over x'. Since $\langle x \mid \hat{H} \mid x' \rangle$ should be some function of x and x', we can write it as $H(x, x')$—which corresponds to H_{ij} in Eq. (16.49). Then Eq. (16.50) is the same as

$$ i\hbar \frac{d}{dt} \psi(x) = \int H(x, x') \psi(x') \, dx' $$

with
$$ H(x, x') \equiv \langle x \mid \hat{H} \mid x' \rangle. \qquad (16.51) $$

According to Eq. (16.51), the rate of change of the ψ at x would depend on the value of ψ at all other points x'; the factor $H(x, x')$ is the amplitude per unit time that the electron will jump from x' to x. *It turns out in nature, however, that this amplitude is zero except for points x' very close to x.* This means—as we saw in the example of the chain of atoms at the beginning of the chapter, Eq. (16.12)—that the right-hand side of Eq. (16.15) can be expressed completely in terms of ψ and the derivatives of ψ with respect to x, all evaluated at the position x.

For a particle moving freely in space with no forces, no disturbances, the correct law of physics is

$$ \int H(x, x') \psi(x') \, dx' = -\frac{\hbar^2}{2m} \frac{d^2}{dx^2} \psi(x). $$

Where did we get that from? Nowhere. It's not possible to derive it from anything you know. It came out of the mind of Schrödinger, invented in his struggle to find an understanding of the experimental observations of the real world. You can perhaps get some clue of why it should be that way by thinking of our derivation of Eq. (16.12) which came from looking at the propagation of an electron in a crystal.

Of course, free particles are not very exciting. What happens if we put forces on the particle? Well, if the force of a particle can be described in terms of a scalar potential $V(x)$—which means we are thinking of electric forces but not magnetic forces—and if we stick to low energies so that we can ignore complexities which come from relativistic motions, then the Hamiltonian which fits the real world gives

$$ \int H(x, x') \psi(x') \, dx' = -\frac{\hbar^2}{2m} \frac{d^2}{dx^2} \psi(x) + V(x)\psi(x). \qquad (16.52) $$

Again, you can get some clue as to the origin of this equation if you go back to the motion of an electron in a crystal, and see how the equations would have to be modified if the energy of the electron varied slowly from one atomic site to the other—as it might do if there were an electric field across the crystal. Then the term E_0 in Eq. (16.7) would vary slowly with position and would correspond to the new term we have added in (16.52).

[You may be wondering why we went straight from Eq. (16.51) to Eq. (16.52) instead of just giving you the correct function for the amplitude $H(x, x') = \langle x \mid \hat{H} \mid x' \rangle$. We did that because $H(x, x')$ can only be written in terms of strange algebraic functions, although the whole integral on the right-hand side of Eq. (16.51) comes out in terms of things you are used to. If you are really curious, $H(x, x')$ can be written in the following way:

$$H(x, x') = -\frac{\hbar^2}{2m} \delta''(x - x') + V(x)\, \delta(x - x'),$$

where δ'' means the second derivative of the delta function. This rather strange function can be replaced by a somewhat more convenient algebraic differential operator, which is completely equivalent:

$$H(x, x') = \left\{ -\frac{\hbar^2}{2m} \frac{d^2}{dx^2} + V(x) \right\} \delta(x - x).$$

We will *not* be using these forms, but will work directly with the form in Eq. (16.52).]

If we now use the expression we have in (16.52) for the integral in (16.50) we get the following differential equation for $\psi(x) = \langle x \mid \psi \rangle$:

$$i\hbar \frac{\partial \psi}{\partial t} = -\frac{\hbar^2}{2m} \frac{\partial^2}{\partial x^2} \psi(x) + V(x)\psi(x). \tag{16.53}$$

It is fairly obvious what we should use instead of Eq. (16.53) if we are interested in motion in three dimensions. The only changes are that d^2/dx^2 gets replaced by

$$\nabla^2 = \frac{\partial^2}{\partial x^2} + \frac{\partial^2}{\partial y^2} + \frac{\partial^2}{\partial z^2},$$

and $V(x)$ gets replaced by $V(x, y, z)$. The amplitude $\psi(x, y, z)$ for an electron moving in a potential $V(x, y, z)$ obeys the differential equation

$$i\hbar \frac{\partial \psi}{\partial t} = -\frac{\hbar^2}{2m} \nabla^2 \psi + V\psi. \tag{16.54}$$

It is called the Schrödinger equation, and was the first quantum-mechanical equation ever known. It was written down by Schrödinger before any of the other quantum equations we have described in this book were discovered.

Although we have approached the subject along a completely different route, the great historical moment marking the birth of the quantum mechanical description of matter occurred when Schrödinger first wrote down his equation in 1926. For many years the internal atomic structure of matter had been a great mystery. No one had been able to understand what held matter together, why there was chemical binding, and especially how it could be that atoms could be stable. Although Bohr had been able to give a description of the internal motion of an electron in a hydrogen atom which seemed to explain the observed spectrum of light emitted by this atom, the reason that electrons moved in this way remained a mystery. Schrödinger's discovery of the proper equations of motion for electrons on an atomic scale provided a theory from which atomic phenomena could be calculated quantitatively, accurately, and in detail. In principle, Schrödinger's equation is capable of explaining all atomic phenomena except those involving magnetism and relativity. It explains the energy levels of an atom, and all the facts of chemical binding. This is, however, true only in principle—the mathematics soon becomes too complicated to solve exactly any but the simplest problems. Only the hydrogen and helium atoms have been calculated to a high accuracy. However, with various approximations, some fairly sloppy, many of the facts of more complicated atoms and of the chemical binding of molecules can be understood. We have shown you some of these approximations in earlier chapters.

The Schrödinger equation as we have written it does not take into account any magnetic effects. It is possible to take such effects into account in an approximate way by adding some more terms to the equation. However, as we have seen in Volume II, magnetism is essentially a relativistic effect, and so a correct description of the motion of an electron in an arbitrary electromagnetic field can only be discussed in a proper relativistic equation. The correct relativistic equation for the motion of an electron was discovered by Dirac a year after Schrödinger brought forth his equation, and takes on quite a different form. We will not be able to discuss it at all here.

Before we go on to look at some of the consequences of the Schrödinger equation, we would like to show you what it looks like for a system with a large number of particles. We will not be making any use of the equation, but just want to show it to you to emphasize that the wave function ψ is not simply an ordinary wave in space, but is a function of many variables. If there are many particles, the equation becomes

$$-i\hbar \frac{\partial \psi(r_1, r_2, r_3, \ldots)}{\partial t} = \sum_i \frac{\hbar^2}{2m_i} \left\{ \frac{\partial^2 \psi}{\partial x_i} + \frac{\partial^2 \psi}{\partial y_i} + \frac{\partial^2 \psi}{\partial x_i} \right\} + V(r_1, r_1, \ldots)\psi. \quad (16.55)$$

The potential function V is what corresponds classically to the total potential energy of all the particles. If there are no external forces acting on the particles, the function V is simply the electrostatic energy of interaction of all the particles. That is, if the ith particle carries the charge $Z_i q_e$, then the function V is simply†

$$V(r_1, r_2, r_3, \ldots) = \sum_{\substack{\text{all} \\ \text{pairs}}} \frac{Z_i Z_j}{r_{ij}} e^2. \quad (16.56)$$

16–6 Quantized energy levels

In a later chapter we will look in detail at a solution of Schrödinger's equation for a particular example. We would like now, however, to show you how one of the most remarkable consequence of Schrödinger's equation comes about—namely, the surprising fact that a differential equation involving only continuous functions of continuous variables in space can give rise to quantum effects such as the discrete energy levels in an atom. The essential fact to understand is how it can be that an electron which is confined to a certain region of space by some kind of a potential "well" must necessarily have only one or another of a certain well-defined set of discrete energies.

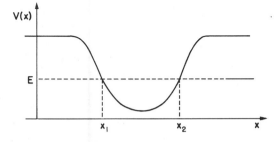

Fig. 16–3. A potential well for a particle moving along x.

Suppose we think of an electron in a one-dimensional situation in which its potential energy varies with x in a way described by the graph in Fig. 16–3. We will assume that this potential is static—it doesn't vary with time. As we have done so many times before, we would like to look for solutions corresponding to states of definite energy, which means, of definite frequency. Let's try a solution of the form

$$\psi = a(x)e^{-iEt/\hbar}. \quad (16.57)$$

---

† We are using the convention of the earlier volumes according to which $e^2 \equiv q_e^2/4\pi\epsilon_0$.

If we substitute this function into the Schrödinger equation, we find that the function $a(x)$ must satisfy the following differential equation:

$$\frac{d^2 a(x)}{dx^2} = \frac{2m}{\hbar}[V(x) - E]a(x). \qquad (16.58)$$

This equation says that at each x the second derivative of $a(x)$ with respect to x is proportional to $a(x)$, the coefficient of proportionality being given by the quantity $(V - E)$. The second derivative of $a(x)$ is the rate of change of its slope. If the potential V is greater than the energy E of the particle, the rate of change of the slope of $a(x)$ will have the same sign as $a(x)$. That means that the curve of $a(x)$ will be concave away from the axis. That is, it will have, more or less, the character of the positive or negative exponential function, $e^{\pm x}$. This means that in the region to the left of x_1, in Fig. 16–3, where V is greater than the assumed energy E, the function $a(x)$ would have to look like one or another of the curves

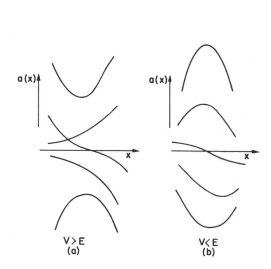

Fig. 16–4. Possible shapes of the wave function $a(x)$ for $V > E$ and for $V < E$.

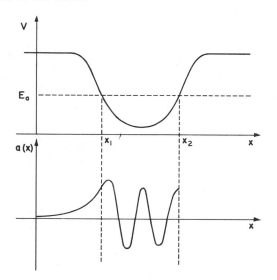

Fig. 16–5. A wave function for the energy E_a which goes to zero for negative x.

shown in part (a) of Fig. 16–4.

If, on the other hand, the potential function V is less than the energy E, the second derivative of $a(x)$ with respect to x has the opposite sign from $a(x)$ itself, and the curve of $a(x)$ will always be concave toward the axis like one of the pieces shown in part (b) of Fig. 16–4. The solution in such a region has, piece-by-piece, roughly the form of a sinusoidal curve.

Now let's see if we can construct graphically a solution for the function $a(x)$ which corresponds to a particle of energy E_a in the potential V shown in Fig. 16–3. Since we are trying to describe a situation in which a particle is bound *inside* the potential well, we want to look for solutions in which the wave amplitude takes on very small values when x is way outside the potential well. We can easily imagine a curve like the one shown in Fig. 16–5 which tends toward zero for large negative values of x, and grows smoothly as it approaches x_1. Since V is equal to E_a at x_1, the curvature of the function becomes zero at this point. Between x_1 and x_2, the quantity $V - E_a$ is always a negative number, so the function $a(x)$ is always concave toward the axis, and the curvature is larger the larger the difference between E_a and V. If we continue the curve into the region between x_1 and x_2, it should go more or less as shown in Fig. 16–5.

Now let's continue this curve into the region to the right of x_2. There it curves away from the axis and takes off toward large positive values, as drawn in Fig. 16–6. For the energy E_a we have chosen, the solution for $a(x)$ gets larger and larger with increasing x. In fact, its *curvature* is also increasing (if the potential continues to stay flat). The amplitude rapidly grows to immense proportions.

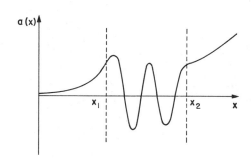

Fig. 16–6. The wave function $a(x)$ of Fig. 16–5 continued beyond x_2.

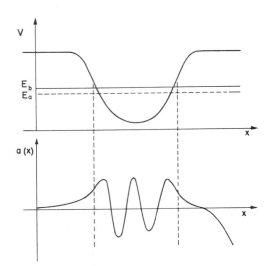

Fig. 16–7. The wave function $a(x)$ for an energy E_b greater than E_a.

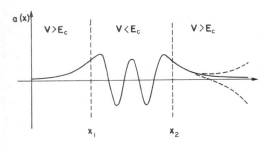

Fig. 16–8. A wave function for the energy E_c between E_a and E_b.

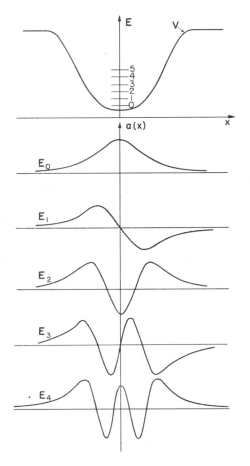

Fig. 16–9. The function $a(x)$ for the five lowest energy bound states.

What does this mean? It simply means that the particle is not "bound" in the potential well. It is infinitely more likely to be found *outside* of the well, than *inside*. For the solution we have manufactured, the electron is more likely to be found at $x = +\infty$ than anywhere else. We have failed to find a solution for a bound particle.

Let's try another energy, say one a little bit higher than E_a—say the energy E_b in Fig. 16–7. If we start with the same conditions on the left, we get the solution drawn in the lower half of Fig. 16–7. It looked at first as though it were going to be better, but it ends up just as bad as the solution for E_a—except that now $a(x)$ is getting more and more *negative* as we go toward large values of x.

Maybe that's the clue. Since changing the energy a little bit from E_a to E_b causes the curve to flip from one side of the axis to the other, perhaps there is some energy lying between E_a and E_b for which the curve will approach zero for large values of x. There is, indeed, and we have sketched how the solution might look in Fig. 16–8.

You should appreciate that the solution we have drawn in the figure is a very special one. If we were to raise or lower the energy ever so slightly, the function would go over into curves like one or the other of the two broken-line curves shown in Fig. 16–8, and we would not have the proper conditions for a bound particle. We have obtained a result that if a particle is to be bound in a potential well, it can do so only if it has a very definite energy.

Does that mean that there is only one energy for a particle bound in a potential well? No. Other energies are possible, but not energies too close to E_c. Notice that the wave function we have drawn in Fig. 16–8 crosses the axis four times in the region between x_1 and x_2. If we were to pick an energy quite a bit lower than E_c, we could have a solution which crosses the axis only three times, only two times, only once, or not at all. The possible solutions are sketched in Fig. 16–9. (There may also be other solutions corresponding to values of the energy higher than the ones shown.) Our conclusion is that if a particle is bound in a potential well, its energy can take on only the certain special values in a discrete energy spectrum. You see how a differential equation can describe the basic fact of quantum physics.

We might remark one other thing. If the energy E is above the top of the potential well, then there are no longer any discrete solutions, and any possible energy is permitted. Such solutions correspond to the scattering of free particles by a potential well. We have seen an example of such solutions when we considered the effects of impurity atoms in a crystal.

17

Symmetry and Conservation Laws

17–1 Symmetry

In classical physics there are a number of quantities which are *conserved*—such a momentum, energy, and angular momentum. Conservation theorems about corresponding quantities also exist in quantum mechanics. The most beautiful thing of quantum mechanics is that the conservation theorems can, in a sense, be derived from something else, whereas in classical mechanics they are practically the starting points of the laws. (There are ways in classical mechanics to do an analogous thing to what we will do in quantum mechanics, but it can be done only at a very advanced level.) In quantum mechanics, however, the conservation laws are very deeply related to the principle of superposition of amplitudes, and to the symmetry of physical systems under various changes. This is the subject of the present chapter. Although we will apply these ideas mostly to the conservation of angular momentum, the essential point is that the theorems about the conservation of all kinds of quantities are—in the quantum mechanics—related to the symmetries of the system.

We begin, therefore, by studying the question of symmetries of systems. A very simple example is the hydrogen molecular ion—we could equally well take the ammonia molecule—in which there are two states. For the hydrogen molecular ion we took as our base states one in which the electron was located near proton number 1, and another in which the electron was located near proton number 2. The two states—which we called $|1\rangle$ and $|2\rangle$—are shown again in Fig. 17–1(a). Now, so long as the two nuclei are both exactly the same, then there is a certain *symmetry* in this physical system. That is to say, if we were to *reflect* the system in the plane halfway between the two protons—by which we mean that everything on one side of the plane gets moved to the symmetric position on the other side—we would get the situations in Fig. 17–1(b). Since the protons are identical, the *operation of reflection* changes $|1\rangle$ into $|2\rangle$ and $|2\rangle$ into $|1\rangle$. We'll call this reflection operation $\hat{P}$ and write

$$\hat{P}\,|1\rangle = |2\rangle, \qquad \hat{P}\,|2\rangle = |1\rangle. \qquad (17.1)$$

So our $\hat{P}$ is an operator in the sense that it "*does something*" to a state to make a new state. The interesting thing is that $\hat{P}$ operating on *any* state produces some *other* state of the system.

Now $\hat{P}$, like any of the other operators we have described, has matrix elements which can be defined by the usual obvious notation. Namely,

$$P_{11} = \langle 1\,|\,\hat{P}\,|\,1\rangle \qquad \text{and} \qquad P_{12} = \langle 1\,|\,\hat{P}\,|\,2\rangle$$

are the matrix elements we get if we multiply $\hat{P}\,|1\rangle$ and $\hat{P}\,|2\rangle$ on the left by $\langle 1\,|$. From Eq. (17.1) they are

$$\langle 1\,|\,\hat{P}\,|\,1\rangle = P_{11} = \langle 1\,|\,2\rangle = 0,$$
$$\langle 1\,|\,\hat{P}\,|\,2\rangle = P_{12} = \langle 1\,|\,1\rangle = 1. \qquad (17.2)$$

In the same way we can get P_{21} and P_{22}. The matrix of $\hat{P}$—*with respect to the base system* $|1\rangle$ *and* $|2\rangle$—is

$$P = \begin{pmatrix} 0 & 1 \\ 1 & 0 \end{pmatrix}. \qquad (17.3)$$

We see once again that the words *operator* and *matrix* in quantum mechanics are

17–1 Symmetry

17–2 Symmetry and conservation

17–3 The conservation laws

17–4 Polarized light

17–5 The disintegration of the Λ^0

17–6 Summary of the rotation matrices

Review: Chapter 52, Vol. I, *Symmetry in Physical Laws*

Reference: Angular Momentum in Quantum Mechanics: A. R. Edmonds, Princeton University Press, 1957

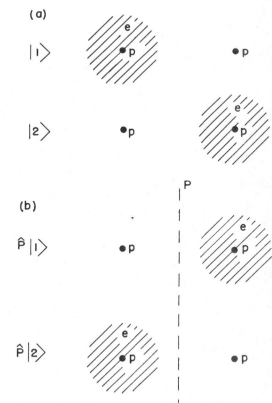

Fig. 17–1. If the states $|1\rangle$ and $|2\rangle$ are reflected in the plane P-P, they go into $|2\rangle$ and $|1\rangle$, respectively.

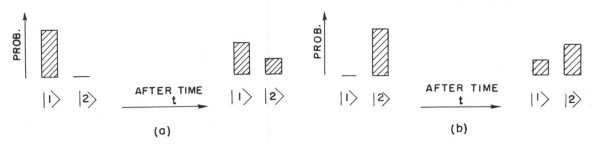

Fig. 17–2. In a symmetric system, if a pure $|1\rangle$ state develops as shown in part (a), a pure $|2\rangle$ state will develop as in part (b).

practically interchangeable. There are slight technical differences—like the difference between a "numeral" and a "number"—but the distinction is something pedantic that we don't have to worry about. So whether $\hat{P}$ defines an operation, or is actually used to define a matrix of numbers, we will call it interchangeably an operator or a matrix.

Now we would like to point out something. We will *suppose* that the *physics* of the whole hydrogen molecular ion system is *symmetrical*. It doesn't have to be —it depends, for instance, on what else is near it. But if the system is symmetrical, the following idea should certainly be true. Suppose we start at $t = 0$ with the system in the state $|1\rangle$ and find after an interval of time t that the system turns out to be in a more complicated situation—in some linear combination of the two base states. Remember that in Chapter 8 we used to represent "going for a period of time" by multiplying by the operator $\hat{U}$. That means that the system would after a while—say 15 seconds to be definite—be in some other state. For example, it might be $\sqrt{2/3}$ parts of the state $|1\rangle$ and $i\sqrt{1/3}$ parts of the state $|2\rangle$, and we would write

$$|\psi \text{ at } 15 \text{ sec}\rangle = \hat{U}(15,0)|1\rangle = \sqrt{2/3}\,|1\rangle + i\sqrt{1/3}\,|2\rangle. \qquad (17.4)$$

Now we ask what happens if we start the system in the *symmetric* state $|2\rangle$ and wait for 15 seconds under the *same conditions*? It is clear that if the world is symmetric—as we are supposing—we should get the state symmetric to (17.4):

$$|\psi \text{ at } 15 \text{ sec}\rangle = \hat{U}(15,0)|2\rangle = \sqrt{2/3}\,|2\rangle + i\sqrt{1/3}\,|1\rangle. \qquad (17.5)$$

The same ideas are sketched diagrammatically in Fig. 17–2. So if the *physics* of a system is symmetrical with respect to some plane, and we work out the behavior of a particular state, we also know the behavior of the state we would get by reflecting the original state in the symmetry plane.

We would like to say the same things a litte bit more generally—which means a little more abstractly. Let $\hat{Q}$ be any one of a number of operations that you could perform on a system *without changing the physics*. For instance, for $\hat{Q}$ we might be thinking of $\hat{P}$, the operation of a *reflection* in the plane between the two atoms in the hydrogen molecule. Or, in a system with two electrons, we might be thinking of the operation of *interchanging* the two electrons. Another possibility would be, in a spherically symmetric system, the operation of a *rotation* of the whole system through a finite angle around some axis—which wouldn't change the physics. Of course, we would normally want to give each special case some special notation for $\hat{Q}$. Specifically, we will normally define the $\hat{R}_y(\theta)$ to be the operation "rotate the system about the y-axis by the angle θ". By $\hat{Q}$ we mean just any one of the operators we have described or any other one—which leaves the basic physical situation unchanged.

Let's think of some more examples. If we have an atom with *no external magnetic* field or *no external electric* field, and if we were to turn the coordinates around any axis, it would be the same physical system. Again, the ammonia molecule is symmetrical with respect to a reflection in a plane parallel to that of the three hydrogens—so long as there is no electric field. When there is an electric field, when we make a reflection we would have to change the electric field also, and that changes the physical problem. But if we have no external field, the molecule is symmetrical.

Now we consider a general situation. Suppose we start with the state $|\psi_1\rangle$ and after some time or other under given physical conditions it has become the state $|\psi_2\rangle$. We can write

$$|\psi_2\rangle = \hat{U}|\psi_1\rangle. \qquad (17.6)$$

[You can be thinking of Eq. (17.4).] Now imagine we perform the operation $\hat{Q}$ on the whole system. The state $|\psi_1\rangle$ will be transformed to a state $|\psi_1'\rangle$, which we can also write as $\hat{Q}|\psi_1\rangle$. Also the state $|\psi_2\rangle$ is changed into $|\psi_2'\rangle = \hat{Q}|\psi_2\rangle$. Now *if the physics* is symmetrical under $\hat{Q}$ (don't forget the *if*; it is not a general property of systems), then, waiting for the same time under the same conditions, we should have

$$|\psi_2'\rangle = \hat{U}|\psi_1'\rangle. \qquad (17.7)$$

[Like Eq. (17.5).] But we can write $\hat{Q}|\psi_1\rangle$ for $|\psi_1'\rangle$ and $\hat{Q}|\psi_2\rangle$ for $|\psi_2'\rangle$ so (17.7) can also be written

$$\hat{Q}|\psi_2\rangle = \hat{U}\hat{Q}|\psi_1\rangle. \qquad (17.8)$$

If we now replace $|\psi_2\rangle$ by $\hat{U}|\psi_1\rangle$—Eq. (17.6)—we get

$$\hat{Q}\hat{U}|\psi_1\rangle = \hat{U}\hat{Q}|\psi_1\rangle. \qquad (17.9)$$

It's not hard to understand what this means. Thinking of the hydrogen ion it says that: "making a reflection and waiting a while"—the expression on the right of Eq. (17.9)—is the same as "waiting a while and then making a reflection"—the expression on the left of (17.9). These should be the same so long as U doesn't change under the reflection.

Since (17.9) is true for *any* starting state $|\psi_1\rangle$, it is really an equation about the operators:

$$\hat{Q}\hat{U} = \hat{U}\hat{Q}. \qquad (17.10)$$

This is what we wanted to get—*it is a mathematical statement of symmetry*. When Eq. (17.10) is true, we say that the operators $\hat{U}$ and $\hat{Q}$ *commute*. We can then *define* "symmetry" in the following way: A physical system is *symmetric* with respect to the operation $\hat{Q}$ when $\hat{Q}$ commutes with $\hat{U}$, the operation of the passage of time. [In terms of matrices, the product of two operators is equivalent to the matrix product, so Eq. (17.10) also holds for the matrices Q and U for a system which is symmetric under the transformation Q.]

Incidentally, since for infinitesimal times ϵ we have $\hat{U} = 1 - i\hat{H}\epsilon/\hbar$—where $\hat{H}$ is the usual Hamiltonian (see Chapter 8)—you can see that if (17.10) is true, it is also true that

$$\hat{Q}\hat{H} = \hat{H}\hat{Q}. \qquad (17.11)$$

So (17.11) is the mathematical statement of the condition for the symmetry of a physical situation under the operator $\hat{Q}$. It *defines* a symmetry.

17–2 Symmetry and conservation

Before applying the result we have just found, we would like to discuss the idea of symmetry a little more. Suppose that we have a very special situation: after we operate on a state with $\hat{Q}$, we get the same state. This is a very special case, but let's suppose it happens to be true for a state $|\psi_0\rangle$ that $|\psi'\rangle = \hat{Q}|\psi_0\rangle$ is physically the same state as $|\psi_0\rangle$. That means that $|\psi'\rangle$ is equal to $|\psi_0\rangle$ except for some phase factor.† How can that happen? For instance, suppose that we

---

† Incidentally, you can show that $\hat{Q}$ is necessarily a *unitary operator*—which means that if it operates on $|\psi\rangle$ to give some number times $|\psi\rangle$, the number must be of the form $e^{i\delta}$, where δ is real. It's a small point, and the proof rests on the following observation. Any operation like a reflection or a rotation doesn't lose any particles, so the normalization of $|\psi'\rangle$ and $|\psi\rangle$ must be the same; they can only differ by a pure imaginary phase factor.

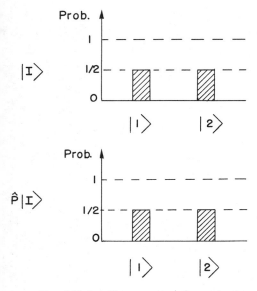

Fig. 17–3. The state $|I\rangle$ and the state $\hat{P} \, |I\rangle$ obtained by reflecting $|I\rangle$ in the central plane.

have an H_2^+ ion in the state which we once called $|I\rangle$. For this state there is equal amplitude to be in the base states $|1\rangle$ and $|2\rangle$. The probabilities are shown as a bar graph in Fig. 17–3(a). If we operate on $|I\rangle$ with the reflection operator $\hat{P}$, it flips the state over changing $|1\rangle$ to $|2\rangle$ and $|2\rangle$ to $|1\rangle$—we get the probabilities shown in Fig. 17–3(b). But that's just the state $|I\rangle$ all over again. If we start with state $|II\rangle$ the probabilities before and after reflection look just the same. However, there is a difference if we look at the *amplitudes*. For the state $|I\rangle$ the amplitudes are the *same* after the reflection, but for the state $|II\rangle$ the amplitudes have the opposite sign. In other words,

$$\hat{P} \, | \, I\rangle = \hat{P} \left\{ \frac{|\,1\rangle + |\,2\rangle}{\sqrt{2}} \right\} = \frac{|\,2\rangle + |\,1\rangle}{\sqrt{2}} = |\,I\rangle,$$

$$\hat{P} \, | \, II\rangle = \hat{P} \left\{ \frac{|\,1\rangle - |\,2\rangle}{\sqrt{2}} \right\} = \frac{|\,2\rangle - |\,1\rangle}{\sqrt{2}} = -\,|\,II\rangle. \tag{17.12}$$

If we write $\hat{P} \, | \, \psi_0\rangle = e^{i\delta} \, |\,\psi_0\rangle$, we have that $e^{i\delta} = 1$ for the state $|\,I\rangle$ and $e^{i\delta} = -1$ for the state $|\,II\rangle$.

Let's look at another example. Suppose we have a RHC polarized photon propagating in the z-direction. If we do the operation of a rotation around the z-axis, we know that this just multiplies the amplitude by $e^{i\phi}$ when ϕ is the angle of the rotation. So for the rotation operation in this case, δ is just equal to the angle of rotation.

Now it is clear that *if it happens to be true* that an operator $\hat{Q}$ just changes the phase of a state at some time, say $t = 0$, *it is true forever*. In other words, if the state $|\,\psi_1\rangle$ goes over into the state $|\,\psi_2\rangle$ after a time t, or

$$\hat{U}(t, 0) \, |\,\psi_1\rangle = |\,\psi_2\rangle \tag{17.13}$$

and if the symmetry of the situation makes it so that

$$\hat{Q} \, |\,\psi_1\rangle = e^{i\delta} \, |\,\psi_1\rangle, \tag{17.14}$$

then it is also true that

$$\hat{Q} \, |\,\psi_2\rangle = e^{i\delta} \, |\,\psi_2\rangle. \tag{17.15}$$

This is clear, since

$$\hat{Q} \, |\,\psi_2\rangle = \hat{Q}\hat{U} \, |\,\psi_1\rangle = \hat{U}\hat{Q} \, |\,\psi_1\rangle,$$

and if $\hat{Q} \, |\,\psi_1\rangle = e^{i\delta} \, |\,\psi_1\rangle$, then

$$\hat{Q} \, |\,\psi_2\rangle = \hat{U}e^{i\delta} \, |\,\psi_1\rangle = e^{i\delta}\hat{U} \, |\,\psi_1\rangle = e^{i\delta} \, |\,\psi_2\rangle.$$

[The sequence of equalities follows from (17.13) and (17.10) for a symmetrical system, from (17.14), and from the fact that a number like $e^{i\delta}$ commutes with an operator.]

So with certain symmetries something which is true initially is true for all times. But isn't that just a *conservation law*? Yes! It says that if you look at the original state and by making a little computation on the side discover that an operation *which is a symmetry operation of the system* produces only a multiplication by a certain phase, then you know that the same property will be true of the final state—the same operation multiplies the final state by the same phase factor. This is always true even though we may not know anything else about the inner mechanism of the universe which changes a system from the initial to the final state. Even if we do not care to look at the details of the machinery by which the system gets from one state to another, we can still say that if a thing is in a state with a certain symmetry character originally, and if the Hamiltonian for this thing is symmetrical under that symmetry operation, then the state will have the same symmetry character for all times. That's the basis of all the conservation laws of quantum mechanics.

Let's look at a special example. Let's go back to the $\hat{P}$ operator. We would like first to modify a little our definition of $\hat{P}$. We want to take for $\hat{P}$ not just a mirror reflection, because that requires defining the plane in which we put the mirror. There is a special kind of a reflection that doesn't require the specification of a plane. Suppose we redefine the operation $\hat{P}$ this way: First you reflect in a mirror in the z-plane so that z goes to $-z$, x stays x, and y stays y; then you turn the system $180°$ about the z-axis so that x is made to go to $-x$ and y to $-y$. The whole thing is called an *inversion*. Every point is projected *through the origin* to the diametrically opposite position. All the coordinates of everything are reversed. We will still use the symbol $\hat{P}$ for this operation. It is shown in Fig. 17–4. It is a little more convenient than a simple reflection because it doesn't require that you specify which coordinate plane you used for the reflection—you need specify only the point which is at the center of symmetry.

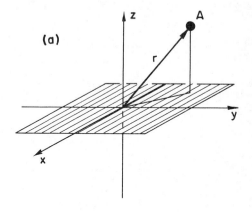

Now let's suppose that we have a state $|\psi_0\rangle$ which under the inversion operation goes into $e^{i\delta}|\psi_0\rangle$—that is,

$$|\psi_0'\rangle = \hat{P}\,|\psi_0\rangle = e^{i\delta}\,|\psi_0\rangle. \tag{17.16}$$

Then suppose that we invert again. After *two* inversions we are right back where we started from—nothing is changed at all. We must have that

$$\hat{P}\,|\psi_0'\rangle = \hat{P}\cdot\hat{P}\,|\psi_0\rangle = |\psi_0\rangle.$$

But

$$\hat{P}\cdot\hat{P}\,|\psi_0\rangle = \hat{P}e^{i\delta}\,|\psi_0\rangle = e^{i\delta}\,\hat{P}\,|\psi_0\rangle = (e^{i\delta})^2\,|\psi_0\rangle.$$

It follows that

$$(e^{i\delta})^2 = 1.$$

So *if the inversion operator is a symmetry operation* of a state, there are only two possibilities for δ:

$$e^{i\delta} = \pm1,$$

which means that

$$\hat{P}\,|\psi_0\rangle = |\psi_0\rangle \qquad \text{or} \qquad \hat{P}\,|\psi_0\rangle = -\,|\psi_0\rangle. \tag{17.17}$$

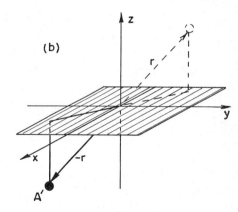

Fig. 17–4. The operation of inversion, $\hat{P}$. Whatever is at the point A at (x, y, z) is moved to the point A' at $(-x, -y, -z)$.

Classically, if a state is symmetric under an inversion, the operation gives back the same state. In quantum mechanics, however, there are the two possibilities: we get the *same state* or *minus* the same state. When we get the *same* state, $\hat{P}\,|\psi_0\rangle = |\psi_0\rangle$, we say that the state $|\psi_0\rangle$ has *even parity*. When the sign is reversed so that $\hat{P}\,|\psi_0\rangle = -\,|\psi_0\rangle$, we say that the state has *odd parity*. (The inversion operator $\hat{P}$ is also known as the parity operator.) The state $|I\rangle$ of the H_2^+ ion has even parity; and the state $|II\rangle$ has odd parity—see Eq. (17.12). There are, of course, states which are not symmetric under the operation $\hat{P}$; these are states with no definite parity. For instance, in the H_2^+ system the state $|I\rangle$ has even parity, the state $|II\rangle$ has odd parity, and the state $|1\rangle$ has no definite parity.

When we speak of an operation like inversion being performed "*on a physical system*" we can think about it in two ways. We can think of *physically moving* whatever is at r to the inverse point at $-r$, or we can think of *looking* at the same system from a new frame of reference x', y', z' related to the old by $x' = -x$, $y' = -y$, and $z' = -z$. Similarly, when we think of rotations, we can think of rotating bodily a physical system, or of rotating the coordinate frame with respect to which we measure the system, keeping the "system" fixed in space. Generally, the two points of view are essentially equivalent. For rotation they are equivalent *except* that rotating a *system* by the angle θ is like rotating the reference frame by the *negative* of θ. In these lectures we have usually considered what happens when a projection is made into a new set of axes. What you get that way is the same as what you get if you leave the axes fixed and rotate the system *backwards* by the same amount. When you do that, the signs of the angles are reversed.†

---

† In other books you may find formulas with different signs; they are probably using a different definition of the angles.

Many of the *laws* of physics—but not all—are unchanged by a reflection or an inversion of the coordinates. They are *symmetric* with respect to an inversion. The laws of electrodynamics, for instance, are unchanged if we change x to $-x$, y to $-y$, and z to $-z$ in *all* the equations. The same is true for the laws of gravity, and for the strong interactions of nuclear physics. Only the weak interactions—responsible for β-decay—do not have this symmetry. (We discussed this in some detail in Chapter 52, Vol. I.) We will for now leave out any consideration of the β-decays. Then in any physical system where β-decays are not expected to produce any appreciable effect—an example would be the emission of light by an atom—the Hamiltonian $\hat{H}$ and the operator $\hat{P}$ will commute. Under these circumstances we have the following proposition. If a state originally has even parity, and if you look at the physical situation at some later time, it will again have even parity. For instance, suppose an atom about to emit a photon is in a state known to have even parity. You look at the whole thing—including the photon—after the emission; it will again have even parity (likewise if you start with odd parity). This principle is called the *conservation of parity*. You can see why the words "conservation of parity" and "reflection symmetry" are closely intertwined in the quantum mechanics. Although until a few years ago it was thought that nature always conserved parity, it is now known that this is *not* true. It has been discovered to be false because the β-decay reaction does not have the inversion symmetry which is found in the other laws of physics.

Now we can prove an interesting theorem (which is true so long as we can disregard weak interactions): Any state of definite energy which is not degenerate must have a definite parity. It must have either even parity or odd parity. (Remember that we have sometimes seen systems in which several states have the same energy—we say that such states are *degenerate*. Our theorem will not apply to them.)

For a state $|\psi_0\rangle$ of definite energy, we know that

$$\hat{H}|\psi_0\rangle = E|\psi_0\rangle, \tag{17.18}$$

where E is just a number—the energy of the state. If we have *any* operator $\hat{Q}$ which is a symmetry operator of the system we can prove that

$$\hat{Q}|\psi_0\rangle = e^{i\delta}|\psi_0\rangle \tag{17.19}$$

so long as $|\psi_0\rangle$ is a unique state of definite energy. Consider the new state $|\psi_0'\rangle$ that you get from operating with $\hat{Q}$. If the physics is symmetric, then $|\psi_0'\rangle$ must have the same energy as $|\psi_0\rangle$. But we have taken a situation in which there *is only one* state of that energy, namely $|\psi_0\rangle$, so $|\psi_0'\rangle$ must be the same state—it can only differ by a phase. That's the physical argument.

The same thing comes out of our mathematics. Our definition of symmetry is Eq. (17.10) or Eq. (17.11) (good for any state ψ),

$$\hat{H}\hat{Q}|\psi\rangle = \hat{Q}\hat{H}|\psi\rangle. \tag{17.20}$$

But we are considering only a state $|\psi_0\rangle$ which is a definite energy state, so that $\hat{H}|\psi_0\rangle = E|\psi_0\rangle$. Since E is just a number that floats through $\hat{Q}$ if we want, we have

$$\hat{Q}\hat{H}|\psi_0\rangle = \hat{Q}E|\psi_0\rangle = E\hat{Q}|\psi_0\rangle.$$

So

$$\hat{H}\{\hat{Q}|\psi_0\rangle\} = E\{\hat{Q}|\psi_0\rangle\}. \tag{17.21}$$

So $|\psi_0'\rangle = \hat{Q}|\psi_0\rangle$ is also a definite energy state of $\hat{H}$—and with the same E. But by our hypothesis, there is only one such state; it must be that $|\psi_0'\rangle = e^{i\delta}|\psi_0\rangle$.

What we have just proved is true for any operator $\hat{Q}$ that is a symmetry operator of the physical system. Therefore, in a situation in which we consider only electrical forces and strong interactions—and no β-decay—so that inversion symmetry is an allowed approximation, we have that $\hat{P}|\psi\rangle = e^{i\delta}|\psi\rangle$. But we have also seen that $e^{i\delta}$ must be either $+1$ or -1. So any state of a definite energy (which is not degenerate) has got either an even parity or an odd parity.

We turn now to another interesting example of an operation: a rotation. We consider the special case of an operator that rotates an atomic system by angle ϕ around the z-axis. We will call this operator† $\hat{R}_z(\phi)$. We are going to suppose that we have a physical situation where we have no influences lined up along the x- and y-axes. Any electric field or magnetic field is taken to be parallel to the z-axis‡ so that there will be no change in the *external* conditions if we rotate the whole physical system about the z-axis. For example, if we have an atom in empty space and we turn the atom around the z-axis by an angle ϕ, we have the same physical system.

Now then, there are *special states* which have the property that such an operation produces a new state which is the original state multiplied by some phase factor. Let us make a quick side remark to show you that when this is true the phase change must always be proportional to the angle ϕ. Suppose that you would rotate twice by the angle ϕ. That's the same thing as rotating by the angle 2ϕ. If a rotation by ϕ has the effect of multiplying the state $|\psi_0\rangle$ by a phase $e^{i\delta}$ so that

$$\hat{R}_z(\phi)\,|\psi_0\rangle = e^{i\delta}\,|\psi_0\rangle,$$

two such rotations in succession would multiply the state by the factor $(e^{i\delta})^2 = e^{i2\delta}$, since

$$\hat{R}_z(\phi)\hat{R}_z(\phi)\,|\psi_0\rangle = \hat{R}_z(\phi)e^{i\delta}\,|\psi_0\rangle = e^{i\delta}\hat{R}_z(\phi)\,|\psi_0\rangle = e^{i\delta}e^{i\delta}\,|\psi_0\rangle.$$

The phase change δ must be proportional to ϕ.¶ We are considering then those special states $|\psi_0\rangle$ for which

$$\hat{R}_z(\phi)\,|\psi_0\rangle = e^{im\phi}\,|\psi_0\rangle, \tag{17.22}$$

where m is some real number.

We also know the remarkable fact that *if* the system is symmetrical for a rotation around z *and if* the original state happens to have the property that (17.22) is true, then it will also have the same property later on. So this number m is a very important one. If we know its value initially, we know its value at the end of the game. It is a number which is *conserved*—m is a *constant of the motion*. The reason that we pull out m is because it hasn't anything to do with any special angle ϕ, and also because it corresponds to something in classical mechanics. In *quantum* mechanics we *choose* to call $m\hbar$—for such states as $|\psi_0\rangle$—the *angular momentum about the z-axis*. If we do that we find that in the limit of large systems the same quantity is equal to the z-component of the angular momentum of classical mechanics. So if we have a state for which a rotation about the z-axis just produces a phase factor $e^{im\phi}$, then we have a state of definite angular momentum about that axis—and the angular momentum is conserved. It is $m\hbar$ now and forever. Of course, you can rotate about any axis, and you get the conservation of angular momentum for the various axes. You see that the conservation of angular momentum is related to the fact that when you turn a system you get the same state with only a new phase factor.

We would like to show you how general this idea is. We will apply it to two other conservation laws which have exact correspondence in the physical ideas to the conservation of angular momentum. In classical physics we also have conservation of momentum and conservation of energy, and it is interesting to see that both of these are related in the same way to some physical symmetry.

---

† Very precisely, we will define $\hat{R}_z(\phi)$ as a rotation of the physical system by $-\phi$ about the z-axis, which is the same as rotating the coordinate frame by $+\phi$.

‡ We can always choose z along the direction of the field provided there is only one field at a time, and its direction doesn't change.

¶ For a fancier proof we should make this argument for small rotations ϵ. Since any angle ϕ is the sum of a suitable n number of these, $\phi = n\epsilon$, $\hat{R}_z(\phi) = [\hat{R}_z(\epsilon)]^n$ and the total phase change is n times that for the small angle ϵ, and is, therefore, proportional to ϕ.

Suppose that we have a physical system—an atom, some complicated nucleus, or a molecule, or something—and it doesn't make any difference if we take the whole system and move it over to a different place. So we have a Hamiltonian which has the property that it depends only on the *internal coordinates* in some sense, and does not depend on the *absolute position* in space. Under those circumstances there is a special symmetry operation we can perform which is a translation in space. Let's define $\hat{D}_x(a)$ as the operation of a displacement by the distance a along the x-axis. Then for any state we can make this operation and get a new state. But again there can be very special states which have the property that when you displace them by a along the x-axis you get the same state except for a phase factor. It's also possible to prove, just as we did above, that when this happens, the phase must be proportional to a. So we can write for these special states $|\psi_0\rangle$

$$\hat{D}_x(a)\,|\psi_0\rangle = e^{ika}\,|\psi_0\rangle. \tag{17.23}$$

The coefficient k, when multiplied by $\hbar$, is called *the x-component of the momentum*. And the reason it is called that is that this number is numerically equal to the classical momentum p_x when we have a large system. The general statement is this: If the Hamiltonian is unchanged when the system is displaced, and if the state starts with a definite momentum in the x-direction, then the momentum in the x-direction will remain the same as time goes on. The total momentum of a system before and after collisions—or after explosions or what not—will be the same.

There is another operation that is quite analogous to the displacement in space: a delay in time. Suppose that we have a physical situation where there is *nothing external* that depends on time, and we start something off at a certain moment in a given state and let it roll. Now if we were to start the same thing off again (in another experiment) two seconds later—or/say, delayed by a time τ—and if nothing in the external conditions depends on the absolute time, the development would be the same and the final state would be the same as the other final state, except that it will get there later by the time τ. Under those circumstances we can also find special states which have the property that the development in time has the special characteristic that the delayed state is just the old, multiplied by a phase factor. Once more it is clear that for these special states the phase change must be proportional to τ. We can write

$$\hat{D}_t(\tau)\,|\psi_0\rangle = e^{-i\omega\tau}\,|\psi_0\rangle. \tag{17.24}$$

It is conventional to use the negative sign in defining ω; with this convention $\omega\hbar$ is the *energy* of the system, *and it is conserved*. So a system of definite energy is one which when displaced τ in time reproduces itself multiplied by $e^{-i\omega\tau}$. (That's what we have said before when we defined a quantum state of definite energy, so we're consistent with ourselves.) It means that if a system is in a state of definite energy, and if the Hamiltonian doesn't depend on t, then no matter what goes on, the system will have the same energy at all later times.

You see, therefore, the relation between the conservation laws and the symmetry of the world. Symmetry with respect to displacements in time implies the conservation of energy; symmetry with respect to position in x, y, or z implies the conservation of that component of momentum. Symmetry with respect to rotations around the x-, y-, and z-axes implies the conservation of the x-, y-, and z-components of angular momentum. Symmetry with respect to reflection implies the conservation of parity. Symmetry with respect to the interchange of two electrons implies the conservation of something we don't have a name for, and so on. Some of these principles have classical analogs and others do not. There are more conservation laws in quantum mechanics than are useful in classical mechanics—or, at least, than are usually made use of.

In order that you will be able to read other books on quantum mechanics, we must make a small technical aside—to describe the notation that people use. The operation of a displacement with respect to time is, of course, just the operation $\hat{U}$ that we talked about before:

$$\hat{D}_t(\tau) = \hat{U}(t + \tau, t). \tag{17.25}$$

Most people like to discuss everything in terms of *infinitesimal* displacements in time, or in terms of infinitesimal displacements in space, or in terms of rotations through infinitesimal angles. Since any finite displacement or angle can be accumulated by a succession of infinitesimal displacements or angles, it is often easier to analyze first the infinitesimal case. The operator of an infinitesimal displacement Δt in time is—as we have defined it in Chapter 8—

$$\hat{D}_t(\Delta t) = 1 - \frac{i}{\hbar} \Delta t \hat{H}. \tag{17.26}$$

Then $\hat{H}$ is analogous to the classical quantity we call energy, because if $\hat{H} | \psi \rangle$ happens to be a constant times $| \psi \rangle$ namely, $\hat{H} | \psi \rangle = E | \psi \rangle$, then that constant is the energy of the system.

The same thing is done for the other operations. If we make a small displacement in x, say by the amount Δx, a state $| \psi \rangle$ will, *in general*, go over into some other state $| \psi' \rangle$. We can write

$$| \psi' \rangle = \hat{D}_x(\Delta x) | \psi \rangle = \left(1 + \frac{i}{\hbar} \hat{p}_x \Delta x \right) | \psi \rangle, \tag{17.27}$$

since as Δx goes to zero, the $| \psi' \rangle$ should become just $| \psi \rangle$ or $\hat{D}_x(0) = 1$, and for small Δx the change of $\hat{D}_x(\Delta x)$ from 1 should be proportional to Δx. Defined this way, the operator $\hat{p}_x$ is called the momentum operator—for the x-component, of course.

For identical reasons, people usually write for small rotations

$$\hat{R}_z(\Delta \phi) | \psi \rangle = \left(1 + \frac{i}{\hbar} \hat{J}_z \Delta \phi \right) | \psi \rangle \tag{17.28}$$

and call $\hat{J}_z$ the operator of the z-component of angular momentum. For those special states for which $\hat{R}_z(\phi) | \psi_0 \rangle = e^{im\phi} | \psi_0 \rangle$, we can for any small angle—say $\Delta \phi$—expand the right-hand side to first order in $\Delta \phi$ and get

$$\hat{R}_z(\Delta \phi) = e^{im\Delta\phi} | \psi_0 \rangle = (1 + im\Delta\phi) | \psi_0 \rangle.$$

Comparing this with the definition of $\hat{J}_z$ in Eq. (17.28), we get that

$$\hat{J}_z | \psi_0 \rangle = m\hbar | \psi_0 \rangle. \tag{17.29}$$

In other words, if you operate with $\hat{J}_z$ on a state *with a definite angular momentum* about the z-axis, you get $m\hbar$ times the same state, where $m\hbar$ is the amount of z-component of angular momentum. It is quite analogous to operating on a definite energy state with $\hat{H}$ to get $E | \psi \rangle$.

We would now like to make some applications of the ideas of the conservation of angular momentum—to show you how they work. The point is that they are really very simple. You knew before that angular momentum is conserved. The only thing you really have to remember from this chapter is that if a state $| \psi_0 \rangle$ has the property that upon a rotation through an angle ϕ about the z-axis, it becomes $e^{im\phi} | \psi_0 \rangle$; it has a z-component of angular momentum equal to $m\hbar$. That's all we will need to do a number of interesting things.

17–4 Polarized light

First of all we would like to check on one idea. In Section 11–4 we showed that when RHC polarized light is viewed in a frame rotated by the angle ϕ about the z-axis† it gets multiplied by $e^{i\phi}$. Does that mean then that the photons of light

---

† Sorry! This angle is the negative of the one we used in Section 11–4.

that are right circularly polarized carry an angular momentum of *one* unit† along the z-axis? *Indeed it does.* It also means that if we have a beam of light containing a large number of photons all circularly polarized the same way—as we would have in a classical beam—it will carry angular momentum. If the total energy carried by the beam in a certain time is W, then there are $N = W/\hbar\omega$ photons. Each one carries the angular momentum $\hbar$, so there is a total angular momentum of

$$J_z = N\hbar = \frac{W}{\omega}. \qquad (17.30)$$

Can we prove classically that light which is right circularly polarized carries an energy and angular momentum in proportion to W/ω? That should be a classical proposition if everything is right. Here we have a case where we can go from the quantum thing to the classical thing. We should see if the classical physics checks. It will give us an idea whether we have a right to call m the angular momentum. Remember what right circularly polarized light is, classically. It's described by an electric field with an oscillating x-component and an oscillating y-component 90° out of phase so that the resultant electric vector ε goes in a circle—as drawn in Fig. 17–5(a). Now suppose that such light shines on a wall which is going to absorb it—or at least some of it—and consider an atom in the wall according to the classical physics. We have often described the motion of the electron in the atom as a harmonic oscillator which can be driven into oscillation by an external electric field. We'll suppose that the atom is isotropic, so that it can oscillate equally well in the x- or y-directions. Then in the circularly polarized light, the x-displacement and the y-displacement are the same, but one is 90° behind the other. The net result is that the electron moves in a circle, as shown in Fig. 17–5(b). The electron is displaced at some displacement r from its equilibrium position at the origin and goes around with some phase lag with respect to the vector ε. The relation between ε and r might be as shown in Fig. 17–5(b). As time goes on, the electric field rotates and the displacement rotates with the same frequency, so their relative orientation stays the same. Now let's look at the work being done on this electron. The rate that energy is being put into this electron is v, its velocity, times the component of $q\varepsilon$ parallel to the velocity:

$$\frac{dW}{dt} = q\varepsilon_t v. \qquad (17.31)$$

But look, there is angular momentum being poured into this electron, because there is always a torque about the origin. The torque is $q\varepsilon_t r$, which must be equal to the rate of change of angular momentum dJ_z/dt:

$$\frac{dJ_z}{dt} = q\varepsilon_t r. \qquad (17.32)$$

Remembering that $v = \omega r$, we have that

$$\frac{dJ_z}{dW} = \frac{1}{\omega}.$$

Therefore, if we integrate the total angular momentum which is absorbed, it is proportional to the total energy—the constant of proportionality being $1/\omega$, which agrees with Eq. (17.30). Light does carry angular momentum—1 unit (times $\hbar$) if it is right circularly polarized along the z-axis, and -1 unit along the z-axis if it is left circularly polarized.

Now let's ask the following question: If light is linearly polarized in the x-direction, what is its angular momentum? Light polarized in the x-direction can be represented as the superposition of RHC and LHC polarized light. Therefore, there is a certain amplitude that the angular momentum is $+\hbar$ and another

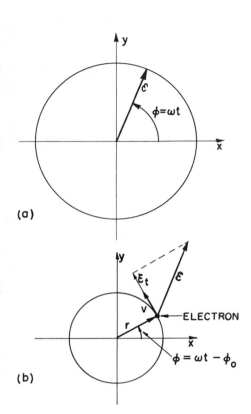

(a)

(b)

Fig. 17–5. (a) The electric field ε in a circularly polarized light wave. (b) The motion of an electron being driven by the circularly polarized light.

---

† It is usually very convenient to measure angular momentum of atomic systems in units of $\hbar$. Then you can say that a spin one-half particle has angular momentum $\pm 1/2$ with respect to any axis. Or, in general, that the z-component of angular momentum is m. You don't need to repeat the $\hbar$ all the time.

amplitude that the angular momentum is $-\hbar$, so it doesn't have a *definite* angular momentum. It has an amplitude to appear with $+\hbar$ and an equal amplitude to appear with $-\hbar$. The interference of these two amplitudes produces the linear polarization, but it has *equal* probabilities to appear with plus or minus one unit of angular momentum. Macroscopic measurements made on a beam of linearly polarized light will show that it carries zero angular momentum, because in a large number of photons there are nearly equal numbers of RHC and LHC photons contributing opposite amounts of angular momentum—the average angular momentum is zero. And in the classical theory you don't find the angular momentum unless there is some circular polarization.

We have said that any spin-one particle can have three values of J_z, namely $+1$, 0, -1 (the three states we saw in the Stern-Gerlach experiment). But light is screwy; it has only two states. It does not have the zero case. This strange lack is related to the fact that light cannot stand still. For a particle of spin j which is standing still, there must be the $2j + 1$ possible states with values of j_z going in steps of 1 from $-j$ to $+j$. But it turns out that for something of spin j with zero mass only the states with the components $+j$ and $-j$ along the direction of motion exist. For example, light does not have three states, but only two—although a photon is still an object of spin one. How is this consistent with our earlier proofs—based on what happens under rotations in space—that for spin-one particles three states are necessary? For a particle at rest, rotations can be made about any axis without changing the momentum state. Particles with zero rest mass (like photons and neutrinos) cannot be at rest; only rotations about the axis along the direction of motion do not change the momentum state. Arguments about rotations around one axis only are insufficient to prove that three states are required, given that one of them varies as $e^{i\phi}$ under rotations by the angle ϕ.†

One further side remark. For a zero rest mass particle, in general, *only one* of the two spin states with respect to the line of motion $(+j, -j)$ is really necessary. For neutrinos—which are spin one-half particles—only the states with the component of angular momentum *opposite* to the direction of motion $(-\hbar/2)$ exist in nature [and only *along* the motion $(+\hbar/2)$ for antineutrinos]. When a system has inversion symmetry (so that parity is conserved, as it is for light) both components $(+j$, and $-j)$ are required.

17–5 The disintegration of the Λ^0

Now we want to give an example of how we use the theorem of conservation of angular momentum in a specifically quantum physical problem. We look at break-up of the lambda particle (Λ^0), which disintegrates into a proton and a π^- meson by a "weak" interaction:

$$\Lambda^0 \to p + \pi^-.$$

Assume we know that the pion has spin zero, that the proton has spin one-half, and that the Λ^0 has spin one-half. We would like to solve the following problem: Suppose that a Λ^0 were to be produced in a way that caused it to be completely polarized—by which we mean that its spin is, say "up," with respect to some suitably chosen z-axis—see Fig. 17–6(a). The question is, with what probability will it disintegrate so that the proton goes off at an angle θ with respect to the z-axis—as in Fig. 17–6(b)? In other words, what is the angular distribution of the disintegrations? We will look at the disintegration in the coordinate system in which the Λ^0 is at rest—we will measure the angles in this rest frame; then they can always be transformed to another frame if we want.

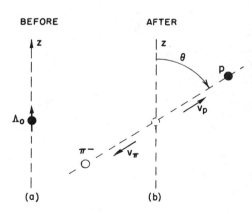

Fig. 17–6. A Λ^0 with spin "up" decays into a proton and a pion (in the CM system). What is the probability that the proton will go off at the angle θ?

---

† We have tried to find at least a proof that the component of angular momentum along the direction of motion must for a zero mass particle be an integral multiple of $\hbar/2$—and not something like $\hbar/3$. Even using all sorts of properties of the Lorentz transformation and what not, we failed. Maybe it's not true. We'll have to talk about it with Prof. Wigner, who knows all about such things.

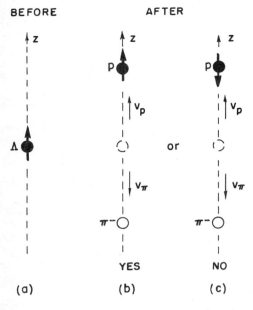

BEFORE   AFTER

YES   NO

(a)   (b)   (c)

Fig. 17-7. Two possibilities for the decay of a spin "up" Λ^0 with the proton going along the $+z$-axis. Only (b) conserves angular momentum.

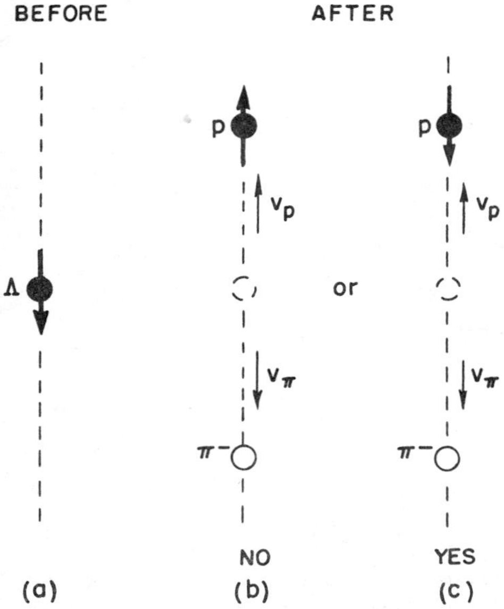

We begin by looking at the special circumstance in which the proton is emitted into a small solid angle $\Delta\Omega$ along the z-axis (Fig. 17-7). Before the disintegration we have a Λ^0 with its spin "up," as in part (a) of the figure. After a short time—for reasons unknown to this day, except that they are connected with the weak decays—the Λ^0 explodes into a proton and a pion. Suppose the proton goes up along the $+z$-axis. Then, from the conservation of momentum, the pion must go down. Since the proton is a spin one-half particle, its spin must be either "up" or "down"—there are, in principle, the two possibilities shown in parts (b) and (c) of the figure. The conservation of angular momentum, however, requires that the proton have spin "up." This is most easily seen from the following argument. A particle moving along the z-axis cannot contribute any angular momentum about this axis by virtue of its motion; therefore, only the spins can contribute to J_z. The spin angular momentum about the z-axis is $+\hbar/2$ before the disintegration, so it must also be $+\hbar/2$ afterward. We can say that since the pion has no spin, the proton spin must be "up."

If you are worried that arguments of this kind may not be valid in quantum mechanics, we can take a moment to show you that they are. The initial state (before the disintegration), which we can call $|\,\Lambda^0$, spin $+z\rangle$ has the property that if it is rotated about the z-axis by the angle ϕ, the state vector gets multiplied by the phase factor $e^{i\phi/2}$. (In the rotated system the state vector is $e^{i\phi/2}\,|\,\Lambda^0$, spin $+z\rangle$.) That's what we mean by spin "up" for a spin one-half particle. Since nature's behavior doesn't depend on our choice of axes, the final state (the proton plus pion) must have the same property. We could write the final state as, say,

$$|\text{ proton going }+z\text{, spin }+z\text{; pion going }-z\rangle.$$

But we really do not need to specify the pion motion, since in the frame we have chosen the pion always moves opposite the proton; we can simplify our description of the final state to

$$|\text{ proton going }+z\text{, spin }+z\rangle.$$

Now what happens to this state vector if we rotate the coordinates about the z-axis by the angle ϕ?

Since the proton and pion are moving along the z-axis, their motion isn't changed by the rotation. (That's why we picked this special case; we couldn't make the argument otherwise.) Also, nothing happens to the pion, because it is spin zero. The proton, however, has spin one-half. If its spin is "up" it will contribute a phase change of $e^{i\phi/2}$ in response to the rotation. (If its spin were "down" the phase change due to the proton would be $e^{-i\phi/2}$.) But the phase change with rotation before and after the excitement must be the same if angular momentum is to be conserved. (And it will be, since there are no outside influences in the Hamiltonian.) So the only possibility is that the proton spin will be "up." If the proton goes up, its spin must also be "up."

We conclude, then, that the conservation of angular momentum permits the process shown in part (b) of Fig. 17-7, but does not permit the process shown in part (c). Since we know that the disintegration occurs, there is some amplitude for process (b)—proton going up with spin "up." We'll let a stand for the amplitude that the disintegration occurs in this way in any infinitesimal interval of time.†

Now let's see what would happen if the Λ^0 spin were initially "down." Again we ask about the decays in which the proton goes up along the z-axis, as shown in Fig. 17-8. You will appreciate that in this case the proton must have spin "down" if angular momentum is conserved. Let's say that the amplitude for such a disintegration is b.

We can't say anything more about the two amplitudes a and b. They depend on the inner machinery of Λ^0, and the weak decays, and nobody yet knows how to

BEFORE   AFTER

NO   YES

(a)   (b)   (c)

Fig. 17-8. The decay along the z-axis for a Λ^0 with spin "down."

---

† We are now assuming that the machinery of the quantum mechanics is sufficiently familiar to you that we can speak about things in a physical way without taking the time to write down all the mathematical details. In case what we are saying here is not clear to you, we have put some of the missing details in a note at the end of the section.

calculate them. We'll have to get them from experiment. But with just these two amplitudes we *can* find out all we want to know about the angular distribution of the disintegration. We only have to be careful always to define completely the states we are talking about.

We want to know the probability that the proton will go off at the angle θ with respect to the z-axis (into a small solid angle $\Delta\Omega$) as drawn in Fig. 17–6. Let's put a new z-axis in this direction and call it the z'-axis. We know how to analyze what happens along this axis. With respect to this new axis, the Λ° no longer has its spin "up," but has a certain amplitude to have its spin "up" and another amplitude to have its spin "down." We have already worked these out in Chapter 6, and again in Chapter 10, Eq. (10.30). The amplitude to be spin "up" is $\cos \theta/2$, and the amplitude to be spin "down" is† $-\sin \theta/2$. When the Λ° spin is "up" along the z'-axis it will emit a proton in the $+z'$-direction with the amplitude a. So the amplitude to find an "up"-spinning proton coming out along the z'-direction is

$$a \cos \frac{\theta}{2}. \tag{17.33}$$

Similarly, the amplitude to find a "down"-spinning proton coming along the positive z'-axis is

$$-b \sin \frac{\theta}{2}. \tag{17.34}$$

The two processes that these amplitudes refer to are shown in Fig. 17–9.

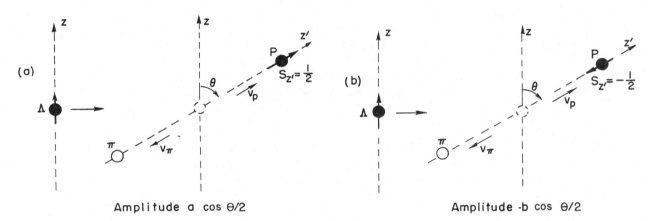

Fig. 17–9. Two possible decay states for the Λ°.

Let's now ask the following easy question. If the Λ° has spin up along the z-axis, what is the probability that the decay proton will go off at the angle θ? The two spin states ("up" or "down" along z') are distinguishable even though we are not going to look at them. So to get the probability we square the amplitudes and add. The probability $f(\theta)$ of finding a proton in a small solid angle $\Delta\Omega$ at θ is

$$f(\theta) = |a|^2 \cos^2 \frac{\theta}{2} + |b|^2 \sin^2 \frac{\theta}{2}. \tag{17.35}$$

Remembering that $\sin^2 \theta/2 = \frac{1}{2}(1 - \cos \theta)$ and that $\cos^2 \theta/2 = \frac{1}{2}(1 + \cos \theta)$, we can write $f(\theta)$ as

$$f(\theta) = \left(\frac{|a|^2 + |b|^2}{2}\right) + \left(\frac{|a|^2 - |b|^2}{2}\right) \cos \theta. \tag{17.36}$$

---

† We have chosen to let z' be in the xz-plane and use the matrix elements for $R_y(\theta)$. You would get the same answer for any other choice.

The angular distribution has the form

$$f(\theta) = \beta(1 + \alpha \cos \theta). \tag{17.37}$$

The probability has one part that is independent of θ and one part that varies linearly with $\cos \theta$. From measuring the angular distribution we can get α and β, and therefore, $|a|$ and $|b|$.

Now there are many other questions we can answer. Are we interested only in protons with spin "up" along the *old z*-axis? Each of the terms in (17–33) and (17–34) will give an amplitude to find a proton with spin "up" and with spin "down" with respect to the z'-axis ($+z'$ and $-z'$). Spin "up" with respect to the old axis $|+z\rangle$ can be expressed in terms of the base states $|+z'\rangle$ and $|-z'\rangle$. We can then combine the two amplitudes (17.33) and (17.34) with the proper coefficients ($\cos \theta/2$ and $-\sin \theta/2$) to get the total amplitude

$$\left(a \cos^2 \frac{\theta}{2} + b \sin^2 \frac{\theta}{2} \right).$$

Its square is the probability that the proton comes out at the angle θ with its spin the same as the Λ^0 ("up" along the z-axis).

If parity were conserved, we could say one more thing. The disintegration of Fig. 17–8 is just the reflection—in say, the $y z$-plane of the disintegration of Fig. 17–7.† If parity were conserved, b would have to be equal to a or to $-a$. Then the coefficient α of (17.37) would be zero, and the disintegration would be equally likely to occur in all directions.

The experimental results show, however, that there *is* an asymmetry in the disintegration. The measured angular distribution does go as $\cos \theta$ as we predict—and not as $\cos^2 \theta$ or any other power. In fact, since the angular distribution has this form, we can deduce from these measurements that the spin of the Λ^0 is $1/2$. Also, we see that parity is not conserved. In fact, the coefficient α is found experimentally to be -0.62 ± 0.05, so b is about twice as large as a. The lack of symmetry under a reflection is quite clear.

You see how much we can get from the conservation of angular momentum. We will give some more examples in the next chapter.

Parenthetical note. By the amplitude a in this section we mean the amplitude that the state $|$ proton going $+z$, spin $+z\rangle$ is generated in an infinitesimal time dt from the state $|\Lambda$, spin $+z\rangle$, or, in other words, that

$$\langle \text{proton going } +z, \text{ spin } +z \,|\, H \,|\, \Lambda, \text{ spin } +z \rangle = i\hbar a, \tag{17.38}$$

where H is the Hamiltonian of the world—or, at least, of whatever is responsible for the Λ-decay. The conservation of angular momentum means that the Hamiltonian must have the property that

$$\langle \text{proton going } +z, \text{ spin } -z \,|\, H \,|\, \Lambda, \text{ spin } +z \rangle = 0. \tag{17.39}$$

By the amplitude b we mean that

$$\langle \text{proton going } +z, \text{ spin } -z \,|\, H \,|\, \Lambda, \text{ spin } -z \rangle = i\hbar b. \tag{17.40}$$

Conservation of angular momentum implies that

$$\langle \text{proton going } +z, \text{ spin } +z \,|\, H \,|\, \Lambda, \text{ spin } -z \rangle = 0. \tag{17.41}$$

If the amplitudes written in (17.33) and (17.34) are not clear, we can express them more mathematically as follows. By (17.33) we intend the amplitude that the Λ with spin along $+z$ will disintegrate into a proton moving along the $+z'$-direction with its spin also in the $+z'$-direction, namely the amplitude

$$\langle \text{proton going } +z', \text{ spin } +z' \,|\, H \,|\, \Lambda, \text{ spin } +z \rangle. \tag{17.42}$$

By the general theorems of quantum mechanics, this amplitude can be written as

$$\sum_i \langle \text{proton going } +z', \text{ spin } +z' \,|\, H \,|\, \Lambda, i \rangle \langle \Lambda, i \,|\, \Lambda, \text{ spin } +z \rangle, \tag{17.43}$$

---

† Remembering that the spin is an axial vector and flips over in the reflection.

where the sum is to be taken over the base states $|\Lambda, i\rangle$ of the Λ-particle at rest. Since the Λ-particle is spin one-half, there are two such base states which can be in any reference base we wish. If we use for base states spin "up" and spin "down" *with respect to z'* $(+z', -z')$, the amplitude of (17.43) is equal to the sum

$$\langle\text{proton going } +z', \text{ spin } +z' \mid H \mid \Lambda, +z'\rangle\langle\Lambda, +z' \mid \Lambda, +z\rangle$$

$$+\langle\text{proton going } +z', \text{ spin } +z' \mid H \mid \Lambda, -z'\rangle\langle\Lambda, -z' \mid \Lambda, +z\rangle. \qquad (17.44)$$

The first factor of the first term is a, and the first factor of the second term is zero—from the definition of (17.38), and from (17.41), which in turn follows from angular momentum conservation. The remaining factor $\langle\Lambda, +z' \mid \Lambda, +z\rangle$ of the first term is just the amplitude that a spin one-half particle which has spin "up" along one axis will also have spin "up" along an axis tilted at the angle θ, which is $\cos\theta/2$—see Table 6-2. So (17.44) is just $a\cos\theta/2$, as we wrote in (17.33). The amplitude of (17.34) follows from the same kind of arguments for a spin "down" Λ-particle.

17–6 Summary of the rotation matrices

We would like now to bring together in one place the various things we have learned about the rotations for particles of spin one-half and spin one—so they will be convenient for future reference. On the next page you will find tables of the two rotation matrices $R_z(\phi)$ and $R_y(\theta)$ for spin one-half particles, for spin-one particles, and for photons (spin-one particles with zero rest mass). For each spin we will give the terms of the matrix $\langle j \mid R \mid i\rangle$ for rotations about the z-axis or the y-axis. They are, of course, exactly equivalent to the amplitudes like $\langle+T \mid 0 S\rangle$ we have used in earlier chapters. We mean by $R_z(\phi)$ that the state is projected into a new coordinate system which is rotated through the angle ϕ about the z-axis—using always the right-hand rule to define the positive sense of the rotation. By $R_y(\theta)$ we mean that the reference axes are rotated by the angle θ about the y-axis. Knowing these two rotations, you can, of course, work out any arbitrary rotation. As usual, we write the matrix elements so that the state on the *left* is a base state of the *new* (rotated) frame and the state on the right is a base state of the old (un-rotated) frame. You can interpret the entries in the tables in many ways. For instance, the entry $e^{-i\phi/2}$ in Table 17–1 means that the matrix element $\langle- \mid R \mid -\rangle = e^{-i\phi/2}$. It also means that $\hat{R} \mid -\rangle = e^{-i\phi/2} \mid -\rangle$, or that $\langle- \mid \hat{R} = \langle- \mid e^{-i\phi/2}$. It's all the same thing.

Table 17–1

Rotation matrices for spin one-half

Two states: $|+\rangle$, "up" along the z-axis, $m = +1/2$
$|-\rangle$, "down" along the z-axis, $m = -1/2$

$R_z(\phi)$	$|+\rangle$	$|-\rangle$
$\langle+|$	$e^{+i\phi/2}$	0
$\langle-|$	0	$e^{-i\phi/2}$

$R_y(\theta)$	$|+\rangle$	$|-\rangle$
$\langle+|$	$\cos\theta/2$	$\sin\theta/2$
$\langle-|$	$-\sin\theta/2$	$\cos\theta/2$

Table 17–2

Rotation matrices for spin one

Three states: $|+\rangle$, $m = +1$
$|0\rangle$, $m = 0$
$|-\rangle$, $m = -1$

$R_z(\phi)$	$|+\rangle$	$|0\rangle$	$|-\rangle$
$\langle+|$	$e^{+i\phi}$	0	0
$\langle0|$	0	1	0
$\langle-|$	0	0	$e^{-i\phi}$

$R_y(\theta)$	$|+\rangle$	$|0\rangle$	$|-\rangle$
$\langle+|$	$\frac{1}{2}(1 + \cos\theta)$	$+\dfrac{1}{\sqrt{2}}\sin\theta$	$\frac{1}{2}(1 - \cos\theta)$
$\langle0|$	$-\dfrac{1}{\sqrt{2}}\sin\theta$	$\cos\theta$	$+\dfrac{1}{\sqrt{2}}\sin\theta$
$\langle-|$	$\frac{1}{2}(1 - \cos\theta)$	$-\dfrac{1}{\sqrt{2}}\sin\theta$	$\frac{1}{2}(1 + \cos\theta)$

Table 17–3

Photons

Two states: $|R\rangle = \dfrac{1}{\sqrt{2}}(|x\rangle + i|y\rangle)$, $m = +1$ (RHC polarized)

$|L\rangle = \dfrac{1}{\sqrt{2}}(|x\rangle - i|y\rangle)$, $m = -1$ (LHC polarized)

$R_z(\phi)$	$|R\rangle$	$|L\rangle$
$\langle R|$	$e^{+i\phi}$	0
$\langle L|$	0	$e^{-i\phi}$

18

Angular Momentum

18–1 Electric dipole radiation

In the last chapter we developed the idea of the conservation of angular momentum in quantum mechanics, and showed how it might be used to predict the angular distribution of the proton from the disintegration of the Λ-particle. We want now to give you a number of other, similar, illustrations of the consequences of momentum conservation in atomic systems. Our first example is the radiation of light from an atom. The conservation of angular momentum (among other things) will determine the polarization and angular distribution of the emitted photons.

Suppose we have an atom which is in an excited state of definite angular momentum—say with a spin of one—and it makes a transition to a state of angular momentum zero at a lower energy, emitting a photon. The problem is to figure out the angular distribution and polarization of the photons. (This problem is almost exactly the same as the Λ^0 disintegration, except that we have spin-one instead of spin one-half particles.) Since the upper state of the atom is spin one, there are three possibilities for its z-component of angular momentum. The value of m could be $+1$, or 0, or -1. We will take $m = +1$ for our example. Once you see how it goes, you can work out the other cases. We suppose that the atom is sitting with its angular momentum along the $+z$-axis—as in Fig. 18–1(a)—and ask with what amplitude it will emit right circularly polarized light upward along the z-axis, so that the atom ends up with zero angular momentum—as shown in part (b) of the figure. Well, we don't know the answer to that. But we do know that right circularly polarized light has one unit of angular momentum about its direction of propagation. So after the photon is emitted, the situation would have to be as shown in Fig. 18–1(b)—the atom is left with zero angular momentum

18–1 Electric dipole radiation

18–2 Light scattering

18–3 The annihilation of positronium

18–4 Rotation matrix for any spin

18–5 Measuring a nuclear spin

18–6 Composition of angular momentum

Added Note 1: Derivation of the rotation matrix

Added Note 2: Conservation of parity in photon emission

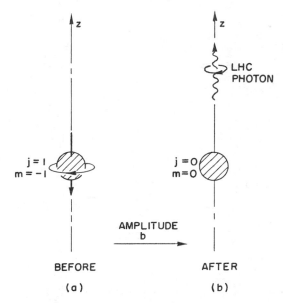

Fig. 18–1. An atom with $m = +1$ emits a RHC photon along the $+z$-axis.

Fig. 18–2. An atom with $m = -1$ emits a LHC photon along the $+z$-axis.

about the z-axis, since we have assumed an atom whose lower state is spin zero. We will let a stand for the amplitude for such an event. More precisely, we let a be the amplitude to emit a photon into a certain small solid angle $\Delta\Omega$, centered on the z-axis, during a time dt. Notice that the amplitude to emit a LHC photon in the same direction is zero. The net angular momentum about the z-axis would be -1 for such a photon and zero for the atom for a total of -1, which would not conserve angular momentum.

Similarly, if the spin of the atom is initially "down" (-1 along the z-axis), it can emit only a LHC polarized photon in the direction of the $+z$-axis, as shown in Fig. 18–2. We will let b stand for the amplitude for this event—meaning again the amplitude that the photon goes into a certain solid angle $\Delta\Omega$. On the other hand, if the atom is in the $m = 0$ state, it cannot emit a photon in the $+z$-direction at all, because a photon can have only the angular momentum $+1$ or -1 along its direction of motion.

Next, we can show that b is related to a. Suppose we perform an inversion of the situation in Fig. 18–1, which means that we should imagine what the system would look like if we were to move each part of the system to an equivalent point on the opposite side of the origin. This does *not* mean that we should reflect the angular momentum vectors, because they are artificial. We should, rather, invert the actual character of the motion that would correspond to such an angular momentum. In Fig. 18–3(a) and (b) we show what the process of Fig. 18–1 looks like before and after an inversion with respect to the center of the atom. Notice that the sense of rotation of the atom is unchanged.† In the inverted system of Fig. 18–3(b) we have an atom with $m = +1$ emitting a LHC photon downward.

If we now rotate the system of Fig. 18–3(b) by 180° about the x- or y-axis, it becomes identical to Fig. 18–2. The combination of the inversion and rotation turns the second process into the first. Using Table 17–2, we see that a rotation of 180° about the y-axis just throws an $m = -1$ state into an $m = +1$ state, so the amplitude b must be equal to the amplitude a *except for a possible sign change due to the inversion.* The sign change in the inversion will depend on the parities of the initial and final state of the atom.

In atomic processes, parity is conserved, so the parity of the whole system must be the same before and after the photon emission. What happens will depend on whether the parities of the initial and final states of the atom are even or odd—the angular distribution of the radiation will be different for different cases. We will take the common case of *odd* parity for the initial state and *even* parity for the final state; it will give what is called "electric dipole radiation." (If the initial and final states have the same parity we say there is "magnetic dipole radiation," which has the character of the radiation from an oscillating current in a loop.) If the parity of the initial state is odd, its amplitude reverses its sign in the inversion which takes the system from (a) to (b) of Fig. 18–3. The final state of the atom has even parity, so its amplitude doesn't change sign. If the reaction is going to conserve parity, the amplitude b must be equal to a in magnitude but of the opposite sign.

We conclude that if the amplitude is a that an $m = +1$ state will emit a photon upward, then for the assumed parities of the initial and final states the amplitude that an $m = -1$ state will emit a LHC photon upward is $-a$.‡

We have all we need to know to find the amplitude for a photon to be emitted at any angle θ with respect to the z-axis. Suppose we have an atom originally polarized with $m = +1$. We can resolve this state into $+1$, 0, and -1 states with respect to a new z'-axis in the direction of the photon emission. The amplitudes for these three states are just the ones given in the lower half of Table 17–2.

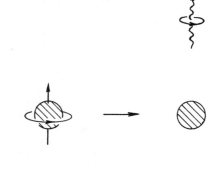

Fig. 18–3. If the process of (a) is transformed by an inversion through the center of the atom, it appears as in (b).

---

† When we change x, y, z into $-x$, $-y$, $-z$, you might think that all vectors get reversed. That is true for *polar* vectors like displacements and velocities, but *not* for an *axial* vector like angular momentum—or any vector which is derived from a cross product of two polar vectors. Axial vectors have the same components after an inversion.

‡ Some of you may object to the argument we have just made, on the basis that the final states we have been considering do not have a definite parity. You will find in Added Note 2 at the end of this chapter another demonstration, which you may prefer.

The amplitude that a RHC photon is emitted in the direction θ is then a times the amplitude to have $m = +1$ in that direction, namely,

$$a \langle + \mid R_y(\theta) \mid + \rangle = \frac{a}{2} (1 + \cos \theta). \tag{18.1}$$

The amplitude that a LHC photon is emitted in the same direction is $-a$ times the amplitude to have $m = -1$ in the new direction. Using Table 17–2, it is

$$-a \langle - \mid R_y(\theta) \mid + \rangle = \frac{a}{-2} (1 - \cos \theta). \tag{18.2}$$

If you are interested in other polarizations you can find out the amplitude for them from the superposition of these two amplitudes. To get the intensity of any component as a function of angle, you must, of course, take the absolute square of the amplitudes.

18–2 Light scattering

Let's use these results to solve a somewhat more complicated problem— but also one which is somewhat more real. We suppose that the same atoms are sitting in their ground state ($j = 0$), and *scatter* an incoming beam of light. Let's say that the light is going initially in the $+z$-direction, so that we have photons coming up to the atom *from* the $-z$-direction, as shown in Fig. 18–4(a). We can consider the scattering of light as a two-step process: The photon is absorbed, and then is re-emitted. If we start with a RHC photon as in Fig. 18–4(a), and angular momentum is conserved, the atom will be in an $m = +1$ state after the absorption—as shown in Fig. 18–4(b). We call the amplitude for this process c. The atom can then emit a RHC photon in the direction θ—as in Fig. 18–4(c). The total amplitude that a RHC photon is scattered in the direction θ is just c times (18.1). Let's call this scattering amplitude $\langle R' \mid S \mid R \rangle$; we have

$$\langle R' \mid S \mid R \rangle = \frac{ac}{2} (1 + \cos \theta). \tag{18.3}$$

There is also an amplitude that a RHC photon will be absorbed and that a LHC photon will be emitted. The product of the two amplitudes is the amplitude $\langle L' \mid S \mid R \rangle$ that a RHC photon is scattered as a LHC photon. Using (18.2), we have

$$\langle L' \mid S \mid R \rangle = -\frac{ac}{2} (1 - \cos \theta). \tag{18.4}$$

Now let's ask about what happens if a LHC photon comes in. When it is absorbed, the atom will go into an $m = -1$ state. By the same kind of arguments we used in the preceding section, we can show that this amplitude must be $-c$. The amplitude that an atom in the $m = -1$ state will emit a RHC photon at the angle θ is a times the amplitude $\langle + \mid R_y(\theta) \mid - \rangle$, which is $\frac{1}{2}(1 - \cos \theta)$. So we have

$$\langle R' \mid S \mid L \rangle = -\frac{ac}{2} (1 - \cos \theta). \tag{18.5}$$

Finally, the amplitude for a LHC photon to be scattered as a LHC photon is

$$\langle L' \mid S \mid L \rangle = \frac{ac}{2} (1 + \cos \theta). \tag{18.6}$$

(There are two minus signs which cancel.)

If we make a measurement of the scattered *intensity* for any given combination of circular polarizations it will be proportional to the square of one of our four amplitudes. For instance, with an incoming beam of RHC light the intensity of the RHC light in the scattered radiation will vary as $(1 + \cos \theta)^2$.

That's all very well, but suppose we start out with *linearly* polarized light. What then? If we have x-polarized light, it can be represented as a superposition

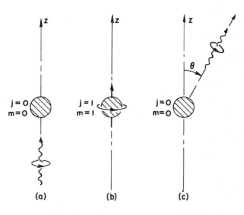

Fig. 18–4. The scattering of light by an atom seen as a two-step process.

18–3

of RHC and LHC light. We write (see Section 11–4)

$$|x\rangle = \frac{1}{\sqrt{2}}(|R\rangle + |L\rangle). \tag{18.7}$$

Or, if we have y-polarized light, we would have

$$|y\rangle = -\frac{i}{\sqrt{2}}(|R\rangle - |L\rangle). \tag{18.8}$$

Now what do you want to know? Do you want the amplitude that an x-polarized photon will scatter into a RHC photon at the angle θ? You can get it by the usual rule for combining amplitudes. First, multiply (18.7) by $\langle R' | S$ to get

$$\langle R' | S | x\rangle = \frac{1}{\sqrt{2}}(\langle R' | S | R\rangle + \langle R' | S | L\rangle), \tag{18.9}$$

and then use (18.3) and (18.5) for the two amplitudes. You get

$$\langle R' | S | x\rangle = \frac{ac}{\sqrt{2}}\cos\theta. \tag{18.10}$$

If you wanted the amplitude that an x-photon would scatter into a LHC photon, you would get

$$\langle L' | S | x\rangle = \frac{ac}{\sqrt{2}}\cos\theta. \tag{18.11}$$

Finally, suppose you wanted to know the amplitude that an x-polarized photon will scatter while keeping its x-polarization. What you want is $\langle x' | S | x\rangle$. This can be written as

$$\langle x' | S | x\rangle = \langle x' | R'\rangle\langle R' | S | x\rangle + \langle x' | L'\rangle\langle L' | S | x\rangle. \tag{18.12}$$

If you then use the relations

$$|R'\rangle = \frac{1}{\sqrt{2}}(|x'\rangle + i|y'\rangle), \tag{18.13}$$

$$|L'\rangle = \frac{1}{\sqrt{2}}(|x'\rangle - i|y'\rangle), \tag{18.14}$$

it follows that

$$\langle x' | R'\rangle = \frac{1}{\sqrt{2}}, \tag{18.15}$$

$$\langle x' | L'\rangle = \frac{1}{\sqrt{2}}. \tag{18.16}$$

So you get that

$$\langle x' | S | x\rangle = ac\cos\theta. \tag{18.17}$$

The answer is that a beam of x-polarized light will be scattered at the direction θ (in the xz-plane) with an *intensity* proportional to $\cos^2\theta$. If you ask about y-polarized light, you find that

$$\langle y' | S | x\rangle = 0. \tag{18.18}$$

So the scattered light is completely polarized in the x-direction.

Now we notice something interesting. The results (18.17) and (18.18) correspond exactly to the classical theory of light scattering we gave in Vol. 1, Section 32–6, where we imagined that the electron was bound to the atom by a linear restoring force—so that it acted like a classical oscillator. Perhaps you are thinking: "It's so much easier in the classical theory; if it gives the right answer why bother with the quantum theory?" For one thing, we have considered so far only the special—though common—case of an atom with a $j = 1$ excited state and a $j = 0$ ground state. If the excited state had spin two, you would get a different result. Also, there is no reason why the model of an electron attached to a spring and driven by an oscillating electric field should work for a single photon. But we have found that it does in fact work, and that the polarization and intensities come out right. So in a certain sense we are bringing the whole course around to the real truth. Whereas we have, in Vol. I, done the theory of the index of refraction, and of light scattering, by the classical theory, we have now shown that the quantum theory gives the same result for the most common case. In effect we have now done the polarization of sky light, for instance, by quantum mechanical arguments, which is the only truly legitimate way.

It should be, of course, that all the classical theories which work are supported ultimately by legitimate quantum arguments. Naturally, those things which we have spent a great deal of time in explaining to you were selected from just those parts of classical physics which still maintain validity in quantum mechanics. You'll notice that we did not discuss in great detail any model of the atom which has electrons going around in orbits. That's because such a model doesn't give results which agree with the quantum mechanics. But the electron on a spring—which is not, in a sense, at all the way an atom "looks"—does work, and so we used that model for the theory of the index of refraction.

18–3 The annihilation of positronium

We would like next to take an example which is very pretty. It is quite interesting and, although somewhat complicated, we hope not too much so. Our example is the system called *positronium*, which is an "atom" made up of an electron and a positron—a bound state of an e^+ and an e^-. It is like a hydrogen atom, except that a positron replaces the proton. This object has—like the hydrogen atom—many states. Also like the hydrogen, the ground state is split into a "hyperfine structure" by the interaction of the magnetic moments. The spins of the electron and positron are each one-half, and they can be either parallel or antiparallel to any given axis. (In the ground state there is no other angular momentum due to orbital motion.) So there are four states: three are the substates of a spin-one system, all with the same energy; and one is a state of spin zero with a different energy. The energy splitting is, however, much larger than the 1420 megacycles of hydrogen because the positron magnetic moment is so much stronger—1000 times stronger—than the proton moment.

The most important difference, however, is that positronium cannot last forever. The position is the antiparticle of the electron; they can annihilate each other. The two particles disappear completely—converting their rest energy into radiation, which appears as γ-rays (photons). In the disintegration, two particles with a finite rest mass go into two or more objects which have zero rest mass.†

We begin by analyzing the disintegration of the spin-zero state of the positronium. It disintegrates into two γ-rays with a lifetime of about 10^{-10} second. Initially, we have a positron and an electron close together and with spins antiparallel, making the positronium system. After the disintegration there are two photons going out with equal and opposite momenta (Fig. 18–5). The momenta must be equal and opposite, because the total momentum after the disintegration must be zero, as it was before, if we are taking the case of annihilation at rest. If the positronium is not at rest, we can ride with it, solve the problem, and then transform everything back to the lab system. (See, we can do anything now; we have all the tools.)

First, we note that the angular distribution is not very interesting. Since the initial state has spin zero, it has no special axis—it is symmetric under all rotations. The final state must then also be symmetric under all rotations. That means that all angles for the disintegration are equally likely—the amplitude is the same for a photon to go in any direction. Of course, once we find *one* of the photons in some direction the *other* must be opposite.

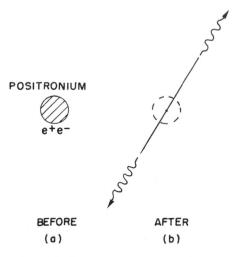

POSITRONIUM

$e^+ e^-$

BEFORE
(a)

AFTER
(b)

Fig. 18–5. The two-photon annihilation of positronium.

---

† In the deeper understanding of the world today, we do not have an easy way to distinguish whether the energy of a photon is less "matter" than the energy of an electron, because as you remember all the particles behave very similarly. The only distinction is that the photon has zero rest mass.

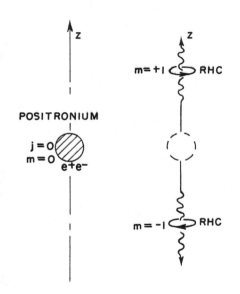

Fig. 18–6. One possibility for positronium annihilation along the z-axis.

The only remaining question, which we now want to look at, is about the polarization of the photons. Let's call the directions of motion of the two photons the plus and minus z-axes. We can use any representations we want for the polarization states of the photons; we will choose for our description right and left circular polarization—always with respect to the directions of motion. Right away, we can see that if the photon going upward is RHC, then angular momentum will be conserved if the downward going photon is also RHC. Each will carry +1 unit of angular momentum *with respect to its momentum direction*, which means plus and minus one unit about the z-axis. The total will be zero, and the angular momentum after the disintegration will be the same as before. See Fig. 18–6.

The same arguments show that if the upward going photon is RHC, the downward cannot be LHC. Then the final state would have two units of angular momentum. This is not permitted if the initial state has spin zero. Note that such a final state is also not possible for the other positronium ground state of spin one, because it can have a maximum of one unit of angular momentum in any direction.

Now we want to show that two-photon annihilation is not possible at all from the spin-one state. You might think that if we took the $j = 1, m = 0$ state—which has zero angular momentum about the z-axis—it should be like the spin-zero state, and could disintegrate into two RHC photons. Certainly, the disintegration sketched in Fig. 18–7(a) conserves angular momentum about the z-axis. But now look what happens if we rotate this system around the y-axis by 180°; we get the picture shown in Fig. 18–7(b). It is exactly the same as in part (a) of the figure. All we have done is interchange the two photons. Now photons are Bose particles; if we interchange them, the amplitude has the same sign, so the amplitude for the disintegration in part (b) must be the same as in part (a). But we have assumed that the initial object is spin one. And when we rotate a spin-one object in a state with $m = 0$ by 180° about the y-axis, its amplitudes change sign (see Table 17–2 for $\theta = \pi$). So the amplitudes for (a) and (b) in Fig. 18–7 should have opposite signs; the spin-one state *cannot disintegrate into two photons*.

When positronium is formed you would expect it to end up in the spin-zero state 1/4 of the time and in the spin-one state (with $m = -1, 0,$ or $+1$)3/4 of the time. So 1/4 of the time you would get two-photon annihilations. The other 3/4

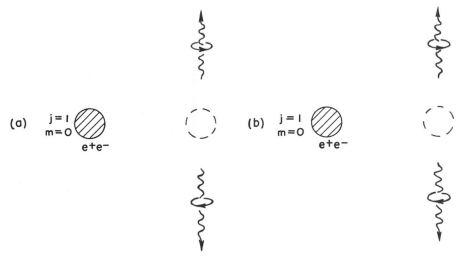

Fig. 18–7. For the $j = 1$ state of positronium, the process (a) and its 180° rotation about y (b) are exactly the same.

† Note that we always analyze the angular momentum about the direction of motion of the particle. If we were to ask about the angular momentum about any other axis, we would have to worry about the possibility of "orbital" angular momentum—from a $p \times r$ term. For instance, we can't say that the photons leave exactly from the center of the positronium. They could leave like two things shot out from the rim of a spinning wheel. We don't have to worry about such possibilities when we take our axis along the direction of motion.

of the time there can be no two-photon annihilations. There is still an annihilation, but it has to go with *three* photons. It is harder for it to do that and the lifetime is 1000 times longer—about 10^{-7} second. This is what is observed experimentally. We will not go into any more of the details of the spin-one annihilation.

So far we have that if we only worry about angular momentum, the spin-zero state of the positronium can go into two RHC photons. There is also another possibility: it can go into two LHC photons as shown in Fig. 18–8. The next question is, what is the relation between the amplitudes for these two possible decay modes? We can find out from the conservation of parity.

To do that, however, we need to know the parity of the positronium. Now theoretical physicists have shown in a way that is not easy to explain that the parity of the electron and the positron—its antiparticle—must be opposite, so that the spin-zero ground state of positronium must be odd. We will just assume that it is odd, and since we will get agreement with experiment, we can take that as sufficient proof.

Let's see then what happens if we make an inversion of the process in Fig. 18–6. When we do that, the two photons reverse directions *and* polarizations. The inverted picture looks just like Fig. 18–8. Assuming that the parity of the positronium is odd, the amplitudes for the two processes in Figs. 18–6 and 18–8 must have the opposite sign. Let's let $|R_1 R_2\rangle$ stand for the final state of Fig. 18–6 in which both photons are RHC, and let $|L_1 L_2\rangle$ stand for the final state of Fig. 18–8, in which both photons are LHC. The true final state—let's call it $|F\rangle$—must be

$$|F\rangle = |R_1 R_2\rangle - |L_1 L_2\rangle. \tag{18.19}$$

Then an inversion changes the R's into L's and gives the state

$$P\,|F\rangle = |L_1 L_2\rangle - |R_1 R_2\rangle = -|F\rangle, \tag{18.20}$$

which is the negative of (18.19). So the final state $|F\rangle$ has negative parity, which is the same as the initial spin-zero state of the positronium. This is the only final state that conserves both angular momentum and parity. There is some amplitude that the disintegration into this state will occur, which we don't need to worry about now, however, since we are only interested in questions about the polarization.

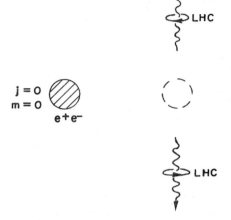

Fig. 18–8. Another possible process for positronium annihilation.

What does the final state of (18.19) mean physically? One thing it means is the following: If we observe the two photons in two detectors which can be set to count separately the RHC or LHC photons, we will always see two RHC photons together, or two LHC photons together. That is, if you stand on one side of the positronium and someone else stands on the opposite side, you can measure the polarization and tell the other guy what polarization he will get. You have a 50–50 chance of catching a RHC photon or a LHC photon; whichever one you get, you can predict that he will get the same.

Since there is a 50–50 chance for RHC or LHC polarization, it sounds as though it might be like linear polarization. Let's ask what happens if we observe the photon in counters that accept only linearly polarized light. For γ-rays it is not as easy to measure the polarization as it is for light; there is no polarizer which works well for such short wavelengths. But let's imagine that there is, to make the discussion easier. Suppose that you have a counter that only accepts light with x-polarization, and that there is a guy on the other side that also looks for linear polarized light with, say, y-polarization. What is the chance you will pick up the two photons from an annihilation? What we need to ask is the amplitude that $|F\rangle$ will be in the state $|x_1 y_2\rangle$. In other words, we want the amplitude

$$\langle x_1 y_2 \,|\, F\rangle,$$

which is, of course, just

$$\langle x_1 y_2 \,|\, R_1 R_2\rangle - \langle x_1 y_2 \,|\, L_1 L_2\rangle. \tag{18.21}$$

Now although we are working with two-particle amplitudes for the two photons, we can handle them just as we did the single particle amplitudes, since each particle acts independently of the other. That means that the amplitude $\langle x_1 y_2 \mid R_1 R_2 \rangle$ is just the product of the two independent amplitudes $\langle x_1 \mid R_1 \rangle$ and $\langle y_2 \mid R_2 \rangle$. Using Table 17–3, these two amplitudes are $1/\sqrt{2}$ and $i/\sqrt{2}$, so

$$\langle x_1 y_2 \mid R_1 R_2 \rangle = +\frac{i}{2}.$$

Similarly, we find that

$$\langle x_1 y_2 \mid L_1 L_2 \rangle = -\frac{i}{2}.$$

Subtracting these two amplitudes according to (18.21), we get that

$$\langle x_1 y_2 \mid F \rangle = + i. \tag{18.22}$$

So there is a *unit* probability† that *if* you get a photon in your x-polarized detector, the other guy will get a photon in his y-polarized detector.

Now suppose that the other guy sets his counter for x-polarization the same as yours. He would never get a count when you got one. If you work it through, you will find that

$$\langle x_1 x_2 \mid F \rangle = 0. \tag{18.23}$$

It will, naturally, also work out that if you set your counter for y-polarization he will get coincident counts only if he is set for x-polarization.

Now this all leads to an interesting situation. Suppose you were to set up something like a piece of calcite which separated the photons into x-polarized and y-polarized beams, and put a counter in each beam. Let's call one the x-counter and the other the y-counter. If the guy on the other side does the same thing, you can always tell him which beam his photon is going to go into. Whenever you and he get simultaneous counts, you can see which of your detectors caught the photon and then tell him which of his counters had a photon. Let's say that in a certain disintegration you find that a photon went into your x-counter; you can tell him that he must have had a count in his y-counter.

Now many people who learn quantum mechanics in the usual (old-fashioned) way find this disturbing. They would like to think that once the photons are emitted it goes along as a wave with a definite character. They would think that since "any given photon" has some "amplitude" to be x-polarized or to be y-polarized, there should be some chance of picking it up in either the x- or y-counter and that this chance shouldn't depend on what some other person finds out about a completely different photon. They argue that "someone else making a measurement shouldn't be able to change the probability that I will find something." Our quantum mechanics says, however, that by making a measurement on photon number one, you *can* predict precisely what the polarization of photon number two is going to be when it is detected. This point was never accepted by Einstein, and he worried about it a great deal—it became known as the "Einstein-Podalsky-Rosen paradox." But when the situation is described as we have done it here, there doesn't seem to be any paradox at all; it comes out quite naturally that what is measured in one place is correlated with what is measured somewhere else. The argument that the result is paradoxical runs something like this:

(1) If you have a counter which tells you whether your photon is RHC or LHC, you can predict exactly what kind of a photon (RHC or LHC) he will find.
(2) The photons he receives must, therefore, each be purely RHC or purely LHC, some of one kind and some of the other.
(3) Surely you cannot alter the physical nature of *his* photons by changing the kind of observation you make on *your* photons. No matter what measurements you make on yours, his must still be either RHC or LHC.

---

† We have not normalized our amplitudes, or multiplied them by the amplitude for the disintegration into any particular final state, but we can see that this result is correct because we get zero probability when we look at the other alternative—see Eq. (18.23).

(4) Now suppose he changes his apparatus to split his photons into two linearly polarized beams with a piece of calcite so that all of his photons go either into an *x*-polarized beam or into a *y*-polarized beam. There is absolutely no way, according to quantum mechanics, to tell into which beam any particular RHC photon will go. There is a 50% probability it will go into the *x*-beam and a 50% probability it will go into the *y*-beam. And the same goes for a LHC photon.

(5) Since each photon is RHC or LHC—according to (2) and (3)—each one must have a 50–50 chance of going into the *x*-beam or the *y*-beam and there is no way to predict which way it will go.

(6) Yet the theory predicts that if *you* see your photon go through an *x*-polarizer you can predict *with certainty* that his photon will go into his *y*-polarized beam. This is in contradiction to (5) so there is a paradox.

Nature apparently doesn't see the "paradox," however, because experiment shows that the prediction in (6) is, in fact, true. We have already discussed the key to this "paradox" in our very first lecture on quantum mechanical behavior in Chapter 35, Vol. I. In the argument above, steps (1), (2), (4), and (6) are all correct, but (3), and its consequence (5), are wrong; they are not a true description of nature. Argument (3) says that by *your* measurement (seeing a RHC or a LHC photon) you can determine which of two alternative events occurs for him (seeing a RHC or a LHC photon), *and* that even if you do *not* make your measurement you can still say that his event will occur either by one alternative or the other. But it was precisely the point of Chapter 35, Vol. I, to point out right at the beginning that this is not so in Nature. *Her* way requires a description in terms of interfering amplitudes, one amplitude for each alternative. A measurement of which alternative actually occurs destroys the interference, but if a measurement is *not* made you cannot still say that "one alternative or the other is still occurring."

If you could determine for each one of your photons whether it was RHC and LHC, and *also* whether it was *x*-polarized (all for the same photon) there would indeed be a paradox. But you cannot do that—it is an example of the uncertainty principle.

Do you still think there is a "paradox"? Make sure that it is, in fact, a paradox about the behavior of Nature, by setting up an imaginary experiment for which the theory of quantum mechanics would predict inconsistent results via two different arguments. Otherwise the "paradox" is only a conflict between reality and your feeling of what reality "ought to be."

Do you think that it is *not* a "paradox," but that it is still very peculiar? On that we can all agree. It is what makes physics fascinating.

18–4 Rotation matrix for any spin

By now you can see, we hope, how important the idea of the angular momentum is in understanding atomic processes. So far, we have considered only systems with spins—or "total angular momentum"—of zero, one-half, or one. There are, of course, atomic systems with higher angular momenta. For analyzing such systems we would need to have tables of rotation amplitudes like those in Section 17–6. That is, we would need the matrix of amplitudes for spin $\frac{3}{2}$, 2, $\frac{5}{2}$, 3, etc. Although we will not work out these tables in detail, we would like to show you how it is done, so that you can do it if you ever need to.

As we have seen earlier, any system which has the spin or "total angular momentum" j can exist in any one of $(2j + 1)$ states for which the z-component of angular momentum can have any one of the discrete values in the sequence j, $j - 1, j - 2, \ldots, -(j - 1), -j$ (all in units of $\hbar$). Calling the z-component of angular momentum of any particular state $m\hbar$, we can define a particular angular momentum state by giving the numerical values of the two "angular momentum quantum numbers" j and m. We can indicate such a state by the state vector $|j, m\rangle$. In the case of a spin one-half particle, the two states are then $|\frac{1}{2}, \frac{1}{2}\rangle$ and $|\frac{1}{2}, -\frac{1}{2}\rangle$; or for a spin-one system, the states would be written in this notation as $|1, +1\rangle, |1, 0\rangle, |1, -1\rangle$. A spin-zero particle has, of course, only the one state $|0, 0\rangle$.

Now we want to know what happens when we project the general state $|j, m\rangle$ into a representation with respect to a rotated set of axes. First, we know that j is a number which characterizes *the system*, so it doesn't change. If we rotate the axes, all we do is get a mixture of the various m-values for the same j. In general, there will be some amplitude that in the rotated frame the system will be in the state $|j, m'\rangle$, where m' gives the new z-component of angular momentum. So what we want are all the matrix elements $\langle j, m' | R | j, m\rangle$ for various rotations. We already know what happens if we rotate by an angle ϕ about the z-axis. The new state is just the old one multiplied by $e^{im\phi}$—it still has the same m-value. We can write this by

$$R_z(\phi)|j, m\rangle = e^{im\phi}|j, m\rangle. \tag{18.24}$$

Or, if you prefer,

$$\langle j, m' | R_z(\phi) | j, m\rangle = \delta_{m,m'}e^{im\phi} \tag{18.25}$$

(where $\delta_{m,m'}$ is 1 if $m' = m$, or zero otherwise).

For a rotation about any other axis there will be a mixing of the various m-states. We could, of course, try to work out the matrix elements for an arbitrary rotation described by the Euler angles β, α, and γ. But it is easier to remember that the most general such rotation can be made up of the three rotations $R_z(\gamma)$, $R_y(\alpha)$, $R_z(\beta)$; so if we know the matrix elements for a rotation about the y-axis, we will have all we need.

How can we find the rotation matrix for a rotation by the angle θ about the y-axis for a particle of spin j? We can't tell you how to do it in a basic way (with what we have had). We did it for spin one-half by a complicated symmetry argument. We then did it for spin one by taking the special case of a spin-one system which was made up of two spin one-half particles. If you will go along with us and accept the fact that in the general case the answers depend only on the spin j, and are independent of how the inner guts of the object of spin j are put together, we can extend the spin-one argument to an arbitrary spin. We can, for example, cook up an artificial system of spin $\frac{3}{2}$ out of three spin one-half objects. We can even avoid complications by imagining that they are all distinct particles—like a proton, an electron, and a muon. By transforming each spin one-half object, we can see what happens to the whole system—remembering that the three amplitudes are multiplied for the combined state. Let's see how it goes in this case.

Suppose we take the three spin one-half objects all with spins "up"; we can indicate this state by $|+ + +\rangle$. If we look at this system in a frame rotated about the z-axis by the angle ϕ, each plus stays a plus, but gets multiplied by $e^{i\phi/2}$. We have three such factors, so

$$R_z(\phi)|+ + +\rangle = e^{i(3\phi/2)}|+ + +\rangle. \tag{18.26}$$

Evidently the state $|+ + +\rangle$ is just what we mean by the $m = +\frac{3}{2}$ state, or the state $|\frac{3}{2}, +\frac{3}{2}\rangle$.

If we now rotate this system about the y-axis, each of the spin one-half objects will have some amplitude to be plus or to be minus, so the system will now be a mixture of the *eight* possible combinations $|+ + +\rangle$, $|+ + -\rangle$, $|+ - +\rangle$, $|- + +\rangle$, $|+ - -\rangle$, $|- + -\rangle$, $|- - +\rangle$, or $|- - -\rangle$. It is clear, however, that these can be broken up into four sets, each set corresponding to a particular value of m. First, we have $|+ + +\rangle$, for which $m = \frac{3}{2}$. Then there are the three states $|+ + -\rangle$, $|+ - +\rangle$, and $|- + +\rangle$—each with two plusses and one minus. Since each spin one-half object has the same chance of coming out minus under the rotation, the amounts of each of these three combinations should be equal. So let's take the combination

$$\frac{1}{\sqrt{3}}\{|+ + -\rangle + |+ - +\rangle + |- + +\rangle\} \tag{18.27}$$

with the factor $1/\sqrt{3}$ put in to normalize the state. If we rotate this state about the z-axis, we get a factor $e^{i\phi/2}$ for each plus, and $e^{-i\phi/2}$ for each minus. Each term in (18.27) is multiplied by $e^{i\phi/2}$, so there is the common factor $e^{i\phi/2}$. This state satisfies our idea of an $m = +\frac{1}{2}$ state; we can conclude that

$$\frac{1}{\sqrt{3}}\{|++-\rangle+|+-+\rangle+|-++\rangle\} = |\tfrac{3}{2},+\tfrac{1}{2}\rangle. \qquad (18.28)$$

Similarly, we can write

$$\frac{1}{\sqrt{3}}\{|+--\rangle+|-+-\rangle+|--+\rangle\} = |\tfrac{3}{2},-\tfrac{1}{2}\rangle, \qquad (18.29)$$

which corresponds to a state with $m = -\frac{1}{2}$. Notice that we take only the *symmetric* combinations—we do not take any combinations with minus signs. They would correspond to states of the same m but a different j. (It's just like the spin-one case, where we found that $(1/\sqrt{2})\{|+-\rangle+|-+\rangle\}$ was the state $|1,0\rangle$, but the state $(1/\sqrt{2})\{|+-\rangle-|-+\rangle\}$ was the state $|0,0\rangle$.) Finally, we would have that

$$|\tfrac{3}{2},-\tfrac{3}{2}\rangle = |---\rangle. \qquad (18.30)$$

We summarize our four states in Table 18–1.

Table 18–1

$$
\begin{array}{ll}
|+++\rangle & = |\tfrac{3}{2},+\tfrac{3}{2}\rangle \\[2mm]
\dfrac{1}{\sqrt{3}}\{|++-\rangle+|+-+\rangle+|-++\rangle\} & = |\tfrac{3}{2},+\tfrac{1}{2}\rangle \\[4mm]
\dfrac{1}{\sqrt{3}}\{|+--\rangle+|-+-\rangle+|--+\rangle\} & = |\tfrac{3}{2},-\tfrac{1}{2}\rangle \\[4mm]
|---\rangle & = |\tfrac{3}{2},-\tfrac{3}{2}\rangle
\end{array}
$$

Now all we have to do is take each state and rotate it about the y-axis and see how much of the other states it gives—using our known rotation matrix for the spin one-half particles. We can proceed in exactly the same way we did for the spin-one case in Section 12–6. (It just takes a little more algebra.) We will follow directly the ideas of Chapter 12, so we won't repeat all the explanations in detail. The states in the system S will be labelled $|\tfrac{3}{2},+\tfrac{3}{2},S\rangle = |+++\rangle$, $|\tfrac{3}{2},+\tfrac{1}{2},S\rangle = (1/\sqrt{3})\{|++-\rangle+|+-+\rangle+|-++\rangle\}$, and so on. The T-system will be one rotated about the y-axis of S by the angle θ. States in T will be labelled $|\tfrac{3}{2},+\tfrac{3}{2},T\rangle$, $|\tfrac{3}{2},+\tfrac{1}{2},T\rangle$, and so on. Of course, $|\tfrac{3}{2},+\tfrac{3}{2},T\rangle$ is the same as $|+'+'+'\rangle$, the primes referring always to the T-system. Similarly, $|\tfrac{3}{2},+\tfrac{1}{2},T\rangle$ will be equal to $(1/\sqrt{3})\{|+'+'-'\rangle+|+'-'+'\rangle+|-'+'+'\rangle\}$, and so on. Each $|+'\rangle$ state in the T-frame comes from both the $|+\rangle$ and $|-\rangle$ states in S via the matrix elements of Table 12–4.

When we have three spin one-half particles, Eq. (12.47) gets replaced by

$$
\begin{aligned}
|+++\rangle = {} & a^3|+'+'+'\rangle + a^2b\{|+'+'-'\rangle+|+'-'+'\rangle+|-'+'+'\rangle\} \\
& + ab^2\{|+'-'-'\rangle+|-'+'-'\rangle+|-'-'+'\rangle\} \\
& + b^3|-'-'-'\rangle. \qquad (18.31)
\end{aligned}
$$

Using the transformation of Table 12–4, we get instead of (12.48) the equation

$$
\begin{aligned}
|\tfrac{3}{2},+\tfrac{3}{2},S\rangle = {} & a^3|\tfrac{3}{2},+\tfrac{3}{2},T\rangle + \sqrt{3}\,a^2b\,|\tfrac{3}{2},+\tfrac{1}{2},T\rangle \\
& + \sqrt{3}\,a^2b\,|\tfrac{3}{2},-\tfrac{1}{2},T\rangle + b^3|\tfrac{3}{2},-\tfrac{3}{2},T\rangle. \qquad (18.32)
\end{aligned}
$$

This already gives us several of our matrix elements $\langle jT \mid iS\rangle$. To get the expression for $(\tfrac{3}{2},+\tfrac{1}{2},S)$ we begin with the transformation of a state with two "+" and one "−" pieces. For instance,

$$|++-\rangle = a^2c \,|+'+'+'\rangle + a^2d \,|+'+'-'\rangle + abc \,|+'-'+'\rangle$$
$$+ bac \,|-'+'+'\rangle + abd \,|+'-'-'\rangle + bad \,|-'+'-'\rangle$$
$$+ b^2c \,|-'-'+'\rangle + b^2d \,|-'-'-'\rangle. \tag{18.33}$$

Adding two similar expressions for $|+-+\rangle$ and $|-++\rangle$ and dividing by $\sqrt{3}$, we find

$$|\tfrac{3}{2},+\tfrac{1}{2},S\rangle = \sqrt{3}\, a^2c \,|\tfrac{3}{2},+\tfrac{3}{2},T\rangle$$
$$+(a^2d + 2abc)\,|\tfrac{3}{2},+\tfrac{1}{2},T\rangle$$
$$+(2bad + b^2c)\,|\tfrac{3}{2},-\tfrac{1}{2},T\rangle$$
$$+\sqrt{3}\, b^2d \,|\tfrac{3}{2},-\tfrac{3}{2},T\rangle. \tag{18.34}$$

Continuing the process we find all the elements $\langle jT \mid iS\rangle$ of the transformation matrix as given in Table 18–2. The first column comes from Eq. (18.32); the second from (18.34). The last two columns were worked out in the same way.

Table 18–2

Rotation matrix for a spin $\tfrac{3}{2}$ particle

(The coefficients a, b, c, and d are given in Table 12–4.)

$\langle jT \mid iS\rangle$	$\lvert\tfrac{3}{2},+\tfrac{3}{2},S\rangle$	$\lvert\tfrac{3}{2},+\tfrac{1}{2},S\rangle$	$\lvert\tfrac{3}{2},-\tfrac{1}{2},S\rangle$	$\lvert\tfrac{3}{2},-\tfrac{3}{2},S\rangle$
$\langle\tfrac{3}{2},+\tfrac{3}{2},T\rvert$	a^3	$\sqrt{3}\, a^2c$	$\sqrt{3}\, ac^2$	c^3
$\langle\tfrac{3}{2},+\tfrac{1}{2},T\rvert$	$\sqrt{3}\, a^2b$	$a^2d + 2abc$	$c^2b + 2dac$	$\sqrt{3}\, c^2d$
$\langle\tfrac{3}{2},-\tfrac{1}{2},T\rvert$	$\sqrt{3}\, ab^2$	$2bad + b^2c$	$2cdb + d^2a$	$\sqrt{3}\, cd^2$
$\langle\tfrac{3}{2},-\tfrac{3}{2},T\rvert$	b^3	$\sqrt{3}\, b^2d$	$\sqrt{3}\, bd^2$	d^3

Now suppose the T-frame were rotated with respect to S by the angle θ about their y-axes. Then a, b, c, and d have the values [see (12.54)] $a = d = \cos\theta/2$, and $c = -b = \sin\theta/2$. Using these values in Table 18–2 we get the forms which correspond to the second part of Table 17–2, but now for a spin $\tfrac{3}{2}$ system.

The arguments we have just gone through are readily generalized to a system of any spin j. The states $|j, m\rangle$ can be put together from $2j$ particles, each of spin one-half. (There are $j + m$ of them in the $|+\rangle$ state and $j - m$ in the $|-\rangle$ state.) Sums are taken over all the possible ways this can be done, and the state is normalized by multiplying by a suitable constant. Those of you who are mathematically inclined may be able to show that the following result comes out†:

$$\langle j, m' \mid R_y(\theta) \mid j, m\rangle = [(j + m)!(j - m)!(j + m')!(j - m')!]^{1/2}$$
$$\times \sum_k \frac{(-1)^k (\cos\theta/2)^{2j+m'-m-2k}(\sin\theta/2)^{m-m'+2k}}{(m - m' + k)!(j + m' - k)!(j - m - k)!k!}, \tag{18.35}$$

where k is to go over all values which give terms ≥ 0 in all the factorials.

This is quite a messy formula, but with it you can check Table 17–2 for $j = 1$ and prepare tables of your own for larger j. Several special matrix elements are of extra importance and have been given special names. For example the matrix elements for $m = m' = 0$ and integral j are known as the Legendre polynomials and are called $P_j (\cos\theta)$:

$$\langle j, 0 \mid R_y(\theta) \mid j, 0\rangle = P_j(\cos\theta). \tag{18.36}$$

---

† If you want details, they are given in an appendix to this chapter.

The first few of these polynomials are:

$$P_0 \, (\cos \theta) \; = \; 1, \tag{18.37}$$

$$P_1 \, (\cos \theta) \; = \; \cos \theta, \tag{18.38}$$

$$P_2 \, (\cos \theta) \; = \; \tfrac{1}{2}(3 \cos^2 \theta - 1), \tag{18.39}$$

$$P_3 \, (\cos \theta) \; = \; \tfrac{1}{2}(5 \cos^3 \theta - 3 \cos \theta). \tag{18.40}$$

18–5 Measuring a nuclear spin

We would like to show you one example of the application of the coefficients we have just described. It has to do with a recent, interesting experiment which you will now be able to understand. Some physicists wanted to find out the spin of a certain excited state of the Ne^{20} nucleus. To do this, they bombarded a carbon target with a beam of accelerated carbon ions, and produced the desired excited state of Ne^{20}—called Ne^{20*}—in the reaction

$$C^{12} + C^{12} \rightarrow Ne^{20*} + \alpha_1,$$

where α_1 is the α-particle, or He^4. Several of the excited states of Ne^{20} produced this way are unstable and disintegrate in the reaction

$$Ne^{20*} \rightarrow O^{16} + \alpha_2.$$

So experimentally there are two α-particles which come out of the reaction. We call them α_1 and α_2; since they come off with different energies, they can be distinguished from each other. Also, by picking a particular energy for α_1 we can pick out any particular excited state of the Ne^{20}.

The experiment was set up as shown in Fig. 18–9. A beam of 16-Mev carbon ions was directed onto a thin foil of carbon. The first α-particle was counted in a silicon diffused junction detector marked α_1—set to accept α-particles of the proper energy moving in the forward direction (with respect to the incident C^{12} beam). The second α-particle was picked up in the counter α_2 at the angle θ with respect to α_1. The counting rate of coincidence signals from α_1 and α_2 were measured as a function of the angle θ.

The idea of the experiment is the following. First, you need to know that the spins of C^{12}, O^{16}, and the α-particle are all zero. If we call the direction of motion of the initial C^{12} the $+z$-direction, then we know that the Ne^{20*} must have zero angular momentum about the z-axis. None of the other particles has any spin; the C^{12} arrives along the z-axis and the α_1 leaves along the z-axis so they can't have any angular momentum about it. So whatever the spin j of the Ne^{20*} is, we know that it is in the state $|j, 0\rangle$. Now what will happen when the Ne^{20*} disintegrates into an O^{16} and the second α-particle? Well, the α-particle is picked up in the counter α_2 and to conserve momentum the O^{16} must go off in the opposite direction.† *About the new axis* through α_2, there can be no component of angular momentum. The final state has zero angular momentum about the new axis, so the Ne^{20*} can disintegrate this way only if it has some amplitude to have m' equal to zero, where m' is the quantum number of the component of angular momentum about the new axis. In fact, the probability of observing α_2 at the angle θ is just the square of the amplitude (or matrix element)

$$\langle j, 0 \mid R_y(\theta) \mid j, 0 \rangle. \tag{18.41}$$

To find the spin of the Ne^{20*} state in question, the intensity of the second α-particle was plotted as a function of angle and compared with the theoretical

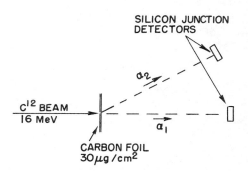

Fig. 18–9. Experimental arrangement for determining the spin of certain states of Ne^{20}.

---

† We can neglect the recoil given to the Ne^{20*} in the first collision. Or better still, we can calculate what it is and make a correction for it.

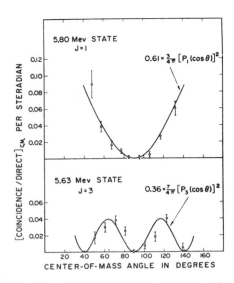

Fig. 18–10. Experimental results for the angular distribution of the α-particles from two excited states of Ne²⁰ produced in the setup of Fig. 18–9. [From J. A. Kuehner, *Physical Review*, Vol. 125, p. 1653, 1962.]

curves for various values of j. As we said in the last section, the amplitudes $\langle j, 0 \mid R_y(\theta) \mid j, 0 \rangle$ are just the functions $P_j(\cos\theta)$. So the possible angular distributions are curves of $[P_j(\cos\theta)]^2$. The experimental results are shown in Fig. 18–10 for two of the excited states. You can see that the angular distribution for the 5.80-Mev state fits very well the curve for $[P_1(\cos\theta)]^2$, and so it must be a spin-one state. The data for the 5.63-Mev state, on the other hand, are quite different; they fit the curve $[P_3(\cos\theta)]^2$. The state has a spin of 3.

From this experiment we have been able to find out the angular momentum of two of the excited states of Ne²⁰*. This information can then be used for trying to understand what the configuration of protons and neutrons is inside this nucleus—one more piece of information about the mysterious nuclear forces.

18–6 Composition of angular momentum

When we studied the hyperfine structure of the hydrogen atom in Chapter 12 we had to work out the internal states of a system composed of two particles—the electron and the proton—each with a spin of one-half. We found that the four possible spin states of such a system could be put together into two groups—a group with one energy that looked to the external world like a spin-one particle, and one remaining state that behaved like a particle of zero spin. That is, putting together two spin one-half particles we can form a system whose "total spin" is one, or zero. In this section we want to discuss in more general terms the spin states of a *system* which is made up of two particles of arbitrary spin. It is another important problem about angular momentum in quantum mechanical systems.

Let's first rewrite the results of Chapter 12 for the hydrogen atom in a form that will be easier to extend to the more general case. We began with two particles which we will now call particle a (the electron) and particle b (the proton). Particle a had the spin j_a $(=\frac{1}{2})$, and its z-component of angular momentum m_a could have one of several values (actually 2, namely $m_a = +\frac{1}{2}$ or $m_a = -\frac{1}{2}$). Similarly, the spin state of particle b is described by its spin j_b and its z-component of angular momentum m_b. Various combinations of the spin states of the two particles could be formed. For instance, we could have particle a with $m_a = \frac{1}{2}$ and particle b with $m_b = -\frac{1}{2}$, to make a state $\mid a, +\frac{1}{2}; b, -\frac{1}{2}\rangle$. In general, the combined states formed a system whose "system spin," or "total spin," or "total angular momentum" J could be 1, or 0. And the system could have a z-component of angular momentum M, which was $+1$, 0, or -1 when $J = 1$, or 0 when $J = 0$. In this new language we can rewrite the formulas in (12.41) and (12.42) as shown in Table 18–3.

In the table the left-hand column describes the compound state in terms of its total angular momentum J and the z-component M. The right-hand column shows how these states are made up in terms of the m-values of the two particles a and b.

We want now to generalize this result to states made up of two objects a and b of arbitrary spins j_a and j_b. We start by considering an example for which $j_a = \frac{1}{2}$

Table 18–3

Composition of angular momenta for two spin $\frac{1}{2}$ particles $(j_a = \frac{1}{2}, j_b = \frac{1}{2})$

$$\mid J = 1, M = +1\rangle = \mid a, +\tfrac{1}{2}; b, +\tfrac{1}{2}\rangle$$

$$\mid J = 1, M = 0\rangle = \frac{1}{\sqrt{2}}\{\mid a, +\tfrac{1}{2}; b, -\tfrac{1}{2}\rangle + \mid a, -\tfrac{1}{2}; b, +\tfrac{1}{2}\rangle\}$$

$$\mid J = 1, M = -1\rangle = \mid a, -\tfrac{1}{2}; b, -\tfrac{1}{2}\rangle$$

$$\mid J = 0, M = 0\rangle = \frac{1}{\sqrt{2}}\{\mid a, +\tfrac{1}{2}; b, -\tfrac{1}{2}\rangle - \mid a, -\tfrac{1}{2}; b, +\tfrac{1}{2}\rangle\}$$

and $j_b = 1$, namely, the deuterium atom in which particle a is an electron (e) and particle b is the nucleus—a deuteron (d). We have then that $j_a = j_e = \frac{1}{2}$. The deuteron is formed of one proton and one neutron in a state whose total spin is one, so $j_b = j_d = 1$. We want to discuss the hyperfine states of deuterium—just the way we did for hydrogen. Since the deuteron has three possible states $m_b = m_d = +1, 0, -1$, and the electron has two, $m_a = m_e = +\frac{1}{2}, -\frac{1}{2}$, there are *six* possible states as follows (using the notation $|\,e, m_e; d, m_d\rangle$):

$$
\begin{aligned}
&|\,e,+\tfrac{1}{2}; d,+1\rangle, \\
&|\,e,+\tfrac{1}{2}; d,0\rangle; |\,e,-\tfrac{1}{2}; d,+1\rangle, \\
&|\,e,+\tfrac{1}{2}; d,-1\rangle; |e,-\tfrac{1}{2}; d,0\rangle, \\
&|\,e,-\tfrac{1}{2}; d,-1\rangle.
\end{aligned}
\tag{18.42}
$$

You will notice that we have grouped the states according to the values of the sum of m_e and m_d—arranged in descending order.

Now we ask: What happens to these states if we project into a different coordinate system? If the new system is just rotated about the z-axis by the angle ϕ, then the state $|\,e, m_e; d, m_d\rangle$ gets multiplied by

$$
e^{im_e\phi}e^{im_d\phi} = e^{i(m_e+m_d)\phi}.
\tag{18.43}
$$

(The state may be thought of as the product $|\,e, m_e\rangle\,|\,d, m_d\rangle$, and each state vector contributes independently its own exponential factor.) The factor (18.43) is of the form $e^{iM\phi}$, so the state $|\,e, m_e; d, m_d\rangle$ has a z-component of angular momentum equal to

$$
M = m_e + m_d.
\tag{18.44}
$$

The z-component of the total angular momentum is the sum of the z-components of angular momentum of the parts.

In the list of (18.42), therefore, the state in the top line has $M = +\frac{3}{2}$, the two in the second line have $M = +\frac{1}{2}$, the next two have $M = -\frac{1}{2}$, and the last state has $M = -\frac{3}{2}$. We see immediately one possibility for the spin J of the combined state (the total angular momentum) must be $\frac{3}{2}$, and this will require four states with $M = +\frac{3}{2}, +\frac{1}{2}, -\frac{1}{2}$, and $-\frac{3}{2}$.

There is only one candidate for $M = \frac{3}{2}$, so we know already that

$$
|\,J = \tfrac{3}{2}, M = +\tfrac{3}{2}\rangle = |\,e,+\tfrac{1}{2}; d, +1\rangle.
\tag{18.45}
$$

But what is the state $|\,J = \frac{3}{2}, M = \frac{1}{2}\rangle$? We have two candidates in the second line of (18.42), and, in fact, any linear combination of them would also have $M = \frac{1}{2}$. So, in general, we must expect to find that

$$
|\,J = \tfrac{3}{2}, M = +\tfrac{1}{2}\rangle = \alpha\,|\,e,+\tfrac{1}{2}; d,0\rangle + \beta\,|\,e,-\tfrac{1}{2}; d,+1\rangle,
\tag{18.46}
$$

where α and β are two numbers. They are called the *Clebsch-Gordon coefficients*. Our next problem is to find out what they are.

We can find out easily if we just remember that the deuteron is made up of a neutron and a proton, and write the deuteron states out more explicitly using the rules of Table 18–3. If we do that, the states listed in (18.42) then look as shown in Table 18–4.

We want to form the four states of $J = \frac{3}{2}$, using the states in the table. But we already know the answer, because in Table 18–1 we have states of spin $\frac{3}{2}$ formed from three spin one-half particles. The first state in Table 18–1 has $|\,J = \frac{3}{2}, M = +\frac{3}{2}\rangle$ and it is $|\,{+}\,{+}\,{+}\rangle$, which—in our present notation—is the same as $|\,e,+\frac{1}{2}; n,+\frac{1}{2}, p,+\frac{1}{2}\rangle$, or the first state in Table 18–4. But this state is also the same as the first in the list of (18.42), confirming our statement in (18.45). The second line of Table 18–1 says—changing to our present notation—that

$$
\begin{aligned}
|\,J = \tfrac{3}{2}; M = +\tfrac{1}{2}\rangle = \frac{1}{\sqrt{3}}\,\{&|\,e,+\tfrac{1}{2}; n,+\tfrac{1}{2}; p,-\tfrac{1}{2}\rangle \\
+ &|\,e,+\tfrac{1}{2}; n,-\tfrac{1}{2}; p,+\tfrac{1}{2}\rangle + |\,e,-\tfrac{1}{2}; n,+\tfrac{1}{2}; p,+\tfrac{1}{2}\rangle\}.
\end{aligned}
\tag{18.47}
$$

Table 18-4

Angular momentum states for a deuterium atom

$$m = \tfrac{3}{2}$$

$$| e,+\tfrac{1}{2}; d,+1 \rangle = | e,+\tfrac{1}{2}; n,+\tfrac{1}{2}; p,+\tfrac{1}{2} \rangle$$

$$m = \tfrac{1}{2}$$

$$| e,+\tfrac{1}{2}; d,0 \rangle = \frac{1}{\sqrt{2}} \{| e,+\tfrac{1}{2}; n,+\tfrac{1}{2}; p,-\tfrac{1}{2} \rangle + | e,+\tfrac{1}{2}; n,-\tfrac{1}{2}; p,+\tfrac{1}{2} \rangle\}$$

$$| e,-\tfrac{1}{2}; d,+1 \rangle = | e,-\tfrac{1}{2}; n,+\tfrac{1}{2}; p,+\tfrac{1}{2} \rangle$$

$$m = -\tfrac{1}{2}$$

$$| e,+\tfrac{1}{2}; d,-1 \rangle = | e,+\tfrac{1}{2}; n,-\tfrac{1}{2}; p,-\tfrac{1}{2} \rangle$$

$$| e,-\tfrac{1}{2}; d,0 \rangle = \frac{1}{\sqrt{2}} \{| e,-\tfrac{1}{2}; n,+\tfrac{1}{2}; p,-\tfrac{1}{2} \rangle + | e,-\tfrac{1}{2}; n,-\tfrac{1}{2}; p,+\tfrac{1}{2} \rangle\}$$

$$m = -\tfrac{3}{2}$$

$$| e,-\tfrac{1}{2}; d,-1 \rangle = | e,-\tfrac{1}{2}; n,-\tfrac{1}{2}; p,-\tfrac{1}{2} \rangle$$

The right side can evidently be put together from the two entries in the second line of Table 18-4 by taking $\sqrt{2/3}$ of the first term with $\sqrt{1/3}$ of the second. That is, Eq. (18-47) is equivalent to

$$| J = \tfrac{3}{2}, M = \tfrac{1}{2} \rangle = \sqrt{2/3}\, | e,+\tfrac{1}{2}; d,0 \rangle + \sqrt{1/3}\, | e,-\tfrac{1}{2}; d,1 \rangle. \qquad (18.48)$$

We have found our two Clebsch-Gordon coefficients α and β in Eq. (18.46):

$$\alpha = \sqrt{2/3}, \qquad \beta = \sqrt{1/3}. \qquad (18.49)$$

Following the same procedure we can find that

$$| J = \tfrac{3}{2}, M = -\tfrac{1}{2} \rangle = \sqrt{1/3}\, | e,+\tfrac{1}{2}; d,-1 \rangle + \sqrt{2/3}\, | e,-\tfrac{1}{2}; d,0 \rangle. \qquad (18.50)$$

And, also, of course,

$$| J = \tfrac{3}{2}, M = -\tfrac{3}{2} \rangle = | e,-\tfrac{1}{2}; d,-1 \rangle. \qquad (18.51)$$

These are the rules for the composition of spin 1 and spin $\tfrac{1}{2}$ to make a total $J = \tfrac{3}{2}$. We summarize (18.45), (18.48), and (18.50) in Table 18-5.

We have, however, only four states here while the system we are considering has six possible states. Of the two states in the second line of (18.42) we have used only one linear combination to form $| J = \tfrac{3}{2}, M = +\tfrac{1}{2} \rangle$. There is another linear combination orthogonal to the one we have taken which also has $M = +\tfrac{1}{2}$, namely

$$\sqrt{1/3}\, | e,+\tfrac{1}{2}; d,0 \rangle - \sqrt{2/3}\, | e,-\tfrac{1}{2}; d,+1 \rangle. \qquad (18.52)$$

Table 18-5

The $J = \tfrac{3}{2}$ states of the deuterium atom

$$| J = \tfrac{3}{2}, M = +\tfrac{3}{2} \rangle = | e,+\tfrac{1}{2}; d,+1 \rangle$$

$$| J = \tfrac{3}{2}, M = +\tfrac{1}{2} \rangle = \sqrt{2/3}\, | e,+\tfrac{1}{2}; d,0 \rangle + \sqrt{1/3}\, | e,-\tfrac{1}{2}; d,1 \rangle$$

$$| J = \tfrac{3}{2}, M = -\tfrac{1}{2} \rangle = \sqrt{1/3}\, | e,+\tfrac{1}{2}; d,-1 \rangle + \sqrt{2/3}\, | e,-\tfrac{1}{2}; d,0 \rangle$$

$$| J = \tfrac{3}{2}, M = -\tfrac{3}{2} \rangle = | e,-\tfrac{1}{2}; d,-1 \rangle$$

Similarly, the two states in the third line of (18.42) can be combined to give two orthogonal states, each with $M = -\frac{1}{2}$. The one orthogonal to (18.52) is

$$\sqrt{2/3}\,|\,e,+\tfrac{1}{2};\,d,-1\rangle - \sqrt{1/3}\,|\,e,-\tfrac{1}{2};\,d,0\rangle. \qquad (18.53)$$

These are the two remaining states. They have $M = m_e + m_d = \pm\frac{1}{2}$; and must be the two states corresponding to $J = \frac{1}{2}$. So we have

$$|\,J = \tfrac{1}{2},\,M = \tfrac{1}{2}\rangle = \sqrt{1/3}\,|\,e,+\tfrac{1}{2};\,d,0\rangle - \sqrt{2/3}\,|\,e,-\tfrac{1}{2};\,d,+1\rangle,$$
$$\qquad\qquad\qquad\qquad\qquad\qquad\qquad\qquad\qquad\qquad (18.54)$$
$$|\,J = \tfrac{1}{2},\,M = -\tfrac{1}{2}\rangle = \sqrt{2/3}\,|\,e,+\tfrac{1}{2};\,d,-1\rangle - \sqrt{1/3}\,|\,e,-\tfrac{1}{2};\,d,0\rangle.$$

We can verify that these two states do indeed behave like the states of a spin one-half object by writing out the deuterium parts in terms of the neutron and proton states—using Table 18–3. The first state in (18.53) is

$$\sqrt{1/6}\{|\,e,+\tfrac{1}{2};\,n,+\tfrac{1}{2};\,p,-\tfrac{1}{2}\rangle + |\,e,+\tfrac{1}{2};\,n,-\tfrac{1}{2};\,p,+\tfrac{1}{2}\rangle\}$$
$$- \sqrt{2/3}\,|\,e,-\tfrac{1}{2};\,n,+\tfrac{1}{2};\,p,+\tfrac{1}{2}\rangle, \qquad (18.55)$$

which can also be written

$$\sqrt{1/3}[\sqrt{1/2}\,\{|\,e,+\tfrac{1}{2};\,n,+\tfrac{1}{2};\,p,-\tfrac{1}{2}\rangle - |\,e,-\tfrac{1}{2};\,n,+\tfrac{1}{2};\,p,+\tfrac{1}{2}\rangle\}$$
$$+ \sqrt{1/2}\,\{|\,e,+\tfrac{1}{2};\,n,-\tfrac{1}{2};\,p,+\tfrac{1}{2}\rangle - |\,e,-\tfrac{1}{2};\,n,+\tfrac{1}{2};\,p,+\tfrac{1}{2}\rangle\}].$$
$$\qquad\qquad\qquad\qquad\qquad\qquad\qquad\qquad\qquad\qquad (18.56)$$

Now look at the terms in the first curly brackets, and think of the e and p taken together. Together they form a spin-zero state (see the bottom line of Table 18–3), and contribute no angular momentum. Only the neutron is left, so the whole of the *first* curly bracket of (18.56) behaves under rotations like a neutron, namely as a state with $J = \frac{1}{2}$, $M = +\frac{1}{2}$. Following the same reasoning, we see that in the *second* curly bracket of (18.56) the electron and neutron team up to produce zero angular momentum, and only the proton contribution—with $m_p = \frac{1}{2}$—is left. The terms behave like an object with $J = \frac{1}{2}$, $M = +\frac{1}{2}$. So the whole expression of (18.56) transforms like $|\,J = +\frac{1}{2},\,M = +\frac{1}{2}\rangle$ as it should. The $M = -\frac{1}{2}$ state which corresponds to (18.57) can be written down (by changing the proper $+\frac{1}{2}$'s to $-\frac{1}{2}$'s) to get

$$\sqrt{1/3}[\sqrt{1/2}\,\{|\,e,+\tfrac{1}{2};\,n,-\tfrac{1}{2};\,p,-\tfrac{1}{2}\rangle - |\,e,-\tfrac{1}{2};\,n,-\tfrac{1}{2};\,p,+\tfrac{1}{2}\rangle\}$$
$$+ \sqrt{1/2}\,\{|\,e,+\tfrac{1}{2};\,n,-\tfrac{1}{2};\,p,-\tfrac{1}{2}\rangle - |\,e,-\tfrac{1}{2};\,n,+\tfrac{1}{2};\,p,-\tfrac{1}{2}\rangle\}].$$
$$\qquad\qquad\qquad\qquad\qquad\qquad\qquad\qquad\qquad\qquad (18.57)$$

You can easily check that this is equal to the second line of (18.54), as it should be if the two terms of that pair are to be the two states of a spin one-half system. So our results are confirmed. A deuteron and an electron can exist in six spin states, four of which act like the states of a spin $\frac{3}{2}$ object (Table 18–5) and two of which act like an object of spin one-half (18.54).

The results of Table 18–5 and of Eq. (18.54) were obtained by making use of the fact that the deuteron is made up of a neutron and a proton. The truth of the equations does not depend on that special circumstance. For *any* spin-one object put together with any spin one-half object the composition laws (and the coefficients) are the same. The set of equations in Table 18–5 means that if the coordinates are rotated about, say, the y-axis—so that the states of the spin one-half particle and of the spin-one particle change according to Table 18–1 and Table 18–2—the linear combinations on the right-hand side will change in the proper way for a spin $\frac{3}{2}$ object. Under the same rotation the states of (18.54) will change as the states of a spin one-half object. The results depend only on the

Table 18–6

**Composition of a spin one-half particle $\left(j_a = \frac{1}{2}\right)$
and a spin-one particle $\left(j_b = 1\right)$.**

$$
\begin{aligned}
\left| J = \tfrac{3}{2}, M = \tfrac{3}{2} \right\rangle &= \left| a, +\tfrac{1}{2}; b, +1 \right\rangle \\[4pt]
\left| J = \tfrac{3}{2}, M = \tfrac{1}{2} \right\rangle &= \sqrt{2/3}\,\left| a, +\tfrac{1}{2}; b, 0 \right\rangle + \sqrt{1/3}\,\left| a, -\tfrac{1}{2}; b, +1 \right\rangle \\[4pt]
\left| J = \tfrac{3}{2}, M = -\tfrac{1}{2} \right\rangle &= \sqrt{1/3}\,\left| a, +\tfrac{1}{2}; b, -1 \right\rangle + \sqrt{2/3}\,\left| a, -\tfrac{1}{2}; b, 0 \right\rangle \\[4pt]
\left| J = \tfrac{3}{2}, M = -\tfrac{3}{2} \right\rangle &= \left| a, -\tfrac{1}{2}; b, -1 \right\rangle \\[10pt]
\left| J = \tfrac{1}{2}, M = +\tfrac{1}{2} \right\rangle &= \sqrt{1/3}\,\left| a, +\tfrac{1}{2}; b, 0 \right\rangle - \sqrt{2/3}\,\left| a, -\tfrac{1}{2}; b, +1 \right\rangle \\[4pt]
\left| J = \tfrac{1}{2}, M = -\tfrac{1}{2} \right\rangle &= \sqrt{2/3}\,\left| a, +\tfrac{1}{2}; b, -1 \right\rangle - \sqrt{1/3}\,\left| a, -\tfrac{1}{2}; b, 0 \right\rangle
\end{aligned}
$$

rotation properties (that is, the spin states) of the two original particles but not in any way on the origins of their angular momenta. We have only made use of this fact to work out the formulas by choosing a special case in which one of the component parts is itself made up of two spin one-half particles in a symmetric state. We have put all our results together in Table 18–6, changing the notation "e" and "d" to "a" and "b" to emphasize the generality of the conclusions.

Suppose we have the general problem of finding the states which can be formed when two objects of arbitrary spins are combined. Say one has j_a (so its z-component m_a runs over the $2j_a + 1$ values from $-j_a$ to $+j_a$) and the other has j_b (with z-component m_b running over the values from $-j_b$ to $+j_b$). The combined states are $\left| a, m_a; b, m_a \right\rangle$, and there are $(2j_a + 1)(2j_b + 1)$ different ones. Now what states of total spin J can be found?

The total z-component of angular momentum M is equal to $m_a + m_b$, and the states can all be listed according to M [as in (18.42)]. The largest M is unique; it corresponds to $m_a = j_a$ and $m_b = j_b$, and is, therefore, just $j_a + j_b$. That means that the largest total spin J is also equal to the sum $j_a + j_b$:

$$ J = (M)_{\text{max}} = j_a + j_b. $$

For the first M value smaller than $(M)_{\text{max}}$, there are two states (either m_a or m_b is one unit less than its maximum). They must contribute one state to the set that goes with $J = j_a + j_b$, and the one left over will belong to a new set with $J = j_a + j_b - 1$. The next M-value—the third from the top of the list—can be formed in *three* ways. (From $m_a = j_a - 2$, $m_b = j_b$; from $m_a = j_a - 1$, $m_b = j_b - 1$; and from $m_a = j_a$, $m_b = j_b - 2$.) Two of these belong to groups already started above; the third tells us that states of $J = j_a + j_b - 2$ must also be included. This argument continues until we reach a stage where in our list we can no longer go one more step down in one of the m's to make new states.

Let j_b be the smaller of j_a and j_b (if they are equal take either one); then only $2j_b$ values of J are required—going in integer steps from $j_a + j_b$ down to $j_a - j_b$. That is, when two objects of spin j_a and j_b are combined, the system can have a total angular momentum J equal to any one of the values

$$ J = \begin{cases} j_a + j_b \\ j_a + j_b - 1 \\ j_a + j_b - 2 \\ \vdots \\ |j_a - j_b|. \end{cases} \tag{18.58} $$

(By writing $|j_a - j_b|$ instead of $j_a - j_b$ we can avoid the extra admonition that $j_a \geq j_b$.)

For *each* of these J values there are the $2J + 1$ states of different M-values—with M going from $+J$ to $-J$. Each of these is formed from linear combinations of the original states $\left| a, m_a; b, m_b \right\rangle$ with appropriate factors—the Clebsch-Gordon coefficients for each particular term. We can consider that these coefficients give the "amount" of the state $|j_a, m_a; j_b, m_b\rangle$ which appears in the state $|J, M\rangle$. So each of the Clebsch-Gordon coefficients has, if you wish, *six* indices identifying its position in the formulas like those of Tables 18–3 and 18–6. That is, calling these coefficients $C(J, M; j_a, m_a; j_b, m_b)$, we could express the equality of the second line of Table 18–6 by writing

$$C(\tfrac{3}{2}, +\tfrac{1}{2}; \tfrac{1}{2}, +\tfrac{1}{2}; 1, 0) = \sqrt{2/3},$$

$$C(\tfrac{3}{2}, +\tfrac{1}{2}; \tfrac{1}{2}, -\tfrac{1}{2}; 1, +1) = \sqrt{1/3}.$$

We will not calculate here the coefficients for any other special cases.† You can, however, find tables in many books. You might wish to try another special case for yourself. The next one to do would be the composition of two spin-one particles. We give just the final result in Table 18–7.

These laws of the composition of angular momenta are very important in particle physics—where they have innumerable applications. Unfortunately, we have no time to look at more examples here.

Table 18–7

Composition of two spin-one particles ($j_a = 1, j_b = 1$)

$|J = 2, M = +2\rangle = |a, +1; b, +1\rangle$

$|J = 2, M = +1\rangle = \dfrac{1}{\sqrt{2}} |a, +1; b, 0\rangle + \dfrac{1}{\sqrt{2}} |a, 0; b, +1\rangle$

$|J = 2, M = 0\rangle = \dfrac{1}{\sqrt{6}} |a, +1; b, -1\rangle + \dfrac{1}{\sqrt{6}} |a, -1; b, +1\rangle + \dfrac{2}{\sqrt{6}} |a, 0; b, 0\rangle$

$|J = 2, M = -1\rangle = \dfrac{1}{\sqrt{2}} |a, 0; b, -1\rangle + \dfrac{1}{\sqrt{2}} |a, -1; b, 0\rangle$

$|J = 2, M = -2\rangle = |a, -1; b, -1\rangle$

$|J = 1, M = +1\rangle = \dfrac{1}{\sqrt{2}} |a, +1; b, 0\rangle - \dfrac{1}{\sqrt{2}} |a, 0; b, +1\rangle$

$|J = 1, M = 0\rangle = \dfrac{1}{\sqrt{2}} |a, +1; b, -1\rangle - \dfrac{1}{\sqrt{2}} |a, -1; b, +1\rangle$

$|J = 1, M = -1\rangle = \dfrac{1}{\sqrt{2}} |a, 0; b, -1\rangle - \dfrac{1}{\sqrt{2}} |a, -1; b, 0\rangle$

$|J = 0, M = 0\rangle = \dfrac{1}{\sqrt{3}} \{|a, +1; b, -1\rangle + |a, -1; b, +1\rangle - |a, 0; b, 0\rangle\}$

Added Note 1: Derivation of the rotation matrix‡

For those who would like to see the details, we work out here the general rotation matrix for a system with spin (total angular momentum) j. It is really not very important to work out the general case; once you have the idea, you can find the general results in tables in many books. On the other hand, after coming this far you might like to see that you can indeed understand even the very complicated formulas of quantum mechanics, such as Eq. (18.35), that come into the description of angular momentum.

---

† A large part of the work is done now that we have the general rotation matrix Eq. (18.35).

‡ The material of this appendix was originally included in the body of the lecture. We now feel that it is unnecessary to include such a detailed treatment of the general case.

We extend the arguments of Section 18–4 to a system with spin j, which we consider to be made up of $2j$ spin one-half objects. The state with $m = j$ would be $|+++\cdots+\rangle$ (with j plus signs). For $m = j - 1$, there will be $2j$ terms like $|++\cdots++-\rangle, |++\cdots+-+\rangle$, and so on. Let's consider the general case in which there are r plusses and s minuses—with $r + s = 2j$. Under a rotation about the z-axis each of the r plusses will contribute $e^{+i\phi/2}$. The result is a phase change of $i(r/2 - s/2)\phi$. You see that

$$m = \frac{r - s}{2}. \tag{18.59}$$

Just as for $J = \frac{3}{2}$, each state of definite m must be the linear combination with plus signs of all the states with the same r and s—that is, states corresponding to every possible arrangement which has r plusses and s minuses. We assume that you can figure out that there are $(r + s)!/r!s!$ such arrangements. To normalize each state, we should divide the sum by the square root of this number. We can write

$$\left[\frac{(r + s)!}{r!s!}\right]^{-1/2} \underbrace{\{|+++\cdots++}_{r}\underbrace{---\cdots--\rangle}_{s}$$

$$+ \text{ (all rearrangements of order)}\} = |j, m\rangle \tag{18.60}$$

with

$$j = \frac{r + s}{2}, \qquad m = \frac{r - s}{2}. \tag{18.61}$$

It will help our work if we now go to still another notation. Once we have defined the states by Eq. (18.60), the two numbers r and s define a state just as well as j and m. It will help us keep track of things if we write

$$|j, m\rangle = |{}^r_s\rangle, \tag{18.62}$$

where, using the equalities of (18.67)

$$r = j + m, \qquad s = j - m.$$

Next, we would like to write Eq. (18.60) with a new *special notation* as

$$|j, m\rangle = |{}^r_s\rangle = \left[\frac{(r + s)!}{r!s!}\right]^{+1/2}\{|+\rangle^r\,|-\rangle^s\}_{\text{perm}}. \tag{18.63}$$

Note that we have changed the exponent of the factor in front to *plus* $\frac{1}{2}$. We do that because there are just $N = (r + s)!/r!s!$ terms inside the curly brackets. Comparing (18.63) with (18.60) it is clear that

$$\{|+\rangle^r\,|-\rangle^s\}\ _{\text{perm}}$$

is just a shorthand way of writing

$$\frac{\{|++\cdots--\rangle + \text{ all rearrangements}\}}{N},$$

where N is the number of different terms in the bracket. The reason that this notation is convenient is that each time we make a rotation, all of the plus signs contribute the same factor, so we get this factor to the rth power. Similarly, all together the s minus terms contribute a factor to the sth power no matter what the sequence of the terms is.

Now suppose we rotate our system by the angle θ about the y-axis. What we want is $R_y(\theta)\,|{}^r_s\rangle$. When $R_y(\theta)$ operates on each $|+\rangle$ it gives

$$R_y(\theta)\,|+\rangle = |+\rangle C + |-\rangle S, \tag{18.64}$$

where $C = \cos\theta/2$ and $S = \sin\theta/2$. When $R_y(\theta)$ operates on each $|-\rangle$ it gives

$$R_y(\theta)\,|-\rangle = |-\rangle C - |+\rangle S.$$

So what we want is

$$R_y(\theta)\,|\,{}^r_s\rangle = \left[\frac{(r+s)!}{r!s!}\right]^{1/2} R_y(\theta)\{|+\rangle^r\,|-\rangle^s\}_{\text{perm}}$$

$$= \left[\frac{(r+s)!}{r!s!}\right]^{1/2} \{(R_y(\theta)\,|+\rangle)^r(R_y(\theta)\,|-\rangle)^s\}_{\text{perm}}$$

$$= \left[\frac{(r+s)!}{r!s!}\right]^{1/2} \{|+\rangle C + |-\rangle S)^r(|-\rangle C - |+\rangle S^s\}_{\text{perm}}. \qquad (18.65)$$

Now each binomial has to be expanded out to its appropriate power and the two expressions multiplied together. There will be terms with $|+\rangle$ to all powers from zero to $(r+s)$. Let's look at all of the terms which have $|+\rangle$ to the r' power. They will appear always multiplied with $|-\rangle$ to the s' power, where $s' = 2j - r'$. Suppose we collect all such terms. For each permutation they will have some numerical coefficient involving the factors of the binomial expansion as well as the factors C and S. Suppose we call that factor $A_{r'}$. Then Eq. (18.65) will look like

$$R_y(\theta)\,|\,{}^r_s\rangle = \sum_{r'=0}^{r+s} \{A_{r'}\,|+\rangle^{r'}\,|-\rangle^{s'}\}_{\text{perm}}. \qquad (18.66)$$

Now let's say that we divide $A_{r'}$ by the factor $[(r'+s')!/r'!s'!]^{1/2}$ and call the quotient $B_{r'}$. Equation (18.66) is then equivalent to

$$R_y(\theta)\,|\,{}^r_s\rangle = \sum_{r'=0}^{r+s} B_{r'}\left[\frac{r'+s'}{r'!s'!}\right]^{1/2} \{|+\rangle^{r'}\,|-\rangle^{s'}\}_{\text{perm}}. \qquad (18.67)$$

(We could just say that this equation defines $B_{r'}$ by the requirement that (18.67) gives the same expression that appears in (18.65).)

With this definition of $B_{r'}$ the remaining factors on the right-hand side of Eq. (18.67) are just the states $|\,{}^{r'}_{s'}\rangle$. So we have that

$$R_y(\theta)\,|\,{}^r_s\rangle = \sum_{r'=0}^{r+s} B_{r'}\,|\,{}^{r'}_{s'}\rangle, \qquad (18.68)$$

with s' always equal to $r + s - r'$. This means, of course, that the coefficients $B_{r'}$ are just the matrix elements we want, namely

$$\langle {}^{r'}_{s'}|\,R_y(\theta)\,|\,{}^r_s\rangle = B_{r'}. \qquad (18.69)$$

Now we just have to push through the algebra to find the various $B_{r'}$. Comparing (18.39) with (18.37)—and remembering that $r' + s' = r + s$—we see that $B_{r'}$ is just the coefficient of $a^r b^s$ in the following expression:

$$\left(\frac{r'!s'!}{r!s!}\right)^{1/2} (aC + bS)^r(bC - aS)^s. \qquad (18.70)$$

It is now only a dirty job to make the expansions by the binomial theorem, and collect the terms with the given power of a and b. If you work it all out, you find that the coefficient of $a^r b^{s'}$ in (18.70) is

$$\left[\frac{r'!s'!}{r!s!}\right]^{1/2} \sum_k (-1)^k S^{r-r'+2k} C^{s+r'-2k} \cdot \frac{r!}{(r-r'+k)!(r'-k)!} \cdot \frac{s!}{(s-k)!k!} \cdot$$
$$(18.71)$$

The sum is to be taken over all integers k which give terms of zero or greater in the factorials. This expression is then the matrix element we wanted.

Finally, we can return to our original notation in terms of j, m, and m' using

$$r = j + m, \qquad r' = j + m', \qquad s = j - m, \qquad s' = j - m'.$$

Making these substitutions, we get Eq. (18.34) in Section 18–4.

Added Note 2: Conservation of parity in photon emission

In Section 1 of this chapter we considered the emission of light by an atom that goes from an excited state of spin 1 to a ground state of spin 0. If the excited state has its spin up ($m = +1$), it can emit a RHC photon along the $+z$-axis or a LHC photon along the $-z$-axis. Let's call these two states of the photon $|R_{\text{up}}\rangle$ and $|L_{\text{dn}}\rangle$. Neither of these states has a definite parity. Letting $\hat{P}$ be the parity operator, $\hat{P}|R_{\text{up}}\rangle = |L_{\text{dn}}\rangle$ and $\hat{P}|L_{\text{dn}}\rangle = |R_{\text{up}}\rangle$.

What about our earlier proof that an atom in a state of definite energy must have a definite parity, and our statement that parity is conserved in atomic processes? Shouldn't the final state in this problem (the state after the emission of a photon) have a definite parity? It *does* if we consider the *complete* final state which contains amplitudes for the emission photons into all sorts of angles. In Section 1 we chose to consider only a part of the complete final state.

If we wish we can look only at final states that do have a definite parity. For example, consider a final state $|\psi_F\rangle$ which has some amplitude α to be a RHC photon going along $+z$ and some amplitude β to be a LHC photon going along $-z$. We can write

$$|\psi_F\rangle = \alpha|R_{\text{up}}\rangle + \beta|L_{\text{dn}}\rangle. \tag{18.72}$$

The parity operation on this state gives

$$\hat{P}|\psi_F\rangle = \alpha|L_{\text{dn}}\rangle + \beta|R_{\text{up}}\rangle. \tag{18.73}$$

This state will be $\pm|\psi_F\rangle$ if $\beta = \alpha$ or if $\beta = -\alpha$. So a final state of even parity is

$$|\psi_F^+\rangle = \alpha\{R_{\text{up}}\rangle + |L_{\text{dn}}\rangle\}, \tag{18.74}$$

and a state of odd parity is

$$|\psi_F^-\rangle = \alpha\{|R_{\text{up}}\rangle - |L_{\text{dn}}\rangle\}. \tag{18.75}$$

Next, we wish to consider the decay of an excited state of odd parity to a ground state of even parity. If parity is to be conserved, the final state of the photon must have odd parity. It must be the state in (18.75). If the amplitude to find $|R_{\text{up}}\rangle$ is α, the amplitude to find $|L_{\text{dn}}\rangle$ is $-\alpha$.

Now notice what happens when we perform a rotation of 180° about the y-axis. The initial excited state of the atom becomes an $m = -1$ state (with no change in sign, according to Table 17-2). And the rotation of the final state gives

$$R_y(180°)|\psi_F^-\rangle = \alpha\{|R_{\text{dn}}\rangle - |L_{\text{up}}\rangle\}. \tag{18.76}$$

Comparing this equation with (18.75), you see that for the assumed parity of the final state, the amplitude to get a LHC photon along $+z$ from the $m = -1$ initial state is the negative of the amplitude to get a RHC photon from the $m = +1$ initial state. This agrees with the result we found in Section 1.

19

The Hydrogen Atom and The Periodic Table

19–1 Schrödinger's equation for the hydrogen atom

19–2 Spherically symmetric solutions

19–3 States with an angular dependence

19–4 The general solution for hydrogen

19–5 The hydrogen wave functions

19–6 The periodic table

19–1 Schrödinger's equation for the hydrogen atom

The most dramatic success in the history of the quantum mechanics was the understanding of the details of the spectra of some simple atoms and the understanding of the periodicities which are found in the table of chemical elements. In this chapter we will at last bring our quantum mechanics to the point of this important achievement, specifically to an understanding of the spectrum of the hydrogen atom. We will at the same time arrive at a qualitative explanation of the mysterious properties of the chemical elements. We will do this by studying in detail the behavior of the electron in a hydrogen atom—for the first time making a detailed calculation of a distribution-in-space according to the ideas we developed in Chapter 16.

For a complete description of the hydrogen atom we should describe the motions of both the proton and the electron. It is possible to do this in quantum mechanics in a way that is analogous to the classical idea of describing the motion of each particle relative to the center of gravity, but we will not do so. We will just discuss an approximation in which we consider the proton to be very heavy, so we can think of it as fixed at the center of the atom.

We will make another approximation by forgetting that the electron has a spin and should be described by relativistic laws of mechanics. Some small corrections to our treatment will be required since we will be using the nonrelativistic Schrödinger equation and will disregard magnetic effects. Small magnetic effects occur because from the electron's point-of-view the proton is a circulating charge which produces a magnetic field. In this field the electron will have a different energy with its spin up than with it down. The energy of the atom will be shifted a little bit from what we will calculate. We will ignore this small energy shift. Also we will imagine that the electron is just like a gyroscope moving around in space always keeping the same direction of spin. Since we will be considering a free atom in space the total angular momentum will be conserved. In our approximation we will assume that the angular momentum of the electron spin stays constant, so all the rest of the angular momentum of the atom—what is usually called "orbital" angular momentum—will also be conserved. To an excellent approximation the electron moves in the hydrogen atom like a particle without spin—the angular momentum of the motion is a constant.

With these approximations the amplitude to find the electron at different places in space can be represented by a function of position in space and time. We let $\psi(x, y, z, t)$ be the amplitude to find the electron somewhere at the time t. According to the quantum mechanics the rate of change of this amplitude with time is given by the Hamiltonian operator working on the same function. From Chapter 16,

$$i\hbar \frac{\partial \psi}{\partial t} = \hat{\mathcal{H}}\psi, \tag{19.1}$$

with

$$\hat{\mathcal{H}} = -\frac{\hbar^2}{2m} \nabla^2 + V(r). \tag{19.2}$$

Here, m is the electron mass, and $V(r)$ is the potential energy of the electron in the electrostatic field of the proton. Taking $V = 0$ at large distances from the proton we can write†

$$V = -\frac{e^2}{r}.$$

The wave function ψ must then satisfy the equation

$$i\hbar \frac{\partial \psi}{\partial t} = -\frac{\hbar^2}{2m} \nabla^2 \psi - \frac{e^2}{r} \psi. \qquad (19.3)$$

We want to look for definite energy states, so we try to find solutions which have the form

$$\psi(\mathbf{r}, t) = e^{-(i/\hbar)Et} \psi(\mathbf{r}). \qquad (19.4)$$

The function $\psi(\mathbf{r})$ must then be a solution of

$$-\frac{\hbar^2}{2m} \nabla^2 \psi = \left(E + \frac{e^2}{r} \right) \psi, \qquad (19.5)$$

where E is some constant—the energy of the atom.

Since the potential energy term depends only on the radius, it turns out to be much more convenient to solve this equation in polar coordinates rather than rectangular coordinates. The Laplacian is defined in rectangular coordinates by

$$\nabla^2 = \frac{\partial^2}{\partial x^2} + \frac{\partial^2}{\partial y^2} + \frac{\partial^2}{\partial z^2}.$$

We want to use instead the coordinates r, θ, ϕ shown in Fig. 19–1. These coordinates are related to x, y, z by

$$x = r \sin\theta \cos\phi; \qquad y = r \sin\theta \sin\phi; \qquad z = r \cos\theta.$$

It's a rather tedious mess to work through the algebra, but you can eventually show that for any function $f(\mathbf{r}) = f(r, \theta, \phi)$,

$$\nabla^2 f(r, \theta, \phi) = \frac{1}{r} \frac{\partial^2}{\partial r^2}(rf) + \frac{1}{r^2} \left\{ \frac{1}{\sin\theta} \frac{\partial}{\partial \theta} \left(\sin\theta \frac{\partial f}{\partial \theta} \right) + \frac{1}{\sin^2\theta} \frac{\partial^2 f}{\partial \phi^2} \right\}. \qquad (19.6)$$

So in terms of the polar coordinates, the equation which is to be satisfied by $\psi(r, \theta, \phi)$ is

$$\frac{1}{r} \frac{\partial^2}{\partial r^2}(r\psi) + \frac{1}{r^2} \left\{ \frac{1}{\sin\theta} \frac{\partial}{\partial \theta} \left(\sin\theta \frac{\partial \psi}{\partial \theta} \right) + \frac{1}{\sin^2\theta} \frac{\partial^2 \psi}{\partial \phi^2} \right\} = -\frac{2m}{\hbar^2} \left(E + \frac{e^2}{r} \right) \psi. \qquad (19.7)$$

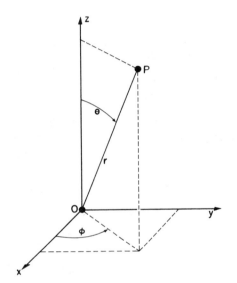

Fig. 19–1. The spherical polar coordinates r, θ, ϕ of the point P.

19–2 Spherically symmetric solutions

Let's first try to find some very simple function that satisfies the horrible equation in (19.7). Although the wave function ψ will, in general, depend on the angles θ and ϕ as well as on the radius r, we can see whether there might be a special situation in which ψ does *not* depend on the angles. For a wave function that doesn't depend on the angles, none of the amplitudes will change in any way if you rotate the coordinate system. That means that all of the components of the angular momentum are zero. Such a ψ must correspond to a state whose total angular momentum is zero. (Actually, it is only the orbital angular momentum which is zero because we still have the spin of the electron, but we are ignoring that part.) A state with zero orbital angular momentum is called by a special name. It is called an "*s*-state"—you can remember "*s* for spherically symmetric."‡

---

† As usual, $e^2 = q_e^2/4\pi\epsilon_0$.

‡ Since these special names are part of the common vocabulary of atomic physics, you will just have to learn them. We will help out by putting them together in a short "dictionary" later in the chapter.

Now if ψ is not going to depend on θ and ϕ then the entire Laplacian contains only the first term and Eq. (19.7) becomes much simpler:

$$\frac{1}{r}\frac{d^2}{dr^2}(r\psi) = -\frac{2m}{\hbar^2}\left(E + \frac{e^2}{r}\right)\psi. \qquad (19.8)$$

Before you start to work on solving an equation like this, it's a good idea to get rid of all excess constants like e^2, m, and $\hbar$, by making some scale changes. Then the algebra will be easier. If we make the following substitutions:

$$r = \frac{\hbar^2}{me^2}\rho, \qquad (19.9)$$

and

$$E = \frac{me^4}{2\hbar^2}\epsilon, \qquad (19.10)$$

then Eq. (19.8) becomes (after multiplying through by ρ)

$$\frac{d^2(\rho\psi)}{d\rho^2} = -\left(\epsilon + \frac{2}{\rho}\right)\rho\psi. \qquad (19.11)$$

These scale changes mean that we are measuring the distance r and energy E as multiples of "natural" atomic units. That is, $\rho = r/r_B$, where $r_B = \hbar^2/me^2$, is called the "Bohr radius" and is about 0.528 angstroms. Similarly, $\epsilon = E/E_R$, with $E_R = me^4/2\hbar^2$. This energy is called the "Rydberg" and is about 13.6 electron volts.

Since the product $\rho\psi$ appears on both sides, it is convenient to work with it rather than with ψ itself. Letting

$$\rho\psi = f, \qquad (19.12)$$

we have the more simple-looking equation

$$\frac{d^2f}{d\rho^2} = -\left(\epsilon + \frac{2}{\rho}\right)f. \qquad (19.13)$$

Now we have to find some function f which satisfies Eq. (19.13)—in other words, we just have to solve a differential equation. Unfortunately, there is no very useful, general method for solving any given differential equation. You just have to fiddle around. Our equation is not easy, but people have found that it can be solved by the following procedure. First, you replace f, which is some function of ρ, by a product of two functions

$$f(\rho) = e^{-\alpha\rho}g(\rho). \qquad (19.14)$$

This just means that you are factoring $e^{-\alpha\rho}$ out of $f(\rho)$. You can certainly do that for any $f(\rho)$ at all. This just shifts our problem to finding the right function $g(\rho)$.

Sticking (19.14) into (19.13), we get the following equation for g:

$$\frac{d^2g}{d\rho^2} - 2\alpha\frac{dg}{d\rho} + \left(\frac{2}{\rho} + \epsilon + \alpha^2\right)g = 0. \qquad (19.15)$$

Since we are free to choose α, let's make

$$\alpha^2 = -\epsilon, \qquad (19.16)$$

and get

$$\frac{d^2g}{d\rho^2} - 2\alpha\frac{dg}{d\rho} + \frac{2}{\rho}g = 0. \qquad (19.17)$$

You may think we are no better off than we were at Eq. (19.13), but the happy thing about our new equation is that it can be solved easily in terms of a power series in ρ. (It is possible, in principle, to solve (19.13) that way too, but it is much harder.) We are saying that Eq. (19.17) can be satisfied by some $g(\rho)$ which can be written as a series,

$$g(\rho) = \sum_{k=1}^{\infty} a_k \rho^k, \qquad (19.18)$$

in which the a_k are constant coefficients. Now all we have to do is find a suitable infinite set of coefficients! Let's check to see that such a solution will work. The first derivative of this $g(\rho)$ is

$$\frac{dg}{d\rho} = \sum_{k=1}^{\infty} a_k k \rho^{k-1},$$

and the second derivative is

$$\frac{d^2 g}{d\rho^2} = \sum_{k=1}^{\infty} a_k k(k-1)\rho^{k-2}.$$

Using these expressions in (19.17) we have

$$\sum_{k=1}^{\infty} k(k-1)a_k \rho^{k-2} - \sum_{k=1}^{\infty} 2\alpha k a_k \rho^{k-1} + \sum_{k=1}^{\infty} 2 a_k \rho^{k-1} = 0. \qquad (19.19)$$

It's not obvious that we have succeeded; but we forge onward. It will all look better if we replace the first sum by an equivalent. Since the first term of the sum is zero, we can replace each k by $k + 1$ without changing anything in the infinite series; with this change the first sum can equally well be written as

$$\sum_{k=1}^{\infty} (k+1)k a_{k+1}\rho^{k-1}.$$

Now we can put all the sums together to get

$$\sum_{k=1}^{\infty} [(k+1)k a_{k+1} - 2\alpha k a_k + 2 a_k]\rho^{k-1} = 0. \qquad (19.20)$$

This power series must vanish for all possible values of ρ. It can do that only if the coefficient of each power of ρ is separately zero. We will have a solution for the hydrogen atom if we can find a set a_k for which

$$(k+1)k a_{k+1} - 2(\alpha k - 1)a_k = 0 \qquad (19.21)$$

for all $k > 1$. That is certainly easy to arrange. Pick any a_1 you like. Then generate all of the other coefficients from

$$a_{k+1} = \frac{2(\alpha k - 1)}{k(k+1)} a_k. \qquad (19.22)$$

With this you will get a_2, a_3, a_4, and so on, and each pair will certainly satisfy (19.21). We get a series for $g(\rho)$ which satisfies (19.17). With it we can make a ψ that satisfies Schrödinger's equation. Notice that the solutions depend on the assumed energy (through α), but for each value of ϵ, there is a corresponding series.

We have a solution, but what does it represent physically? We can get an idea by seeing what happens far from the proton—for large values of ρ. Out there, the high-order terms of the series are the most important, so we should look at what happens for large k. When $k \gg 1$, Eq. (19.22) is approximately the same as

$$a_{k+1} = \frac{2\alpha}{k} a_k,$$

which means that

$$a_{k+1} \approx \frac{(2\alpha)^k}{k!}. \qquad (19.23)$$

But these are just the coefficients of the series for $e^{+2\alpha\rho}$. The function of g is a rapidly increasing exponential. Even coupled with $e^{-\alpha\rho}$ to produce $f(\rho)$—see

Eq. (19.14)—it still gives a solution for $f(\rho)$ which goes like $e^{\alpha\rho}$ for large ρ. We have found a mathematical solution but not a physical one. It represents a situation in which the electron is *least* likely to be near the proton! It is always more likely to be found at a very large radius ρ. A wave function for a *bound* electron must go to zero for large ρ.

We have to think whether there is some way to beat the game, and there is. Observe! If it just happened by luck that α were equal to $1/n$, where n is any integer, then Eq. (19.22) would make $a_{n+1} = 0$. All higher terms would also be zero. We wouldn't have an infinite series but a finite polynomial. Any polynomial increases more slowly than $e^{\alpha\rho}$, so the term $e^{-\alpha\rho}$ will eventually beat it down, and the function f will go to zero for large ρ. The only *bound-state* solutions are those for which $\alpha = 1/n$, with $n = 1, 2, 3, 4$, and so on.

Looking back to Eq. (19.16), we see that the bound-state solutions to the spherically symmetric wave equation can exist only when

$$-\epsilon = 1, \frac{1}{4}, \frac{1}{9}, \frac{1}{16}, \ldots, \frac{1}{n^2}, \ldots$$

The allowed energies are just these fractions times the Rydberg, $E_R = me^4/2\hbar^2$, or the energy of the nth energy level is

$$E_n = -E_R \frac{1}{n^2}. \tag{19.24}$$

There is, incidentally, nothing mysterious about negative numbers for the energy. The energies are negative because when we chose to write $V = -e^2/r$, we picked our zero point as the energy of an electron located far from the proton. When it is close to the proton, its energy is less, so somewhat below zero. The energy is lowest (most negative) for $n = 1$, and increases toward zero with increasing n.

Before the discovery of quantum mechanics, it was known from experimental studies of the spectrum of hydrogen that the energy levels could be described by Eq. (19.24), where E_R was found from the observations to be about 13.6 electron volts. Bohr then devised a model which gave the same equation and predicted that E_R should be $me^4/2\hbar^2$. But it was the first great success of the Schrödinger theory that it could reproduce this result from a basic equation of motion for the electron.

Now that we have solved our first atom, let's look at the nature of the solution we got. Pulling all the pieces together, each solution looks like this:

$$\psi_n = \frac{f_n(\rho)}{\rho} = \frac{e^{-\rho/n}}{\rho} g_n(\rho), \tag{19.25}$$

where

$$g_n(\rho) = \sum_{k=1}^{n} a_k \rho^k \tag{19.26}$$

and

$$a_{k+1} = \frac{2(k/n - 1)}{k(k + 1)} a_k. \tag{19.27}$$

So long as we are mainly interested in the relative probabilities of finding the electron at various places we can pick any number we wish for a_1. We may as well set $a_1 = 1$. (People often choose a_1 so that the wave function is "normalized," that is, so that the integrated probability of finding the electron anywhere in the atom is equal to 1. We have no need to do that just now.)

For the lowest energy state, $n = 1$, and

$$\psi_1(\rho) = e^{-\rho}. \tag{19.28}$$

For a hydrogen atom in its ground (lowest-energy) state, the amplitude to find the electron at any point drops off exponentially with the distance from the proton. It is most likely to be found right at the proton, and the characteristic spreading distance is about one unit in ρ, or about one Bohr radius, r_B.

Putting $n = 2$ gives the next higher level. The wave function for this state will have two terms. It is

$$\psi_2(\rho) = \left(1 - \frac{\rho}{2}\right)e^{-\rho/2}. \qquad (19.29)$$

The wave function for the next level is

$$\psi_3(\rho) = \left(1 - \frac{2\rho}{3} + \frac{2}{27}\rho^2\right)e^{-\rho/3}. \qquad (19.30)$$

The wave functions for these first three levels are plotted in Fig. 19–2. You can see the general trend. All of the wave functions approach zero rapidly for large ρ after oscillating a few times. In fact, the number of "bumps" is just equal to n—or, if you prefer, the number of zero-crossings of ψ_n is $n - 1$.

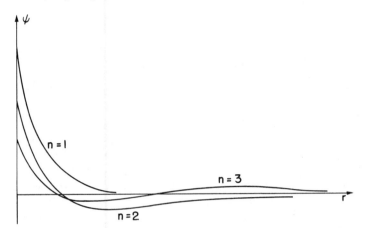

Fig. 19–2. The wave functions for the first three $l = 0$ states of the hydrogen atom. (The scales are chosen so that the total probabilities are equal.)

19–3 States with an angular dependence

In the states described by the $\psi_n(r)$ we have found that the probability amplitude for finding the electron is spherically symmetric—depending only on r, the distance for the proton. Such states have zero orbital angular momentum. We should now inquire about states which may have some angular dependences.

We could, if we wished, just investigate the strictly mathematical problem of finding the functions of r, θ, and ϕ which satisfy the differential equation (19.7)—putting in the additional physical conditions that the only acceptable functions are ones which go to zero for large r. You will find this done in many books. We are going to take a short cut by using the knowledge we already have about how amplitudes depend on angles in space.

The hydrogen atom in any particular state is a particle with a certain "spin" j—the quantum number of the total angular momentum. Part of this spin comes from the electron's intrinsic spin, and part from the electron's motion. Since each of these two components acts independently (to an excellent approximation) we will again ignore the spin part and think only about the "orbital" angular momentum. This orbital motion behaves, however, just like a spin. For example, if the orbital quantum number is l, the z-component of angular momentum can be $l, l - 1, l - 2, \dots, -l$. (We are, as usual, measuring in units of $\hbar$.) Also, all the rotation matrices and other properties we have worked out still apply. (From now on we will *really* ignore the electron's spin; when we speak of "angular momentum" we will mean only the orbital part.)

Since the potential V in which the electron moves depends only on r and not on θ or ϕ, the Hamiltonian is symmetric under all rotations. It follows that the angular momentum and all its components are conserved. (This is true for motion in *any* "central field"—one which depends only on r—so is not a special feature of the Coulomb e^2/r potential.)

Now let's think of some possible state of the electron; its internal angular structure will be characterized by the quantum number l. Depending on the "orientation" of the total angular momentum with respect to the z-axis, the z-component of angular momentum will be m, which is one of the $2l + 1$ possibilities between $+l$ and $-l$. Let's say $m = 1$. With what amplitude will the electron be found on the z-axis at some distance r? Zero. An electron on the z-axis *cannot* have any orbital angular momentum around that axis. Alright, suppose m is zero, then there can be some nonzero amplitude to find the electron at each distance from the proton. We'll call this amplitude $F_l(r)$. It is the amplitude to find the electron at the distance r up along the z-axis, when the atom is in the state $|l, 0\rangle$, by which we mean orbital spin l and z-component $m = 0$.

If we know $F_l(r)$ everything is known. For any state $|l, m\rangle$, we know the amplitude $\psi_{l,m}(\mathbf{r})$ to find the electron *anywhere* in the atom. How? Watch. Suppose we have the atom in the state $|l, m\rangle$, what is the amplitude to find the electron at the angle θ, ϕ and the distance r from the origin? Put a new z-axis, say z', at that angle (see Fig. 19–3), and ask what is the amplitude that the electron will be at the distance r along the new axis z'? We know that it cannot be found along z' unless its z'-component of angular momentum, say m', is zero. When m' *is* zero, however, the amplitude to find the electron along z' is $F_l(r)$. Therefore, the result is the product of two factors. The first is the amplitude that an atom in the state $|l, m\rangle$ along the z-axis will be in the state $|l, m' = 0\rangle$ *with respect to the z'-axis*. Multiply that amplitude by $F_l(r)$ and you have the amplitude $\psi_{l,m}(\mathbf{r})$ to find the electron at (r, θ, ϕ) with respect to the original axes.

Let's write it out. We have worked out earlier the transformation matrices for rotations. To go from the frame x, y, z to the frame x', y', z' of Fig. 19–3, we can rotate first around the z-axis by the angle ϕ, and then rotate about the *new* y-axis (y') by the angle θ. This combined rotation is the product

$$R_y(\theta)R_z(\phi).$$

The amplitude to find the state l, $m' = 0$ after the rotation is

$$\langle l, 0 \mid R_y(\phi)R_z(\phi) \mid l, m\rangle. \tag{19.31}$$

Our result, then, is

$$\psi_{l,m}(\mathbf{r}) = \langle l, 0 \mid R_y(\theta)R_z(\phi) \mid l, m\rangle F_l(r). \tag{19.32}$$

The orbital motion can have only integral values of l. (If the electron can be found anywhere at $r \neq 0$, there is some amplitude to have $m = 0$ in that direction. And $m = 0$ states exist only for integral spins.) The rotation matrices for $l = 1$ are given in Table 17–2. For larger l you can use the general formulas we worked out in Chapter 18. The matrices for $R_z(\phi)$ and $R_y(\theta)$ appear separately, but you know how to combine them. For the general case you would start with the state $|l, m\rangle$ and operate with $R_z(\phi)$ to get the new state $R_z(\phi) | l, m\rangle$. Then you operate on this state with $R_y(\theta)$ to get the state $R_y(\theta)R_z(\phi) | l, m\rangle$ (which is just $e^{im\phi} | l, m\rangle$). Multiplying by $\langle l, 0 |$ gives the matrix element (19.31).

The matrix elements of the rotation operation are algebraic functions of θ and ϕ. The particular functions which appear in (19.31) also show up in many kinds of problems which involve waves in spherical geometries and so has been given a special name. Not everyone uses the same convention; but one of the most common ones is

$$\langle l, 0 \mid R_y(\theta)R_z(\phi) \mid l, m\rangle \equiv a Y_{l,m}(\theta, \phi). \tag{19.33}$$

The functions $Y_{l,m}(\theta, \phi)$ are called the *spherical harmonics*, and a is just a numerical factor which depends on the definition chosen for $Y_{l,m}$. For the usual definition,

$$a = \sqrt{\frac{4}{2l + 1}}. \tag{19.34}$$

With this notation, the hydrogen wave functions can be written

$$\psi_{l,m}(\mathbf{r}) = Y_{l,m}(\theta, \phi)F_l(r). \tag{19.35}$$

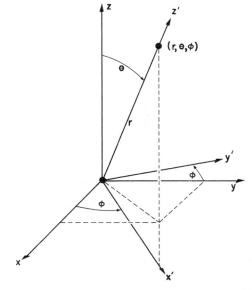

Fig. 19–3. The point (r, θ, ϕ) is on the z'-axis of the x'y'z' coordinate frame.

The angle functions $Y_{l,m}(\theta, \phi)$ are important not only in many quantum-mechanical problems, but also in many areas of classical physics in which the ∇^2 operator appears, such as electromagnetism. As another example of their use in quantum mechanics, consider the disintegration of an excited state of Ne^{20} (such as we discussed in the last chapter) which decays by emitting an α-particle and going into O^{16}:

$$Ne^{20*} \rightarrow O^{16} + He^4.$$

Suppose that the excited state has some spin l (necessarily an integer) and that the z-component of angular momentum is m. We might now ask the following: given l and m, what is the amplitude that we will find the α-particle going off in a direction which makes the angle θ with respect to the z-axis and the angle ϕ with respect to the xz-plane—as shown in Fig. 19–4.

To solve this problem we make, first, the following observation. A decay in which the α-particle goes straight up along z must come from a state with $m = 0$. This is so because both O^{16} and the α-particle have spin zero, and because their motion cannot have any angular momentum about the z-axis. Let's call this amplitude a (per unit solid angle). Then, to find the amplitude for a decay at the arbitrary angle of Fig. 19–4, all we need to know is what amplitude the given initial state has zero angular momentum about the decay direction. The amplitude for the decay at θ and ϕ is then a times the amplitude that a state $| l, m \rangle$ with respect to the z-axis will be in the state $| l, 0 \rangle$ with respect to z'—the decay direction. This latter amplitude is just what we have written in (19.31). The probability to see the α-particle at θ, ϕ is

$$P(\theta, \phi) = a^2 \, |\langle l, 0 \mid R_y(\theta)R_z(\phi) \mid l, m \rangle|^2.$$

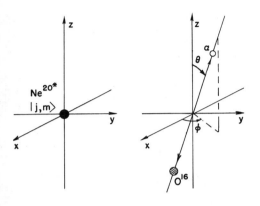

Fig. 19–4. The decay of an excited state of Ne^{20}.

As an example, consider an initial state with $l = 1$ and various values of m. From Table 17–2 we know the necessary amplitudes. They are

$$\langle 1, 0 \mid R_y(\theta)R_z(\phi) \mid 1, +1 \rangle = -\frac{1}{\sqrt{2}} \sin \theta e^{i\phi},$$

$$\langle 1, 0 \mid R_y(\theta)R_z(\phi) \mid 1, 0 \rangle = \cos \theta, \qquad (19.36)$$

$$\langle 1, 0 \mid R_y(\theta)R_z(\phi) \mid 1, -1 \rangle = -\frac{1}{\sqrt{2}} \sin \theta e^{-i\phi}.$$

These are the three possible angular distribution amplitudes—depending on the m-value of the initial nucleus.

Amplitudes such as the ones in (19.36) appear so often and are sufficiently important that they are given *several* names. If the angular distribution amplitude is proportional to any one of the three functions or any linear combination of them, we say, "The system has an orbital angular momentum of one." Or we may say, "The Ne^{20*} emits a p-wave α-particle." Or we say, "The α-particle is emitted in an $l = 1$ state." Because there are so many ways of saying the same thing it is useful to have a dictionary. If you are going to understand what other physicists are talking about, you will just have to memorize the language. In Table 19–1 we give a dictionary of orbital angular momentum.

If the orbital angular momentum is zero, then there is no change when you rotate the coordinate system and there is no variation with angle—the "dependence" on angle is as a constant, say 1. This is also called an "s-state", and there is only one such state—as far as angular dependence is concerned. If the orbital angular momentum is 1, then the amplitude of the angular variation may be any one of the three functions given—depending on the value of m—or it may be a linear combination. These are called "p-states," and there are three of them. If the orbital angular momentum is 2 then there are the five functions shown. Any linear combination is called an "$l = 2$," or a "d-wave" amplitude. Now you can immediately guess what the next letter is—what should come after s, p, d? Well, of course, f, g, h, and so on down the alphabet! The letters don't mean anything. (They did once mean something—they meant "sharp" lines, "principal" lines, "diffuse" lines and

Table 19–1

Dictionary of orbital angular momentum

($l = j =$ an integer)

Orbital angular momentum, l	z-component, m	Angular dependence of amplitudes	Name	Number of states	Orbital parity
0	0	1	s	1	$+$
1	$+1$	$-\dfrac{1}{\sqrt{2}} \sin\theta\, e^{i\phi}$	p	3	$-$
	0	$\cos\theta$			
	-1	$\dfrac{1}{\sqrt{2}} \sin\theta\, e^{-i\phi}$			
2	$+2$	$\dfrac{\sqrt{6}}{4} \sin^2\theta\, e^{2i\phi}$	d	5	$+$
	$+1$	$\dfrac{\sqrt{6}}{2} \sin\theta \cos\theta\, e^{i\phi}$			
	0	$\tfrac{1}{2}(3\cos^2\theta - 1)$			
	-1	$-\dfrac{\sqrt{6}}{2} \sin\theta \cos\theta\, e^{-i\phi}$			
	-2	$\dfrac{\sqrt{6}}{4} \sin^2\theta\, e^{-2i\phi}$			
3		$\langle l, 0 \mid R_y(\theta)R_z(\phi) \mid l, m \rangle$	f		
4		$= Y_{l,m}(\theta, \phi)$	g	$2l + 1$	$(-1)^l$
5		$= P_l^m(\cos\theta)e^{im\phi}$	h		
$\vdots$			$\vdots$		

"fundamental" lines of the optical spectra of atoms. But those were in the days when people did not know where the lines came from. After f there were no special names, so we now just continue with g, h, and so on.)

The angular functions in the table go by several names—and are sometimes defined with slightly different conventions about the numerical factors that appear out in front. Sometimes they are called "spherical harmonics," and written as $Y_{l,m}(\theta, \phi)$. Sometimes they are written $P_l^m(\cos\theta)e^{im\phi}$, and if $m = 0$, simply as $P_l(\cos\theta)$. The functions $P_l(\cos\theta)$ are called the "Legendre polynomials" in $\cos\theta$, and the functions $P_l^m(\cos\theta)$ are called the "associated Legendre functions." You will find tables of these functions in many books.

Notice, incidentally, that all the functions for a given l have the property that that they have the same parity—for odd l they change sign under an inversion and for even l they don't change. So we can write that the parity of a state of orbital angular momentum l is $(-1)^l$.

As we have seen, these angular distributions may refer to a nuclear disintegration or some other process, or to the distribution of the amplitude to find an electron at some place in the hydrogen atom. For instance, if an electron is in a p-state ($l = 1$) the amplitude to find it can depend on the angle in many possible ways—but all are linear combinations of the three functions for $l = 1$ in Table 19–1. Let's take the case $\cos\theta$. That's interesting. That means that the amplitude is positive, say, in the upper part ($\theta < \pi/2$), is negative in the lower part ($\theta > \pi/2$), and is zero when θ is 90°. Squaring this amplitude we see that the probability of finding the electron varies with θ as shown in Fig. 19–5—and is independent of ϕ. This angular distribution is responsible for the fact that in molecular binding the attraction of an electron in an $l = 1$ state for another atom depends on direction—it is the origin of the directed valences of chemical attraction.

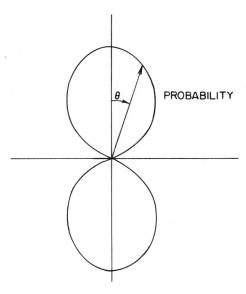

Fig. 19–5. A polar graph of $\cos^2\theta$, which is the relative probability of finding an electron at various angles from the z-axis (for a given r) in an atomic state with $l = 1$ and $m = 0$.

19–4 The general solution for hydrogen

In Eq. (19.35) we have written the wave functions for the hydrogen atom as

$$\psi_{l,m}(\mathbf{r}) = Y_{l,m}(\theta, \phi)F_l(r). \tag{19.37}$$

These wave functions must be solutions of the differential equation (19.7). Let's see what that means. Put (19.37) into (19.7); you get

$$\frac{Y_{l,m}}{r}\frac{\partial^2}{\partial r^2}(rF_l) + \frac{F_l}{r^2 \sin\theta}\frac{\partial}{\partial\theta}\left(\sin\theta\,\frac{\partial Y_{l,m}}{\partial\theta}\right) + \frac{F_l}{r^2 \sin^2\theta}\frac{\partial^2 Y_{l,m}}{\partial\phi^2}$$
$$= -\frac{2m}{\hbar^2}\left(E + \frac{e^2}{r}\right)Y_{l,m}F_l. \tag{19.38}$$

Now multiply through by r^2/F_l and rearrange terms. The result is

$$\frac{1}{\sin\theta}\frac{\partial}{\partial\theta}\left(\sin\theta\,\frac{\partial Y_{l,m}}{\partial\theta}\right) + \frac{1}{\sin^2\theta}\frac{\partial^2 Y_{l,m}}{\partial\phi^2}$$
$$= -\left[\frac{r^2}{F_l}\left\{\frac{1}{r}\frac{d^2}{dr^2}(rF_l) + \frac{2m}{\hbar^2}\left(E + \frac{e^2}{r}\right)\right\}\right]Y_{l,m}. \tag{19.39}$$

The left-hand side of this equation depends on θ and ϕ, *but not on* r. No matter what value we choose for r, the left side doesn't change. *This must also be true for the right-hand side.* Although the quantity in the square brackets has r's all over the place, the whole quantity cannot depend on r, otherwise we wouldn't have an equation good for all r. As you can see, the bracket also does not depend on θ or ϕ. It must be some constant. Its value may well depend on the l-value of the state we are studying, since the function F_l must be the one appropriate to that state; we'll call the constant K_l. Equation (19.35) is therefore equivalent to *two* equations:

$$\frac{1}{\sin\theta}\frac{\partial}{\partial\theta}\left(\sin\theta\,\frac{\partial Y_{l,m}}{\partial\theta}\right) + \frac{1}{\sin^2\theta}\frac{\partial^2}{\partial\theta^2} = -K_l Y_{l,m}, \tag{19.40}$$

$$\frac{1}{r}\frac{\partial^2}{\partial r^2}(rF_l) + \frac{2m}{\hbar^2}\left(E + \frac{e^2}{r}\right)F_l = K_l\frac{F_l}{r^2}. \tag{19.41}$$

Now look at what we've done. For any state described by l and m, we know the functions $Y_{l,m}$; we can use Eq. (19.40) to determine the constant K_l. Putting K_l into Eq. (19.41) we have a differential equation for the function $F_l(r)$. If we can solve that equation for $F_l(r)$, we have all of the pieces to put into (19.37) to give $\psi(r)$.

What is K_l? First, notice that it must be the same for all m (which go with a particular l), so we can pick any m we want for $Y_{l,m}$ and plug it into (19.40) to solve for K_l. Perhaps the easiest one to use is $Y_{l,l}$. From Eq. (18.24),

$$R_z(\phi)\,|\,l, l\rangle = e^{il\phi}\,|\,l, l\rangle. \tag{19.42}$$

The matrix element for $R_y(\theta)$ is also quite simple:

$$\langle l, 0\,|\,R_y(\theta)\,|\,l, l\rangle = b\,(\sin\theta)^l, \tag{19.43}$$

where b is some number.† Combining the two, we obtain

$$Y_{l,l} \propto e^{il\phi}\sin^l\theta. \tag{19.44}$$

---

† You can with some work show that this comes out of Eq. (18.35), but it is also easy to work out from first principles following the ideas of Section 18–4. A state $|\,l, l\rangle$ can be made out of $2l$ spin one-half particles all with spins up; while the state $|\,l, 0\rangle$ would have l up and l down. Under the rotation the amplitude that an up-spin remains up is $\cos\theta/2$, and that an up-spin goes down is $\sin\theta/2$. We are asking for the amplitude that l up-spins stay up, while the other l up-spins go down. The amplitude for that is $(\cos\theta/2\sin\theta/2)^l$ which is the same as $\sin^l\theta$.

Putting this function into (19.40) gives

$$K_l = l(l+1). \tag{19.45}$$

Now that we have determined K_l, Eq. (19.41) tells us about the radial function $F_l(r)$. It is, of course, just the Schrödinger equation with the angular part replaced by its equivalent $K_l F_l/r^2$. Let's rewrite (19.41) in the form we had in Eq. (19.8), as follows:

$$\frac{1}{r}\frac{d^2}{dr^2}(rF_l) = -\frac{2m}{\hbar^2}\left\{E + \frac{e^2}{r} - \frac{l(l+1)\hbar^2}{2mr^2}\right\}F_l. \tag{19.46}$$

A mysterious term has been added to the potential energy. Although we got this term by some mathematical shenanigan, it has a simple physical origin. We can give you an idea about where it comes from in terms of a semi-classical argument. Then perhaps you will not find it quite so mysterious.

Think of a classical particle moving around some center of force. The total energy is conserved and is the sum of the potential and kinetic energies

$$U = V(r) + \tfrac{1}{2}mv^2 = \text{constant}.$$

In general, v can be resolved into a radial component v_r and a tangential component $r\dot\theta$; then

$$v^2 = v_r^2 + (r\dot\theta)^2.$$

Now the angular momentum $mr^2\dot\theta$ is also conserved; say it is equal to L. We can then write

$$mr^2\dot\theta = L, \quad \text{or} \quad r\dot\theta = \frac{L}{mr},$$

and the energy is

$$U = \tfrac{1}{2}mv_r^2 + V(r) + \frac{L^2}{2mr^2}.$$

If there were no angular momentum we would have just the first two terms. Adding the angular momentum L does to the energy just what adding a term $L^2/2mr^2$ to the potential energy would do. But this is almost exactly the extra term in (19.46). The only difference is that $l(l+1)\hbar$ appears for the angular momentum instead of $l^2\hbar^2$ as we might expect. But we have seen before (for example, Volume II, Section 34–7)† that this is just the substitution that is usually required to make a quasi-classical argument agree with a correct quantum-mechanical calculation. We can, then, understand the new term as a "pseudo-potential" which gives the "centrifugal force" term that appears in the equations of radial motion for a rotating system. (See the discussion of "pseudo-forces" in Volume I, Section 12–5.)

We are now ready to solve Eq. (19.46) for $F_l(r)$. It is very much like Eq. (19.8), so the same technique will work again. Everything goes as before until you get to Eq. (19.19) which will have the additional term

$$-l(l+1)\sum_{k=1}^{\infty} a_k\rho^{k-2}. \tag{19.47}$$

This term can also be written as

$$-l(l+1)\left\{\frac{a_1}{\rho} - \sum_{k=1}^{\infty} a_{k+1}\rho^{k-1}\right\}. \tag{19.48}$$

(We have taken out the first term and then shifted the running index k down by 1.) Instead of Eq. (19.20) we have

$$\sum_{k=1}^{\infty}[\{k(k+1) - l(l+1)\}a_{k+1} - 2(\alpha k - 1)a_k]\rho^{k-1}$$
$$-\frac{l(l+1)a_1}{\rho} = 0. \tag{19.49}$$

———
† See Appendix to this volume.

There is only one term in ρ^{-1}, so it must be zero. The coefficient a_1 must be zero (unless $l = 0$ and we have our previous solution). Each of the other terms is made zero by having the square bracket come out zero for every k. This condition replaces Eq. (19.21) by

$$a_{k+1} = \frac{2(\alpha k - 1)}{k(k + 1) - l(l + 1)} a_k. \tag{19.50}$$

This is the only significant change from the spherically symmetric case.

As before the series must terminate if we are to have solutions which can represent bound electrons. The series will end at $k = n$ if $\alpha n = 1$. We get again the same condition on α, that it must be equal to $1/n$, where n is some integer. However, Eq. (19.50) also gives a new restriction. The index k cannot be equal to l, the denominator becomes zero and a_{l+1} is infinite. That is, since $a_1 = 0$, Eq. (19.50) implies that all successive a_k are zero until we get to a_{l+1}, which can be nonzero. This means that k must start at $l + 1$ and end at n.

Our final result is that for any l there are many possible solutions which we can call $F_{n,l}$ where $n \geq l + 1$. Each solution has the energy

$$E_n = -\frac{me^4}{2\hbar^2}\left(\frac{1}{n^2}\right). \tag{19.51}$$

The wave function for the state of this energy with the angular quantum numbers l and m is

$$\psi_{n,l,m} = Y_{l,m}(\theta, \phi) F_{n,l}(\rho), \tag{19.52}$$

with

$$\rho F_{n,l}(\rho) = e^{-\alpha\rho} \sum_{k=l+1}^{n} a_k \rho^k. \tag{19.53}$$

The coefficients a_k are obtained from (19.50). We have, finally, a complete description of the states of a hydrogen atom.

19–5 The hydrogen wave functions

Let's review what we have discovered. The states which satisfy Schrödinger's equation for an electron in a Coulomb field are characterized by three quantum numbers n, l, m, all integers. The angular distribution of the electron amplitude can have only certain forms which we call $Y_{l,m}$. They are labeled by l, *the quantum number of total angular momentum*, and m, *the "magnetic" quantum number*, which can range from $-l$ to $+l$. For each angular configuration, various possible radial distributions $F_{n,l}(r)$ of the electron amplitude are possible; they are labeled by the *principle quantum number n* —which can range from $l + 1$ to ∞. The energy of the state depends only on n, and increases with increasing n.

The lowest energy, or ground, state is an s-state. It has $l = 0$, $n = 0$, and $m = 0$. It is a "nondegenerate" state—there is only one with this energy, and its wave function is spherically symmetric. The amplitude to find the electron is a maximum at the center, and falls off monatonically with increasing distance from the center. We can visualize the electron amplitude as a blob as shown in Fig. 19–6(a).

There are other s-states with higher energies, for $n = 2, 3, 4, \ldots$ For each energy there is only one version ($m = 0$), and they are all spherically symmetric. These states have amplitudes which alternate in sign one or more times with increasing r. There are $n - 1$ spherical nodal surfaces—the places where ψ goes through zero. The $2s$-state ($l = 0$, $n = 2$), for example, will look as sketched in Fig. 19–6(b). (The dark areas indicate regions where the amplitude is large, and the plus and minus signs indicate the relative phases of the amplitude.) The energy levels of the s-states are shown in the first column of Fig. 19–7.

Then there are the p-states—with $l = 1$. For each n, which must be 2 or greater, there are three states of the same energy, one each for $m = +1$, $m = 0$, and $m = -1$. The energy levels are as shown in Fig. 19–7. The angular dependences of these states are given in Table 19–1. For instance, for $m = 0$, if the

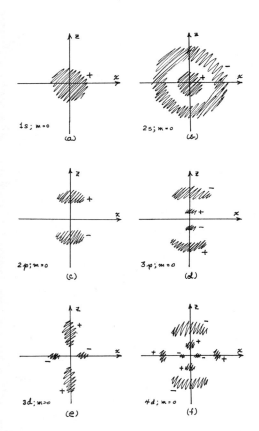

Fig. 19–6. Rough sketches showing the general nature of some of the hydrogen wave functions. The shaded regions show where the amplitudes are large. The plus and minus signs show the relative sign of the amplitude in each region.

amplitude is positive for θ near zero, it will be negative for θ near 180°. There is a nodal plane coincident with the xy-plane. For $n > 2$ there are also spherical nodes. The $n = 2$, $m = 0$ amplitude is sketched in Fig. 19–6(c), and the $n = 3$, $m = 0$ wave function is sketched in Fig. 19–6(d).

You might think that since m represents a kind of "orientation" in space, there should be similar distributions with the peaks of amplitude along the x-axis or along the y-axis. Are these perhaps the $m = +1$ and $m = -1$ states? No. But since we have three states with equal energies, any linear combinations of the three will also be stationary states of the same energy. It turns out that the "x"-state—which corresponds to the "z"-state, or $m = 0$ state, of Fig. 19–6(c)—is a linear combination of the $m = +1$ and $m = -1$ states. The corresponding "y"-state is another combination. Specifically, we mean that

$$\text{"}z\text{"} = |1, 0\rangle,$$

$$\text{"}x\text{"} = \frac{|1, +1\rangle + |1, -1\rangle}{\sqrt{2}},$$

$$\text{"}y\text{"} = \frac{|1, +1\rangle - |1, -1\rangle}{i\sqrt{2}}.$$

These states all look the same when referred to their particular axes.

The d-states ($l = 2$) have five possible values of m for each energy, the lowest energy has $n = 3$. The levels go as shown in Fig. 19–7. The angular dependences get more complicated. For instance the $m = 0$ states have two conical nodes, so the wave function reverses phase from $+$, to $-$, to $+$ as you go around from the north pole to the south pole. The rough form of the amplitude is sketched in (e) and (f) of Fig. 19–6 for the $m = 0$ states with $n = 3$ and $n = 4$. Again, the larger n's have spherical nodes.

We will not try to describe any more of the possible states. You will find the hydrogen wave functions described in more detail in many books. Two good references are L. Pauling and E. B. Wilson, *Introduction to Quantum Mechanics*, McGraw-Hill (1935); and R. B. Leighton, *Principles of Modern Physics*, McGraw-Hill (1959). You will find in them graphs of some of the functions and pictorial representations of many states.

We would like to mention one particular feature of the wave functions for higher l: for $l > 0$ the amplitudes are zero at the center. That is not surprising, since it's hard for an electron to have angular momentum when its radius arm is very small. For this reason, the higher the l, the more the amplitudes are "pushed away" from the center. If you look at the way the radial functions $F(r)$ vary for small r, you find from (19.53) that

$$F_{n,l}(r) \approx r^l.$$

Such a dependence on r means that for larger l's you have to go farther from $r = 0$ before you get an appreciable amplitude. This behavior is, incidentally, determined by the centrifugal force term in the radial equation, so the same thing will apply for any potential that varies slower than $1/r^2$ for small r—which most atomic potentials do.

19–6 The periodic table

We would like now to apply the theory of the hydrogen atom in an approximate way to get some understanding of the chemist's periodic table of the elements. For an element with atomic number Z there are Z electrons held together by the electric attraction of the nucleus but with mutual repulsion of the electrons. To get an exact solution we would have to solve Schrödinger's equation for Z electrons in a Coulomb field. For helium the equation is

$$-\frac{\hbar}{i}\frac{\partial \psi}{\partial t} = -\frac{\hbar^2}{2m}(\nabla_1^2 \psi + \nabla_2^2 \psi) + \left(-\frac{2e^2}{r_1} - \frac{2e^2}{r_2} + \frac{e^2}{r_{12}}\right)\psi,$$

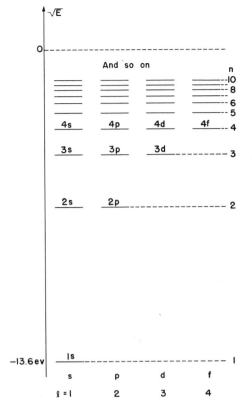

Fig. 19–7. The energy level diagram for hydrogen.

where ∇_1^2 is a Laplacian which operates on r_1, the coordinate of one electron; ∇_2^2 operates on r_2; and $r_{12} = |r_1 - r_2|$. (We are again neglecting the spin of the electrons.) To find the stationary states and energy levels we would have to find solutions of the form

$$\psi = f(r_1, r_2)e^{-(i/\hbar)Et}.$$

The geometrical dependence is contained in f, which is a function of six variables —the simultaneous positions of the two electrons. No one has found an analytic solution, although solutions for the lowest energy states have been obtained by numerical methods.

With 3, 4, or 5 electrons it is hopeless to try to obtain exact solutions, and it is going too far to say that quantum mechanics has given a precise understanding of the periodic table. It is possible, however, even with a sloppy approximation—and some fixing—to understand, at least qualitatively, many chemical properties which show up in the periodic table.

The chemical properties of atoms are determined primarily by their lowest energy states. We can use the following approximate theory to find these states and their energies. First, we neglect the electron spin, *except* that we adopt the exclusion principle and say that any particular electronic state can be occupied by only one electron. This means that any particular orbital configuration can have up to *two* electrons—one with spin up, the other with spin down. Next we disregard the *details* of the interactions between the electrons in our first approximation, and say that each electron moves in a *central field* which is the combined field of the nucleus and all the other electrons. For neon, which has 10 electrons, we say that one electron sees an average potential due to the nucleus plus the other nine electrons. We imagine then that in the Schrödinger equation for each electron we put a $V(r)$ which is a $1/r$ field modified by a spherically symmetric charge density coming from the other electrons.

In this model each electron acts like an independent particle. The angular dependence of its wave function will be just the same as the ones we had for the hydrogen atom. There will be s-states, p-states, and so on; and they will have the various possible m-values. Since $V(r)$ no longer goes as $1/r$, the radial part of the wave functions will be somewhat different, but it will be qualitatively the same, so we will have the same radial quantum numbers, n. The energies of the states will also be somewhat different.

H

With these ideas, let's see what we get. The ground state of hydrogen has $l = m = 0$ and $n = 1$; we say the electron configuration is $1s$. The energy is -13.6 ev. This means that it takes 13.6 electron volts to pull the electron off the atom. We call this the "ionization energy", W_I. A large ionization energy means that it is harder to pull the electron off and, in general, that the material is chemically less active.

He

Now take helium. Both electrons can be in the same lowest state (one spin up and the other spin down). In this lowest state the electron moves in a potential which is for small r like a Coulomb field for $z = 2$ and for large r like a Coulomb field for $z = 1$. The result is a "hydrogen-like" $1s$ state with a somewhat lower energy. Both electrons occupy identical $1s$ states ($l = 0, m = 0$). The observed ionization energy (to remove *one* electron) is 24.6 electron volts. Since the $1s$ "shell" is now filled—we allow only two electrons—there is practically no tendency for an electron to be attracted from another atom. Helium is chemically inert.

Li

The lithium nucleus has a charge of 3. The electron states will again be hydrogen-like, and the three electrons will occupy the lowest three energy levels. Two will go into $1s$ states and the third will go into an $n = 2$ state. But with $l = 0$ or $l = 1$? In hydrogen these states have the same energy, but in other atoms they don't, for the following reason. Remember that a 2s state has some amplitude to be near the nucleus while the 2p state does not. That means that a 2s electron will feel some of the triple electric charge of the Li nucleus, but that a 2p electron will stay out where the field looks like the Coulomb field of a single charge. The extra attraction lowers the energy of the 2s state relative to the 2p state. The energy levels will be roughly as shown in Fig. 19–8—which you should compare with the corresponding diagram for hydrogen in Fig. 19–7. So the lithium atom will have two electrons in 1s states and one in a 2s. Since the 2s electron has a higher energy than a 1s electron it is relatively easily removed. The ionization energy of lithium is only 5.4 electron volts, and it is quite active chemically.

So you can see the patterns which develop; we have given in Table 19–2 a list of the first 36 elements, showing the states occupied by the electrons in the ground state of each atom. The Table gives the ionization energy for the most loosely bound electron, and the number of electrons occupying each "shell"— by which we mean states with the same n. Since the different l-states have different

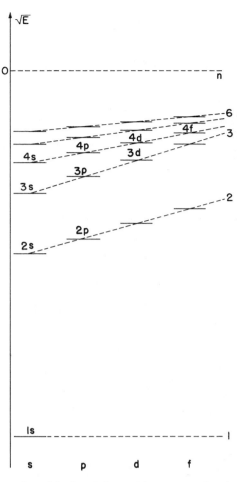

Fig. 19–8. Schematic energy level diagram for an atomic electron with other electrons present. (The scale is *not* the same as Fig. 19–7.)

Table 19–2

The electron configurations of the first 36 elements

Z	Element		W_I(ev)	Electron Configuration									
				1s	2s	2p	3s	3p	3d	4s	4p	4d	4f
1	H	hydrogen	13.6	1									
2	He	helium	24.6	2									
3	Li	lithium	5.4		1								
4	Be	beryllium	9.3		2								
5	B	boron	8.3		2	1							
6	C	carbon	11.3	FILLED	2	2	Number of electrons						
7	N	nitrogen	14.5	(2)	2	3	in each state						
8	O	oxygen	13.6		2	4							
9	F	fluorine	17.4		2	5							
10	Ne	neon	21.6		2	6							
11	Na	sodium	5.1				1						
12	Mg	magnesium	7.6				2						
13	Al	aluminum	6.0				2	1					
14	Si	silicon	8.1	—FILLED—			2	2					
15	P	phosphorus	10.5				2	3					
16	S	sulfur	10.4	(2)		(8)	2	4					
17	Cl	chlorine	13.0				2	5					
18	A	argon	15.8				2	6					
19	K	potassium	4.3							1			
20	Ca	calcium	6.1							2			
21	Sc	scandium	6.5						1	2			
22	Ti	titanium	6.8						2	2			
23	V	vanadium	6.7				——FILLED——		3	2			
24	Cr	chromium	6.8						5	1			
25	Mn	manganese	7.4	(2)		(8)	(8)		5	2			
26	Fe	iron	7.9						6	2			
27	Co	cobalt	7.9						7	2			
28	Ni	nickel	7.6						8	2			
29	Cu	copper	7.7						10	1			
30	Zn	zinc	9.4						10	2			
31	Ga	gallium	6.0							2	1		
32	Ge	germanium	7.9				——FILLED——			2	2		
33	As	arsenic	9.8							2	3		
34	Se	selenium	9.7	(2)		(8)	(18)			2	4		
35	Br	bromine	11.8							2	5		
36	Kr	krypton	14.0							2	6		

energies, each *l*-value corresponds to a sub-shell of $2(2l + 1)$ possible states (of different *m* and electron spin). These all have the same energy—except for some very small effects we are neglecting.

Be

Beryllium is like lithium except that it has two electrons in the 2*s* state as well as two in the filled 1*s* shell.

B to Ne

Boron has 5 electrons. The fifth must go into a 2*p* state. There are $2 \times 3 = 6$ different 2*p* states, so we can keep adding electrons until we get to a total of 8. This takes us to neon. As we add these electrons we are also increasing *Z*, so the whole electron distribution gets pulled in closer and closer to the nucleus and the energy of the 2*p* states goes down. By the time we get to neon the ionization energy is up to 21.6 volts. Neon does not easily give up an electron. Also there are no more low-energy slots to be filled, so it won't try to grab an extra electron. Neon is chemically inert. Fluorine, on the other hand, *does* have an empty position where an electron can drop into a state of low energy, so it is quite active in chemical reactions.

Na to A

With sodium the eleventh electron must start a new shell—going into a 3*s* state. The energy level of this state is much higher; the ionization energy jumps down; and sodium is an active chemical. From sodium to argon the *s* and *p* states with $n = 3$ are occupied in exactly the same sequence as for lithium to neon. Angular configurations of the electrons in the outer unfilled shell have the same sequence, and the progression of ionization energies is quite similar. You can see why the chemical properties repeat with increasing atomic number. Magnesium acts chemically much like beryllium, silicon like carbon, and chlorine like fluorine. Argon is inert like neon.

You may have noticed that there is a slight peculiarity in the sequence of ionization energies between lithium and neon, and a similar one between sodium and argon. The last electron is bound to the oxygen atom somewhat less than we might expect. And sulphur is similar. Why should that be? We can understand it if we put in just a little bit of the effects of the interactions between individual electrons. Think of what happens when we put the first 2*p* electron onto the boron atom. It has six possibilities—three possible *p*-states, each with two spins. Imagine that the electron goes with spin up into the $m = 0$ state, which we have also called the "*z*" state because it hugs the *z*-axis. Now what will happen in carbon? There are now two 2*p* electrons. If one of them goes into the "*z*" state, where will the second one go? It will have lower energy if it stays away from the first electron, which it can do by going into, say, the "*x*" state of the 2*p* shell. (This state is, remember, just a linear combination of the $m = +1$ and $m = -1$ states.) Next, when we go to nitrogen, the three 2*p* electrons will have the lowest energy of mutual repulsion if they go one each into the "*x*," "*y*," and "*z*" configurations. For oxygen, however, the jig is up. The fourth electron must go into one of the filled states—with opposite spin. It is strongly repelled by the electron already in that state, so its energy will not be as low as it might otherwise be, and it is more easily removed. That explains the break in the sequence of binding energies which appears between nitrogen and oxygen, and between phosphorus and silicon.

K to Zn

After argon, you would, at first, think that the new electrons would start to fill up the 3*d* states. But they don't. As we described earlier—and illustrated in Fig. 19–7—the higher angular momentum states get pushed up in energy. By the time we get to the 3*d* states they are pushed to an energy a little bit above the energy of the 4*s* state. So in potassium the last electron goes into the 4*s* state. After this shell is filled (with two electrons) at calcium, the $3d$ states begin to be filled for scandium, titanium, and vanadium.

The energies of the $3p$ and $4s$ states are so close together that small effects can shift the balance either way. By the time we get to put four electrons into the $3d$ states, their repulsion raises the energy of the $4s$ state just enough that its energy is slightly above the $3d$ energy, so one electron shifts over. For chromium we don't get a 4, 2 combination as we would have expected, but instead a 5, 1 combination. The new electron added to get manganese fills up the $4s$ shell again, and the states of the $3d$ shell are then occupied one by one until we reach copper.

Since the outermost shell of manganese, iron, cobalt, and nickel have the same configurations, however, they all tend to have similar chemical properties. (This effect is much more pronounced in the rare-earth elements which all have the same outer shell but a progressively filling inner shell which has much less influence on their chemical properties.)

In copper an electron is robbed from the $4s$ shell, finally completing the $3d$ shell. The energy of the 10, 1 combination is, however, so close to the 9, 2 configuration for copper that just the presence of another atom nearby can shift the balance. For this reason the two last electrons of copper are nearly equivalent, and copper can have a valence of either 1 or 2. (It sometimes acts as though its electrons were in the 9, 2 combination.) Similar things happen at other places and account for the fact that other metals, such as iron, combine chemically with either of two valences. By zinc, both the $3d$ and $4s$ shells are filled once and for all.

Ga to Kr

From gallium to krypton the sequence proceeds normally again, filling the $4p$ shell. The outer shells, the energies, and the chemical properties repeat the pattern of boron to neon and aluminum to argon.

Krypton, like argon and neon, is known as "noble" gas. All three are chemically "inert." This means only that, having filled shells of relatively low energy, there are few situations in which there is an energy advantage for them to join in a simple combination with other elements. Having a filled shell is not enough. Beryllium and magnesium have filled s-shells, but the energy of these shells is too high to lead to stability. Similarly, one would have expected another "noble" element at nickel, if the energy of the $3d$ shell had been lower (or the $4s$, higher). On the other hand, krypton is not completely inert; it will form a weakly-bound compound with chlorine.

Since our sample has turned up most of the main features of the periodic table, we stop our examination at element number 36—there are still seventy or so more!

We would like to bring up only one more point—that we not only can understand the valences to some extent but also can say something about the directional properties of the chemical bonds. Take an atom like oxygen which has four $2p$ electrons. The first three go into "x," "y," and "z" states and the fourth will double one of these states, leaving two—say "x" and "y"—vacant. Consider then what happens in H_2O. Each of the two hydrogens are willing to share an electron with the oxygen, helping the oxygen to fill a shell. These electrons will tend to go into the "x" and "y" vacancies. So the water molecule should have the two hydrogen atoms making a right angle with respect to the center of the oxygen. The angle is actually 105°. We can even understand why the angle is larger than 90°. In sharing their electrons the hydrogens end up with a net positive charge. The electric repulsion "strains" the wave functions and pushes the angle out to 105°. The same situation occurs in H_2S. But because the sulphur atom is larger, the two hydrogen atoms are farther apart, there is less repulsion, and the angle is only pushed out to about 93°. Selenium is even larger, so in H_2Se the angle is very nearly 90°.

We can use the same arguments to understand the geometry of ammonia, H_3N. Nitrogen has room for three more $2p$ electrons, on each for the "x," "y," and "z" type states. The three hydrogens should join on at right angles to each other. The angles come out a little larger than 90°—again from the electric repulsion—but at least we see why the molecule of H_3N is not flat. The angles in phosphene, H_3P, are close to 90°, and in H_3As are still closer. We assumed that NH_3 was not flat when we described it as a two-state system. And the nonflatness is what makes the ammonia maser possible. Now we see that also that shape can be understood from our quantum mechanics.

The Schrödinger equation has been one of the great triumphs of physics. By providing the key to the underlying machinery of atomic structure it has given an explanation for atomic spectra, for chemistry, and for the nature of matter.

20

Operators

20–1 Operations and operators

All the things we have done so far in quantum mechanics could be handled with ordinary algebra, although we did from time to time show you some special ways of writing quantum-mechanical quantities and equations. We would like now to talk some more about some interesting and useful mathematical ways of describing quantum-mechanical things. There are many ways of approaching the subject of quantum mechanics, and most books use a different approach from the one we have taken. As you go on to read other books you might not see right away the connections of what you will find in them to what we have been doing. Although we will also be able to get a few useful results, the main purpose of this chapter is to tell you about some of the different ways of writing the same physics. Knowing them you should be able to understand better what other people are saying. When people were first working out classical mechanics they always wrote all the equations in terms of x-, y-, and z-components. Then someone came along and pointed out that all of the writing could be made much simpler by inventing the vector notation. It's true that when you come down to figuring something out you often have to convert the vectors back to their components. But it's generally much easier to see what's going on when you work with vectors and also easier to do many of the calculations. In quantum mechanics we were able to write many things in a simpler way by using the idea of the "state vector." The state vector $|\psi\rangle$ has, of course, nothing to do with geometric vectors in three dimensions but is an abstract symbol that *stands for a physical state*, identified by the "label," or "name," ψ. The idea is useful because the laws of quantum mechanics can be written as algebraic equations in terms of these symbols. For instance, our fundamental law that any state can be made up from a linear combination of base states is written as

$$|\psi\rangle = \sum_i C_i |i\rangle, \tag{20.1}$$

where the C_i are a set of ordinary (complex) numbers—the amplitudes $C_i = \langle i | \psi \rangle$ —while $|1\rangle$, $|2\rangle$, $|3\rangle$, and so on, stand for the base states in some base, or *representation*.

If you take some physical state and do something to it—like rotating it, or like waiting for the time Δt—you get a different state. We say, "performing an operation on a state produces a new state." We can express the same idea by an equation:

$$|\phi\rangle = \hat{A} |\psi\rangle. \tag{20.2}$$

An operation on a state produces another state. The *operator* $\hat{A}$ stands for some particular operation. When this operation is performed on any state, say $|\psi\rangle$, it produces some other state $|\phi\rangle$.

What does Eq. (20.2) mean? We *define* it this way. If you multiply the equation by $\langle i |$ and expand $|\psi\rangle$ according to Eq. (20.1), you get

$$\langle i | \phi \rangle = \sum_j \langle i | \hat{A} | j \rangle \langle j | \psi \rangle. \tag{20.3}$$

(The states $|j\rangle$ are from the same set as $|i\rangle$.) This is now just an algebraic equation. The numbers $\langle i | \phi \rangle$ give the amount of each base state you will find in $|\phi\rangle$, and it is given in terms of a linear superposition of the amplitudes $\langle j | \psi \rangle$ that you find

20–1 Operations and operators

20–2 Average energies

20–3 The average energy of an atom

20–4 The position operator

20–5 The momentum operator

20–6 Angular momentum

20–7 The change of averages with time

$|\psi\rangle$ in each base state. The numbers $\langle i \,|\, \hat{A} \,|\, j \rangle$ are just the coefficients which tell how much of $\langle j \,|\, \psi \rangle$ goes into each sum. The operator $\hat{A}$ is described numerically by the set of numbers, or "matrix,"

$$A_{ij} \equiv \langle i \,|\, \hat{A} \,|\, j \rangle. \tag{20.4}$$

So Eq. (20.2) is a high-class way of writing Eq. (20.3). Actually it is a little more than that; something more is implied. In Eq. (20.2) we do not make any reference to a set of base states. Equation (20.3) is an image of Eq. (20.2) in terms of some set of base states. But, as you know, you may use any set you wish. And this idea is implied in Eq. (20.2). The operator way of writing avoids making any particular choice. Of course, when you want to get definite you have to choose *some* set. When you make your choice, you use Eq. (20.3). So the *operator* equation (20.2) is a more abstract way of writing the *algebraic* equation (20.3). It's similar to the difference between writing

$$\boldsymbol{c} = \boldsymbol{a} \times \boldsymbol{b}$$

instead of

$$c_x = a_y b_z - a_z b_y,$$

$$c_y = a_z b_x - a_x b_y,$$

$$c_z = a_x b_y - a_y b_x.$$

The first way is much handier. When you want *results*, however, you will eventually have to give the components with respect to some set of axes. Similarly, if you want to be able to say what you really mean by $\hat{A}$, you will have to be ready to give the matrix A_{ij} in terms of *some* set of base states. So long as you have in mind some set A_{ij}, Eq. (20.2) means just the same as Eq. (20.3). (You should remember also that once you know a matrix for one particular set of base states you can always calculate the corresponding matrix that goes with any other base. You can transform the matrix from one "representation" to another.)

The operator equation in (20.2) also allows a new way of thinking. If we imagine some operator $\hat{A}$, we can use it with any state $|\psi\rangle$ to create a new state $\hat{A} |\psi\rangle$. Sometimes a "state" we get this way may be very peculiar—it may not represent any *physical* situation we are likely to encounter in nature. (For instance, we may get a state that is not normalized to represent one electron.) In other words, we may at times get "states" that are mathematically artificial. Such artificial "states" may still be useful, perhaps as the mid-point of some calculation.

We have already shown you many examples of quantum-mechanical operators. We have had the rotation operator $\hat{R}_y(\theta)$ which takes a state $|\psi\rangle$ and produces a new state, which is the old state as seen in a rotated coordinate system. We have had the parity (or inversion) operator $\hat{P}$, which makes a new state by reversing all coordinates. We have had the operators $\hat{\sigma}_x$, $\hat{\sigma}_y$, and $\hat{\sigma}_z$ for spin one-half particles.

The operator $\hat{J}_z$ was defined in Chapter 17 in terms of the rotation operator for a small angle ϵ.

$$\hat{R}_z(\epsilon) = 1 + \frac{i}{\hbar} \, \epsilon \, \hat{J}_z. \tag{20.5}$$

This just means, of course, that

$$\hat{R}_z(\epsilon) \,|\, \psi \rangle = |\, \psi \rangle + \frac{i}{\hbar} \, \epsilon \, \hat{J}_z \,|\, \psi \rangle. \tag{20.6}$$

In this example, $\hat{J}_z \,|\, \psi \rangle$ is $\hbar/i\epsilon$ times the state you get if you rotate $|\psi\rangle$ by the small angle ϵ and then subtract the original state. It represents a "state" which is the *difference* of two states.

One more example. We had an operator $\hat{p}_x$—called the momentum operator (x-component) defined in an equation like (20.6). If $\hat{D}_x(L)$ is the operator which displaces a state along x by the distance L, then $\hat{p}_x$ is defined by

$$\hat{D}_x(\delta) = 1 + \frac{i}{\hbar}\,\delta\hat{p}_x, \tag{20.7}$$

where δ is a small displacement. Displacing the state $|\psi\rangle$ along x by a small distance δ gives a new state $|\psi'\rangle$. We are saying that this new state is the old state plus a small new piece

$$\frac{i}{\hbar}\,\delta\hat{p}_x\,|\psi\rangle.$$

The operators we are talking about work on a state vector like $|\psi\rangle$, which is an abstract description of a physical situation. They are quite different from *algebraic* operators which work on mathematical functions. For instance, d/dx is an "operator" that works on $f(x)$ by changing it to a new function $f'(x) = df/dx$. Another example is the algebraic operator ∇^2. You can see why the same word is used in both cases, but you should keep in mind that the two kinds of operators are different. A quantum-mechanical operator $\hat{A}$ does *not* work on an algebraic function, but on a state vector like $|\psi\rangle$. Both kinds of operators are used in quantum mechanics and often in similar kinds of equations, as you will see a little later. When you are first learning the subject it is well to keep the distinction always in mind. Later on, when you are more familiar with the subject, you will find that it is less important to keep any sharp distinction between the two kinds of operators. You will, indeed, find that most books generally use the same notation for both!

We'll go on now and look at some useful things you can do with operators. But first, one special remark. Suppose we have an operator $\hat{A}$ whose matrix in some base is $A_{ij} \equiv \langle i\,|\,\hat{A}\,|\,j\rangle$. The amplitude that the state $\hat{A}\,|\psi\rangle$ is also in some other state $|\phi\rangle$ is $\langle\phi\,|\,\hat{A}\,|\,\psi\rangle$. Is there some meaning to the complex conjugate of this amplitude? You should be able to show that

$$\langle\phi\,|\,\hat{A}\,|\,\psi\rangle^* = \langle\psi\,|\,\hat{A}^\dagger\,|\,\phi\rangle, \tag{20.8}$$

where $\hat{A}^\dagger$ (read "A dagger") is an operator whose matrix elements are

$$A^\dagger_{ij} = (A_{ji})^*. \tag{20.9}$$

To get the i, j element of $A^\dagger$ you go to the j, i element of $\hat{A}$ (the indexes are reversed) and take its complex conjugate. The amplitude that the state $\hat{A}^\dagger\,|\phi\rangle$ is in $|\psi\rangle$ is the complex conjugate of the amplitude that $\hat{A}\,|\psi\rangle$ is in $|\phi\rangle$. The operator $\hat{A}^\dagger$ is called the "Hermitian adjoint" of $\hat{A}$. Many important operators of quantum mechanics have the special property that when you take the Hermitian adjoint, you get the same operator back. If $\hat{B}$ is such an operator, then

$$\hat{B}^\dagger = \hat{B},$$

and it is called a "self-adjoint" or "Hermitian," operator.

20–2 Average energies

So far we have reminded you mainly of what you already know. Now we would like to discuss a new question. How would you find the *average* energy of a system—say, an atom? If an atom is in a particular state of definite energy and you measure the energy, you will find a certain energy E. If you keep repeating the measurement on each one of a whole series of atoms which are all selected to be in the same state, all the measurements will give E, and the "average" of your measurements will, of course, be just E.

Now, however, what happens if you make the measurement on some state $|\psi\rangle$ which is *not* a stationary state? Since the system does not have a definite energy, one measurement would give one energy, the same measurement on another atom in the same state would give a different energy, and so on. What would you get for the average of a whole series of energy measurements?

We can answer the question by projecting the state $|\psi\rangle$ onto the set of states of definite energy. To remind you that this is a special base set, we'll call the states $|\eta_i\rangle$. Each of the states $|\eta_i\rangle$ has a definite energy E_i. In this representation,

$$|\psi\rangle = \sum_i C_i \,|\,\eta_i\rangle. \qquad (20.10)$$

When you make an energy measurement and get some number E_i, you have found that the system was in the state η_i. But you may get a different number for each measurement. Sometimes you will get E_1, sometimes E_2, sometimes E_3, and so on. The *probability* that you observe the energy E_1 is just the probability of finding the system in the state $|\eta_1\rangle$, which is, of course, just the absolute square of the amplitude $C_1 = \langle\eta_1\,|\,\psi\rangle$. The probability of finding each of the possible energies E_i is

$$P_i = |\,C_i\,|^2. \qquad (20.11)$$

How are these probabilities related to the mean value of a whole sequence of energy measurements? Let's imagine that we get a series of measurements like this: E_1, E_7, E_{11}, E_9, E_1, E_{10}, E_7, E_2, E_3, E_9, E_6, E_4, and so on. We continue for, say, a thousand measurements. When we are finished we add all the energies and divide by one thousand. That's what we mean by the average. There's also a short-cut to adding all the numbers. You can count up how many times you get E_1, say that is N_1, and then count up the number of times you get E_2, call that N_2, and so on. The sum of all the energies is certainly just

$$N_1E_1 + N_2E_2 + N_3E_3 + \cdots = \sum_i N_iE_i.$$

The average energy is this sum divided by the total number of measurements which is just the sum of all the N_i's, which we can call N;

$$E_{\text{av}} = \frac{\sum_i N_iE_i}{N}. \qquad (20.12)$$

We are almost there. What we *mean* by the probability of something happening is just the number of times we expect it to happen divided by the total number of tries. The ratio N_i/N should—for large N—be very near to P_i, the probability of finding the state $|\eta_i\rangle$, although it will not be exactly P_i because of the statistical fluctuations. Let's write the predicted (or "expected") average energy as $\langle E\rangle_{\text{av}}$; then we can say that

$$\langle E\rangle_{\text{av}} = \sum_i P_iE_i. \qquad (20.13)$$

The same arguments apply for any measurement. The average value of a measured quantity A should be equal to

$$\langle A\rangle_{\text{av}} = \sum_i P_iA_i,$$

where A_i are the various possible values of the observed quantity, and P_i is the probability of getting that value.

Let's go back to our quantum-mechanical state $|\psi\rangle$. It's average energy is

$$\langle E\rangle_{\text{av}} = \sum_i |C_i|^2E_i = \sum_i C_i^*C_iE_i. \qquad (20.14)$$

Now watch this trickery! First, we write the sum as

$$\sum_i \langle\psi\,|\,\eta_i\rangle E_i\langle\eta_i\,|\,\psi\rangle. \qquad (20.15)$$

Next we treat the left-hand $\langle\psi\,|$ as a common "factor." We can take this factor out of the sum, and write it as

$$\langle\psi\,|\,\left\{\sum_i \,|\,\eta_i\rangle E_i\langle\eta_i\,|\,\psi\rangle\right\}.$$

This expression has the form

$$\langle \psi \mid \phi \rangle,$$

where $\mid \phi \rangle$ is some "cooked-up" state defined by

$$\mid \phi \rangle = \sum_i \mid \eta_i \rangle E_i \langle \eta_i \mid \psi \rangle. \qquad (20.16)$$

It is, in other words, the state you get if you take each base state $\mid \eta_i \rangle$ in the amount $E_i \langle \eta_i \mid \psi \rangle$.

Now remember what we mean by the states $\mid \eta_i \rangle$. They are supposed to be the stationary states—by which we mean that for each one,

$$\hat{H} \mid \eta_i \rangle = E_i \mid \eta_i \rangle.$$

Since E_i is just a number, the right-hand side is the same as $\mid \eta_i \rangle E_i$, and the sum in Eq. (20.16) is the same as

$$\sum_i \hat{H} \mid \eta_i \rangle \langle \eta_i \mid \psi \rangle.$$

Now i appears only in the famous combination that contracts to unity, so

$$\sum_i \hat{H} \mid \eta_i \rangle \langle \eta_i \mid \psi \rangle = \hat{H} \sum_i \mid \eta_i \rangle \langle \eta_i \mid \psi \rangle = \hat{H} \mid \psi \rangle.$$

Magic! Equation (20.16) is the same as

$$\mid \phi \rangle = \hat{H} \mid \psi \rangle. \qquad (20.17)$$

The average energy of the state $\mid \psi \rangle$ can be written very prettily as

$$\langle E \rangle_{\text{av}} = \langle \psi \mid \hat{H} \mid \psi \rangle. \qquad (20.18)$$

To get the average energy you operate on $\mid \psi \rangle$ with $\hat{H}$, and then multiply by $\langle \psi \mid$. A simple result.

Our new formula for the average energy is not only pretty. It is also useful, because now we don't need to say anything about any particular set of base states. We don't even have to know all of the possible energy levels. When we go to calculate, we'll need to describe our state in terms of *some* set of base states, but if we know the Hamiltonian matrix H_{ij} for *that* set we can get the average energy. Equation (19.18) says that for *any* set of base states $\mid i \rangle$, the average energy can be calculated from

$$\langle E \rangle_{\text{av}} = \sum_{ij} \langle \psi \mid i \rangle \langle i \mid \hat{H} \mid j \rangle \langle j \mid \psi \rangle, \qquad (20.19)$$

where the amplitudes $\langle i \mid H \mid j \rangle$ are just the elements of the matrix H_{ij}.

Let's check this result for the special case that the states $\mid i \rangle$ are the definite energy states. For them, $\hat{H} \mid j \rangle = E_j \mid j \rangle$, so $\langle i \mid \hat{H} \mid j \rangle = E_j \delta_{ij}$ and

$$\langle E \rangle_{\text{av}} = \sum_{ij} \langle \psi \mid i \rangle E_i \delta_{ij} \langle j \mid \psi \rangle = \sum_i E_i \langle \psi \mid i \rangle \langle i \mid \psi \rangle,$$

which is right.

Equation (20.19) can, incidentally, be extended to other physical measurements which you can express as an operator. For instance, $\hat{L}_z$ is the operator of the z-component of the angular momentum $\mathbf{L}$. The average of the z-component for the state $\mid \psi \rangle$ is

$$\langle L_z \rangle_{\text{av}} = \langle \psi \mid \hat{L}_z \mid \psi \rangle.$$

One way to prove it is to think of some situation in which the energy is proportional to the angular momentum. Then all the arguments go through in the same way.

In summary, if a physical observable A is related to a suitable quantum-mechanical operator $\hat{A}$, the average value of A for the state $|\psi\rangle$ is given by

$$\langle A \rangle_{av} = \langle \psi \mid \hat{A} \mid \psi \rangle. \tag{20.20}$$

By this we mean that

$$A_{av} = \langle \psi \mid \phi \rangle, \tag{20.21}$$

with

$$\mid \phi \rangle = \hat{A} \mid \psi \rangle. \tag{20.22}$$

20–3 The average energy of an atom

Suppose we want the average energy of an atom in a state described by a wave function $\psi(r)$; How do we find it? Let's first think of a one-dimensional situation with a state $|\psi\rangle$ defined by the amplitude $\langle x \mid \psi \rangle = \psi(x)$. We are asking for the special case of Eq. (20.19) applied to the coordinate representation. Following our usual procedure, we replace the states $|i\rangle$ and $|j\rangle$ by $|x\rangle$ and $|x'\rangle$, and change the sums to integrals. We get

$$\langle E \rangle_{av} = \iint \langle \psi \mid x \rangle \langle x \mid \hat{H} \mid x' \rangle \langle x' \mid \psi \rangle \, dx \, dx'. \tag{20.23}$$

This integral can, if we wish, be written in the following way:

$$\int \langle \psi \mid x \rangle \langle x \mid \phi \rangle \, dx, \tag{20.24}$$

with

$$\langle x \mid \phi \rangle = \int \langle x \mid \hat{H} \mid x' \rangle \langle x' \mid \psi \rangle \, dx'. \tag{20.25}$$

The integral over x' in (20.25) is the same one we had in Chapter 16—see Eq. (16.50) and Eq. (16.52)—and is equal to

$$-\frac{\hbar^2}{2m} \frac{d^2}{dx^2} \psi(x) + V(x)\psi(x).$$

We can therefore write

$$\langle x \mid \phi \rangle = \left\{ -\frac{\hbar^2}{2m} \frac{d^2}{dx^2} + V(x) \right\} \psi(x). \tag{20.26}$$

Remember that $\langle \psi \mid x \rangle = \langle x \mid \psi \rangle^* = \psi^*(x)$; using this equality, the average energy in Eq. (20.23) can be written as

$$\langle E \rangle_{av} = \int \psi^*(x) \left\{ -\frac{\hbar^2}{2m} \frac{d^2}{dx^2} + V \right\} \psi(x) \, dx. \tag{20.27}$$

Given a wave function $\psi(x)$, you can get the average energy by doing this integral. You can begin to see how we can go back and forth from the state-vector ideas to the wave-function ideas.

The quantity in the braces of Eq. (20.27) is an *algebraic* operator.‡ We will write it as $\hat{\mathcal{H}}$

$$\hat{\mathcal{H}} = -\frac{\hbar^2}{2m} \frac{d^2}{dx^2} + V.$$

With this notation Eq. (20.23) becomes

$$\langle E \rangle_{av} = \int \psi^*(x) \hat{\mathcal{H}} \psi(x) \, dx. \tag{20.28}$$

The algebraic operator $\hat{\mathcal{H}}$ defined here is, of course, not identical to the quantum-mechanical operator $\hat{H}$. The new operator works on a function of position $\psi(x) = \langle x \mid \psi \rangle$ to give a new function of x, $\phi(x) = \langle x \mid \phi \rangle$; while $\hat{H}$

---

‡ The "operator" $V(x)$ means "multiply by $V(x)$."

operates on a state vector $|\psi\rangle$ to give another state vector $|\phi\rangle$, without implying the coordinate representation or any particular representation at all. Nor is $\hat{\mathcal{H}}$ strictly the same as $\hat{H}$ even in the coordinate representation. If we choose to work in the coordinate representation, we would interpret $\hat{H}$ in terms of a matrix $\langle x \mid \hat{H} \mid x'\rangle$ which depends somehow on the two "indices" x and x'; that is, we expect—according to Eq. (20.25)—that $\langle x \mid \phi \rangle$ is related to all the amplitudes $\langle x \mid \psi \rangle$ by an integration. On the other hand, we find that $\hat{\mathcal{H}}$ is a differential operator. We have already worked out in Section 16–5 the connection between $\langle x \mid \hat{H} \mid x'\rangle$ and the algebraic operator $\hat{\mathcal{H}}$.

We should make one qualification on our results. We have been assuming that the amplitude $\psi(x) = \langle x \mid \psi \rangle$ is normalized. By this we mean that the scale has been chosen so that

$$\int |\psi(x)|^2\, dx = 1;$$

so the probability of finding the electron *somewhere* is unity. If you should choose to work with a $\psi(x)$ which is not normalized you should write

$$\langle E \rangle_{\mathrm{av}} = \frac{\int \psi^*(x)\hat{\mathcal{H}}\psi(x)\, dx}{\int \psi^*(x)\psi(x)\, dx}. \qquad (20.29)$$

It's the same thing.

Notice the similarity in form between Eq. (20.28) and Eq. (20.18). These two ways of writing the same result appear often when you work with the x-representation. You can go from the first form to the second with any $\hat{A}$ which is a *local* operator, where a local operator is one which in the integral

$$\int \langle x \mid \hat{A} \mid x'\rangle\langle x' \mid \psi \rangle\, dx'$$

can be written as $\hat{\alpha}\,\psi(x)$, where $\hat{\alpha}$ is a differential algebraic operator. There are, however, operators for which this is not true. For them you must work with the basic equations in (20.21) and (20.22).

You can easily extend the derivation to three dimensions. The result is that‡

$$\langle E \rangle_{\mathrm{av}} = \int \psi(\mathbf{r})\hat{\mathcal{H}}\psi(\mathbf{r})d\,\mathrm{Vol}, \qquad (20.30)$$

with

$$\hat{\mathcal{H}} = -\frac{\hbar^2}{2m}\nabla^2 + V(\mathbf{r}), \qquad (20.31)$$

and with the understanding that

$$\int |\psi|^2 d\,\mathrm{Vol} = 1. \qquad (20.32)$$

The same equations can be extended to systems with several electrons in a fairly obvious way, but we won't bother to write down the results.

With Eq. (20.30) we can calculate the average energy of an atomic state even without knowing its energy levels. All we need is the wave function. It's an important law. We'll tell you about one interesting application. Suppose you want to know the ground-state energy of some system—say the helium atom, but it's too hard to solve Schrödinger's equation for the wave function, because there are too many variables. Suppose, however, that you take a guess at the wave function—pick any function you like—and calculate the average energy. That is, you use Eq. (20.29)—generalized to three dimensions—to find what the average energy would be if the atom were really in the state described by this wave function. This energy will certainly be higher than the ground-state energy which is the lowest

---

‡ We write d Vol for the element of volume. It is, of course, just $dx\, dy\, dz$, and the integral goes from $-\infty$ to $+\infty$ in all three coordinates.

possible energy the atom can have.‡ Now pick another function and calculate its average energy. If it is lower than your first choice you are getting closer to the true ground-state energy. If you keep on trying all sorts of artificial states you will be able to get lower and lower energies, which come closer and closer to the ground-state energy. If you are clever, you will try some functions which have a few adjustable parameters. When you calculate the energy it will be expressed in terms of these parameters. By varying the parameters to give the lowest possible energy, you are trying out a whole class of functions at once. Eventually you will find that it is harder and harder to get lower energies and you will begin to be convinced that you are fairly close to the lowest possible energy. The helium atom has been solved in just this way—not by solving a differential equation, but by making up a special function with a lot of adjustable parameters which are eventually chosen to give the lowest possible value for the average energy.

20–4 The position operator

What is the average value of the position of an electron in an atom? For any particular state $|\psi\rangle$ what is the average value of the coordinate x? We'll work in one dimension and let you extend the ideas to three dimensions or to systems with more than one particle. We have a state described by $\psi(x)$, and we keep measuring x over and over again. What is the average? It is

$$\int xP(x)\,dx,$$

where $P(x)$ is the probability of finding the electron in a little element dx at x. Suppose the probability density $P(x)$ varies with x as shown in Fig. 20–1. The electron is most likely to be found near the peak of the curve. The average value of x is also somewhere near the peak. It is, in fact, just the center of gravity of the area under the curve.

We have seen earlier that $P(x)$ is just $|\psi(x)|^2 = \psi^*(x)\psi(x)$, so we can write the average of x as

$$\langle x\rangle_{\text{av}} = \int \psi^*(x)x\psi(x)\,dx. \tag{20.33}$$

Our equation for $\langle x\rangle_{\text{av}}$ has the same form as Eq. (20.33). For the average energy, the energy operator $\hat{\mathcal{H}}$ appears between the two ψ's, for the average position there is just x. (If you wish you can consider x to be the algebraic operator "multiply by x.") We can carry the parallelism still further, expressing the average position in a form which corresponds to Eq. (20.18). Suppose we just write

$$\langle x\rangle_{\text{av}} = \langle \psi \mid \alpha\rangle \tag{20.34}$$

with

$$|\alpha\rangle = \hat{x}\mid\psi\rangle, \tag{20.35}$$

and then see if we can find the operator $\hat{x}$ which generates the state $|\alpha\rangle$, which will make Eq. (20.34) agree with Eq. (20.33). That is, we must find a $|\alpha\rangle$, so that

$$\langle\psi\mid\alpha\rangle = \langle x\rangle_{\text{av}} = \int \langle\psi\mid x\rangle x\langle x\mid\psi\rangle\,dx. \tag{20.36}$$

First, let's expand $\langle\psi\mid\phi\rangle$ in the x-representation. It is

$$\langle\psi\mid\alpha\rangle = \int \langle\psi\mid x\rangle\langle x\mid\alpha\rangle\,dx. \tag{20.37}$$

Now compare the integrals in the last two equations. You see that in the x-representation

$$\langle x\mid\alpha\rangle = x\langle x\mid\psi\rangle. \tag{20.38}$$

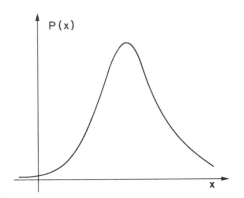

Fig. 20–1. A curve of probability density representing a localized particle.

P(x)

x

---

‡ You can also look at it this way. Any function (that is, state) you choose can be written as a linear combination of the base states which are definite energy states. Since in this combination there is a mixture of higher energy states in with the lowest energy state, the average energy will be higher than the ground-state energy.

Operating on $|\psi\rangle$ with $\hat{x}$ to get $|\alpha\rangle$ is equivalent to multiplying $\psi(x) = \langle x\,|\,\psi\rangle$ by x to get $\alpha(x) = \langle x\,|\,\alpha\rangle$. We have a definition of $\hat{x}$ in the coordinate representation.‡

[We have not bothered to try to get the x-representation of the matrix of the operator $\hat{x}$. If you are ambitious you can try to show that

$$\langle x\,|\,\hat{x}\,|\,x'\rangle = x\,\delta(x - x'). \tag{20.39}$$

You can then work out the amusing result that

$$\hat{x}\,|\,x\rangle = x\,|\,x\rangle. \tag{20.40}$$

The operator $\hat{x}$ has the interesting property that when it works on the base states $|\,x\rangle$ it is equivalent to multiplying by x.]

Do you want to know the average value of x^2? It is

$$\langle x^2\rangle_{\mathrm{av}} = \int \psi^*(x) x^2 \psi(x)\,dx. \tag{20.41}$$

Or, if you prefer you can write

$$\langle x^2\rangle_{\mathrm{av}} = \langle \psi\,|\,\alpha'\rangle$$

with

$$|\,\alpha'\rangle = \hat{x}^2\,|\,\psi\rangle. \tag{20.42}$$

By $\hat{x}^2$ we mean $\hat{x}\hat{x}$—the two operators are used one after the other. With the second form you can calculate $\langle x^2\rangle_{\mathrm{av}}$ using any representation (base-states) you wish. If you want the average of x^n, or of any polynomial in x, you can see how to get it.

20–5 The momentum operator

Now we would like to calculate the mean *momentum* of an electron—again, we'll stick to one dimension. Let $P(p)\,dp$ be the probability that a measurement will give a momentum between p and $p + dp$. Then

$$\langle p\rangle_{\mathrm{av}} = \int p\,P(p)\,dp. \tag{20.43}$$

Now we let $\langle p\,|\,\psi\rangle$ be the amplitude that the state $|\,\psi\rangle$ is in a definite momentum state $|\,p\rangle$. This is the same amplitude we called $\langle \mathrm{mom}\,p\,|\,\psi\rangle$ in Section 16–3 and is a function of p just as $\langle x\,|\,\psi\rangle$ is a function of x. There we chose to normalize the amplitude so that

$$P(p) = \frac{1}{2\pi\hbar}\,|\langle p\,|\,\psi\rangle|^2. \tag{20.44}$$

We have, then,

$$\langle p\rangle_{\mathrm{av}} = \int \langle \psi\,|\,p\rangle\,p\,\langle p\,|\,\psi\rangle\,\frac{dp}{2\pi\hbar}. \tag{20.45}$$

The form is quite similar to what we had for $\langle x\rangle_{\mathrm{av}}$.

If we want, we can play exactly the same game we did with $\langle x\rangle_{\mathrm{av}}$. First, we can write the integral above as

$$\int \langle \psi\,|\,p\rangle\langle p\,|\,\beta\rangle\,\frac{dp}{2\pi\hbar}. \tag{20.46}$$

You should now recognize this equation as just the expanded form of the amplitude $\langle \psi\,|\,\beta\rangle$—expanded in terms of the base states of definite momentum. From Eq.

---

‡ Equation (20.38) does *not* mean that $|\,\alpha\rangle = x\,|\,\psi\rangle$. You cannot "factor out" the $\langle x\,|$, because the multiplier x in front of $\langle x\,|\,\psi\rangle$ is a number which is different for each state $\langle x\,|$. It is the value of the coordinate of the electron in the state $|\,x\rangle$. See Eq. (20.40).

(20.45) the state $|\beta\rangle$ is defined *in the momentum representation* by

$$\langle p \,|\, \beta\rangle = p\langle p \,|\, \psi\rangle \tag{20.47}$$

That is, we can now write

$$\langle p\rangle_{\text{av}} = \langle \psi \,|\, \beta\rangle \tag{20.48}$$

with

$$|\beta\rangle = \hat{p}\,|\,\psi\rangle, \tag{20.49}$$

where the operator $\hat{p}$ is defined in terms of the p-representation by Eq. (20.47).

[Again, you can if you wish show that the matrix form of $\hat{p}$ is

$$\langle p \,|\, \hat{p} \,|\, p'\rangle = p\,\delta(p - p'), \tag{20.50}$$

and that

$$\hat{p}\,|\,p\rangle = p\,|\,p\rangle. \tag{20.51}$$

It works out the same as for x.]

Now comes an interesting question. We can write $\langle p\rangle_{\text{av}}$ as we have done in Eqs. (20.45) and (20.48), and we know the meaning of the operator $\hat{p}$ *in the momentum representation.* But how should we interpret $\hat{p}$ in the *coordinate* representation? That is what we will need to know if we have some wave function $\psi(x)$, and we want to compute its average momentum. Let's make clear what we mean. If we start by saying that $\langle p\rangle_{\text{av}}$ is given by Eq. (20.48), we can expand that equation in terms of the p-representation to get back to Eq. (20.45). If we are given the p-description of the state—namely the amplitude $\langle p \,|\, \psi\rangle$, which is an algebraic function of the momentum p—we can get $\langle p \,|\, \phi\rangle$ from Eq. (20.47) and proceed to evaluate the integral. The question now is: What do we do if we are given a description of the state in the x-representation, namely the wave function $\psi(x) = \langle x \,|\, \psi\rangle$?

Well, let's start by expanding Eq. (20.48) in the x-representation. It is

$$\langle p\rangle_{\text{av}} = \int \langle \psi \,|\, x\rangle\langle x \,|\, \beta\rangle\,dx. \tag{20.52}$$

Now, however, we need to know what the state $|\beta\rangle$ is in the x-representation. If we can find it, we can carry out the integral. So our problem is to find the function $\beta(x) = \langle x \,|\, \beta\rangle$.

We can find it in the following way. In Section 16–3 we saw how $\langle p \,|\, \beta\rangle$ was related to $\langle x \,|\, \beta\rangle$. According to Eq. (16.24),

$$\langle p \,|\, \beta\rangle = \int e^{-ipx/\hbar}\langle x \,|\, \beta\rangle\,dx. \tag{20.53}$$

If we know $\langle p \,|\, \beta\rangle$ we can solve this equation for $\langle x \,|\, \beta\rangle$. What we want, of course, is to express the result somehow in terms of $\psi(x) = \langle x \,|\, \psi\rangle$, which we are assuming to be known. Suppose we start with Eq. (20.47) and again use Eq. (16.24) to write

$$\langle p \,|\, \beta\rangle = p\langle p \,|\, \psi\rangle = p\int e^{-ipx/\hbar}\psi(x)\,dx. \tag{20.54}$$

Since the integral is over x we can put the p inside the integral and write

$$\langle p \,|\, \beta\rangle = \int e^{-ipx/\hbar}p\psi(x)\,dx. \tag{20.55}$$

Compare this with (20.53). You would say that $\langle x \,|\, \beta\rangle$ is equal to $p\psi(x)$. No, No! The wave function $\langle x \,|\, \beta\rangle = \beta(x)$ can depend only on x—not on p. That's the whole problem.

However, some ingenious fellow discovered that the integral in (20.55) could be integrated by parts. The derivative of $e^{-ipx/\hbar}$ with respect to x is $(-i/\hbar)pe^{-ipx/\hbar}$, so the integral in (20.55) is equivalent to

$$-\frac{\hbar}{i}\int \frac{d}{dx}\,(e^{-ipx/\hbar})\psi(x)\,dx.$$

If we integrate by parts, it becomes

$$-\frac{\hbar}{i}\left[e^{-ipx/\hbar}\psi(x)\right]_{-\infty}^{+\infty} + \frac{\hbar}{i}\int e^{-ipx/\hbar}\frac{d\psi}{dx}\,dx.$$

So long as we are considering bound states, so that $\psi(x)$ goes to zero at $x = \pm\infty$, the bracket is zero and we have

$$\langle p \mid \beta \rangle = \frac{\hbar}{i}\int e^{-ipx/\hbar}\frac{d\psi}{dx}\,dx. \qquad (20.56)$$

Now compare this result with Eq. (20.53). You see that

$$\langle x \mid \beta \rangle = \frac{\hbar}{i}\frac{d}{dx}\psi(x). \qquad (20.57)$$

We have the necessary piece to be able to complete Eq. (20.52). The answer is

$$\langle p \rangle_{\mathrm{av}} = \int \psi^*(x)\frac{\hbar}{i}\frac{d}{dx}\psi(x)\,dx. \qquad (20.58)$$

We have found how Eq. (20.48) looks in the coordinate representation.

Now you should begin to see an interesting pattern developing. When we asked for the average energy of the state $|\psi\rangle$ we said it was

$$\langle E \rangle_{\mathrm{av}} = \langle \psi \mid \phi \rangle, \quad \text{with} \quad |\phi_1\rangle = \hat{H}\,|\psi\rangle.$$

The same thing is written in the coordinate world as

$$\langle E \rangle_{\mathrm{av}} = \int \psi^*(x)\phi(x)\,dx \quad \text{with} \quad \phi(x) = \hat{\mathcal{H}}\psi(x).$$

Here $\hat{\mathcal{H}}$ is an *algebraic* operator which works a function of x. When we asked about the average value of x, we found that it could also be written

$$\langle x \rangle_{\mathrm{av}} = \langle \psi \mid \alpha \rangle, \quad \text{with} \quad |\alpha\rangle = \hat{x}\,|\psi\rangle.$$

In the coordinate world the corresponding equations are

$$\langle x \rangle_{\mathrm{av}} = \int \psi^*(x)\alpha(x)\,dx, \quad \text{with} \quad \alpha(x) = x\psi(x).$$

When we asked about the average value of p, we wrote

$$\langle p \rangle_{\mathrm{av}} = \langle \psi \mid \beta \rangle, \quad \text{with} \quad |\beta\rangle = \hat{p}\,|\psi\rangle.$$

In the coordinate world the equivalent equations were

$$\langle p \rangle_{\mathrm{av}} = \int \psi(x)\beta(x)\,dx, \quad \text{with} \quad \beta(x) = \frac{\hbar}{i}\frac{d}{dx}\psi(x).$$

In each of our three examples we start with the state $|\psi\rangle$ and produce another (hypothetical) state by a *quantum-mechanical* operator. In the coordinate representation we generate the corresponding wave function by operating on the wave function $\psi(x)$ with an *algebraic* operator. There are the following one-to-one correspondences (for one-dimensional problems):

$$\hat{H} \rightarrow \hat{\mathcal{H}} = -\frac{\hbar^2}{2m}\frac{d^2}{dx^2} + V(x),$$

$$\hat{x} \rightarrow x, \qquad (20.59)$$

$$\hat{p}_x \rightarrow \hat{\mathcal{P}}_x = \frac{\hbar}{i}\frac{\partial}{\partial x}.$$

Table 20-1

Physical Quantity	Operator	Coordinate Form
Energy	$\hat{H}$	$\hat{\mathcal{H}} = -\dfrac{\hbar^2}{2m} \nabla^2 + V(r)$
Position	$\hat{x}$	x
	$\hat{y}$	y
	$\hat{z}$	z
Momentum	$\hat{p}_x$	$\hat{\mathcal{P}}_x = \dfrac{\hbar}{i} \dfrac{\partial}{\partial x}$
	$\hat{p}_y$	$\hat{\mathcal{P}}_y = \dfrac{\hbar}{i} \dfrac{\partial}{\partial y}$
	$\hat{p}_z$	$\hat{\mathcal{P}}_z = \dfrac{\hbar}{i} \dfrac{\partial}{\partial z}$

In this list, we have introduced the symbol $\mathcal{P}_x$ for the algebraic operator $(\hbar/i)\partial/\partial x$:

$$\hat{\mathcal{P}}_x = \frac{\hbar}{i} \frac{\partial}{\partial x}, \tag{20.60}$$

and we have inserted the x subscript on $\hat{\mathcal{P}}$ to remind you that we have been working only with the x-component of momentum.

You can easily extend the results to three dimensions. For the other components of the momentum,

$$\hat{p}_y \rightarrow \hat{\mathcal{P}}_y = \frac{\hbar}{i} \frac{\partial}{\partial y},$$

$$\hat{p}_z \rightarrow \hat{\mathcal{P}}_z = \frac{\hbar}{i} \frac{\partial}{\partial z}.$$

If you want, you can even think of an operator of the *vector* momentum and write

$$\hat{p} \rightarrow \hat{\mathcal{P}} = \frac{\hbar}{i} \left(e_x \frac{\partial}{\partial x} + e_y \frac{\partial}{\partial y} + e_z \frac{\partial}{\partial z} \right),$$

where e_x, e_y, and e_z are the unit vectors in the three directions. It looks even more elegant if we write

$$\hat{p} \rightarrow \hat{\mathcal{P}} = \frac{\hbar}{i} \nabla. \tag{20.61}$$

Our general result is that for at least some quantum-mechanical operators, there are corresponding algebraic operators in the coordinate representation. We summarize our results so far—extended to three dimensions—in Table 20–1. For each operator we have the two equivalent forms:‡

$$| \phi \rangle = \hat{A} | \psi \rangle \tag{20.62}$$

or

$$\phi(r) = \hat{\alpha}\psi(r). \tag{20.63}$$

We will now give a few illustrations of the use of these ideas. The first one is just to point out the relation between $\hat{\mathcal{P}}$ and $\hat{\mathcal{H}}$. If we use $\hat{\mathcal{P}}_x$ twice, we get

$$\hat{\mathcal{P}}_x \hat{\mathcal{P}}_x = -\hbar^2 \frac{\partial^2}{\partial x^2}.$$

---

‡ In many books the same symbol is used for $\hat{A}$ and $\hat{\alpha}$, because they both stand for the same physics, and because it is convenient not to have to write different kinds of letters. You can usually tell which one is intended by the context.

This means that we can write the equality

$$\hat{\mathcal{H}} = \frac{1}{2m}\{\hat{\mathscr{P}}_x\hat{\mathscr{P}}_x + \hat{\mathscr{P}}_y\hat{\mathscr{P}}_y + \hat{\mathscr{P}}_z\hat{\mathscr{P}}_z\} + V(r).$$

Or, using the vector notation,

$$\hat{\mathcal{H}} = \frac{1}{2m}\,\hat{\mathbf{P}}\cdot\hat{\mathbf{P}} + V(r). \tag{20.64}$$

(In an algebraic operator, any term without the operator symbol ($\hat{\ }$) means just a straight multiplication.) This equation is nice because it's easy to remember if you haven't forgotten your classical physics. Everyone knows that the energy is (nonrelativistically) just the kinetic energy $p^2/2m$ plus the potential energy, and $\hat{\mathcal{H}}$ is the operator of the total energy.

This result has impressed people so much that they try to teach students all about classical physics before quantum mechanics. (We think differently!) But such parallels are often misleading. For one thing, when you have operators, the *order* of various factors is important; but that is not true for the factors in a classical equation.

In Chapter 17 we defined an operator $\hat{p}_x$ in terms of the displacement operator $\hat{D}_x$ by [see Eq. (17.27)]

$$|\psi'\rangle = \hat{D}_x(\delta)|\psi\rangle = \left(1 + \frac{i}{\hbar}\,\hat{p}_x\delta\right)|\psi\rangle, \tag{20.65}$$

where δ is a *small* displacement. We should show you that this is equivalent to our new definition. According to what we have just worked out, this equation should mean the same as

$$\psi'(x) = \psi(x) + \frac{\partial\psi}{\partial x}\,\delta.$$

But the right-hand side is just the Taylor expansion of $\psi(x + \delta)$, which is certainly what you get if you displace the state to the left by δ (or shift the coordinates to the right by the same amount). Our two definitions of $\hat{p}$ agree!

Let's use this fact to show something else. Suppose we have a bunch of particles which we label $1, 2, 3, \ldots$, in some complicated system. (To keep things simple we'll stick to one dimension.) The wave function describing the state is a function of all the coordinates $x_1, x_2, x_3, \ldots$ We can write it as $\psi(x_1, x_2, x_3, \ldots)$. Now displace the system (to the left) by δ. The new wave function

$$\psi'(x_1, x_2, x_3, \ldots) = \psi(x_1 + \delta, x_2 + \delta, x_3 + \delta, \ldots)$$

can be written as

$$\psi'(x_1, x_2, x_3, \ldots) = \psi(x_1, x_2, x_3, \ldots)$$
$$+ \left\{\delta\,\frac{\partial\psi}{\partial x_1} + \delta\,\frac{\partial\psi}{\partial x_2} + \delta\,\frac{\partial\psi}{\partial x_3} + \cdots\right\}. \tag{20.66}$$

According to Eq. (20.65) the operator of the momentum of the state $|\psi\rangle$ (let's call it the *total* momentum) is equal to

$$\hat{\mathscr{P}}_{\text{total}} = \frac{\hbar}{i}\left\{\frac{\partial}{\partial x_1} + \frac{\partial}{\partial x_2} + \frac{\partial}{\partial x_3} + \cdots\right\}.$$

But this is just the same as

$$\hat{\mathscr{P}}_{\text{total}} = \hat{\mathscr{P}}_{x1} + \hat{\mathscr{P}}_{x2} + \hat{\mathscr{P}}_{x3} + \cdots. \tag{20.67}$$

The operators of momentum obey the rule that the total momentum is the sum of the momenta of all the parts. Everything holds together nicely, and many of the things we have been saying are consistent with each other.

20–6 Angular momentum

Let's for fun look at another operation—the operation of orbital angular momentum. In Chapter 17 we defined an operator $\hat{J}_z$ in terms of $\hat{R}_z(\varphi)$, the operator of a rotation by the angle φ about the z-axis. We consider here a system described simply by a single wave function $\psi(r)$, which is a function of coordinates only, and does not take into account the fact that the electron may have its spin either up or down. That is, we want for the moment to disregard *intrinsic* angular momentum and think about only the *orbital* part. To keep the distinction clear, we'll call the orbital operator $\hat{L}_z$, and define it in terms of the operator of a rotation by an infinitesimal angle ϵ by

$$\hat{R}_z(\epsilon)\,|\,\psi\rangle = \left(1 + \frac{i}{\hbar}\,\epsilon\,\hat{L}_z\right)|\,\psi\rangle.$$

(Remember, this definition applies only to a state $|\,\psi\rangle$ which has no internal spin variables, but depends only on the coordinates $r = x, y, x$.) If we look at the state $|\,\psi\rangle$ in a new coordinate system, rotated about the z-axis by the small angle ϵ, we see a new state

$$|\,\psi'\rangle = \hat{R}_z(\epsilon)|\,\psi\rangle.$$

If we choose to describe the state $|\,\psi\rangle$ in the coordinate representation—that is, by its wave function $\psi(r)$, we would expect to be able to write

$$\psi'(r) = \left(1 + \frac{i}{\hbar}\,\epsilon\,\hat{\mathcal{L}}_z\right)\psi(x). \tag{20.68}$$

What is $\hat{\mathcal{L}}$? Well, a point P at x and y in the *new* coordinate system (really x' and y', but we will drop the primes) was formerly at $x - \epsilon y$ and $y + \epsilon x$, as you can see from Fig. 20–2. Since the amplitude for the electron to be at P isn't changed by the rotation of the coordinates we can write

$$\psi'(x, y, z) = \psi(x + \epsilon y, y - \epsilon x, z) = \psi(x, y, z) + \epsilon y\,\frac{\partial\psi}{\partial x} - \epsilon x\,\frac{\partial\psi}{\partial y}$$

(remembering that ϵ is a small angle). This means that

$$\hat{\mathcal{L}}_z = \frac{\hbar}{i}\left(x\,\frac{\partial}{\partial y} - y\,\frac{\partial}{\partial x}\right). \tag{20.69}$$

That's our answer. But notice. It is equivalent to

$$\hat{\mathcal{L}}_z = x\hat{\mathcal{P}}_y - y\hat{\mathcal{P}}_x. \tag{20.70}$$

Returning to our quantum-mechanical operators, we can write

$$\hat{L}_z = x\hat{p}_y - y\hat{p}_x. \tag{20.71}$$

This formula is easy to remember because it looks like the familiar formula of classical mechanics; it is the z-component of

$$\boldsymbol{L} = \boldsymbol{r} \times \boldsymbol{p}. \tag{20.72}$$

One of the fun parts of this operator business is that many classical equations get carried over into a quantum-mechanical form. Which ones don't? There had better be some that don't come out right, because if everything did, then there would be nothing different about quantum mechanics. There would be no new physics. Here is one equation which is different. In classical physics

$$xp_x - p_x x = 0.$$

What is it in quantum mechanics?

$$\hat{x}\hat{p}_x - \hat{p}_x\hat{x} = \;?$$

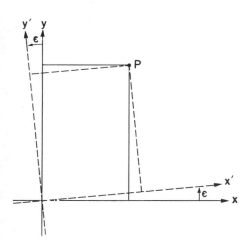

Fig. 20–2. Rotation of the axes around the z-axis by the small angle ϵ.

Let's work it out in the x-representation. So that we'll know what we are doing we put in some wave function $\psi(x)$. We have

$$x \mathcal{P}_x \psi(x) - \hat{\mathcal{P}}_x x \psi(x),$$

or

$$x \frac{\hbar}{i} \frac{\partial}{\partial \psi} \psi(x) - \frac{\hbar}{i} \frac{\partial}{\partial x} x \psi(x).$$

Remember now that the derivatives operate on everything to the right. We get

$$x \frac{\hbar}{i} \frac{\partial \psi}{\partial x} - \frac{\hbar}{i} \psi(x) - \frac{\hbar}{i} x \frac{\partial \psi}{\partial x} = -\frac{\hbar}{i} \psi(x). \tag{20.73}$$

The answer is *not* zero. The whole operation is equivalent simply to multiplication by $-\hbar/i$:

$$\hat{x} \hat{p}_x - \hat{p}_x \hat{x} = -\frac{\hbar}{i}. \tag{20.74}$$

If Plank's constant were zero, the classical and quantum results would be the same, and there would be no quantum mechanics to learn!

Incidentally, if any two operators $\hat{A}$ and $\hat{B}$, when taken together like this:

$$\hat{A}\hat{B} - \hat{B}\hat{A},$$

do *not* give zero, we say that "the operators do not commute." And an equation such as (20.74) is called a "commutation rule." You can see that the commutation rule for p_x and y is

$$\hat{p}_x \hat{y} - \hat{y} \hat{p}_x = 0.$$

There is another very important commutation rule that has to do with angular momenta. It is

$$\hat{L}_x \hat{L}_y - \hat{L}_y \hat{L}_x = i\hbar \hat{L}_z. \tag{20.75}$$

You can get some practice with $\hat{x}$ and $\hat{p}$ operators by proving it for yourself.

It is interesting to notice that operators which do not commute can also occur in classical physics. We have already seen this when we have talked about rotation in space. If you rotate something, such as a book, by 90° around x and then 90° around y, you get something different from rotating first by 90° around y and then by 90° around x. It is, in fact, just this property of space that is responsible for Eq. (20.75).

20–7 The change of averages with time

Now we want to show you something else. How do averages change with time? Suppose for the moment that we have an operator $\hat{A}$, which does not itself have time in it in any obvious way. We mean an operator like $\hat{x}$ or $\hat{p}$. (We exclude things like, say, the operator of some external potential that was being varied with time, such as $V(x, t)$.) Now suppose we calculate $\langle A \rangle_{\text{av}}$, in some state $| \psi \rangle$, which is

$$\langle A \rangle_{\text{av}} = \langle \psi | \hat{A} | \psi \rangle. \tag{20.76}$$

How will $\langle A \rangle_{\text{av}}$ depend on time? Why should it? One reason might be that the operator itself depended explicitly on time—for instance, if it had to do with a time-varying potential like $V(x, t)$. But even if the operator does not depend on t, say, for example, the operator $\hat{A} = \hat{x}$, the corresponding average may depend on time. Certainly the average position of a particle could be moving. How does such a motion come out of Eq. (20.76) if $\hat{A}$ has no time dependence? Well, the state $| \psi \rangle$ might be changing with time. For nonstationary states we have often shown a time dependence explicitly by writing a state as $| \psi(t) \rangle$. We want to show that the rate of change of $\langle A \rangle_{\text{av}}$ is given by a new operator we will call $\dot{\hat{A}}$. Remember that $\hat{A}$ is an operator, so that putting a dot over the A does not here mean taking the time derivative, but is just a way of writing a *new* operator $\hat{A}$ which is defined by

$$\frac{d}{dt}\langle A\rangle_{\mathrm{av}} = \langle\psi\,|\,\hat{A}\,|\,\psi\rangle. \tag{20.77}$$

Our problem is to find the operator $\hat{A}$.

First, we know that the rate of change of a state is given by the Hamiltonian. Specifically,

$$i\hbar\frac{d}{dt}\,|\,\psi(t)\rangle = \hat{H}\,|\,\psi(t)\rangle. \tag{20.78}$$

This is just the abstract way of writing our original definition of the Hamiltonian:

$$i\hbar\frac{dC_i}{dt} = \sum_{ij} H_{ij}C_j. \tag{20.79}$$

If we take the complex conjugate of this equation, it is equivalent to

$$-i\hbar\frac{d}{dt}\langle\psi(t)\,| = \langle\psi(t)\,|\,\hat{H}. \tag{20.80}$$

Next, see what happens if we take the derivatives with respect to t of Eq. (20.76). Since each ψ depends on t, we have

$$\frac{d}{dt}\langle A\rangle_{\mathrm{av}} = \left(\frac{d}{dt}\langle\psi\,|\right)\hat{A}\,|\,\psi\rangle + \langle\psi\,|\,\hat{A}\left(\frac{d}{dt}\,|\,\psi\rangle\right). \tag{20.81}$$

Finally, using the two equations in (20.78) and (20.79) to replace the derivatives, we get

$$\frac{d}{dt}\langle A\rangle_{\mathrm{av}} = \frac{i}{\hbar}\,\{\langle\psi\,|\,\hat{H}\hat{A}\,|\,\psi\rangle - \langle\psi\,|\,\hat{A}\hat{H}\,|\,\psi\rangle\}.$$

This equation is the same as

$$\frac{d}{dt}\langle A\rangle_{\mathrm{av}} = \frac{i}{\hbar}\langle\psi\,|\,(\hat{H}\hat{A} - \hat{A}\hat{H})\,|\,\psi\rangle.$$

Comparing this equation with Eq. (20.77), you see that

$$\hat{A} = \frac{i}{\hbar}\,(\hat{H}\hat{A} - \hat{A}\hat{H}). \tag{20.82}$$

That is our interesting proposition, and it is true for any operator $\hat{A}$.

Incidentally, if the operator $\hat{A}$ should *itself* be time dependent, we would have had

$$\hat{A} = \frac{i}{\hbar}\,(\hat{H}\hat{A} - \hat{A}H) + \frac{\partial\hat{A}}{\partial t}. \tag{20.83}$$

Let us try out Eq. (20.82) on some example to see whether it really makes sense. For instance, what operator corresponds to $\hat{x}$? We say it should be

$$\hat{x} = \frac{i}{\hbar}\,(\hat{H}\hat{x} - \hat{x}\hat{H}). \tag{20.84}$$

What is this? One way to find out is to work it through in the coordinate representation using the algebraic operator for $\mathcal{H}$. In this representation the commutator is

$$\mathcal{H}x - x\mathcal{H} = \left\{\frac{\hbar^2}{2m}\frac{d^2}{dx^2} + V(x)\right\}x - x\left\{\frac{\hbar^2}{2m}\frac{d^2}{dx^2} + V(x)\right\}.$$

If you operate with this or any wave function $\psi(x)$ and work out all of the derivatives where you can, you end up after a little work with

$$-\frac{\hbar^2}{2m}\frac{d\psi}{dx}.$$

But this is just the same as

$$-i \frac{\hbar}{m} \hat{\mathscr{O}}_x \psi,$$

so we find that

$$\hat{H}\hat{x} = \hat{x}\hat{H} = -i \frac{\hbar}{m} p_x \qquad (20.85)$$

or that

$$\dot{x} = \frac{\hat{p}_x}{m}. \qquad (20.86)$$

A pretty result. It means that if the mean value of x is changing with time the drift of the center of gravity is the same as the mean momentum divided by m. Exactly like classical mechanics.

Another example. What is the rate of change of the average momentum of a state? Same game. Its operator is

$$\dot{\hat{p}} = \frac{i}{\hbar} (\hat{H}\hat{p} - \hat{p}\hat{H}). \qquad (20.87)$$

Again you can work it out in the x representation. Remember that $\hat{p}$ becomes d/dx, and this means that you will be taking the derivative of the potential energy V (in the $\hat{\mathscr{H}}$)—but only in the second term. It turns out that it is the only term which does not cancel, and you find that

$$\hat{\mathscr{H}}\hat{\mathscr{P}} - \hat{\mathscr{P}}\hat{\mathscr{H}} = -i\hbar \frac{dV}{dx}$$

or that

$$\dot{\hat{p}} = - \frac{dV}{dx}. \qquad (20.88)$$

Again the classical result. The right-hand side is the force, so we have derived Newton's law! But remember—these are the laws for the *operators* which give the *average* quantities. They do not describe what goes on in detail inside an atom.

Quantum mechanics has the essential difference that $\hat{p}\hat{x}$ is not equal to $\hat{x}\hat{p}$. They differ by a little bit—by the small number $\hbar$. But the whole wondrous complications of interference, waves, and all, result from the little fact that $\hat{x}\hat{p} - \hat{p}\hat{x}$ is not quite zero.

The history of this idea is also interesting. Within a period of a few months in 1926, Heisenberg and Schrödinger independently found correct laws to describe atomic mechanics. Schrödinger invented his wave function $\psi(x)$ and found his equation. Heisenberg, on the other hand, found that nature could be described by classical equations, except that $xp - px$ should be equal to $\hbar/i$, which he could make happen by defining them in terms of special kinds of matrices. In our language he was using the energy-representation, with its matrices. Both Heisenberg's matrix algebra and Schrödinger's differential equation explained the hydrogen atom. A few months later Schrödinger was able to show that the two theories were equivalent—as we have seen here. But the two different mathematical forms of quantum mechanics were discovered independently.

The Schrödinger Equation in a Classical Context: A Seminar on Superconductivity

21–1 Schrödinger's equation in a magnetic field

This lecture is only for entertainment. I would like to give the lecture in a somewhat different style—just to see how it works out. It's not a part of the course —in the sense that it is not supposed to be a last minute effort to teach you something new. But, rather, I imagine that I'm giving a seminar or research report on the subject to a more advanced audience, to people who have already been educated in quantum mechanics. The main difference between a seminar and a regular lecture is that the seminar speaker does not carry out all the steps, or all the algebra. He says: "If you do such and such, this is what comes out," instead of showing all of the details. So in this lecture I'll describe the ideas all the way along but just give you the *results* of the computations. You should realize that you're not supposed to understand everything immediately, but believe (more or less) that things would come out if you went through the steps.

All that aside, this is a subject I *want* to talk about. It is recent and modern and would be a perfectly legitimate talk to give at a research seminar. My subject is the Schrödinger equation in a classical setting—the case of superconductivity.

Ordinarily, the wave function which appears in the Schrödinger equation applies to only one or two particles. And the wave function itself is not something that has a classical meaning—unlike the electric field, or the vector potential, or things of that kind. The wave function for a single particle *is* a "field"—in the sense that it is a function of position—but it does not generally have a classical significance. Nevertheless, there are some situations in which a quantum mechanical wave function *does* have classical significance, and they are the ones I would like to take up. The peculiar quantum mechanical behavior of matter on a small scale doesn't usually make itself felt on a large scale except in the standard way that it produces Newton's laws—the laws of the so-called classical mechanics. But there are certain situations in which the peculiarities of quantum mechanics can come out in a special way on a large scale.

At low temperatures, when the energy of a system has been reduced very, very low, instead of a large number of states being involved, only a very, very small number of states near the ground state are involved. Under those circumstances the quantum mechanical character of that ground state can appear on a macroscopic scale. It is the purpose of this lecture to show a connection between quantum mechanics and large-scale effects—not the usual discussion of the way that quantum mechanics reproduces Newtonian mechanics on the average, but a special situation in which quantum mechanics will produce its own characteristic effects on a large or "macroscopic" scale.

I will begin by reminding you of some of the properties of the Schrödinger equation.† I want to describe the behavior of a particle in a magnetic field using the Schrödinger equation, because the superconductive phenomena are involved with magnetic fields. An external magnetic field is described by a vector potential, and the problem is: what are the laws of quantum mechanics in a vector potential? The principle that describes the behavior of quantum mechanics in a vector potential is very simple. The amplitude that a particle goes from one place to another along a certain route when there's a field present is the same as the ampli-

21–1 Schrödinger's equation in a magnetic field

21–2 The equation of continuity for probabilities

21–3 Two kinds of momentum

21–4 The meaning of the wave function

21–5 Superconductivity

21–6 The Meissner effect

21–7 Flux quantization

21–8 The dynamics of superconductivity

21–9 The Josephson junction

---

† I'm not really reminding you, because I haven't shown you some of these equations before; but remember the spirit of this seminar.

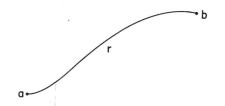

Fig. 21–1. The amplitude to go from a to b along the path Γ is proportional to $\exp (iq/\hbar) \int_a^b \mathbf{A} \cdot d\mathbf{s}$.

tude that it would go along the same route when there's no field, multiplied by the exponential of the line integral of the vector potential, times the electric charge divided by Planck's constant[1] (see Fig. 21–1):

$$\langle b \mid a \rangle_{\text{in } A} = \langle b \mid a \rangle_{A=0} \cdot \exp \left\{ \frac{iq}{\hbar} \int_a^b A \cdot ds \right\}. \tag{21.1}$$

It is a basic statement of quantum mechanics.

Now without the vector potential the Schrödinger equation of a charged particle (nonrelativistic, no spin) is

$$-\frac{\hbar}{i} \frac{\partial \psi}{\partial t} = \hat{\mathcal{H}} \psi = \frac{1}{2m} \left(\frac{\hbar}{i} \nabla \right) \cdot \left(\frac{\hbar}{i} \nabla \right) \psi + q\phi\psi, \tag{21.2}$$

where ϕ is the electric potential so that $q\phi$ is the potential energy.† Equation (21.1) is equivalent to the statement that in a magnetic field the gradients in the Hamiltonian are replaced in each case by the gradient minus qA, so that Eq. (21.2) becomes

$$-\frac{\hbar}{i} \frac{\partial \psi}{\partial t} = \hat{\mathcal{H}} \psi = \frac{1}{2m} \left(\frac{\hbar}{i} \nabla - qA \right) \cdot \left(\frac{\hbar}{i} \nabla - qA \right) \psi + q\phi\psi. \tag{21.3}$$

This is the Schrödinger equation for a particle with charge q moving in an electromagnetic field A, ϕ (nonrelativistic, no spin).

To show that this is true I'd like to illustrate by a simple example in which instead of having a continuous situation we have a line of atoms along the x-axis with the spacing b and we have an amplitude $-K$ for an electron to jump from one atom to another when there is no field.‡ Now according to Eq. (21.1) if there's a vector potential in the x-direction $A_x(x, t)$, the amplitude to jump will be altered from what it was before by a factor $\exp (iq/\hbar A_x b)$, the exponent being $iq/\hbar$ times the vector potential integrated from one atom to the next. For simplicity we will write $(q/\hbar)A_x \equiv f(x)$, since A_x will, in general, depend on x. If the amplitude to find the electron at the atom "n" located at x is called $C(x) \equiv C_n$, then the rate of change of that amplitude is given by the following equation:

$$-\frac{\hbar}{i} \frac{\partial}{\partial t} C(x) = E_0 C(x) - Ke^{-ibf(x+b/2)} C(x + b)$$
$$- Ke^{+ibf(x-b/2)} C(x - b). \tag{21.4}$$

There are three pieces. First, there's some energy E_0 if the electron is located at x. As usual, that gives the term $E_0 C(x)$. Next, there is the term $-KC(x + b)$, which is the amplitude for the electron to have jumped backwards one step from atom "$n + 1$," located at $x + b$. However, in doing so in a vector potential, the phase of the amplitude must be shifted according to the rule in Eq. (21.1). If A_x is not changing appreciably in one atomic spacing, the integral can be written as just the value of A_x at the midpoint, times the spacing b. So $(iq/\hbar)$ times the integral is just $bf(x + b/2)$. Since the electron is jumping backwards, I showed this phase shift with a minus sign. That gives the second piece. In the same manner there's a certain amplitude to have jumped from the other side, but this time we need the vector potential at a distance $(b/2)$ on the other side of x, times the distance b. That gives the third piece. The sum gives the equation for the amplitude to be at x in a vector potential.

Now we know that if the function $C(x)$ is smooth enough (we take the long wavelength limit), and if we let the atoms get closer together, Eq. (16.4) will approach the behavior of an electron in free space. So the next step is to expand both sides of (21.4) in powers of b, assuming b is very small. For example, if b is zero the right-hand side is just $(E_0 - 2K)C(x)$, so in the zeroth approximation

---

[1] Volume II, Section 15–5.

† Not to be confused with our earlier use of ϕ for a state label!

‡ K is the same quantity that was called A in the problem of a linear lattice with no magnetic field. See Chapter 13.

the energy is $E_0 - 2K$. Next comes the terms in b. But because the two exponentials have opposite signs, only even powers of b remain. So if you make a Taylor expansion of $C(x)$, of $f(x)$, and of the exponentials, and then collect the terms in b^2, you get

$$-\frac{\hbar}{i}\frac{\partial C(x)}{\partial t} = E_0 C(x) - 2KC(x)$$
$$- Kb^2\{C''(x) - 2if(x)C'(x) - if'(x)C(x) - f^2(x)C(x)\}. \quad (21.5)$$

(The "primes" mean differentiation with respect to x.)

Now this horrible combination of things looks quite complicated. But mathematically it's exactly the same as

$$-\frac{\hbar}{i}\frac{\partial C(x)}{\partial t} = (E_0 - 2K)C(x) - Kb^2\left[\frac{\partial}{\partial x} - if(x)\right]\left[\frac{\partial}{\partial x} - if(x)\right]C(x). \quad (21.6)$$

The second bracket operating on $C(x)$ gives $C'(x)$ plus $if(x)C(x)$. The first bracket operating on these two terms gives the C'' term and terms in the first derivative of $f(x)$ and the first derivative of $C(x)$. Now remember that the solutions for zero magnetic field[2] represent a particle with an effective mass m_{eff} given by

$$Kb^2 = \frac{\hbar}{m_{eff}}.$$

If you then set $E_0 = -2K$, and put back $f(x) = (q/\hbar)A_x$, you can easily check that Eq. (21.6) is the same as the first part of Eq. (21.3). (The origin of the potential energy term is well known, so I haven't bothered to include it in this discussion.) The proposition of Eq. (21.1) that the vector potential changes all the amplitudes by the exponential factor is the same as the rule that the momentum operator, $(\hbar/i)\nabla$ gets replaced by

$$\frac{\hbar}{i}\nabla - qA,$$

as you see in the Schrödinger equation of (21.3).

21–2 The equation of continuity for probabilities

Now I turn to a second point. An important part of the Schrödinger equation for a single particle is the idea that the probability to find the particle at a position is given by the absolute square of the wave function. It is also characteristic of the quantum mechanics that probability is conserved in a local sense. When the probability of finding the electron somewhere decreases, while the probability of the electron being elsewhere increases (keeping the total probability unchanged), something must be going on in between. In other words, the electron has a continuity in the sense that if the probability decreases at one place and builds up at another place, there must be some kind of flow between. If you put a wall, for example, in the way, it will have an influence and the probabilities will not be the same. So the conservation of probability alone is not the complete statement of the conservation law, just as the conservation of energy alone is not as deep and important as the *local* conservation of energy.[3] If energy is disappearing, there must be a flow of energy to correspond. In the same way, we would like to find a "current" of probability such that if there is any change in the probability density (the probability of being found in a unit volume), it can be considered as coming from an inflow or an outflow due to some current. This current would be a vector which could be interpreted this way—the x component would be the net probability per second and per unit area that a particle passes in the x direction across a plane parallel to the y-z plane. Passage toward $+x$ is considered a positive flow, and passage in the opposite direction, a negative flow.

---

[2] Section 13–3.
[3] Volume II, Section 27–1.

Is there such a current? Well, you know that the probability density $P(r, t)$ is given in terms of the wave function by

$$P(r, t) = \psi^*(r, t)\psi(r, t). \qquad (21.7)$$

I am asking: Is there a current J such that

$$\frac{\partial P}{\partial t} = -\nabla \cdot J? \qquad (21.8)$$

If I take the time derivative of Eq. (21.7), I get two terms:

$$\frac{\partial P}{\partial t} = \psi^* \frac{\partial \psi}{\partial t} + \psi \frac{\partial \psi^*}{\partial t}. \qquad (21.9)$$

Now use the Schrödinger equation—Eq. (21.3)—for $\partial\psi/\partial t$; and take the complex conjugate of it to get $\partial\psi^*/\partial t$—each i gets its sign reversed. You get

$$\frac{\partial P}{\partial t} = -\frac{i}{\hbar} \psi^* \frac{1}{2m} \left(\frac{\hbar}{i}\nabla - qA\right) \cdot \left(\frac{\hbar}{i}\nabla - qA\right)\psi + e\phi\psi^*\psi$$
$$- \psi \frac{1}{2m} \left(\frac{\hbar}{i}\nabla + qA\right) \cdot \left(\frac{\hbar}{i}\nabla + qA\right)\psi^* - e\phi\psi\psi^*. \qquad (21.10)$$

The potential terms and a lot of other stuff cancel out. And it turns out that what is left can indeed be written as a perfect divergence. The whole equation is equivalent to

$$\frac{\partial P}{\partial t} = -\nabla \cdot \left\{ \frac{1}{2m} \psi^* \left(\frac{\hbar}{i}\nabla - qA\right)\psi + \psi \left(-\frac{\hbar}{i}\nabla - qA\right)\psi^* \right\}. \qquad (21.11)$$

It is really not as complicated as it seems. It is a symmetrical combination of ψ^* times a certain operation on ψ, plus ψ^* times the complex conjugate operation on ψ. It is some quantity plus its own complex conjugate, so the whole thing is real—as it ought to be. The operation can be remembered this way: it is just the momentum operator $\hat{\mathcal{P}}$ minus qA. I could write the current in Eq. (21.8) as

$$J = \frac{1}{2} \left\{ \left[\frac{\hat{\mathcal{P}} - qA}{m}\psi\right]^* \psi + \psi^* \left[\frac{\hat{\mathcal{P}} - qA}{m}\right]\psi \right\}. \qquad (21.12)$$

There is then a current J which completes Eq. (21.8).

Equation (21.10) shows that the probability is conserved locally. If a particle disappears from one region it cannot appear in another without something going on in between. Imagine that the first region is surrounded by a closed surface far enough out that there is zero probability to find the electron at the surface. The total probability to find the electron somewhere inside the surface is the volume integral of P. But according to Gauss's theorem the volume integral of the divergence J is equal to the surface integral of J. If ψ is zero at the surface, Eq. (21.10) says that J is zero, so the total probability to find the particle inside can't change. Only if some of the probability approaches the boundary can some of it leak out. We can say that it only gets out by moving through the surface—and that is local conservation.

21–3 Two kinds of momentum

The equation for the current is rather interesting, and sometimes causes a certain amount of worry. You would think the current would be something like the density of particles times the velocity. The density should be something like $\psi\psi^*$, which is o.k. And each term in Eq. (21.12) looks like the typical form for the average-value of the operator

$$\frac{\hat{\mathcal{P}} - qA}{m}, \qquad (21.13)$$

so maybe we should think of it as the velocity of flow. It looks as though we have two suggestions for relations of velocity to momentum, because we would also think that momentum divided by mass, $\hat{\mathcal{P}}/m$, should be a velocity. The two possibilities differ by the vector potential.

It happens that these two possibilities were also discovered in classical physics, when it was found that momentum could be defined in two ways.[4] One of them is called "kinematic momentum," but for absolute clarity I will in this lecture call it the "mv-momentum." This is the momentum obtained by multiplying mass by velocity. The other is a more mathematical, more abstract momentum, sometimes called the "dynamical momentum," which I'll call "p-momentum." The two possibilities are

$$mv\text{-momentum} = mv, \qquad (21.14)$$

$$p\text{-momentum} = mv + qA. \qquad (21.15)$$

It turns out that in quantum mechanics with magnetic fields it is the p-momentum which is connected to the gradient operator $\hat{\mathcal{P}}$, so it follows that (21.13) is the operator of a velocity.

I'd like to make a brief digression to show you what this is all about—why there must be something like Eq. (21.15) in the quantum mechanics. The wave function changes with time according to the Schrödinger equation in Eq. (21.3). If I would suddenly change the vector potential, the wave function wouldn't change at the first instant; only its rate of change changes. Now think of what would happen in the following circumstance. Suppose I have a long solenoid, in which I can produce a flux of magnetic field (B-field), as shown in Fig. 21–2. And there is a charged particle sitting nearby. Suppose this flux nearly instantaneously builds up from zero to something. I start with zero vector potential and then I turn on a vector potential. That means that I produce suddenly a circumferential vector potential A. You'll remember that the line integral of A around a loop is the same as the flux of B through the loop.[5] Now what happens if I suddenly turn on a vector potential? According to the quantum mechanical equation the sudden change of A does not make a sudden change of ψ; the wave function is still the same. So the gradient is also unchanged.

But remember what happens electrically when I suddenly turn on a flux. During the short time that the flux is rising, there's an electric field generated whose line integral is the rate of change of the flux with time:

$$E = -\frac{\partial A}{\partial t}. \qquad (21.16)$$

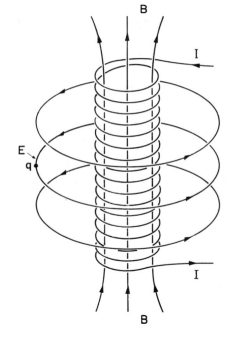

Fig. 21–2. The electric field outside a solenoid with an increasing current.

That electric field is enormous if the flux is changing rapidly, and it gives a force on the particle. The force is the charge times the electric field, and so during the build up of the flux the particle obtains a total impulse (that is, a change in mv) equal to $-qA$. In other words, if you suddenly turn on a vector potential at a charge, this charge immediately picks up an "mv" momentum equal to $-qA$. But there is something that isn't changed immediately and that's the difference between mv and $-qA$. And so the sum $p = mv + qA$ is something which is not changed when you make a sudden change in the vector potential. This quantity p is what we have called the p-momentum and is of importance in classical mechanics in the theory of dynamics, but it also has a direct significance in quantum mechanics. It depends on the character of the wave function, and it is the one to be identified with the operator

$$\hat{\mathcal{P}} = \frac{\hbar}{i}\,\boldsymbol{\nabla}.$$

[4] See, for example, J. D. Jackson, *Classical Electrodynamics*, John Wiley and Sons, Inc. New York (1962), p. 408.

[5] Volume II, Chapter 14, Section 14–1.

When Schrödinger first discovered his equation he discovered the conservation law of Eq. (21.9) as a consequence of his equation. But he imagined incorrectly that P was the electric charge density of the electron and that J was the electric current density, so he thought that the electrons interacted with the electromagnetic field through these charges and currents. When he solved his equations for the hydrogen atom and calculated ψ, he wasn't calculating the probability of anything —there were no amplitudes at that time—the interpretation was completely different. The atomic nucleus was stationary but there were currents moving around; the charges P and currents J would generate electromagnetic fields and the thing would radiate light. He soon found on doing a number of problems that it didn't work out quite right. It was at this point that Born made an essential contribution to our ideas regarding quantum mechanics. It was Born who correctly (as far as we know) interpreted the ψ of the Schrödinger equation in terms of a probability amplitude—that very difficult idea that the square of the amplitude is not the charge density but is only the probability per unit volume of finding an electron there, and that when you do find the electron some place the entire charge is there. That whole idea is due to Born.

The wave function $\psi(r)$ for an electron in an atom does not, then, describe a smeared-out electron with a smooth charge density. The electron is either here, or there, or somewhere else, but wherever it is, it is a point charge. On the other hand, think of a situation in which there are an enormous number of particles in exactly the same state, a very large number of them with exactly the same wave function. Then what? One of them is here and one of them is there, and the probability of finding any one of them at a given place is proportional to $\psi\psi^*$. But since there are so many particles, if I look in any volume $dx\,dy\,dz$ I will generally find a number close to $\psi\psi^*\,dx\,dy\,dz$. So in a situation in which ψ is the wave function for each of an enormous number of particles which are all in the same state, $\psi\psi^*$ *can* be interpreted as the density of particles. If, under these circumstances, each particle carries the same charge q, we can, in fact, go further and interpret $\psi^*\psi$ as the density of *electricity*. Normally, $\psi\psi^*$ is given the dimensions of a probability density, then ψ should be multiplied by q to give the dimensions of a charge density. For our present purposes we can put this constant factor into ψ, and take $\psi\psi^*$ itself as the electric charge density. With this understanding, $\hat{J}$ (the current of probability I have calculated) becomes directly the electric current density.

So in the situation in which we can have very many particles in exactly the same state, there is possible a new physical interpretation of the wave functions. The charge density and the electric current can be calculated directly from the wave functions and the wave functions take on a physical meaning which extends into classical, macroscopic situations.

Something similar can happen with neutral particles. When we have the wave function of a single photon, it is the amplitude to find a photon somewhere. Although we haven't ever written it down there is an equation for the photon wave function analogous to the Schrödinger equation for the electron. The photon equation is just the same as Maxwell's equations for the electromagnetic field, and the wave function is the same as the vector potential A. The wave function turns out to be just the vector potential. The quantum physics is the same thing as the classical physics because photons are noninteracting Bose particles and many of them can be in the same state—as you know, they *like* to be in the same state. The moment that you have billions in the same state (that is, in the same electromagnetic wave), you can measure the wave function, which is the vector potential, directly. Of course, it worked historically the other way. The first observations were on situations with many photons in the same state, and so we were able to discover the correct equation for a single photon by observing directly with our hands on a macroscopic level the nature of wave function.

Now the trouble with the electron is that you cannot put more than one in the same state. Therefore, it was long believed that the wave function of the

Schrödinger equation would never have a macroscopic representation analogous to the macroscopic representation of the amplitude for photons. On the other hand, it is now realized that the phenomena of superconductivity presents us with just this situation.

21–5 Superconductivity

As you know, very many metals become superconducting below a certain temperature[6]—the temperature is different for different metals. When you reduce the temperature sufficiently the metals conduct electricity without any resistance. This phenomenon has been observed for a very large number of metals but not for all, and the theory of this phenomenon has caused a great deal of difficulty. It took a very long time to understand what was going on inside of superconductors, and I will only describe enough of it for our present purposes. It turns out that due to the interactions of the electrons with the vibrations of the atoms in the lattice, there is a small net effective *attraction* between the electrons. The result is that the electrons form together, if I may speak very qualitatively and crudely, bound pairs.

Now you know that a single electron is a Fermi particle. But a bound pair would act as a Bose particle, because if I exchange both electrons in a pair I change the sign of the wave function twice, and that means that I don't change anything. A pair *is* a Bose particle.

The energy of pairing—that is, the net attraction—is very, very weak. Only a tiny temperature is needed to throw the electrons apart by thermal agitation, and convert them back to "normal" electrons. But when you make the temperature sufficiently low that they have to do their very best to get into the absolutely lowest state; then they do collect in pairs.

I don't wish you to imagine that the pairs are really held together very closely like a point particle. As a matter of fact, one of the great difficulties of understanding this phenomena originally was that that is not the way things are. The two electrons which form the pair are really spread over a considerable distance; and the mean distance between pairs is relatively smaller than the size of a single pair. Several pairs are occupying the same space at the same time. Both the reason why electrons in a metal form pairs and an estimate of the energy given up in forming a pair have been a triumph of recent times. This fundamental point in the theory of superconductivity was first explained in the theory of Bardeen, Cooper, and Schrieffer,[7] but that is not the subject of this seminar. We will accept, however, the idea that the electrons do, in some manner or other, work in pairs, that we can think of these pairs as behaving more or less like particles, and that we can therefore talk about the wave function for a "pair."

Now the Schrödinger equation for the pair will be more or less like Eq. (21.3). There will be one difference in that the charge q will be twice the charge of an electron. Also, we don't know the inertia—or effective mass—for the pair in the crystal lattice, so we don't know what number to put in for m. Nor should we think that if we go to very high frequencies (or short wavelengths), this is exactly the right form, because the kinetic energy that corresponds to very rapidly varying wave functions may be so great as to break up the pairs. At finite temperatures there are always a few pairs which are broken up according to the usual Boltzmann theory. The probability that a pair is broken is proportional to $\exp(-E_{pair}/kT)$. The electrons that are not bound in pairs are called "normal" electrons and will move around in the crystal in the ordinary way. I will, however, consider only the situation at essentially zero temperature—or, in any case, I will disregard the complications produced by those electrons which are not in pairs.

---

[6] First discovered by Onnes in 1911; H. K. Onnes, Comm. Phys. Lab., Univ. Leyden, Nos. 119, 120, 122 (1911). You will find a nice up-to-date discussion of the subject in E. A. Lynton, *Superconductivity*, John Wiley and Sons, Inc., New York, 1962.

[7] J. Bardeen, L. N. Cooper, and J. R. Schrieffer, *Phys. Rev.* **108**, 1175 (1957).

Since electron pairs are bosons, when there are a lot of them in a given state there is an especially large amplitude for other pairs to go to the same state. So nearly all of the pairs will be locked down at the lowest energy in *exactly the same state*—it won't be easy to get one of them into another state. There's more amplitude to go into the same state than into an unoccupied state by the famous factor $\sqrt{n}$, where n is the occupancy of the lowest state. So we would expect all the pairs to be moving in the same state.

What then will our theory look like? I'll call ψ the wave function of a pair in the lowest energy state. However, since $\psi\psi^*$ is going to be proportional to the charge density ρ, I can just as well write ψ as the square root of the charge density times some phase factor:

$$\psi(\mathbf{r}) = \rho(\mathbf{r})e^{i\theta(\mathbf{r})}, \qquad (21.17)$$

where ρ and θ are real functions of $\mathbf{r}$. (Any complex function can, of course, be written this way.) It's clear what we mean when we talk about the charge density, but what is the physical meaning of the phase θ of the wave function? Well, let's see what happens if we substitute $\psi(\mathbf{r})$ into Eq. (21.12), and express the current density in terms of these new variables ρ and θ. It's just a change of variables and I won't go through all the algebra, but it comes out

$$\mathbf{J} = \frac{\hbar}{m}\left(\nabla\theta - \frac{q}{\hbar}\mathbf{A}\right)\rho \,. \qquad (21.18)$$

Since both the current density and the charge density have a direct physical meaning for the superconducting electron gas, both ρ and θ are real things. The phase is just as observable as ρ; it is a piece of the current density $\mathbf{J}$. The *absolute* phase is not observable, but if the gradient of the phase is known everywhere, the phase is known except for a constant. You can define the phase at one point, and then the phase everywhere is determined.

Incidentally, the equation for the current can be analyzed a little nicer, when you think that the current density $\mathbf{J}$ is *in fact* the charge density times the velocity of motion of the fluid of electrons, or ρv. Equation (21.18) is then equivalent to

$$m v = \hbar \nabla\theta - qA. \qquad (21.19)$$

Notice that there are two pieces in the mv-momentum; one is a contribution from the vector potential, and the other, a contribution from the behavior of the wave function. In other words, the quantity $\hbar \nabla\theta$ is just what we have called the p-momentum.

21–6 The Meissner effect

Now we can describe some of the phenomena of superconductivity. First, there is no electrical resistance. There's no resistance because all the electrons are collectively in the same state. In the ordinary flow of current you knock one electron or the other out of the regular flow, gradually deteriorating the general momentum. But here to get one electron away from what all the others are doing is very hard because of the tendency of all Bose particles to go in the same state. A current once started, just keeps on going forever.

It's also easy to understand that if you have a piece of metal in the superconducting state and turn on a magnetic field which isn't too strong (we won't go into the details of how strong), the magnetic field can't penetrate the metal. If, as you build up the magnetic field, any of it were to build up inside the metal, there would be a rate of change of flux which would produce an electric field, and an electric field would immediately generate a current which, by Lenz's law, would oppose the flux. Since all the electrons will move together, an infinitesimal electric field will generate enough current to oppose completely any applied magnetic field. So if you turn the field on after you've cooled a metal to the superconducting state, it will be excluded.

Even more interesting is a related phenomenon discovered experimentally by Meissner.[8] If you have a piece of the metal at a high temperature (so that it is a normal conductor) and establish a magnetic field through it, and then you lower the temperature below the critical temperature (where the metal becomes a super-conductor), *the field is expelled.* In other words, it starts up its own current—and in just the right amount to push the field out.

We can see the reason for that in the equations, and I'd like to explain how. Suppose that we take a piece of superconducting material which is in one lump. Then in a steady situation of any kind the divergence of the current must be zero because there's no place for it to go. It is convenient to choose to make the divergence of A equal to zero. (I should explain why choosing this convention doesn't mean any loss of generality, but I don't want to take the time.) Taking the divergence of Eq. (21.18), then gives that the Laplacian of θ is equal to zero. One moment. What about the variation of ρ? I forgot to mention an important point. There is a background of positive charge in this metal due to the atomic ions of the lattice. If the charge density ρ is uniform there is no net charge and no electric field. If there would be any accumulation of electrons in one region the charge wouldn't be neutralized and there would be a terrific repulsion pushing the electrons apart.† So in ordinary circumstances the charge density of the electrons in the superconductor is almost perfectly uniform—I can take ρ as a constant. Now the only way that $\nabla^2\theta$ can be zero everywhere inside the lump of metal is for θ to be a constant. And that means that there is no contribution to J from p-momentum. Equation (21.18) then says that the current is proportional to ρ times A. So everywhere in a lump of superconducting material the current is necessarily proportional to the vector potential:

$$J = -\rho\,\frac{q}{m}\,A. \tag{21.20}$$

Since ρ and q have the same (negative) sign, and since ρ is a constant, I can set $\rho q/m = -(\text{some constant})$; then

$$J = -(\text{some constant})A. \tag{21.21}$$

This equation was originally proposed by London and London[9] to explain the experimental observations of superconductivity—long before the quantum mechanical origin of the effect was understood.

Now we can use Eq. (21.20) in the equations of electromagnetism to solve for the fields. The vector potential is related to the current density by

$$\nabla^2 A = -\frac{1}{\epsilon_0 c^2}\,J. \tag{21.22}$$

If I use Eq. (21.21) for J, I have

$$\nabla^2 A = \lambda^2 A, \tag{21.23}$$

where λ^2 is just a new constant;

$$\lambda^2 = \rho\,\frac{q}{\epsilon_0 m c^2}. \tag{21.24}$$

We can now try to solve this equation for A and see what happens in detail. For example, in one dimension Eq. (21.23) has exponential solutions of the form $e^{-\lambda x}$ and $e^{+\lambda x}$. These solutions mean that the vector potential must *decrease* exponentially as you go from the surface into the material. (It can't increase

---

[8] W. Meissner and R. Ochsenfeld, *Naturwiss.* **21,** 787 (1933).

[9] H. London and F. London, *Proc. Roy. Soc.* (London) **A149,** 71 (1935); *Physica* **2,** 341 (1935).

† Actually if the electric field were too strong, pairs would be broken up and the "normal" electrons created would move in to help neutralize any excess of positive charge. Still, it takes energy to make these normal electrons, so the main point is that a nearly uniform density ρ is highly favored energetically.

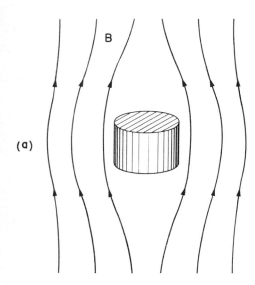

(a)

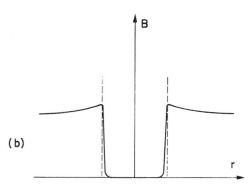

(b)

Fig. 21-3. (a) A superconducting cylinder is a magnetic field; (b) the magnetic field B as a function of r.

because there would be a blow up.) If the piece of metal is very large compared to $1/\lambda$, the field only penetrates to a thin layer at the surface—a layer about $1/\lambda$ in thickness. The entire remainder of the interior is free of field, as sketched in Fig. 21-3. This is the explanation of the Meissner effect.

How big is the distance λ? Well, remember that r_0, the "electromagnetic radius" of the electron (2.8×10^{-13} cm), is given by

$$mc^2 = \frac{q_e^2}{4\pi\epsilon_0 r_0}.$$

Also, remember that q in Eq. (21.24) is twice the charge on an electron, so

$$\frac{q}{\epsilon_0 mc^2} = \frac{8\pi r_0}{q_e}.$$

Writing ρ as $q_e N$, where N is the number of electrons per cubic centimeter, we have

$$\lambda^2 = 8\pi N r_0. \qquad (21.25)$$

For a metal such as lead there are about 3×10^{22} atoms per cm^3, so if each one contributed only one conduction electron, $1/\lambda$ would be about 2×10^{-5} cm. That gives you the order of magnitude.

21-7 Flux quantization

The London equation (21.21) was proposed to account for the observed facts of superconductivity including the Meissner effect. In recent times, however, there have been some even more dramatic predictions. One prediction made by London was so peculiar that nobody paid much attention to it until recently. I will now discuss it. This time instead of taking a single lump, suppose we take a *ring* whose thickness is large compared to $1/\lambda$, and try to see what would happen if we started with a magnetic field through the ring, then cooled it to the super-conducting state, and afterward removed the original source of $\boldsymbol{B}$. The sequence of events is sketched in Fig. 21-4. In the normal state there will be a field in the body of the ring as sketched in part (a) of the figure. When the ring is made super-conducting, the field is forced outside of the *material* (as we have just seen). There will then be some flux through the hole of the ring as sketched in part (b). If the external field is now removed, the lines of field going through the hole are "trapped" as shown in part (c). The flux Φ through the center can't decrease because $\partial\Phi/\partial t$ must be equal to the line integral of E around the ring, which is zero in a superconductor. As the external field is removed a super current starts flowing around the ring to keep the flux through the ring a constant. (It's the old eddy-current idea, only with zero resistance.) These currents will, however, all flow near the surface (down to a depth $1/\lambda$), as can be shown by the same kind of analysis that I made for the solid block. These currents can keep the magnetic field out of the body of the ring, and produce the permanently trapped magnetic field as well.

Now, however, there is an essential difference, and our equations predict a surprising effect. The argument I made above that θ must be a constant in a solid block *does not apply for a ring*, as you can see from the following arguments.

Well inside the body of the ring the current density J is zero; so Eq. (21.18) gives

$$\hbar\nabla\theta = qA. \qquad (21.26)$$

Now consider what we get if we take the line integral of A around a curve Γ, which goes around the ring near the center of its cross-section so that it never gets near the surface, as drawn in Fig. 21-5. From Eq. (21.26),

$$\hbar\oint \nabla\theta \cdot ds = q\oint A \cdot ds. \qquad (21.27)$$

Now you know that the line integral of A around any loop is equal to the flux of B through the loop

$$\oint A \cdot ds = \Phi.$$

Equation (21.27) the becomes

$$\oint \nabla\theta \cdot ds = \frac{q}{\hbar}\Phi. \qquad (21.28)$$

The line integral of a gradient from one point to another (say from point 1 to point 2) is the difference of the values of the function at the two points. Namely,

$$\int_1^2 \nabla\theta \cdot ds = \theta_2 - \theta_1.$$

If we let the two end points 1 and 2 come together to make a closed loop you might at first think that θ_2 would equal θ_1, so that the integral in Eq. (21.28) would be zero. That would be true for a closed loop in a simply-connected piece of super-conductor, but it is not necessarily true for a ring-shaped piece. The only physical requirement we can make is that *there can be only one value of the wave function for each point.* Whatever θ does as you go around the ring, when you get back to the starting point the θ you get must give the same value for the wave function

$$\psi = \sqrt{\rho}\, e^{i\theta}.$$

This will happen if θ changes by $2\pi n$, where n is any integer. So if we make one complete turn around the ring the left-hand side of Eq. (21.27) must be $\hbar \cdot 2\pi n$. Using Eq. (21.28), I get that

$$2\pi n\hbar = q\Phi. \qquad (21.29)$$

The trapped flux must always be an integer times $2\pi\hbar/q$! If you would think of the ring as a classical object with an ideally perfect (that is, infinite) conductivity, you would think that whatever flux was initially found through it would just stay there—any amount of flux at all could be trapped. But the quantum-mechanical theory of superconductivity says that the flux can be zero, or $2\pi\hbar/q$, or $4\pi\hbar/q$, or $6\pi\hbar/q$, and so on, but no value in between. It must be a multiple of a basic quantum mechanical unit.

London[10] predicted that the flux trapped by a superconducting ring would be quantized and said that the possible values of the flux would be given by Eq. (21.29) with q equal to the electronic charge. According to London the basic unit of flux should be $2\pi\hbar/q_e$, which is about 4×10^{-7} gauss = cm^2. To visualize such a flux, think of a tiny cylinder a tenth of a millimeter in diameter; the magnetic field inside it when it contains this amount of flux is about one percent of the earth's magnetic field. It should be possible to observe such a flux by a sensitive magnetic measurement.

In 1961 such a quantized flux was looked for and found by Deaver and Fairbank[11] at Stanford University and at about the same time by Doll and Nabauer[12] in Germany.

In the experiment of Deaver and Fairbank, a tiny cylinder of superconductor was made by electroplating a thin layer of tin on a one-centimeter length of No. 56 (1.3×10^{-3} cm diameter) copper wire. The tin becomes superconducting below 3.8°K, while the copper remains a normal metal. The wire was put in a small controlled magnetic field, and the temperature reduced until the tin became superconducting. Then the exeternal source of field was removed. You would

[10] F. London, *Superfluids;* John Wiley and Sons, Inc., New York, 1950, Vol. I, p. 152.
[11] B. S. Deaver, Jr., and W. M. Fairbank, *Phys. Rev. Letters* **7**, 43 (1961).
[12] R. Doll and M. Nabauer, *Phys. Rev. Letters* **7**, 51 (1961).

Fig. 21-4. A ring in a magnetic field: (a) in the normal state; (b) in the superconducting state; (c) after the external field is removed.

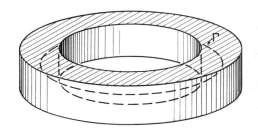

Fig. 21-5. The curve Γ inside a superconducting ring.

expect this to generate a current by Lenz's law so that the flux inside would not change. The little cylinder should now have magnetic moment proportional to the flux inside. The magnetic moment was measured by jiggling the wire up and down (like the needle on a sewing machine, but at the rate of 100 cycles per second) inside a pair of little coils at the ends of the tin cylinder. The induced voltage in the coils was then a measure of the magnetic moment.

When the experiment was done by Deaver and Fairbank, they found that the flux was quantized, *but that the basic unit was only one-half as large as London had predicted.* Doll and Nabauer got the same result. At first this was quite mysterious,† but we now understand why it should be so. According to the Bardeen, Cooper, and Schrieffer theory of superconductivity, the q which appears in Eq. (21.29) is the charge of a *pair* of electrons and so is equal to $2q_e$. The basic flux unit is

$$\Phi_0 = \frac{\pi\hbar}{q_e} \approx 2 \times 10^{-7} \text{ gauss-cm} \tag{21.30}$$

or one-half the amount predicted by London. Everything now fits together, and the measurements show the existence of the predicted purely quantum-mechanical effect on a large scale.

21–8 The dynamics of superconductivity

The Meissner effect and the flux quantization are two confirmations of our general ideas. Just for the sake of completeness I would like to show you what the complete equations of a superconducting fluid would be from this point of view—it is rather interesting. Up to this point I have only put the expression for ψ into equations for charge density and current. If I put it into the complete Schrödinger equation I get equations for ρ and θ. It should be interesting to see what develops, because here we have a "fluid" of electron pairs with a charge density ρ and a mysterious θ—we can try to see what kind of equations we get for such a "fluid"! So we substitute the wave function of Eq. (21.17) into the Schrödinger equation (21.3) and remember that ρ and θ are real functions of x, y, and z. If we separate real and imaginary parts we obtain then two equations. To write them in a shorter form I will—following Eq. (21.19)—write

$$\frac{\hbar}{m}\nabla\theta - \frac{q}{m}A = v. \tag{21.31}$$

One of the equations I get is then

$$\frac{\partial\rho}{\partial t} = \nabla \cdot \rho v. \tag{21.32}$$

Since ρv is first J, this is just the continuity equation once more. The other equation I obtain tells how θ varies; it is

$$\hbar\frac{\partial\theta}{\partial t} = -\frac{m}{2}v^2 + q\phi - \frac{\hbar^2}{2m}\left\{\frac{1}{\sqrt{\rho}}\nabla^2(\sqrt{\rho})\right\}. \tag{21.33}$$

Those who are thoroughly familiar with hydrodynamics (of which I'm sure few of you are) will recognize this as the equation of motion for an electrically charged fluid if we identify $\hbar\theta$ as the "velocity potential"—except that the last term, which should be the energy of compression of the fluid, has a rather strange dependence on the density ρ. In any case, the equation says that rate of change of the quantity $\hbar\theta$ is given by a kinetic energy term, $\frac{1}{2}mv^2$, plus a potential energy term, $q\phi$, with an additional term, containing the factor $\hbar^2$, which we could call a "quantum mechanical energy." We have seen that inside a superconductor ρ is kept very

---

† It has once been suggested by Onsager that this might happen (see F. London, Ref. 10), although no one else ever understood why.

uniform by the electrostatic forces, so this term can almost certainly be neglected in every practical application provided we have only one superconducting region. If we have a boundary between two superconductors (or other circumstances in which the value of ρ may change rapidly) this term can become important.

For those who are not so familiar with the equations of hydrodynamics, I can rewrite Eq. (21.33) in a form that makes the physics more apparent by using Eq. (21.31) to express θ in terms of v. Taking the gradient of the whole of Eq. (21.33) and expressing $\nabla\theta$ in terms of A and v by using (21.31), I get

$$\frac{\partial v}{\partial t} = \frac{q}{m}\left(-\nabla\phi - \frac{\partial A}{\partial t}\right) - v \times (\nabla \times v) - (v \times \nabla)v - \nabla\frac{\hbar^2}{2m}\left(\frac{1}{\sqrt{\rho}}\nabla^2\sqrt{\rho}\right).$$

(21.34)

What does this equation mean? First, remember that

$$-\nabla\phi - \frac{\partial A}{\partial t} = E.$$

(21.35)

Next, notice that if I take the curl of Eq. (21.19), I get

$$\nabla \times v = -\frac{q}{m}\nabla \times A,$$

(21.36)

since the curl of a gradient is always zero. But $\nabla \times A$ is the magnetic field B, so the first two terms can be written as

$$\frac{q}{m}(E + v \times B).$$

Finally, you should understand that $\partial v/\partial t$ stands for the rate of change of the velocity of the fluid at a point. If you concentrate on a particular particle, its acceleration is the *total* derivative of v (or, as it is sometimes called in fluid dynamics, the "comoving acceleration"), which is related to $\partial v/\partial t$ by[13]

$$\left.\frac{dv}{dt}\right|_{\text{comoving}} = \frac{\partial v}{\partial t} + (v \cdot \nabla)v.$$

(21.37)

This extra term also appears as the third term on the right side of Eq. (21.25). Taking it to the left side, I can write Eq. (21.25) in the following way:

$$\left.m\frac{dv}{dt}\right|_{\text{comoving}} = q(E + v \times B) - \nabla\frac{\hbar^2}{2}\left(\frac{1}{\sqrt{\rho}}\nabla^2\sqrt{\rho}\right).$$

(21.38)

We also have from Eq. (21.36) that

$$\nabla \times v = -\frac{q}{m}B.$$

(21.39)

These two equations are the equations of motion of the superconducting electron fluid. The first equation is just Newton's law for a charged fluid in an electromagnetic field. It says that the acceleration of each particle of the fluid whose charge is q comes from the ordinary Lorentz force $q(E + v \times B)$ plus an additional force, which is the gradient of some mystical quantum mechanical potential—a force which is not very big except at the junction between two superconductors. The second equation says that the fluid is "ideal"—the curl of v has zero divergence (the divergence of B is always zero). That means that the velocity can be expressed in terms of velocity potential. Ordinarily one writes that $\nabla \times v = 0$ for an ideal fluid, but for an *ideal charged fluid in a magnetic field*, this gets modified to Eq. (21.40).

So, Schrödinger's equation for the electron pairs in a superconductor gives us the equations of motion of an electrically charged ideal fluid. Superconductivity is the same as the problem of the hydrodynamics of a charged liquid. If you want

---

[13] See Volume II, Section 40–2.

to solve any problem about superconductors you take these equations for the fluid [or the equivalent pair, Eqs. (21.32) and (21.33)], and combine them with Maxwell's equations to get the fields. (The charges and currents you use to get the fields must, of course, include the ones from the superconductor as well as from the external sources.)

Incidentally, I believe that Eq. (21.38) is not quite correct, but ought to have an additional term involving the density. This new term does not depend on quantum mechanics, but comes from the ordinary energy associated with variations of density. Just as in an ordinary fluid there should be a potential energy density proportional to the square of the deviation of ρ from ρ_0, the undisturbed density (which is, here, also equal to the charge density of the crystal lattice). Since there will be forces proportional to the gradient of this energy, there should be another term in Eq. (21.38) of the form: (const) $\nabla(\rho - \rho_0)^2$. This term did not appear from the analysis because it comes from the interactions between particles, which I neglected in using an independent-particle approximation. It is, however, just the force I referred to when I made the qualitative statement that electrostatic forces would tend to keep ρ nearly constant inside a superconductor.

21–9 The Josephson junction

I would like to discuss next a very interesting situation that was noticed by Josephson[14] while analyzing what might happen at a junction between two superconductors. Suppose we have two superconductors which are connected by a thin layer of insulating material as in Fig. 21–6. Such an arrangement is now called a "Josephson junction." If the insulating layer is thick, the electrons can't get through; but if the layer is thin enough, there can be an appreciable quantum mechanical amplitude for electrons to jump across. This is just another example of the quantum-mechanical penetration of a barrier. Josephson analyzed this situation and discovered that a number of strange phenomenon should occur.

In order to analyze such a junction I'll call the amplitude to find an electron on one side, ψ_1, and the amplitude to find it on the other, ψ_2. In the superconducting state the wave function, ψ_1 is the common wave function of all the electrons on one side, and ψ_2 is the corresponding function on the other side. I could do this problem for different kinds of superconductors, but let us take a very simple situation in which the material is the same on both sides so that the junction is symmetrical and simple. Also, for a moment let there be no magnetic field. Then the two amplitudes should be related in the following way:

$$i\hbar \frac{\partial \psi_1}{\partial t} = U_1 \psi_1 + K \psi_2,$$

$$i\hbar \frac{\partial \psi_2}{\partial t} = U_2 \psi_2 + K \psi_1.$$

The constant K is a characteristic of the junction. If K were zero, these two equations would just describe the lowest energy state—with energy U—of each superconductor. But there is coupling between the two sides by the amplitude K that there may be leakage from one side to the other. (It is just the "flip-flop" amplitude of a two-state system.) If the two sides are identical, U_1 would equal U_2 and I could just subtract them off. But now suppose that we connect the two superconducting regions to the two terminals of a battery so that there is a potential difference V across the junction. Then $U_1 - U_2 = qV$. I can, for convenience, define the zero of energy to be halfway between, then the two equations are

$$i\hbar \frac{\partial \psi_1}{\partial t} = \frac{qV}{2} \psi_1 + K \psi_2,$$

$$i\hbar \frac{\partial \psi_2}{\partial t} = -\frac{qV}{2} \psi_2 + K \psi_1.$$

(21.40)

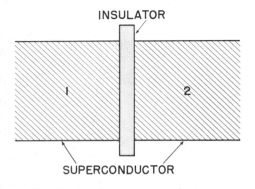

INSULATOR

1   2

SUPERCONDUCTOR

Fig. 21–6. Two superconductors separated by a thin insulator.

---

[14] B. D. Josephson, *Physics Letters* **1**, 251 (1962).

These are the standard equations for two quantum mechanical states coupled together. This time, let's analyze these equations in another way. Let's make the substitutions

$$\psi_1 = \sqrt{\rho_1}e^{i\theta_1},$$
$$\psi_2 = \sqrt{\rho_2}e^{i\theta_2}, \tag{21.41}$$

where θ_1 and θ_2 are the phases on the two sides of the junction and ρ_1 and ρ_2 are the density of electrons at those two points. Remember that in actual practice ρ_1 and ρ_2 are almost exactly the same and are equal to ρ_0, the normal density of electrons in the superconducting material. Now if you substitute these equations for ψ_1 and ψ_2 into (21.40), you get four equations by equating the real and imaginary parts in each case. Letting $(\theta_2 - \theta_1) = \delta$, for short, the result is

$$\dot{\rho}_1 = +\frac{2}{\hbar} K\sqrt{\rho_2\rho_1}\ \sin \delta,$$
$$\dot{\rho}_2 = -\frac{2}{\hbar} K\sqrt{\rho_2\rho_1}\ \sin \delta, \tag{21.42}$$

$$\dot{\theta}_1 = +\frac{K}{\hbar}\sqrt{\frac{\rho_2}{\rho_1}}\ \cos \delta - \frac{qV}{2\hbar},$$
$$\dot{\theta}_2 = +\frac{K}{\hbar}\sqrt{\frac{\rho_1}{\rho_2}}\ \cos \delta + \frac{qV}{2\hbar}. \tag{21.43}$$

The first two equations say that $\dot{\rho}_1 = -\dot{\rho}_2$. "But," you say, "they must both be zero if ρ_1 and ρ_2 are both constant and equal to zero." Not quite. These equations are not the whole story. They say what $\dot{\rho}_1$ and $\dot{\rho}_2$ would be *if there were no extra electric forces* due to an unbalance between the electron fluid and the background of positive ions. They tell how the densities would *start* to change, and therefore describe the kind of current that would begin to flow. This current from side 1 to side 2 would be just $\dot{\rho}_1$(or $-\dot{\rho}_2$), or

$$J = \frac{2K}{\hbar}\sqrt{\rho_1\rho_2}\ \sin \delta. \tag{21.44}$$

Such a current would soon charge up side 2, *except* that we have forgotten that the two sides are connected by wires to the battery. The current that flows will not charge up region 2 (or discharge region 1) because currents will flow to keep the potential constant. These currents from the battery have not been included in our equations. When they are included, ρ_1 and ρ_2 do not in fact change, but the current across the junction is still given by Eq. (21.44).

Since ρ_1 and ρ_2 do remain constant and equal to ρ_0, let's set $2K\rho_0/\hbar = J_0$, and write

$$J = J_0 \sin \delta. \tag{21.45}$$

J_0, like K, is then a number which is a characteristic of the particular junction.

The other pair of equations (21.43) tells us about θ_1 and θ_2. We are interested in the difference $\delta = \theta_2 - \theta_1$ to use Eq. (21.45); what we get is

$$\dot{\delta} = \dot{\theta}_2 - \dot{\theta}_1 = \frac{qV}{\hbar}. \tag{21.46}$$

That means that we can write

$$\delta(t) = \delta_0 + \frac{q}{\hbar}\int V(t)\ dt, \tag{21.47}$$

where δ_0 is the value of δ at $t = 0$. Remember also that q is the charge of a pair, namely, $q = 2q_e$. In Eqs. (21.45) and (21.47) we have an important result, the general theory of the Josephson junction.

Now what are the consequences? First, put on a dc voltage. If you put on a dc voltage, V_0, the argument of the sine becomes $(\delta_0 + (q/\hbar)V_0 t)$. Since $\hbar$ is a small number (compared to ordinary voltage and times), the sine oscillates rather rapidly and the net current is nothing. (In practice, since the temperature is not zero, you would get a small current due to the conduction by "normal" electrons.) On the other hand if you have *zero* voltage across the junction, you can get a current! With no voltage the current can be any amount between $+J_0$ and $-J_0$ (depending on the value of δ_0). But try to put a voltage across it and the current goes to zero. This strange behavior has recently been observed experimentally.[15]

There is another way of getting a current—by applying a voltage at a very high frequency in addition to a dc voltage. Let

$$V = V_0 + v \cos \omega t,$$

where $v \ll V$. Then $\delta(t)$ is

$$\delta_0 + \frac{q}{\hbar} V_0 t + \frac{q}{\hbar} \frac{v}{\omega} \sin \omega t.$$

Now for Δx small,

$$\sin (x + \Delta x) \approx \sin x + \Delta x \cos x.$$

Using this approximation for $\sin \delta$, I get

$$J = J_0 \left[\sin \left(\delta_0 + \frac{q}{\hbar} V_0 t \right) + \frac{q}{\hbar} \frac{v}{\omega} \sin \omega t \cos \left(\delta_0 + \frac{q}{\hbar} V_0 t \right) \right] \cdot$$

The first term is zero on the average, but the second term is not if

$$\omega = \frac{q}{\hbar} V_0.$$

There should be a current if the ac voltage has just this frequency. Shapiro[16] claims to have observed such a resonance effect.

If you look up papers on the subject you will find that they often write the formula for the current as

$$J = J_0 \sin \left(\delta_0 + \frac{2q_e}{\hbar} \int A \cdot ds \right), \tag{21.48}$$

where the integral is to be taken across the junction. The reason for this is that when there's a vector potential across the junction the flip-flop amplitude is modified in phase in the way that we explained earlier. If you chase that extra phase through, it comes out as given above.

Finally, I would like to describe a very dramatic and interesting experiment which has recently been made on the interference of the currents from each of two junctions. In quantum mechanics we're used to the interference between amplitudes from two different slits. Now we're going to do the interference between two junctions caused by the difference in the phase of the arrival of the currents through two different paths. In Fig. 21–7, I show two different junctions, "a" and "b", connected in parallel. The ends, P and Q, are connected to our electrical intruments which measure any current flow. The external current, J_{total}, will be the sum of the currents through the two junctions. Let J_a and J_b be the currents through the two junctions, and let their phases be δ_a and δ_b. Now the phase difference of the wave functions between P and Q must be the same whether you go on one route or the other. Along the route through junction "a", the phase difference between P and Q is δ_a plus the line integral of the vector potential along the upper route:

$$\Delta \text{Phase}_{P \to Q} = \delta_a + \frac{2q_e}{\hbar} \int_{\text{upper}} A \cdot ds. \tag{21.49}$$

[15] P. W. Anderson and J. M. Rowell, *Phys. Rev. Letters* **10**, 230 (1963).
[16] S. Shapiro, *Phys. Rev. Letters* **11**, 80 (1963).

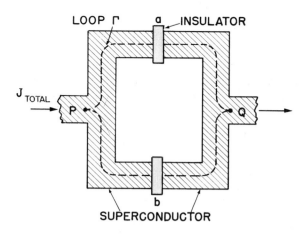

Fig. 21–7. Two Josephson junctions in parallel.

Why? Because the phase θ is related to A by Eq. (21.26). If you integrate that equation along some path, the left-hand side gives the phase change, which is then just proportional to the line integral of A, as we have written here. The phase change along the lower route can be written similarly

$$\Delta\mathrm{Phase}_{P\to Q} = \delta_\mathrm{b} + \frac{2q_e}{\hbar} \int_{\mathrm{lower}} A \cdot ds. \qquad (21.50)$$

These two must be equal; and if I subtract them I get that the difference of the deltas must be the line integral of A around the circuit:

$$\delta_\mathrm{b} - \delta_\mathrm{a} = \frac{2q_e}{\hbar} \oint_\Gamma A \cdot ds.$$

Here the integral is around the closed loop Γ of Fig. 21–7 which circles through both junctions. The integral over A is the magnetic flux Φ through the loop. So the two δ's are going to differ by $2q_e/\hbar$ times the magnetic flux Φ which passes between the two branches of the circuit:

$$\delta_\mathrm{b} - \delta_\mathrm{a} = \frac{2q_e}{\hbar} \Phi. \qquad (21.51)$$

I can control this phase difference by changing the magnetic field on the circuit, so I can adjust the differences in phases and see whether or not the total current that flows through the two junctions shows any interference of the two parts. The total current will be the sum of J_a and J_b. For convenience, I will write

$$\delta_\mathrm{a} = \delta_0 + \frac{q_e}{\hbar} \Phi, \qquad \delta_\mathrm{b} = \delta_0 - \frac{q_e}{\hbar} \Phi.$$

Then,

$$J_\mathrm{total} = J_0 \left\{ \sin\left(\delta_0 + \frac{q_e}{\hbar}\Phi\right) + \sin\left(\delta_0 - \frac{q_e}{\hbar}\Phi\right) \right\}$$

$$= J_0 \sin\delta_0 \cos\frac{q_e\Phi}{\hbar}. \qquad (21.52)$$

Now we don't know anything about δ_0, and nature can adjust that anyway she wants depending on the circumstances. In particular, it will depend on the external voltage we apply to the junction. No matter what we do, however, $\sin\delta_0$ can never get bigger than 1. So the *maximum* current for any given Φ is given by

$$J_\mathrm{max} = J_0 \left| \cos\frac{q_e\Phi}{\hbar} \right|.$$

This maximum current will vary with Φ and will itself have maxima whenever

$$\Phi = n\frac{\pi\hbar}{q_e},$$

with n some integer. That is to say that the current takes on its maximum values where the flux linkage has just those quantized values we found in Eq. (21.30)!

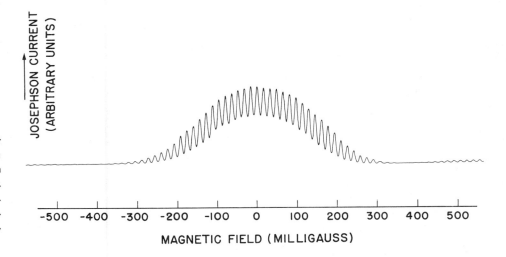

Fig. 21-8. A recording of the current through a pair of Josephson junctions as a function of the magnetic field in the region between the two junctions (see Fig. 21-7). [This recording was provided by R. C. Jaklevic, J. Lambe, A. H. Silver, and J. E. Mercereau of the Scientific Laboratory, Ford Motor Company.]

The Josephson current through a double junction was recently measured[17] as a function of the magnetic field in the area between the junctions. The results are shown in Fig. 21-8. There is a general background of current from various effects we have neglected, but the rapid oscillations of the current with changes in the magnetic field are due to the interference term $\cos q_e \Phi / \hbar$ of Eq. (21.52).

One of the intriguing questions about quantum mechanics is the question of whether the vector potential exists in a place where there's no field.[18] This experiment I have just described has also been done with a tiny solenoid between the two junctions so that the only significant magnetic B field is inside the solenoid and a negligible amount is on the superconducting wires themselves. Yet it is reported that the amount of current depends oscillatorily on the flux of magnetic field inside that solenoid even though that field never touches the wires—another demonstration of the "physical reality" of the vector potential.[19]

I don't know what will come next. But look what can be done. First, notice that the interference between two junctions can be used to make a sensitive magnetometer. If a pair of junctions is made with an enclosed area of, say, 1 mm^2, the maxima in the curve of Fig. 21-8 would be separated by 2×10^{-6} gauss. It is certainly possible to tell when you are 1/10 of the way between two peaks; so it should be possible to use such a junction to measure magnetic fields as small as 2×10^{-7} gauss—or to measure larger fields to such a precision. One should be able to go even farther. Suppose for example we put a set of 10 or 20 junctions close together and equally spaced. Then we can have the interference between 10 or 20 slits and as we change the magnetic field we will get very sharp maxima and minima. Instead of a 2-slit interference we can have a 20- or perhaps even a 100-slit interferometer for measuring the magnetic field. Perhaps we can predict that the measurement of magnetic fields will—by using the effects of quantum-mechanical interference—eventually become almost as precise as the measurement of wavelength of light.

These then are some illustrations of things that are happening in modern times—the transistor, the laser, and now these junctions, whose ultimate practical applications are still not known. The quantum mechanics which was discovered in 1926 has had nearly 40 years of development, and rather suddenly it has begun to be exploited in many practical and real ways. We are really getting control of nature on a very delicate and beautiful level.

I am sorry to say, gentlemen, that to participate in this adventure it is absolutely imperative that you learn quantum mechanics as soon as possible. It was our hope that in this course we would find a way to make comprehensible to you at the earliest possible moment the mysteries of this part of physics.

[17] Jaklevic, Lambe, Silver, and Mercereau, *Phys. Rev. Letters* **12**, 159 (1964).
[18] Jaklevic, Lambe, Silva, and Mercereau, *Phys. Rev. Letters* **12**, 274 (1964).
[19] See Volume II, Chapter 15, Section 15-5.

Feynman's Epilogue

Well, I've been talking to you for two years and now I'm going to quit. In some ways I would like to apologize, and other ways not. I hope—in fact, I know—that two or three dozen of you have been able to follow everything with great excitement, and have had a good time with it. But I also know that "the powers of instruction are of very little efficacy except in those happy circumstances in which they are practically superfluous." So, for the two or three dozen who have understood everything, may I say I have done nothing but shown you the things. For the others, if I have made you hate the subject, I'm sorry. I never taught elementary physics before, and I apologize. I just hope that I haven't caused a serious trouble to you, and that you do not leave this exciting business. I hope that someone else can teach it to you in a way that doesn't give you indigestion, and that you will find someday that, after all, it isn't as horrible as it looks.

Finally, may I add that the main purpose of my teaching has not been to prepare you for some examination—it was not even to prepare you to serve industry or the military. I wanted most to give you some appreciation of the wonderful world and the physicist's way of looking at it, which, I believe, is a major part of the true culture of modern times. (There are probably professors of other subjects who would object, but I believe that they are completely wrong.)

Perhaps you will not only have some appreciation of this culture; it is even possible that you may want to join in the greatest adventure that the human mind has ever begun.

Appendix

Much of the work of this volume assumes a knowledge of the subject of atomic magnetism which is treated in Chapters 34 and 35 of Volume II. For the convenience of those readers who may not have Volume II at hand, these two chapters are reproduced here. Their contents are as follows:

CHAPTER 34 *The Magnetism of Matter*

34–1 Diamagnetism and paramagnetism
34–2 Magnetic moments and angular momentum
34–3 The precession of atomic magnets
34–4 Diamagnetism
34–5 Larmor's theorem
34–6 Classical physics gives neither diamagnetism nor paramagnetism
34–7 Angular momentum in quantum mechanics
34–8 The magnetic energy of atoms

CHAPTER 35 *Paramagnetism and Magnetic Resonance*

35–1 Quantized magnetic states
35–2 The Stern-Gerlach experiment
35–3 The Rabi molecular-beam method
35–4 The paramagnetism of bulk materials
35–5 Cooling by adiabatic demagnetization
35–6 Nuclear magnetic resonance

The Magnetism of Matter

34–1 Diamagnetism and paramagnetism

In this chapter we are going to talk about the magnetic properties of materials. The material which has the most striking magnetic properties is, of course, iron. Similar magnetic properties are shared also by the elements nickel, cobalt, and—at sufficiently low temperatures (below 16°C)—by gadolinium, as well as by a number of peculiar alloys. That kind of magnetism, called *ferromagnetism*, is sufficiently striking and complicated that we will discuss it in a special chapter. However, all ordinary substances do show some magnetic effects, although very small ones—a thousand to a million times less than the effects in ferromagnetic materials. Here we are going to describe ordinary magnetism, that is to say, the magnetism of substances other than the ferromagnetic ones.

This small magnetism is of two kinds. Some materials are *attracted* toward magnetic fields; others are *repelled*. Unlike the electrical effect in matter, which always causes dielectrics to be attracted, there are two signs to the magnetic effect. These two signs can be easily shown with the help of a strong electromagnet which has one sharply pointed pole piece and one flat pole piece, as drawn in Fig. 34–1. The magnetic field is much stronger near the pointed pole than near the flat pole. If a small piece of material is fastened to a long string and suspended between the poles, there will, in general, be a small force on it. This small force can be seen by the slight displacement of the hanging material when the magnet is turned on. The few ferromagnetic materials are attracted very strongly toward the pointed pole; all other materials feel only a very weak force. Some are weakly attracted to the pointed pole; and some are weakly repelled.

34–1 Diamagnetism and paramagnetism

34–2 Magnetic moments and angular momentum

34–3 The precession of atomic magnets

34–4 Diamagnetism

34–5 Larmor's theorem

34–6 Classical physics gives neither diamagnetism nor paramagnetism

34–7 Angular momentum in quantum mechanics

34–8 The magnetic energy of atoms

Review: Section 15–1, "The forces on a current loop; energy of a dipole."

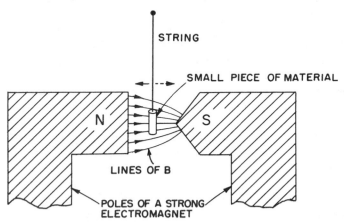

Fig. 34–1. A small cylinder of bismuth is weakly repelled by the sharp pole; a piece of aluminum is attracted.

The effect is most easily seen with a small cylinder of bismuth, which is *repelled* from the high-field region. Substances which are repelled in this way are called *diamagnetic*. Bismuth is one of the strongest diamagnetic materials, but even with it, the effect is still quite weak. Diamagnetism is always very weak. If a small piece of aluminum is suspended between the poles, there is also a weak force, but *toward* the pointed pole. Substances like aluminum are called *paramagnetic*. (In such an experiment, eddy-current forces arise when the magnet is turned on and off, and these can give off strong impulses. You must be careful to look for the net displacement after the hanging object settles down.)

We want now to describe briefly the mechanisms of these two effects. First, in many substances the atoms have no permanent magnetic moments, or rather, all the magnets within each atom balance out so that the *net* moment of the atom is zero. The electron spins and orbital motions all exactly balance out, so that any particular atom has no average magnetic moment. In these circumstances, when you turn on a magnetic field little extra currents are generated inside the atom by induction. According to Lenz's law, these currents are in such a direction as to oppose the increasing field. So the induced magnetic moments of the atoms are directed *opposite* to the magnetic field. This is the mechanism of diamagnetism.

Then there are some substances for which the atoms do have a permanent magnetic moment—in which the electron spins and orbits have a net circulating current that is not zero. So besides the diamagnetic effect (which is always present), there is also the possibility of lining up the individual atomic magnetic moments. In this case, the moments try to line up *with* the magnetic field (in the way the permanent dipoles of a dielectric are lined up by the electric field), and the induced magnetism tends to enhance the magnetic field. These are the paramagnetic substances. Paramagnetism is generally fairly weak because the lining-up forces are relatively small compared with the forces from the thermal motions which try to derange the order. It also follows that paramagnetism is usually sensitive to the temperature. (The paramagnetism arising from the spins of the electrons responsible for conduction in a metal constitutes an exception. We will not be discussing this phenomenon here.) For ordinary paramagnetism, the lower the temperature, the stronger the effect. There is more lining-up at low temperatures when the deranging effects of the collisions are less. Diamagnetism, on the other hand, is more or less independent of the temperature. In any substance with built-in magnetic moments there is a diamagnetic as well as a paramagnetic effect, but the paramagnetic effect usually dominates.

In Chapter 11 we described a *ferroelectric* material, in which all the electric dipoles get lined up by their own mutual electric fields. It is also possible to imagine the magnetic analog of ferroelectricity, in which all the atomic moments would line up and lock together. If you make calculations of how this should happen, you will find that because the magnetic forces are so much smaller than the electric forces, thermal motions should knock out this alignment even at temperatures as low as a few tenths of a degree Kelvin. So it would be impossible at room temperature to have any permanent lining up of the magnets.

On the other hand, this is exactly what does happen in iron—it does get lined up. There is an effective force between the magnetic moments of the different atoms of iron which is much, much greater than the *direct magnetic* interaction. It is an indirect effect which can be explained only by quantum mechanics. It is about ten thousand times stronger than the direct magnetic interaction, and is what lines up the moments in ferromagnetic materials. We discuss this special interaction in a later chapter.

Now that we have tried to give you a qualitative explanation of diamagnetism and paramagnetism, we must correct ourselves and say that *it is not possible* to understand the magnetic effects of materials in any honest way from the point of view of classical physics. Such magnetic effects are a *completely quantum-mechanical phenomenon*. It is, however, possible to make some phoney classical arguments and to get some idea of what is going on. We might put it this way. You can make some classical arguments and get guesses as to the behavior of the material, but these arguments are not "legal" in any sense because it is absolutely essential that quantum mechanics be involved in every one of these magnetic phenomena. On the other hand, there are situations, such as in a plasma or a region of space with many free electrons, where the electrons do obey the laws of classical mechanics. And in those circumstances, some of the theorems from classical magnetism are worth while. Also, the classical arguments are of some value for historical reasons. The first few times that people were able to guess at the meaning and behavior of magnetic materials, they used classical arguments. Finally, as we have already illustrated, classical mechanics can give us some useful guesses as to what might happen—even though the really honest way to study this subject would be to learn quantum mechanics first and then to understand the magnetism in terms of quantum mechanics.

On the other hand, we don't want to wait until we learn quantum mechanics inside out to understand a simple thing like diamagnetism. We will have to lean on the classical mechanics as kind of half showing what happens, realizing, however, that the arguments are really not correct. We therefore make a series of theorems about classical magnetism that will confuse you because they will prove different things. Except for the last theorem, every one of them will be wrong. Furthermore, they will all be wrong as a description of the physical world, because quantum mechanics is left out.

34–2 Magnetic moments and angular momentum

The first theorem we want to prove from classical mechanics is the following: If an electron is moving in a circular orbit (for example, revolving around a nucleus under the influence of a central force), there is a definite ratio between the magnetic moment and the angular momentum. Let's call J the angular momentum and μ the magnetic moment of the electron in the orbit. The magnitude of the angular momentum is the mass of the electron times the velocity times the radius. (See Fig. 34–2.) It is directed perpendicular to the plane of the orbit.

$$J = mvr. \tag{34.1}$$

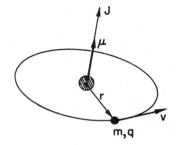

Fig. 34–2. For any circular orbit the magnetic moment μ is q/2m times the angular momentum J.

(This is, of course, a nonrelativistic formula, but it is a good approximation for atoms, because for the electrons involved v/c is generally of the order of $e^2/\hbar c = 1/137$, or about 1 percent.)

The magnetic moment of the same orbit is the current times the area. (See Section 14–5.) The current is the charge per unit time which passes any point on the orbit, namely, the charge q times the frequency of rotation. The frequency is the velocity divided by the circumference of the orbit; so

$$I = q\,\frac{v}{2\pi r}.$$

The area is πr^2, so the magnetic moment is

$$\mu = \frac{qvr}{2}. \tag{34.2}$$

It is also directed perpendicular to the plane of the orbit. So J and μ are in the same direction:

$$\mu = \frac{q}{2m}\,J \ \text{(orbit)}. \tag{34.3}$$

Their ratio depends neither on the velocity nor on the radius. For any particle moving in a circular orbit the magnetic moment is equal to $q/2m$ times the angular momentum. For an electron, the charge is negative—we can call it $-q_e$; so for an electron

$$\mu = -\frac{q_e}{2m}\,J \ \text{(electron orbit)}. \tag{34.4}$$

That's what we would expect classically and, miraculously enough, it is also true quantum-mechanically. It's one of those things. However, if you keep going with the classical physics, you find other places where it gives the wrong answers, and it is a great game to try to remember which things are right and which things are wrong. We might as well give you immediately what is true *in general* in quantum mechanics. First, Eq. (34.4) is true for *orbital motion*, but that's not the only magnetism that exists. The electron also has a spin rotation about its own axis (something like the earth rotating on its axis), and as a result of that spin it has both an angular momentum and a magnetic moment. But for reasons that are purely quantum-mechanical—there is no classical explanation—the ratio of μ

to J for the electron spin is twice as large as it is for orbital motion of the spinning electron:

$$\boldsymbol{\mu} = -\frac{q_e}{m}\, \boldsymbol{J} \text{ (electron spin).} \qquad (34.5)$$

In any atom there are, generally speaking, several electrons and some combination of spin and orbit rotations which builds up a total angular momentum and a total magnetic moment. Although there is no classical reason why it should be so, it is *always true* in quantum mechanics that (for an isolated atom) the direction of the magnetic moment is exactly opposite to the direction of the angular momentum. The ratio of the two is not necessarily either $-q_e/m$ or $-q_e/2m$, but somewhere in between, because there is a mixture of the contributions from the orbits and the spins. We can write

$$\boldsymbol{\mu} = -g\left(\frac{q_e}{2m}\right) \boldsymbol{J}, \qquad (34.6)$$

where g is a factor which is characteristic of the state of the atom. It would be 1 for a pure orbital moment, or 2 for a pure spin moment, or some other number in between for a complicated system like an atom. This formula does not, of course, tell us very much. It says that the magnetic moment is *parallel to* the angular momentum, but can have any magnitude. The form of Eq. (34.6) is convenient, however, because g—called the "Landé g-factor"—is a dimensionless constant whose magnitude is of the order of one. It is one of the jobs of quantum mechanics to predict the g-factor for any particular atomic state.

You might also be interested in what happens in nuclei. In nuclei there are protons and neutrons which may move around in some kind of orbit and at the same time, like an electron, have an intrinsic spin. Again the magnetic moment is parallel to the angular momentum. Only now the order of magnitude of the ratio of the two is what you would expect for a *proton* going around in a circle, with m in Eq. (34.3) equal to the *proton* mass. Therefore it is usual to write for nuclei

$$\boldsymbol{\mu} = g\left(\frac{q_e}{2m_p}\right) \boldsymbol{J}, \qquad (34.7)$$

where m_p is the mass of the proton, and g—called the *nuclear g*-factor—is a number near one, to be determined for each nucleus.

Another important difference for a nucleus is that the *spin* magnetic moment of the proton does *not* have a g-factor of 2, as the electron does. For a proton, $g = 2(2.79)$. Surprisingly enough, the *neutron* also has a spin magnetic moment, and its magnetic moment relative to its angular momentum is $2(-1.93)$. The neutron, in other words, is not exactly "neutral" in the magnetic sense. It is like a little magnet, and it has the kind of magnetic moment that a rotating *negative* charge would have.

34–3 The precession of atomic magnets

One of the consequences of having the magnetic moment proportional to the angular momentum is that an atomic magnet placed in a magnetic field will *precess*. First we will argue classically. Suppose that we have the magnetic moment $\boldsymbol{\mu}$ suspended freely in a uniform magnetic field. It will feel a torque $\boldsymbol{\tau}$, equal to $\boldsymbol{\mu} \times \boldsymbol{B}$, which tries to bring it in line with the field direction. But the atomic magnet is a gyroscope—it has the angular momentum $\boldsymbol{J}$. Therefore the torque due to the magnetic field will not cause the magnet to line up. Instead, the magnet will *precess*, as we saw when we analyzed a gyroscope in Chapter 20 of Volume I. The angular momentum—and with it the magnetic moment—precesses about an axis parallel to the magnetic field. We can find the rate of precession by the same method we used in Chapter 20 of the first volume.

Suppose that in a small time Δt the angular momentum changes from $\boldsymbol{J}$ to $\boldsymbol{J}'$, as drawn in Fig. 34–3, staying always at the same angle θ with respect to the direction of the magnetic field $\boldsymbol{B}$. Let's call ω_p the angular velocity of the precession, so that in the time Δt the angle *of precession* is $\omega_p \Delta t$. From the geometry of the figure, we see that the change of angular momentum in the time Δt is

$$\Delta J = (J \sin \theta)(\omega_p \Delta t).$$

So the rate of change of the angular momentum is

$$\frac{dJ}{dt} = \omega_p J \sin \theta, \tag{34.8}$$

which must be equal to the torque:

$$\tau = \mu B \sin \theta. \tag{34.9}$$

The angular velocity of precession is then

$$\omega_p = \frac{\mu}{J} B. \tag{34.10}$$

Substituting μ/J from Eq. (34.6), we see that for an atomic system

$$\omega_p = g \frac{q_e B}{2m}; \tag{34.11}$$

the precession frequency is proportional to B. It is handy to remember that for an atom (or electron)

$$f_p = \frac{\omega_p}{2\pi} = (1.4 \text{ megacycles/gauss}) gB, \tag{34.12}$$

and that for a nucleus

$$f_p = \frac{\omega_p}{2\pi} = (0.76 \text{ kilocycles/gauss}) gB. \tag{34.13}$$

(The formulas for atoms and nuclei are different only because of the different conventions for g for the two cases.)

According to the *classical* theory, then, the electron orbits—and spins—in an atom should precess in a magnetic field. Is it also true quantum-mechanically? It is essentially true, but the meaning of the "precession" is different. In quantum mechanics one cannot talk about the *direction* of the angular momentum in the same sense as one does classically; nevertheless, there is a very close analogy—so close that we continue to call it "precession." We will discuss it later when we talk about the quantum-mechanical point of view.

34-4 Diamagnetism

Next we want to look at *dia*magnetism from the classical point of view. It can be worked out in several ways, but one of the nice ways is the following. Suppose that we slowly turn on a magnetic field in the vicinity of an atom. As the magnetic field changes an *electric* field is generated by magnetic induction. From Faraday's law, the line integral of E around any closed path is the rate of change of the magnetic flux through the path. Suppose we pick a path Γ which is a circle of radius r concentric with the center of the atom, as shown in Fig. 34-4. The average tangential electric field E around this path is given by

$$E2\pi r = -\frac{d}{dt}(B\pi r^2),$$

and there is a circulating electric field whose strength is

$$E = -\frac{r}{2}\frac{dB}{dt}.$$

The induced electric field acting on an electron in the atom produces a torque equal to $-q_e Er$, which must equal the rate of change of the angular momentum dJ/dt:

$$\frac{dJ}{dt} = \frac{q_e r^2}{2}\frac{dB}{dt}. \tag{34.14}$$

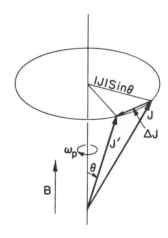

Fig. 34-3. An object with angular momentum J and a parallel magnetic moment μ placed in a magnetic field B precesses with the angular velocity ω_p.

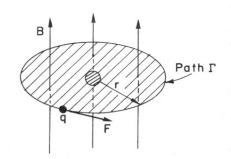

Fig. 34-4. The induced electric forces on the electrons in an atom.

Integrating with respect to time from zero field, we find that the change in angular momentum due to turning on the field is

$$\Delta J = \frac{q_e r^2}{2} B. \tag{34.15}$$

This is the extra angular momentum from the twist given to the electrons as the field is turned on.

This added angular momentum makes an extra magnetic moment which, because it is an *orbital* motion, is just $-q_e/2m$ times the angular momentum. The induced diamagnetic moment is

$$\Delta\mu = -\frac{q_e}{2m}\Delta J = -\frac{q_e^2 r^2}{4m} B. \tag{34.16}$$

The minus sign (as you can see is right by using Lenz's law) means that the added moment is opposite to the magnetic field.

We would like to write Eq. (34.16) a little differently. The r^2 which appears is the radius from an axis through the atom parallel to B, so if B is along the z-direction, it is $x^2 + y^2$. If we consider spherically symmetric atoms (or average over atoms with their natural axes in all directions) the average of $x^2 + y^2$ is 2/3 of the average of the square of the true radial distance from the center *point* of the atom. It is therefore usually more convenient to write Eq. (34.16) as

$$\Delta\mu = -\frac{q_e^2}{6m} \langle r^2 \rangle_{\mathrm{av}} B. \tag{34.17}$$

In any case, we have found an induced atomic moment proportional to the magnetic field B and opposing it. This is diamagnetism of matter. It is this magnetic effect that is responsible for the small force on a piece of bismuth in a nonuniform magnetic field. (You could compute the force by working out the energy of the induced moments in the field and seeing how the energy changes as the material is moved into or out of the high-field region.)

We are still left with the problem: What is the mean square radius, $\langle r^2 \rangle_{\mathrm{av}}$? Classical mechanics cannot supply an answer. We must go back and start over with quantum mechanics. In an atom we cannot really say where an electron is, but only know the probability that it will be at some place. If we interpret $\langle r^2 \rangle_{\mathrm{av}}$ to mean the average of the square of the distance from the center for the probability distribution, the diamagnetic moment given by quantum mechanics is just the same as formula (34.17). This equation, of course, is the moment for one electron. The total moment is given by the sum over all the electrons in the atom. The surprising thing is that the classical argument and quantum mechanics give the same answer, although, as we shall see, the classical argument that gives Eq. (34.17) is not really valid in classical mechanics.

The same diamagnetic effect occurs even when an atom already has a permanent moment. Then the system will precess in the magnetic field. As the whole atom precesses, it takes up an additional small angular velocity, and that slow turning gives a small current which represents a correction to the magnetic moment. This is just the diamagnetic effect represented in another way. But we don't really have to worry about that when we talk about paramagnetism. If the diamagnetic effect is first computed, as we have done here, we don't have to worry about the fact that there is an extra little current from the precession. That has already been included in the diamagnetic term.

34–5 Larmor's theorem

We can already conclude something from our results so far. First of all, in the classical theory the moment μ was always proportional to J, with a given constant of proportionality for a particular atom. There wasn't any spin of the electrons, and the constant of proportionality was always $-q_e/2m$; that is to say, in Eq. (34.6) we should set $g = 1$. The ratio of μ to J was independent of the internal motion of the electrons. Thus, according to the classical theory, all systems of electrons would precess with *the same* angular velocity. (This is *not* true in quantum mechanics.) This result is related to a theorem in classical mechanics that we would now like to prove. Suppose we have a group of electrons which are all held together by attraction toward a central point—as the electrons are attracted by a nucleus. The electrons will also be interacting with each other, and can, in general, have complicated motions. Suppose you have solved for the motions with *no* magnetic field and then want to know what the motions would be *with* a weak magnetic field. The theorem says that the motion with a weak magnetic field is always one of the no-field solutions with an added rotation, about the axis of the field, with the angular velocity $\omega_L = q_e B/2m$. (This is the same as ω_p, if $g = 1$.) There are, of course, many possible motions. The point is that for every motion without the magnetic field there is a corresponding motion in the field, which is the original motion plus a uniform rotation. This is called Larmor's theorem, and ω_L is called the *Larmor frequency*.

We would like to show how the theorem can be proved, but we will let you work out the details. Take, first, one electron in a central force field. The force on it is just $F(r)$, directed toward the center. If we now turn on a uniform magnetic field, there is an additional force, $qv \times B$; so the total force is

$$F(r) + qv \times B. \tag{34.18}$$

Now let's look at the same system from a coordinate system rotating with angular velocity ω about an axis through the center of force and parallel to B. This is no longer an inertial system, so we have to put in the proper pseudoforces—the centrifugal and Coriolis forces we talked about in Chapter 19 of Volume I. We found there that in a frame rotating with angular velocity ω, there is an apparent *tangential* force proportional to v_r, the radial component of velocity:

$$F_t = -2m\omega v_r. \tag{34.19}$$

And there is an apparent radial force which is given by

$$F_r = m\omega^2 r + 2m\omega v_t, \tag{34.20}$$

where v_t is the tangential component of the velocity, measured *in* the rotating frame. (The radial component v_r for rotating and inertial frames is the same.)

Now for small enough angular velocities (that is, if $\omega r \ll v_t$), we can neglect the first term (centrifugal) in Eq. (34.20) in comparison with the second (Coriolis). Then Eqs. (34.19) and (34.20) can be written together as

$$F = -2m\omega \times v. \tag{34.21}$$

If we now *combine* a rotation and a magnetic field, we must add the force in Eq. (34.21) to that in Eq. (34.18). The total force is

$$F(r) + qv \times B + 2mv \times \omega \tag{34.22}$$

[we reverse the cross product and the sign of Eq. (34.21) to get the last term]. Looking at our result, we see that if

$$2m\omega = -qB$$

the two terms on the right cancel, and in the moving frame the only force is $F(r)$. The motion of the electron is just the same as with no magnetic field—and, of course, no rotation. We have proved Larmor's theorem for one electron. Since the proof assumes a small ω, it also means that the theorem is true only for weak magnetic fields. The only thing we could ask you to improve on is to take the case of many electrons mutually interacting with each other, but all in the same central field, and prove the same theorem. So no matter how complex an atom is, if it has a central field the theorem is true. But that's the end of the classical mechanics, because it isn't true in fact that the motions precess in that way. The precession frequency ω_p of Eq. (34.11) is only equal to ω_L if g happens to be equal to 1.

Now we would like to demonstrate that according to classical mechanics there can be no diamagnetism and no paramagnetism at all. It sounds crazy—first, we have proved that there are paramagnetism, diamagnetism, precessing orbits, and so on, and now we are going to prove that it is all wrong. Yes!—We are going to prove that *if* you follow the *classical* mechanics far enough, there are no such magnetic effects—*they all cancel out.* If you start a classical argument in a certain place and don't go far enough, you can get any answer you want. But the only legitimate and correct proof shows that there is no magnetic effect whatever.

It is a consequence of classical mechanics that if you have any kind of system—a gas with electrons, protons, and whatever—kept in a box so that the whole thing can't turn, there will be no magnetic effect. It is possible to have a magnetic effect if you have an isolated system, like a star held together by itself, which can start rotating when you put on the magnetic field. But if you have a piece of material that is held in place so that it can't start spinning, then there will be no magnetic effects. What we mean by holding down the spin is summarized this way: At a given temperature we suppose that there is *only one state* of thermal equilibrium. The theorem then says that if you turn on a magnetic field and wait for the system to get into thermal equilibrium, there will be no paramagnetism or diamagnetism—there will be no induced magnetic moment. Proof: According to statistical mechanics, the probability that a system will have any given state of motion is proportional to $e^{-U/kt}$, where U is the energy of that motion. Now what is the energy of motion? For a particle moving in a constant magnetic field, the energy is the ordinary potential energy plus $mv^2/2$, with nothing additional for the magnetic field. [You know that the forces from electromagnetic fields are $q(E + v \times B)$, and that the rate of work $F \cdot v$ is just $qE \cdot v$, which is not affected by the magnetic field.] So the energy of a system, whether it is in a magnetic field or not, is always given by the kinetic energy plus the potential energy. Since the probability of any motion depends only on the energy—that is, on the velocity and position—it is the same whether or not there is a magnetic field. For *thermal* equilibrium, therefore, the magnetic field has no effect. If we have one system in a box, and then have another system in a second box, this time with a magnetic field, the probability of any particular velocity at any point in the first box is the same as in the second. If the first box has no average circulating current (which it will not have if it is in equilibrium with the stationary walls), there is no average magnetic moment. Since in the second box all the motions are the same, there is no average magnetic moment there either. Hence, if the temperature is kept constant and thermal equilibrium is re-established after the field is turned on, there can be no magnetic moment induced by the field—according to classical mechanics. We can only get a satisfactory understanding of magnetic phenomena from quantum mechanics.

Unfortunately, we cannot assume that you have a thorough understanding of quantum mechanics, so this is hardly the place to discuss the matter. On the other hand, we don't always have to learn something first by learning the exact rules and then by learning how they are applied in different cases. Almost every subject that we have taken up in this course has been treated in a different way. In the case of electricity, we wrote the Maxwell equations on "Page One" and then deduced all the consequences. That's one way. But we will *not* now try to begin a new "Page One," writing the equations of quantum mechanics and deducing everything from them. We will just have to tell you some of the consequences of quantum mechanics, before you learn where they come from. So here we go.

34–7 Angular momentum in quantum mechanics

We have already given you a relation between the magnetic moment and the angular momentum. That's pleasant. But what do the magnetic moment and the angular momentum *mean* in quantum mechanics? In quantum mechanics it turns out to be best to define things like magnetic moments in terms of the other concepts such as energy, in order to make sure that one knows what it means. Now, it is easy to define a magnetic moment in terms of energy, because the energy of a moment in a magnetic field is, in the classical theory, $\boldsymbol{\mu} \cdot \boldsymbol{B}$. Therefore, the following definition has been taken in quantum mechanics: If we calculate the energy of a system in a magnetic field and we find that it is proportional to the field strength (for small field), the coefficient is called the component of magnetic moment in the direction of the field. (We don't have to get so elegant for our work now; we can still think of the magnetic moment in the ordinary, to some extent classical, sense.)

Now we would like to discuss the idea of angular momentum in quantum mechanics—or rather, the characteristics of what, in quantum mechanics, is called angular momentum. You see, when you go to new kinds of laws, you can't just assume that each word is going to mean exactly the same thing. You may think, say, "Oh, I know what angular momentum is. It's that thing that is changed by a torque." But what's a torque? In quantum mechanics we have to have new definitions of old quantities. It would, therefore, be legally best to call it by some other name such as "quantangular momentum," or something like that, because it is the angular momentum as defined in quantum mechanics. But if we can find a quantity in quantum mechanics which is identical to our old idea of angular momentum when the system becomes large enough, there is no use in inventing an extra word. We might as well just call it angular momentum. With that understanding, this odd thing that we are about to describe *is* angular momentum. It is the thing which in a large system we recognize as angular momentum in classical mechanics.

First, we take a system in which angular momentum is conserved, such as an atom all by itself in empty space. Now such a thing (like the earth spinning on its axis) could, in the ordinary sense, be spinning around any axis one wished to choose. And for a given spin, there could be many different "states," all of the same energy, each "state" corresponding to a particular direction of the axis of the angular momentum. So in the classical theory, with a given angular momentum, there is an infinite number of possible states, all of the same energy.

It turns out in quantum mechanics, however, that several strange things happen. First, the number of states in which such a system *can exist* is limited—there is only a finite number. If the system is small, the finite number is very small, and if the system is large, the finite number gets very, very large. Second, we *cannot* describe a "state" by giving the *direction* of its angular momentum, but only by giving the *component* of the angular momentum along some direction—say in the z-direction. Classically, an object with a given total angular momentum J could have, for its z-component, any value from $+J$ to $-J$. But quantum-mechanically, the z-component of angular momentum can have only certain discrete values. Any given system—a particular atom, or a nucleus, or anything—with a given energy, has a characteristic number j, and its z-component of angular momentum can only be one of the following set of values:

$$
\begin{aligned}
& j\hbar \\
& (j-1)\hbar \\
& (j-2)\hbar \\
& \quad\vdots \\
& -(j-2)\hbar \\
& -(j-1)\hbar \\
& -j\hbar
\end{aligned}
\tag{34.23}
$$

The largest z-component is j times $\hbar$; the next smaller is one unit of $\hbar$ less, and so on down to $-j\hbar$. The number j is called "the spin of the system." (Some people call it the "total angular momentum quantum number"; but we'll call it the "spin.")

You may be worried that what we are saying can only be true for some "special" z-axis. But that is not so. For a system whose spin is j, the component of angular momentum along *any* axis can have only one of the values in (34.23). Although it is quite mysterious, we ask you just to accept it for the moment. We will come back and discuss the point later. You may at least be pleased to hear that the z-component goes from some number to minus the *same* number, so that we at least don't have to decide which is the plus direction of the z-axis. (Certainly, if we said that it went from $+j$ to minus a different amount, that would be infinitely mysterious, because we wouldn't have been able to define the z-axis, pointing the other way.)

Now if the z-component of angular momentum must go down by integers from $+j$ to $-j$, then j must be an integer. No! Not quite; twice j must be an integer. It is only the *difference* between $+j$ and $-j$ that must be an integer. So, in general, the spin j is either an integer or a half-integer, depending on whether $2j$ is even or odd. Take, for instance, a nucleus like lithium, which has a spin of three-halves, $j = 3/2$. Then the angular momentum around the z-axis, in units of $\hbar$, is one of the following:

$$+3/2$$
$$+1/2$$
$$-1/2$$
$$-3/2.$$

There are four possible states, each of the same energy, if the nucleus is in empty space with no external fields. If we have a system whose spin is two, then the z-component of angular momentum has only the values, in units of $\hbar$,

$$2$$
$$1$$
$$0$$
$$-1$$
$$-2.$$

If you count how many states there are for a given j, there are $(2j + 1)$ possibilities. In other words, if you tell me the energy and also the spin j, it turns out that there are exactly $(2j + 1)$ states with that energy, each state corresponding to one of the different possible values of the z-component of the angular momentum.

We would like to add one other fact. If you pick out any atom of known j at random and measure the z-component of the angular momentum, then you may get any one of the possible values, and each of the values is *equally* likely. All of the states are in fact single states, and each is just as good as any other. Each one has the same "weight" in the world. (We are assuming that nothing has been done to sort out a special sample.) This fact has, incidentally, a simple classical analog. If you ask the same question classically: What is the likelihood of a particular z-component of angular momentum if you take a random sample of systems, all with the same total angular momentum?—the answer is that all values from the maximum to the minimum are equally likely. (You can easily work that out.) The classical result corresponds to the equal probability of the $(2j + 1)$ possibilities in quantum mechanics.

From what we have so far, we can get another interesting and somewhat surprising conclusion. In certain classical calculations the quantity that appears in the final result is the *square* of the magnitude of the angular momentum $\boldsymbol{J}$—in other words, $\boldsymbol{J} \cdot \boldsymbol{J}$. It turns out that it is often possible to *guess* at the correct quantum-mechanical formula by using the classical calculation and the following simple rule: Replace $J^2 = \boldsymbol{J} \cdot \boldsymbol{J}$ by $j(j + 1)\hbar^2$. This rule is commonly used, and usually gives the correct result, but *not* always. We can give the following argument to show why you might expect this rule to work.

The scalar product $\boldsymbol{J} \cdot \boldsymbol{J}$ can be written as

$$\boldsymbol{J} \cdot \boldsymbol{J} = J_x^2 + J_y^2 + J_z^2.$$

Since it is a scalar, it should be the same for any orientation of the spin. Suppose we pick samples of any given atomic system at random and make measurements of J_x^2, or J_y^2, or J_z^2, the *average value* should be the same for each. (There is no special distinction for any one of the directions.) Therefore, the average of $\boldsymbol{J} \cdot \boldsymbol{J}$ is just equal to three times the average of any component squared, say of J_z^2;

$$\langle \mathbf{J} \cdot \mathbf{J} \rangle_{\text{av}} = 3 \langle J_z^2 \rangle.$$

But since $\mathbf{J} \cdot \mathbf{J}$ is the same for all orientations, its average is, of course, just its constant value; we have

$$\mathbf{J} \cdot \mathbf{J} = 3 \langle J_z^2 \rangle_{\text{av}}. \tag{34.24}$$

If we now say that we will use the same equation for quantum mechanics, we can easily find $\langle J_z^2 \rangle_{\text{av}}$. We just have to take the sum of the $(2j + 1)$ possible values of J_z^2, and divide by the total number;

$$\langle J_z^2 \rangle_{\text{av}} = \frac{j^2 + (j - 1)^2 + \cdots + (-j + 1)^2 + (-j)^2}{2j + 1} \hbar^2. \tag{34.25}$$

For a system with a spin of 3/2, it goes like this:

$$\langle J_z^2 \rangle_{\text{av}} = \frac{(3/2)^2 + (1/2)^2 + (-1/2)^2 + (-3/2)^2}{4} \hbar^2 = \frac{5}{4} \hbar^2.$$

We conclude that

$$\mathbf{J} \cdot \mathbf{J} = 3 \langle J_z^2 \rangle_{\text{av}} = 3 \tfrac{5}{4} \hbar^2 = \tfrac{3}{2}(\tfrac{3}{2} + 1)\hbar^2.$$

We will leave it for you to show that Eq. (34.25), together with Eq. (34.24), gives the general result

$$\mathbf{J} \cdot \mathbf{J} = j(j + 1)\hbar^2. \tag{34.26}$$

Although we would think classically that the largest possible value of the z-component of $\mathbf{J}$ is just the magnitude of $\mathbf{J}$—namely, $\sqrt{\mathbf{J} \cdot \mathbf{J}}$—quantum mechanically the maximum of J_z is always a little less than that, because $j\hbar$ is always less than $\sqrt{j(j + 1)}\, \hbar$. The angular momentum is never "completely along the z-direction."

34–8 The magnetic energy of atoms

Now we want to talk again about the magnetic moment. We have said that in quantum mechanics the magnetic moment of a particular atomic system can be written in terms of the angular momentum by Eq. (34.6);

$$\boldsymbol{\mu} = -g \left(\frac{q_e}{2m} \right) \mathbf{J}, \tag{34.27}$$

where $-q_e$ and m are the charge and mass of the electron.

An atomic magnet placed in an external magnetic field will have an extra magnetic energy which depends on the component of its magnetic moment along the field direction. We know that

$$U_{\text{mag}} = -\boldsymbol{\mu} \cdot \mathbf{B}. \tag{34.28}$$

Choosing our z-axis along the direction of $\mathbf{B}$,

$$U_{\text{mag}} = -\mu_z B. \tag{34.29}$$

Using Eq. (34.27), we have that

$$U_{\text{mag}} = g \left(\frac{q_e}{2m} \right) J_z B.$$

Quantum mechanics says that J_z can have only certain values: $j\hbar$, $(j - 1)\hbar, \ldots,$ $-j\hbar$. Therefore, the magnetic energy of an atomic system is not arbitrary; it can have only certain values. Its maximum value, for instance, is

$$g \left(\frac{q_e}{2m} \right) \hbar j B.$$

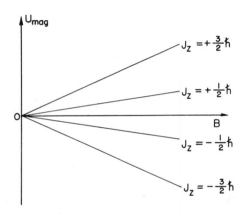

Fig. 34–5. The possible magnetic energies of an atomic system with a spin of 3/2 in a magnetic field **B**.

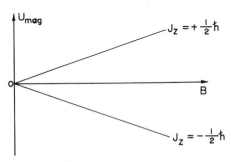

Fig. 34–6. The two possible energy states of an electron in a magnetic field **B**.

The quantity $q_e\hbar/2m$ is usually given the name "the Bohr magneton" and written μ_B:

$$\mu_B = \frac{q_e\hbar}{2m}.$$

The possible values of the magnetic energy are

$$U_{\mathrm{mag}} = g\mu_B B\frac{J_z}{\hbar},$$

where $J_z/\hbar$ takes on the possible values j, $(j-1)$, $(j-2)$, ..., $(-j+1)$, $-j$.

In other words, the energy of an atomic system is changed when it is put in a magnetic field by an amount that is proportional to the field, and proportional to J_z. We say that the energy of an atomic system is "split into $2j+1$ levels" by a magnetic field. For instance, an atom whose energy is U_0 outside a magnetic field and whose j is 3/2, will have four possible energies when placed in a field. We can show these energies by an energy-level diagram like that drawn in Fig. 34–5. Any particular atom can have only one of the four possible energies in any given field B. That is what quantum mechanics says about the behavior of an atomic system in a magnetic field.

The simplest "atomic" system is a single electron. The spin of an electron is 1/2, so there are two possible states: $J_z = \hbar/2$ and $J_z = -\hbar/2$. For an electron at rest (no orbital motion), the spin magnetic moment has a g-value of 2, so the magnetic energy can be either $\pm\mu_B B$. The possible energies in a magnetic field are shown in Fig. 34–6. Speaking loosely we say that the electron either has its spin "up" (along the field) or "down" (opposite the field).

For systems with higher spins, there are more states. We can think that the spin is "up" or "down" or cocked at some "angle" in between, depending on the value of J_z.

We will use these quantum mechanical results to discuss the magnetic properties of materials in the next chapter.

35

Paramagnetism and Magnetic Resonance

35–1 Quantized magnetic states

In the last chapter we described how in quantum mechanics the angular momentum of a thing does not have an arbitrary direction, but its component along a given axis can take on only certain equally spaced, discrete values. It is a shocking and peculiar thing. You may think that perhaps we should not go into such things until your minds are more advanced and ready to accept this kind of an idea. Actually, your minds will never become more advanced—in the sense of being able to accept such a thing easily. There isn't any descriptive way of making it intelligible that isn't so subtle and advanced in its own form that it is more complicated than the thing you were trying to explain. The behavior of matter on a small scale—as we have remarked many times—is different from anything that you are used to and is very strange indeed. As we proceed with classical physics, it is a good idea to try to get a growing acquaintance with the behavior of things on a small scale, at first as a kind of experience without any deep understanding. Understanding of these matters comes very slowly, if at all. Of course, one does get better able to know what is going to happen in a quantum-mechanical situation—if that is what understanding means—but one never gets a comfortable feeling that these quantum-mechanical rules are "natural." Of course they *are*, but they are not natural to our own experience at an ordinary level. We should explain that the attitude that we are going to take with regard to this rule about angular momentum is quite different from many of the other things we have talked about. We are not going to try to "explain" it, but we must at least *tell* you what happens; it would be dishonest to describe the magnetic properties of materials without mentioning the fact that the classical description of magnetism—of angular momentum and magnetic moments—is incorrect.

One of the most shocking and disturbing features about quantum mechanics is that if you take the angular momentum along any particular axis you find that it is always an integer or half-integer times $\hbar$. This is so no matter which axis you take. The subtleties involved in that curious fact—that you can take any other axis and find that the component for it is also locked to the same set of values—we will leave to a later chapter, when you will experience the delight of seeing how this apparent paradox is ultimately resolved.

We will now just accept the fact that for every atomic system there is a number j, called the *spin* of the system—which must be an integer or a half-integer—and that the component of the angular momentum along any particular axis will always have one of the following values between $+j\hbar$ and $-j\hbar$:

$$J_z = \text{one of} \left\{ \begin{matrix} j \\ j-1 \\ j-2 \\ \vdots \\ -j+2 \\ -j+1 \\ -j \end{matrix} \right\} \cdot \hbar. \qquad (35.1)$$

We have also mentioned that every simple atomic system has a magnetic moment which has the same direction as the angular momentum. This is true not only for atoms and nuclei but also for the fundamental particles. Each fundamental particle has its own characteristic value of j and its magnetic moment.

35–1 Quantized magnetic states

35–2 The Stern-Gerlach experiment

35–3 The Rabi molecular-beam method

35–4 The paramagnetism of bulk materials

35–5 Cooling by adiabatic demagnetization

35–6 Nuclear magnetic resonance

Review: Chapter 11, *Inside Dielectrics*

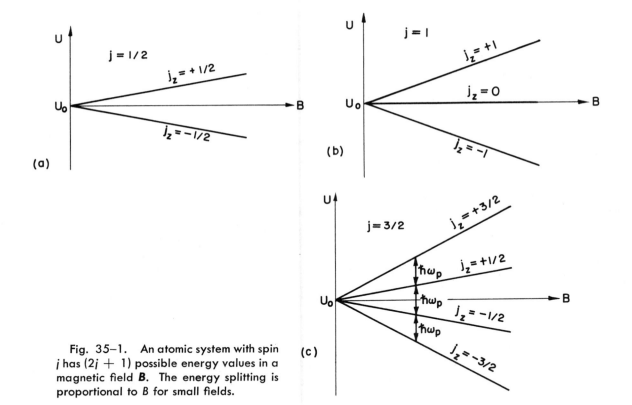

Fig. 35–1. An atomic system with spin j has $(2j + 1)$ possible energy values in a magnetic field **B**. The energy splitting is proportional to B for small fields.

(For some particles, both are zero.) What we mean by "the magnetic moment" in this statement is that the energy of the system in a magnetic field, say in the z-direction, can be written as $-\mu_z B$ for small magnetic fields. We must have the condition that the field should not be too great, otherwise it could disturb the internal motions of the system and the energy would not be a measure of the magnetic moment that was there before the field was turned on. But if the field is sufficiently weak, the field changes the energy by the amount

$$\Delta U = -\mu_z B, \tag{35.2}$$

with the understanding that in this equation we are to replace μ_z by

$$\mu_z = g\left(\frac{q}{2m}\right) J_z, \tag{35.3}$$

where J_z has one of the values in Eq. (35.1).

Suppose we take a system with a spin $j = 3/2$. Without a magnetic field, the system has four different possible states corresponding to the different values of J_z, all of which have exactly the same energy. But the moment we turn on the magnetic field, there is an additional energy of interaction which separates these states into four slightly different energy levels. The energies of these levels are given by a certain energy proportional to B, multiplied by $\hbar$ times $3/2, 1/2, -1/2,$ and $-3/2$—the values of J_z. The splitting of the energy levels for atomic systems with spins of $1/2, 1,$ and $3/2$ are shown in the diagrams of Fig. 35–1. (Remember that for any arrangement of electrons the magnetic moment is always directed opposite to the angular momentum.)

You will notice from the diagrams that the "center of gravity" of the energy levels is the same with and without a magnetic field. Also notice that the spacings from one level to the next are always equal for a given particle in a given magnetic field. We are going to write the energy spacing, for a given magnetic field B, as $\hbar\omega_p$—which is just a definition of ω_p. Using Eqs. (35.2) and (35.3), we have

$$\hbar\omega_p = g\frac{q}{2m}\hbar B$$

or

$$\omega_p = g\frac{q}{2m}B. \tag{35.4}$$

The quantity $g(q/2m)$ is just the ratio of the magnetic moment to the angular momentum—it is a property of the particle. Equation (35.4) is the same formula that we got in Chapter 34 for the angular velocity of precession in a magnetic field, for a gyroscope whose angular momentum is J and whose magnetic moment is μ.

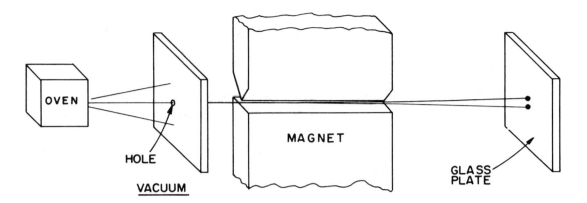

Fig. 35–2. The experiment of Stern and Gerlach.

35–2 The Stern-Gerlach experiment

The fact that the angular momentum is quantized is such a surprising thing that we will talk a little bit about it historically. It was a shock from the moment it was discovered (although it was expected theoretically). It was first observed in an experiment done in 1922 by Stern and Gerlach. If you wish, you can consider the experiment of Stern-Gerlach as a direct justification for a belief in the quantization of angular momentum. Stern and Gerlach devised an experiment for measuring the magnetic moment of individual silver atoms. They produced a beam of silver atoms by evaporating silver in a hot oven and letting some of them come out through a series of small holes. This beam was directed between the pole tips of a special magnet, as shown in Fig. 35–2. Their idea was the following. If the silver atom has a magnetic moment μ, then in a magnetic field B it has an energy $-\mu_z B$, where z is the direction of the magnetic field. In the classical theory, μ_z would be equal to the magnetic moment times the cosine of the angle between the moment and the magnetic field, so the extra energy in the field would be

$$\Delta U = -\mu B \cos\theta. \qquad (35.5)$$

Of course, as the atoms come out of the oven, their magnetic moments would point in every possible direction, so there would be all values of θ. Now if the magnetic field varies very rapidly with z—if there is a strong field gradient—then the magnetic energy will also vary with position, and there will be a force on the magnetic moments whose direction will depend on whether cosine θ is positive or negative. The atoms will be pulled up or down by a force proportional to the derivative of the magnetic energy; from the principle of virtual work,

$$F_z = -\frac{\partial U}{\partial z} = \mu \cos\theta \, \frac{\partial B}{\partial z}. \qquad (35.6)$$

Stern and Gerlach made their magnet with a very sharp edge on one of the pole tips in order to produce a very rapid variation of the magnetic field. The beam of silver atoms was directed right along this sharp edge, so that the atoms would feel a vertical force in the inhomogeneous field. A silver atom with its magnetic moment directed horizontally would have no force on it and would go straight past the magnet. An atom whose magnetic moment was exactly vertical would have a force pulling it up toward the sharp edge of the magnet. An atom whose magnetic moment was pointed downward would feel a downward push. Thus, as they left the magnet, the atoms would be spread out according to their vertical components of magnetic moment. In the classical theory all angles are possible, so that when the silver atoms are collected by deposition on a glass plate, one should expect a smear of silver along a vertical line. The height of the line would be proportional to the magnitude of the magnetic moment. The abject failure of classical ideas was completely revealed when Stern and Gerlach saw what actually happened. They found on the glass plate two distinct spots. The silver atoms had formed two beams.

That a beam of atoms whose spins would apparently be randomly oriented gets split up into two separate beams is most miraculous. How does the magnetic moment *know* that it is only allowed to take on certain components in the direction of the magnetic field? Well, that was really the beginning of the discovery of the quantization of angular momentum, and instead of trying to give you a theoretical explanation, we will just say that you are stuck with the result of this experiment just as the physicists of that day had to accept the result when the experiment was done. It is an *experimental fact* that the energy of an atom in a magnetic field takes on a series of individual values. For each of these values the energy is proportional to the field strength. So in a region where the field varies, the principle of virtual work tells us that the possible magnetic force on the atoms will have a set of separate values; the force is different for each state, so the beam of atoms is split into a small number of separate beams. From a measurement of the deflection of the beams, one can find the strength of the magnetic moment.

35–3 The Rabi molecular-beam method

We would now like to describe an improved apparatus for the measurement of magnetic moments which was developed by I. I. Rabi and his collaborators. In the Stern-Gerlach experiment the deflection of atoms is very small, and the measurement of the magnetic moment is not very precise. Rabi's technique permits a fantastic precision in the measurement of the magnetic moments. The method is based on the fact that the original energy of the atoms in a magnetic field is split up into a finite number of energy levels. That the energy of an atom in the magnetic field can have only certain discrete energies is really not more surprising than the fact that atoms *in general* have only certain discrete energy levels—something we mentioned often in Volume I. Why should the same thing *not* hold for atoms in a magnetic field? It does. But it is the attempt to correlate this with the idea of an *oriented magnetic moment* that brings out some of the strange implications of quantum mechanics.

When an atom has two levels which differ in energy by the amount ΔU, it can make a transition from the upper level to the lower level by emitting a light quantum of frequency ω, where

$$\hbar\omega = \Delta U. \tag{35.7}$$

The same thing can happen with atoms in a magnetic field. Only then, the energy differences are so small that the frequency does not correspond to light, but to microwaves or to radiofrequencies. The transitions from the lower energy level to an upper energy level of an atom can also take place with the absorption of light or, in the case of atoms in a magnetic field, by the absorption of microwave energy. Thus if we have an atom in a magnetic field, we can cause transitions from one state to another by applying an additional electromagnetic field of the proper frequency. In other words, if we have an atom in a strong magnetic field and we "tickle" the atom with a weak varying electromagnetic field, there will be a certain probability of knocking it to another level if the frequency is near to the ω in Eq. (35.7). For an atom in a magnetic field, this frequency is just what we have earlier called ω_p and it is given in terms of the magnetic field by Eq. (35.4). If the atom is tickled with the wrong frequency, the chance of causing a transition is very small. Thus there is a sharp *resonance* at ω_p in the probability of causing a transition. By measuring the frequency of this resonance in a known magnetic field B, we can measure the quantity $g(q/2m)$—and hence the g-factor—with great precision.

It is interesting that one comes to the same conclusion from a classical point of view. According to the classical picture, when we place a small gyroscope with a magnetic moment μ and an angular momentum J in an external magnetic field, the gyroscope will precess about an axis parallel to the magnetic field. (See Fig. 35–3.) Suppose we ask: How can we change the angle of the classical gyroscope with respect to the field—namely, with respect to the z-axis? The magnetic field produces a torque around a *horizontal* axis. Such a torque you would think is *trying* to line up the magnet with the field, but it only causes the precession. If we want to change the angle of the gyroscope with respect to the z-axis, we must exert a torque on it *about the z-axis*. If we apply a torque which goes in the same direction as the precession, the angle of the gyroscope will change to give a smaller component of J in the z-direction. In Fig. 35–3, the angle between J and the z-axis would increase. If we try to hinder the precession, J moves toward the vertical.

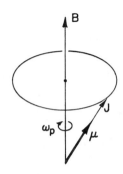

Fig. 35–3.   The classical precession of an atom with the magnetic moment μ and the angular momentum J.

For our precessing atom in a uniform magnetic field, how can we apply the kind of torque we want? The answer is: with a weak magnetic field from the side. You might at first think that the direction of this magnetic field would have to rotate with the precession of the magnetic moment, so that it was always at right angles to the moment, as indicated by the field B' in Fig. 35–4(a). Such a field works very well, but an *alternating* horizontal field is almost as good. If we have a small horizontal field B', which is always in the x-direction (plus or minus) and which oscillates with the frequency ω_p, then on each one-half cycle the torque on the magnetic moment reverses, so that it has a cumulative effect which is almost as effective as a rotating magnetic field. Classically, then, we would expect the component of the magnetic moment along the z-direction to change if we have a very weak oscillating magnetic field at a frequency which is exactly ω_p. Classically, of course, μ_z would change continuously, but in quantum mechanics the z-component of the magnetic moment cannot adjust continuously. It must jump suddenly from one value to another. We have made the comparison between the consequences of classical mechanics and quantum mechanics to give you some clue as to what might happen classically and how it is related to what actually happens in quantum mechanics. You will notice, incidentally, that the expected resonant frequency is the same in both cases.

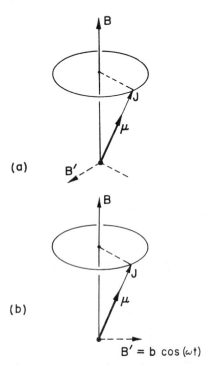

Fig. 35–4.   The angle of precession of an atomic magnet can be changed by a horizontal magnetic field always at right angles to μ, as in (a), or by an oscillating field, as in (b).

One additional remark: From what we have said about quantum mechanics, there is no apparent reason why there couldn't also be transitions at the frequency $2\omega_p$. It happens that there isn't any analog of this in the classical case, and also it doesn't happen in the quantum theory either—at least not for the particular method of inducing the transitions that we have described. With an oscillating horizontal magnetic field, the probability that a frequency $2\omega_p$ would cause a jump of two steps at once is zero. It is only at the frequency ω_p that transitions, either upward or downward, are likely to occur.

Now we are ready to describe Rabi's method for measuring magnetic moments. We will consider here only the operation for atoms with a spin of 1/2. A diagram of the apparatus is shown in Fig. 35–5. There is an oven which gives out a stream of neutral atoms which passes down a line of three magnets. Magnet 1

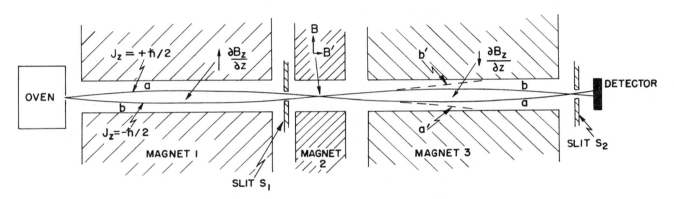

Fig. 35–5.   The Rabi molecular-beam apparatus.

is just like the one in Fig. 35–2, and has a field with a strong field gradient—say, with $\partial B_z/\partial z$ positive. If the atoms have a magnetic moment, they will be deflected downward if $J_z = +\hbar/2$, or upward if $J_z = -\hbar/2$ (since for electrons μ is directed opposite to $\mathbf{J}$). If we consider only those atoms which can get through the slit S_1, there are two possible trajectories, as shown. Atoms with $J_z = +\hbar/2$ must go along curve a to get through the slit, and those with $J_z = -\hbar/2$ must go along curve b. Atoms which start out from the oven along other paths will not get through the slit.

Magnet 2 has a uniform field. There are no forces on the atoms in this region, so they go straight through and enter magnet 3. Magnet 3 is just like magnet 1 but with the field *inverted*, so that $\partial B_z/\partial z$ has the opposite sign. The atoms with $J_z = +\hbar/2$ (we say "with spin up"), that felt a downward push in magnet 1, get an *upward* push in magnet 3; they continue on the path a and go through slit S_2 to a detector. The atoms with $J_z = -\hbar/2$ ("with spin down") also have opposite forces in magnets 1 and 3 and go along the path b, which also takes them through slit S_2 to the detector.

The detector may be made in various ways, depending on the atom being measured. For example, for atoms of an alkali metal like sodium, the detector can be a thin, hot tungsten wire connected to a sensitive current meter. When sodium atoms land on the wire, they are evaporated off as Na^+ ions, leaving an electron behind. There is a current from the wire proportional to the number of sodium atoms arriving per second.

In the gap of magnet 2 there is a set of coils that produces a small horizontal magnetic field $\mathbf{B}'$. The coils are driven with a current which oscillates at a variable frequency ω. So between the poles of magnet 2 there is a strong, constant, vertical field $\mathbf{B}_0$ and a weak, oscillating, horizontal field $\mathbf{B}'$.

Suppose now that the frequency ω of the oscillating field is set at ω_p—the "precession" frequency of the atoms in the field $\mathbf{B}$. The alternating field will cause some of the atoms passing by to make transitions from one J_z to the other. An atom whose spin was initially "up" ($J_z = +\hbar/2$) may be flipped "down" ($J_z = -\hbar/2$). Now this atom has the direction of its magnetic moment reversed, so it will feel a *downward* force in magnet 3 and will move along the path a', shown in Fig. 35–5. It will no longer get through the slit S_2 to the detector. Similarly, some of the atoms whose spins were initially down ($J_z = -\hbar/2$) will have their spins flipped up ($J_z = +\hbar/2$) as they pass through magnet 2. They will then go along the path b' and will not get to the detector.

If the oscillating field $\mathbf{B}'$ has a frequency appreciably different from ω_p, it will not cause any spin flips, and the atoms will follow their undisturbed paths to the detector. So you can see that the "precession" frequency ω_p of the atoms in the field $\mathbf{B}_0$ can be found by varying the frequency ω of the field $\mathbf{B}'$ until a decrease is observed in the current of atoms arriving at the detector. A decrease in the current will occur when ω is "in resonance" with ω_p. A plot of the detector current as a function of ω might look like the one shown in Fig. 35–6. Knowing ω_p, we can obtain the g-value of the atom.

Such atomic-beam or, as they are usually called, "molecular" beam resonance experiments are a beautiful and delicate way of measuring the magnetic properties of atomic objects. The resonance frequency ω_p can be determined with great precision—in fact, with a greater precision than we can measure the magnetic field $\mathbf{B}_0$, which we must know to find g.

35–4 The paramagnetism of bulk materials

We would like now to describe the phenomenon of the paramagnetism of bulk materials. Suppose we have a substance whose atoms have permanent magnetic moments, for example a crystal like copper sulfate. In the crystal there are copper ions whose inner electron shells have a net angular momentum and a net magnetic moment. So the copper ion is an object which has a permanent magnetic moment. Let's say just a word about which atoms have magnetic moments and which ones don't. Any atom, like sodium for instance, which has an *odd* number

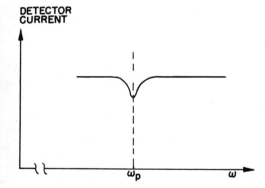

DETECTOR
CURRENT

Fig. 35–6. The current of atoms in the beam decreases when $\omega = \omega_p$.

of electrons, will have a magnetic moment. Sodium has one electron in its unfilled shell. This electron gives the atom a spin and a magnetic moment. Ordinarily, however, when compounds are formed the extra electrons in the outside shell are coupled together with other electrons whose spin directions are exactly opposite, so that all the angular momenta and magnetic moments of the valence electrons usually cancel out. That's why, in general, molecules do not have a magnetic moment. Of course if you have a gas of sodium atoms, there is no such cancellation.* Also, if you have what is called in chemistry a "free radical"—an object with an odd number of valence electrons—then the bonds are not completely satisfied, and there is a net angular momentum.

In most bulk materials there is a net magnetic moment only if there are atoms present whose *inner* electron shell is not filled. Then there can be a net angular momentum and a magnetic moment. Such atoms are found in the "transition element" part of the periodic table—for instance, chromium, manganese, iron, nickel, cobalt, palladium, and platinum are elements of this kind. Also, all of the rare earth elements have unfilled inner shells and permanent magnetic moments. There are a couple of other strange things that also happen to have magnetic moments, such as liquid oxygen, but we will leave it to the chemistry department to explain the reason.

Now suppose that we have a box full of atoms or molecules with permanent moments—say a gas, or a liquid, or a crystal. We would like to know what happens if we apply an external magnetic field. With *no* magnetic field, the atoms are kicked around by the thermal motions, and the moments wind up pointing in all directions. But when there is a magnetic field, it acts to line up the little magnets; then there are more moments lying toward the field than away from it. The material is "magnetized."

We define the *magnetization* M of a material as the net magnetic moment per unit volume, by which we mean the vector sum of all the atomic magnetic moments in a unit volume. If there are N atoms per unit volume and their *average* moment is $\langle \mu \rangle_{\text{av}}$ then M can be written as N times the average atomic moment:

$$M = N\langle \mu \rangle_{\text{av}}. \tag{35.8}$$

The definition of M corresponds to the definition of the electric polarization P of Chapter 10.

The classical theory of paramagnetism is just like the theory of the dielectric constant we showed you in Chapter 11. One assumes that each of the atoms has a magnetic moment μ, which always has the same magnitude but which can point in any direction. In a field B, the magnetic energy is $-\mu \cdot B = -\mu B \cos\theta$, where θ is the angle between the moment and the field. From statistical mechanics, the relative probability of having any angle is $e^{-\text{energy}/kT}$, so angles near zero are more likely than angles near π. Proceeding exactly as we did in Section 11–3, we find that for small magnetic fields M is directed parallel to B and has the magnitude

$$M = \frac{N\mu^2 B}{3kT}. \tag{35.9}$$

[See Eq. (11.20).] This approximate formula is correct only for $\mu B/kT$ much less than one.

We find that the induced magnetization—the magnetic moment per unit volume—is proportional to the magnetic field. This is the phenomenon of paramagnetism. You will see that the effect is stronger at lower temperatures and weaker at higher temperatures. When we put a field on a substance, it develops, for small fields, a magnetic moment proportional to the field. The ratio of M to B (for small fields) is called the magnetic *susceptibility*.

Now we want to look at paramagnetism from the point of view of quantum mechanics. We take first the case of an atom with a spin of $1/2$. In the absence of

---

* Ordinary Na vapor is mostly monatomic, although there are also some molecules of Na_2.

a magnetic field the atoms have a certain energy, but in a magnetic field there are two possible energies, one for each value of J_z. For $J_z = +\hbar/2$, the energy is changed by the magnetic field by the amount

$$\Delta U_1 = +g \left(\frac{q_e \hbar}{2m} \right) \cdot \frac{1}{2} \cdot B. \qquad (35.10)$$

(The energy shift ΔU is positive for an atom because the electron charge is negative.) For $J_z = -\hbar/2$, the energy is changed by the amount

$$\Delta U_2 = -g \left(\frac{q_e \hbar}{2m} \right) \cdot \frac{1}{2} \cdot B. \qquad (35.11)$$

To save writing, let's set

$$\mu_0 = g \left(\frac{q_e \hbar}{2m} \right) \cdot \frac{1}{2} ; \qquad (35.12)$$

then

$$\Delta U = \pm \mu_0 B. \qquad (35.13)$$

The meaning of μ_0 is clear: $-\mu_0$ is the z-component of the magnetic moment in the up-spin case, and $+\mu_0$ is the z-component of the magnetic moment in the down-spin case.

Now statistical mechanics tells us that the probability that an atom is in one state or another is proportional to

$$e^{-(\text{Energy of state})/kT}.$$

With no magnetic field the two states have the same energy; so when there is equilibrium in a magnetic field, the probabilities are proportional to

$$e^{-\Delta U/kT}. \qquad (35.14)$$

The number of atoms per unit volume with spin up is

$$N_{\text{up}} = a e^{-\mu_0 B/kt}, \qquad (35.15)$$

and the number with spin down is

$$N_{\text{down}} = a e^{+\mu_0 B/kt}. \qquad (35.16)$$

The constant a is to be determined so that

$$N_{\text{up}} + N_{\text{down}} = N, \qquad (35.17)$$

the total number of atoms per unit volume. So we get that

$$a = \frac{N}{e^{+\mu_0 B/kT} + e^{-\mu_0 B/kT}}. \qquad (35.18)$$

What we are interested in is the *average* magnetic moment along the z-axis. The atoms with spin up will contribute a moment of $-\mu_0$, and those with spin down will have a moment of $+\mu_0$; so the average moment is

$$\langle \mu \rangle_{\text{av}} = \frac{N_{\text{up}}(-\mu_0) + N_{\text{down}}(+\mu_0)}{N}. \qquad (35.19)$$

The magnetic moment per unit volume M is then $N \langle \mu \rangle_{\text{av}}$. Using Eqs. (35.15), (35.16), and (35.17), we get that

$$M = N\mu_0 \frac{e^{+\mu_0 B/kT} - e^{-\mu_0 B/kT}}{e^{+\mu_0 B/kT} + e^{-\mu_0 B/kT}}. \qquad (35.20)$$

This is the quantum-mechanical formula for M for atoms with $j = 1/2$. Incidentally, this formula can also be written somewhat more concisely in terms of the hyperbolic tangent function:

$$M = N\mu_0 \tanh \frac{\mu_0 B}{kT}. \tag{35.21}$$

A plot of M as a function of B is given in Fig. 35.7. When B gets very large, the hyperbolic tangent approaches 1, and M approaches the limiting value $N\mu_0$. So at high fields, the magnetization *saturates*. We can see why that is; at high enough fields the moments are all lined up in the same direction. In other words, they are all in the spin-down state, and each atom contributes the moment μ_0.

In most normal cases—say, for typical moments, room temperatures, and the fields one can normally get (like 10,000 gauss)—the ratio $\mu_0 B/kT$ is about 0.02. One must go to very low temperatures to see the saturation. For normal temperatures, we can usually replace $\tanh x$ by x, and write

$$M = \frac{N\mu_0^2 B}{kT}. \tag{35.22}$$

Just as we saw in the classical theory, M is proportional to B. In fact, the formula is almost exactly the same, except that there seems to be a factor of 1/3 missing. But we still need to relate the μ_0 in our quantum formula to the μ that appears in the classical result, Eq. (35.9).

In the classical formula, what appears is $\mu^2 = \boldsymbol{\mu} \cdot \boldsymbol{\mu}$, the square of the vector magnetic moment, or

$$\boldsymbol{\mu} \cdot \boldsymbol{\mu} = \left(g \frac{q_e}{2m}\right)^2 \boldsymbol{J} \cdot \boldsymbol{J}. \tag{35.23}$$

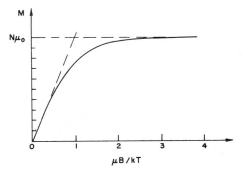

Fig. 35–7. The variation of the paramagnetic magnetization with the magnetic field strength B.

We pointed out in the last chapter that you can very likely get the right answer from a classical calculation by replacing $\boldsymbol{J} \cdot \boldsymbol{J}$ by $j(j + 1)\hbar^2$. In our particular example, we have $j = 1/2$, so

$$j(j + 1)\hbar^2 = \tfrac{3}{4}\hbar^2.$$

Substituting this for $\boldsymbol{J} \cdot \boldsymbol{J}$ in Eq. (35.23), we get

$$\boldsymbol{\mu} \cdot \boldsymbol{\mu} = \left(g \frac{q_e}{2m}\right)^2 \frac{3\hbar^2}{4},$$

or in terms of μ_0, defined in Eq. (35.12), we get

$$\boldsymbol{\mu} \cdot \boldsymbol{\mu} = 3\mu_0^2.$$

Substituting this for μ^2 in the classical formula, Eq. (35.9), does indeed reproduce the correct quantum formula, Eq. (35.22).

The quantum theory of paramagnetism is easily extended to atoms of any spin j. The low-field magnetization is

$$M = Ng^2 \frac{j(j + 1)}{3} \frac{\mu_B^2 B}{kT}, \tag{35.24}$$

where

$$\mu_B = \frac{q_e \hbar}{2m} \tag{35.25}$$

is a combination of constants with the dimensions of a magnetic moment. Most atoms have moments of roughly this size. It is called the *Bohr magneton*. The spin magnetic moment of the electron is almost exactly one Bohr magneton.

35–5 Cooling by adiabatic demagnetization

There is a very interesting special application of paramagnetism. At very low temperatures it is possible to line up the atomic magnets in a strong field. It is then possible to get down to *extremely* low temperatures by a process called *adiabatic demagnetization*. We can take a paramagnetic salt (for example, one containing a number of rare-earth atoms like praseodynium-ammonium-nitrate), and start by cooling it down with liquid helium to one or two degrees absolute in a strong magnetic field. Then the factor $\mu B/kT$ is larger than 1—say more like 2 or 3. Most of the spins are lined up, and the magnetization is nearly saturated. Let's say, to make it easy, that the field is very powerful and the temperature is very low, so that nearly all the atoms are lined up. Then you isolate the salt thermally (say, by removing the liquid helium and leaving a good vacuum) and turn off the magnetic field. The temperature of the salt goes way down.

Now if you were to turn off the field *suddenly*, the jiggling and shaking of the atoms in the crystal lattice would gradually knock all the spins out of alignment. Some of them would be up and some down. But if there is no field (and disregarding the interactions between the atomic magnets, which will make only a slight error), it takes no energy to turn over the atomic magnets. They could randomize their spins without any energy change and, therefore, without any temperature change.

Suppose, however, that while the atomic magnets are being flipped over by the thermal motion there is still some magnetic field present. Then it requires some work to flip them over opposite to the field—*they must do work against the field*. This takes energy from the thermal motions and lowers the temperature. So if the strong magnetic field is not removed too rapidly, the temperature of the salt will decrease—it is cooled by the demagnetization. From the quantum-mechanical view, when the field is strong all the atoms are in the lowest state, because the odds against any being in the upper state are impossibly big. But as the field is lowered, it gets more and more likely that thermal fluctuations will knock an atom into the upper state. When that happens, the atom absorbs the energy $\Delta U = \mu_0 B$. So if the field is turned off slowly, the magnetic transitions can take energy out of the thermal vibrations of the crystal, cooling it off. It is possible in this way to go from a temperature of a few degrees absolute down to a temperature of a few thousandths of a degree.

Would you like to make something even colder than that? It turns out that Nature has provided a way. We have already mentioned that there are also magnetic moments for the atomic nuclei. Our formulas for paramagnetism work just as well for nuclei, except that the moments of nuclei are roughly a *thousand times smaller*. [They are of the order of magnitude of $q\hbar/2m_p$, where m_p is the *proton* mass, so they are smaller by the ratio of the masses of the electron and proton.] With such magnetic moments, even at a temperature of 2°K, the factor $\mu B/kT$ is only a few parts in a thousand. But if we use the paramagnetic demagnetization process to get down to a temperature of a few thousandths of a degree, $\mu B/kT$ becomes a number near 1—at these low temperatures we can begin to saturate the nuclear moments. That is good luck, because we can then use the adiabatic demagnetization of the *nuclear* magnetism to reach still lower temperatures. Thus it is possible to do two stages of magnetic cooling. First we use adiabatic demagnetization of paramagnetic ions to reach a few thousandths of a degree. Then we use the cold paramagnetic salt to cool some material which has a strong nuclear magnetism. Finally, when we remove the magnetic field from this material, its temperature will go down to within a *millionth* of a degree of absolute zero—if we have done everything very carefully.

35–6 Nuclear magnetic resonance

We have said that atomic paramagnetism is very small and that nuclear magnetism is even a thousand times smaller. Yet it is relatively easy to observe the nuclear magnetism by the phenomenon of "nuclear magnetic resonance." Suppose we take a substance like water, in which all of the electron spins are exactly balanced so that their net magnetic moment is zero. The molecules will still have a very, very tiny magnetic moment due to the nuclear magnetic moment of the hydrogen nuclei. Suppose we put a small sample of water in a magnetic field B. Since the protons (of the hydrogen) have a spin of 1/2, they will have two possible energy states. If the water is in thermal equilibrium, there will be slightly more protons in the lower energy states—with their moments directed parallel to the field. There is a small net magnetic moment per unit volume. Since the proton moment is only about one-thousandth of an atomic moment, the magnetization which goes as μ^2—using Eq. (35.22)—is only about one-millionth as strong as typical atomic paramagnetism. (That's why we have to pick a material with no atomic magnetism.) If you work it out, the difference between the number of protons with spin up and with spin down is only one part in 10^8, so the effect is indeed very small! It can still be observed, however, in the following way.

Suppose we surround the water sample with a small coil that produces a small horizontal oscillating magnetic field. If this field oscillates at the frequency ω_p, it will induce transitions between the two energy states—just as we described for the Rabi experiment in Section 35–3. When a proton flips from an upper energy state to a lower one, it will give up the energy $\mu_z B$ which, as we have seen, is equal to $\hbar\omega_p$. If it flips from the lower energy state to the upper one, it will *absorb* the energy $\hbar\omega_p$ from the coil. Since there are slightly more protons in the lower state than in the upper one, there will be a net *absorption* of energy from the coil. Although the effect is very small, the slight energy absorption can be seen with a sensitive electronic amplifier.

Just as in the Rabi molecular-beam experiment, the energy absorption will be seen only when the oscillating field is in resonance, that is, when

$$\omega = \omega_p = g\left(\frac{q_e}{2m_p}\right)B.$$

It is often more convenient to search for the resonance by varying B while keeping ω fixed. The energy absorption will evidently appear when

$$B = \frac{2m_p}{g\,q_e}\,\omega.$$

A typical nuclear magnetic resonance apparatus is shown in Fig. 35–8. A high-frequency oscillator drives a small coil placed between the poles of a large electromagnet. Two small auxiliary coils around the pole tips are driven with a 60-cycle current so that the magnetic field is "wobbled" about its average value by a very small amount. As an example, say that the main current of the magnet is set to give a field of 5000 gauss, and the auxiliary coils produce a variation of ± 1 gauss about this value. If the oscillator is set at 21.2 megacycles per second, it will then be at the proton resonance each time the field sweeps through 5000 gauss [using Eq. (34.13) with $g = 5.58$ for the proton].

The circuit of the oscillator is arranged to give an additional output signal proportional to any *change* in the power being absorbed from the oscillator. This signal is fed to the vertical deflection amplifier of an oscilloscope. The horizontal sweep of the oscilloscope is triggered once during each cycle of the field-wobbling frequency. (More usually, the horizontal deflection is made to follow in proportion to the wobbling field.)

Before the water sample is placed inside the high-frequency coil, the power drawn from the oscillator is some value. (It doesn't change with the magnetic field.) When a small bottle of water is placed in the coil, however, a signal appears on the oscilloscope, as shown in the figure. We see a picture of the power being absorbed by the flipping over of the protons!

In practice, it is difficult to know how to set the main magnet to exactly 5000 gauss. What one does is to adjust the main magnet current until the resonance signal appears on the oscilloscope. It turns out that this is now the most convenient way to make an accurate measurement of the strength of a magnetic field. Of course, at some time *someone* had to measure accurately the magnetic field and frequency to determine the g-value of the proton. But now that this has been done, a proton resonance apparatus like that of the figure can be used as a "proton resonance magnetometer."

We should say a word about the shape of the signal. If we were to wobble the magnetic field very slowly, we would expect to see a normal resonance curve. The energy absorption would read a maximum when ω_p arrived exactly at the

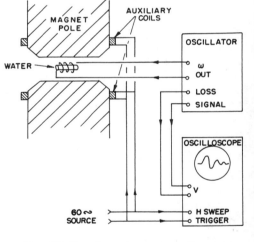

Fig. 35–8. A nuclear magnetic resonance apparatus.

oscillator frequency. There would be some absorption at nearby frequencies because all the protons are not in exactly the same field—and different fields mean slightly different resonant frequencies.

One might wonder, incidentally, whether at the resonance frequency we should see any signal at all. Shouldn't we expect the high-frequency field to equalize the populations of the two states—so that there should be no signal except when the water is first put in? Not exactly, because although we are *trying* to equalize the two populations, the thermal motions on their part are trying to keep the proper ratios for the temperature T. If we sit at the resonance, the power being absorbed by the nuclei is just what is being lost to the thermal motions. There is, however, relatively little "thermal contact" between the proton magnetic moments and the atomic motions. The protons are relatively isolated down in the center of the electron distributions. So in pure water, the resonance signal is, in fact, usually too small to be seen. To increase the absorption, it is necessary to increase the "thermal contact." This is usually done by adding a little iron oxide to the water. The iron atoms are like small magnets; as they jiggle around in their thermal dance, they make tiny jiggling magnetic fields at the protons. These varying fields "couple" the proton magnets to the atomic vibrations and tend to establish thermal equilibrium. It is through this "coupling" that protons in the higher energy states can lose their energy so that they are again capable of absorbing energy from the oscillator.

In practice the output signal of a nuclear resonance apparatus does not look like a normal resonance curve. It is usually a more complicated signal with oscillations—like the one drawn in the figure. Such signal shapes appear because of the changing fields. The explanation should be given in terms of quantum mechanics, but it can be shown that in such experiments the classical ideas of precessing moments always give the correct answer. Classically, we would say that when we arrive at resonance we start driving a lot of the precessing nuclear magnets synchronously. In so doing, we make them precess *together*. These nuclear magnets, all rotating together, will set up an induced emf in the oscillator coil at the frequency ω_p. But because the magnetic field is increasing with time, the precession frequency is increasing also, and the induced voltage is soon at a frequency a little higher than the oscillator frequency. As the induced emf goes alternately in phase and out of phase with the oscillator, the "absorbed" power goes alternately positive and negative. So on the oscilloscope we see the beat note between the proton frequency and the oscillator frequency. Because the proton frequencies are not all identical (different protons are in slightly different fields) and also possibly because of the disturbance from the iron oxide in the water, the freely precessing moments soon get out of phase, and the beat signal disappears.

These phenomena of magnetic resonance have been put to use in many ways as tools for finding out new things about matter—especially in chemistry and nuclear physics. It goes without saying that the numerical values of the magnetic moments of nuclei tell us something about their structure. In chemistry, much has been learned from the structure (or shape) of the resonances. Because of magnetic fields produced by nearby nuclei, the exact position of a nuclear resonance is shifted somewhat, depending on the environment in which any particular nucleus finds itself. Measuring these shifts helps determine which atoms are near which other ones and helps to elucidate the details of the structure of molecules. Equally important is the electron spin resonance of free radicals. Although not present to any very large extent in equilibrium, such radicals are often intermediate states of chemical reactions. A measurement of an electron spin resonance is a delicate test for the presence of free radicals and is often the key to understanding the mechanism of certain chemical reactions.

Index

Aberration, I–27–7, I–34–10
Absolute zero, I–1–5
Absorption, I–31–8 ff
  of light, III–9–14 f
Absorption coefficient, II–32–8
Acceleration, I–8–8 ff
  components of, I–9–3
  of gravity, I–9–4
Accelerator guide field, II–29–4 ff
Acceptor, III–14–5
Activation energy, I–42–7
Active circuit element, II–22–5
Adams, J. C., I–7–5
Adiabatic compression, I–39–5
Adiabatic demagnetization, II–35–9 f
Adiabatic expansion, I–44–5
Adjoint, III–11–22
Affective future, I–17–4
Aharanov, II–15–12
Air trough, I–10–5
Algebra, I–22–1 ff
Algebraic operator, III–20–6
Alternating-current circuits, II–22–1 ff
Alternating-current generator, II–17–6 ff
Alnico V, II–37–10
Amber, II–1–10
Ammeter, II–16–1
Ammonia maser, III–9–1 ff
Ammonia molecule, III–8–11 ff
  states of, III–9–1 ff
Ampere, A., II–13–3
Ampere's law, II–13–4
Amperian current, II–36–2
Amplitudes, III–8–1 f, III–2–1 ff
  interfering, III–5–10 ff
  of oscillation, I–21–3
  probability, III–3–1 ff
  space dependence of, III–54–1 ff,
    III–13–4
  time dependence of, III–7–1 ff
  transformation of, III–6–1 ff
Amplitude modulation, I–48–3
Analog computer, I–25–8
Anderson, C. D., I–52–10
Angle, of incidence, I–26–3
  of precession, II–34–4
  of reflection, I–26–3
Angstrom (unit), I–1–3
Angular frequency, I–21–3, I–29–2
Angular momentum, I–7–7, I–18–5 f,
  I–20–1, III–20–14 ff
  composition of, III–18–4 ff
  conservation of, I–4–7, I–18–6 ff,
    I–20–5
  of rigid body, I–20–8
Anomalous refraction, I–33–9 f
Antiferromagnetic material, II–37–11

Antimatter, I–52–10 f, III–11–16
Antiparticle, I–2–8, III–11–13
Antiproton, III–11–13
Argon, III–19–16
Aristotle, I–5–1
Atom, I–1–2
  metastable, I–42–10
  Rutherford-Bohr model, II–5–3
  stability of, II–5–3
  Thompson model, II–5–3
Atomic clock, I–5–5
Atomic currents, II–13–5 f
Atomic hypothesis, I–1–2
Atomic orbits, II–1–8
Atomic particles, I–2–9 f
Atomic polarizability, II–32–2
Atomic processes, I–1–5 f
Attenuation, I–31–8
Avogadro, A., I–39–2
Avogadro's number, I–41–10
Axial vector, I–52–6 f

Barkhausen effect, II–37–9
Baryons, III–11–13
Base states, III–5–8 ff, III–12–1 ff,
  of the world, III–8–5 ff
Battery, II–22–6
Becquerel, A. H., I–28–3
Bell, A. G., II–16–3
Benzene molecule, III–10–10 ff,
  III–15–7 ff
Bernoulli's theorem, II–40–6 ff
Bessel function, II–23–6
Betatron, II–17–5
Biot-Savart law, II–14–10
Birefringence, I–33–3 ff
Blackbody radiation, I–41–5 f
Blackbody spectrum, III–4–8 ff
Boehm, I–52–10
Bohm, II–7–7, II–15–12
Bohr. N., I–42–9, II–5–3
Bohr magneton, II–34–12
Bohr radius, I–38–6, III–2–6, III–19–3
Boltzmann, L., I–41–2
Boltzmann factor, III–14–4
Boltzmann's law, I–40–2 f
Bopp, II–28–8
Born, M., I–37–1, I–38–9, II–28–7,
  III–1–1, III–2–9, III–21–6
Boron, III–19–16
Bose particles, III–4–1 ff, III–15–6 f
Boundary layer, II–41–9
Boundary-value problems, II–7–1
Boyle's law, I–40–8
"Boys" camera, II–9–10
Bragg, L., II–30–9

Bragg-Nye crystal model, II–30–9 ff
Breaking-drop theory, II–9–9
Bremsstrahlung, I–34–6 f
Brewster's angle, I–33–6
Briggs, H., I–22–6
Brown, R., I–41–1
Brownian motion, I–1–8, I–6–5, I–41–1 ff
Brush discharge, II–9–9
Bulk modulus, II–38–3
Butadiene molecule, III–15–10

Calculus, differential, I–8–4, II–2–1 ff
  integral, II–3–1 ff
  of variations, II–19–3
Cantilever beam, II–38–10
Capacitance, I–23–5
  mutual, II–22–17
Capacitor, I–14–9, I–23–5, II–22–3 ff,
  II–23–2 ff
  parallel-plate, I–14–9, II–6–11 ff, II–8–3
Capacity, II–6–12
  of a condenser, II–8–2
Capillary action, I–51–8
Carnot, S., I–4–2, I–44–2 ff
Carnot cycle, I–44–5 f, I–45–2
Carrier signal, I–48–3
Carriers, negative, III–14–2
  positive, III–14–2
Catalyst, I–42–8
Cavendish, H., I–7–9
Cavendish's experiment, I–7–9
Cavity resonator, II–23–1 ff
Center of mass, I–18–1 f, I–19–1 ff
Centrifugal force, I–7–5, I–12–11
Çerenkov, P. A., I–51–2
Çerenkov radiation, I–51–2
Charge, conservation of, I–4–7, II–13–1 f
  on electron, I–12–7
  line of, II–5–3 f
  motion of, II–29–1 ff
  sheet of, II–5–4
  sphere of, II–5–4 f
Charge density, II–5–4
Charge separation, II–9–7 ff
Charged conductor, II–8–2 ff
Chemical energy, I–4–2
Chemical kinetics, I–42–7 f
Chemical reaction, I–1–6 ff
Chlorophyll molecule, III–15–11
Chromaticity, I–35–6 f
Circuits, alternating-current, II–22–1 ff
  equivalent, II–22–10 f
Circuit elements, II–23–1 f
  active, II–22–5
  passive, II–22–5
Circular motion, I–21–4

Circulation, II–1–5, II–3–8 ff
Classical electron radius, II–28–3
Classical limit, III–7–10
Clausius, R., I–44–2, I–44–3
Clausius-Clapeyron equation,
    I–45–6 ff
Clausius-Mossotti equation, II–11–6 f,
    II–32–7
Cleavage plane, II–30–1
Coaxial line, II–24–1
Coefficient, absorption, II–32–8
    of coupling, II–17–14
    of friction, I–12–4
    gravitational, I–7–9
    of viscosity, II–41–2
Collision, I–16–6
    elastic, I–10–7
Colloidal particles, II–7–8 ff
Color vision, I–35–1 ff
    physiochemistry of, I–35–9 f
Commutation rule, III–20–15
Complex impedance, I–23–7
Complex numbers, I–22–7 ff, I–23–1 ff
Complex variable, II–7–2 ff
Compound eye, I–36–6 ff
Compression, adiabatic, I–39–5
    isothermal, I–44–5
Condenser, parallel-plate, I–14–9,
    II–6–11 ff, II–8–3
Conduction band, III–14–1
Conductivity, II–32–10
    thermal, II–2–8, II–12–2
Conductor, II–1–2
Cones, I–35–1
Conservation, of angular momentum,
    I–4–7, I–18–6 ff, I–20–5
    of charge, I–4–7, II–13–1 f
    of energy, I–3–2, I–4–1 ff, II–27–1 f,
    II–42–10
    of linear momentum, I–4–7, I–10–1 ff
    of potential energy, III–7–6 ff
    of strangeness, III–11–12
Contraction hypothesis, I–15–3
Copernicus, I–7–1
Coriolis force, I–19–8 f
Cornea, I–35–1
Cosmic rays, II–9–2
Couette flow, II–41–10 ff
Coulomb's law, I–28–2, II–4–2 ff,
    II–5–6
Coupling, coefficient of, II–17–14
Covalent bond, II–30–2
Cross product, II–2–8, II–31–8
Cross section for scattering, I–32–7
Crystal, II–30–1 ff
    geometry of, II–30–1 f
Crystal diffraction, I–38–4 f, III–2–4 f
Crystal lattice, II–30–3 f
    propagation, III–13–1 ff
    imperfections, III–13–10 f
Cubic cell, II–30–7
Curie law, II–11–5
Curie temperature, II–36–13
Curie-Weiss law, II–11–9
Curl operator, II–2–8, II–3–1
Current, Amperian, II–36–2
    atomic, II–13–5 f
    eddy, II–16–6
    electric, II–13–1 f
    induced, II–16–1 ff
Current density, II–13–1

Curvature, intrinsic, II–42–5
    mean, II–42–6
    negative, II–42–4
    positive, II–42–4
    in three-dimensional space, II–42–5 f
Curved space, II–42–1 ff
Cutoff frequency, II–22–14

D'Alembertian, II–25–8
Debye length, II–7–9
Dedekind, R., I–22–4
    definite energy, states, III–13–3 ff
Degrees of freedom, I–25–2, I–39–12
Demagnetization, adiabatic, II–35–9 f
Density, I–1–4
Derivative, I–8–5 ff
    partial, I–14–9
Diamagnetism, II–34–1 ff
    diamond lattice, III–14–1
Dicke, R. H., I–7–11
Dielectric, II–10–1 ff, II–11–1 ff
Dielectric constant, II–10–1 f
Differential calculus, I–8–4, II–2–1 ff
Diffraction, I–30–1 ff
    by screen, I–31–10 f
Diffraction grating, I–29–5, I–30–3 ff
Diffusion, I–43–1 ff
    of neutrons, II–12–6 ff
Dipole, II–21–5 ff
    electric, II–6–2 ff
    magnetic, II–14–7 f
Dipole moment, I–12–6, II–6–7
Dipole potential, II–6–4 ff
Dipole radiator, I–28–5 f, I–29–3 ff
Dirac, P., I–52–10, II–2–1, II–28–7,
    III–8–2, III–12–6
Dirac equation, I–20–6
Dislocation, II–30–8, II–30–9
Dispersion, I–31–6 ff
Distance, I–5–5 ff
Distance measurement, color brightness,
    I–5–6
    triangulation, I–5–6
Divergence, II–25–7
Divergence operator, II–2–7, II–3–1
Domain, II–37–6
Donor site, III–14–4
Doppler effect, I–17–8, I–23–9, I–34–7 f,
    I–38–6, II–42–9, III–2–6, III–12–9
Dot product, I–2–4, II–25–3
Double stars, I–7–6
Drag coefficient, II–41–7
"Dry" water, II–40–1 ff
Dyes, III–10–12
Dynamical momentum, III–21–5
Dynamics, I–7–2 f, I–9–1 ff
    relativistic, I–15–9 f

Eddy current, II–16–6
Effective mass, III–13–7
Efficiency of ideal engine, I–44–7 f
Eigenstates, III–11–22
Eigenvalues, III–11–21
Einstein, A., I–2–6, I–7–11, I–12–12,
    I–15–1, I–16–1, I–41–8, I–42–8,
    I–42–9, II–42–1, II–42–6, II–42–8,
    II–42–13 f
Elastic collision, I–10–7
Elastic constants, II–39–6, II–39–10 f

Elastic energy, I–4–2, I–4–6
Elastic materials, II–39–1 ff
Elastica, II–38–12
Elasticity, II–38–1 ff
Elasticity tensor, II–39–4 ff
Electret, II–11–8
Electric charge density, II–2–8, II–4–3,
    III–21–6
Electric current, II–13–1 f
    in the atmosphere, II–9–2 f
Electric current density, II–2–8, III–21–6
Electric dipole, II–6–2 ff
Electric dipole matrix element, III–9–15
Electric field, I–2–4, I–12–7 f, II–1–2,
    II–1–3, II–6–1 ff, II–7–1 ff
    relativity of, II–13–6 ff
Electric flux, II–1–4
Electric force, I–2–3 ff, II–1–1 ff, II–13–1
Electric potential, II–4–4
Electric susceptibility, II–10–4
Electrical energy, I–4–2, II–15–3 ff
Electrical forces, II–1–1 ff, II–13–1
Electrodynamics, II–1–3
    relativistic notation, II–25–1 ff
Electromagnet, II–36–9 ff
Electromagnetic energy, I–29–2
Electromagnetic field, I–2–2, I–2–5, I–10–9
Electromagnetic mass, II–28–3 f
Electromagnetic radiation, I–26–1,
    I–28–1 ff
Electromagnetic waves, II–21–1 f
    cosmic rays, I–2–5
    gamma rays, I–2–5
    infrared, I–2–5, I–23–8, I–26–1
    light, I–2–5
    ultraviolet, I–2–5, I–26–1
    x-rays, I–2–5, I–26–1
Electromagnetism, II–1–1 ff
    laws of, II–1–5 ff
Electromotive force, II–16–2
Electron, I–2–4, I–37–1, I–37–4 ff,
    III–1–1, III–1–4 ff
    charge on, I–12–7
    radius of, classical, I–32–4
Electron cloud, I–6–11
Electron configuration, III–19–15
Electron-hole pairs, III–14–3
Electron microscope, II–29–3 f
Electron-ray tube, I–12–9
Electron volt (unit), I–34–4
Electronic polarization, II–11–1 ff
Electrostatic energy, II–8–1 ff
    of charges, II–8–1 f
    of ionic crystal, II–8–4 ff
    in nuclei, II–8–6 ff
    of a point charge, II–8–12
Electrostatic equations, II–10–6 f
Electrostatic field, II–5–1 ff, II–7–1 f
    energy in, II–8–9 ff
    of a grid, II–7–10 f
Electrostatic lens, II–29–2 f
Electrostatic potential, equations of, II–6–1
Electrostatics, II–4–1 ff, II–5–1
Ellipse, I–7–1
Emissivity, II–6–14
Energy, II–22–11 f
    chemical, I–4–2
    of a condenser, II–8–2 ff
    conservation of, I–3–2, I–4–1 ff,
    II–27–1 f
    elastic, I–4–2, I–4–6

electrical, I–4–2, II–15–3 ff
electromagnetic, I–29–2
electrostatic, II–8–1 ff
  in electrostatic field, II–8–9 ff
  gravitational, I–4–2 ff
  heat, I–4–2, I–4–6, I–10–7, I–10–8
  kinetic, I–1–7, I–4–2, I–4–5 f, I–39–4
  magnetic, II–17–12 ff
  mass, I–4–2, I–4–7
  mechanical, II–15–3 ff
  nuclear, I–4–2
  potential, I–4–4, I–13–1 ff, I–14–1 ff
  radiant, I–4–2
  relativistic, I–16–1 ff
Energy density, II–27–2
Energy diagram, III–14–1
Energy flux, II–27–2
Energy level diagram, III–14–3
Energy levels, I–38–7 f, III–12–7 ff,
  III–2–7 f
Energy theorem, I–50–7 f
Enthalpy, I–45–5
Entropy, I–44–10 ff, I–46–7 ff
Eötvös, L., I–7–11
Equation of motion, II–42–14
Equilibrium, I–1–6
Equipotential surfaces, II–4–11 f
Equivalent circuits, II–22–10 f
Ethylene molecule, III–15–8
Euclid, I–5–6
Euclidean geometry, I–12–3
Euler force, II–38–11
Evaporation, I–1–5 f
  of a liquid, I–40–3 f, I–42–1 ff
Excess radius, II–42–4
Exchange force, II–37–2
Excited state, II–8–7, III–13–9
Exciton, III–13–9
Exclusion principle, III–4–12 ff
Expansion, adiabatic, I–44–5
  isothermal, I–44–5
Exponential atmosphere, I–40–1 f
Eye, compound, I–36–6 ff
  human, I–35–1 f, I–36–3 ff

Farad (unit), I–25–7, II–6–13
Faraday, M., II–10–1
Faraday's law of induction, II–17–2
Fermat, P., I–26–3
Fermi (unit), I–5–10
Fermi, E., I–5–10
Fermi particles, III–4–1 ff, III–15–7 f
Ferrite, II–37–12
Ferroelectricity, II–11–8 ff
Ferromagnetic crystal, III–15–1
Ferromagnetic insulators, II–37–12
Ferromagnetism, II–34–1 f, II–36–1 ff,
  II–37–1 ff
Feynman, R., II–28–8
Fields, I–2–2, I–2–4, I–2–5, I–10–9,
  I–12–7 ff, I–13–8 f, I–14–7 ff
  in a cavity, II–5–8 f
  of a charged conductor, II–6–8
  of a conductor, II–5–7 f
  electric, I–2–4, I–12–7 f, II–1–2,
    II–1–3, II–6–1 ff, II–7–1 ff
  electrostatic, II–5–1 ff, II–7–1 f
  magnetic, II–1–2, II–1–3, II–13–1,
    II–14–1 ff
  magnetizing, II–36–7

scalar, II–2–2 ff
  superposition of, I–12–9
  two-dimensional, II–7–2 ff
  vector, II–1–4 f, II–2–1 ff
Field energy, II–27–1 ff
  of a point charge, II–28–1 f
Field equation, II–42–14
Field index, II–29–5
Field-ion microscope, II–6–14
Field lines, II–4–11
Field momentum, II–27–9 ff
  of a moving charge, II–28–2 f
Field strength, II–1–4
Filter, II–22–14 ff
Flow, fluid, II–12–8 ff
  irrotational, II–40–5
  viscous, II–41–4 f
Fluid flow, II–12–8 ff
Fluorine, III–19–16
Flux, II–4–7 ff
  electric, II–1–4
  of a vector field, II–3–2 ff
Flux quantization, III–21–10
Flux rule, II–17–1 ff
Focal length, I–27–1 ff
Focus, I–26–5
Force, centrifugal, I–7–5, I–12–11
  components of, I–9–3
  conservative, I–14–3 ff
  Coriolis, I–19–8 f
  electrical, I–2–3 ff, II–1–1 ff, II–13–1
  electromotive, II–16–2
  gravitational, I–2–3
  Lorentz, II–13–1, II–15–14
  magnetic, II–1–2, II–13–1
  molecular, I–1–3, I–12–6 f
  moment of, I–18–5
  nonconservative, I–14–6 f
  nuclear, I–12–12, III–10–6 ff
  pseudo, I–12–10 ff
Fourier, J., I–50–2 f
Fourier analysis, I–50–2 ff
Fourier theorem, II–7–11
Fourier transform, I–25–4
Four-vectors, I–15–8 f, I–17–5 ff,
  II–25–1 ff
Fovea, I–35–1
Frank, I., I–51–2
Franklin, B., II–5–6
Frequency, angular, I–21–3, I–29–2
  of oscillation, I–2–5
  plasma, II–7–6, II–32–12
Fresnel's reflection formulas, I–33–8
Friction, I–10–5, I–12–3 ff
  coefficient of, I–12–4

Galileo, I–5–1, I–7–2, I–9–1, I–52–3
Galilean relativity, I–10–3
Galilean transformation, I–12–11
Gallium, III–19–17
Galvanometer, II–1–8, II–16–1
Garnet, II–37–12
Gauss (unit), I–34–4
Gauss, K., II–16–2
Gauss' law, II–4–9 f, II–5–1 ff
Gauss' theorem, II–3–5, III–21–4
Gaussian surface, II–10–1
Geiger, II–5–3
Gell-Mann, M., I–2–9
Generator, alternating-current, II–17–6 ff

electric, II–16–1 ff, II–22–5 ff
  van de Graaff, II–5–9, II–8–7
Geometrical optics, I–26–1, I–27–1 f
Gerlach, II–35–3
Gradient operator, II–2–4, II–3–1
Gravitation, I–2–3, I–7–1 ff, I–12–2,
  II–42–1
Gravitational acceleration, I–9–4
Gravitational coefficient, I–7–9
Gravitational energy, I–4–2 ff
Gravitational field, I–12–8 ff, I–13–8 f
Gravity, I–13–3 ff, II–42–8 ff
  acceleration of, I–9–4
Green's function, I–25–4
Ground state, II–8–7, III–7–2
Gyroscope, I–20–5 ff

Hall effect, III–14–7
Hamiltonian matrix, III–8–10 f
Hamilton's first principal function,
  II–19–8
Harmonic motion, I–21–4, I–23–1 ff
Harmonic oscillator, I–10–1, I–21–1 ff
  forced, I–21–5 f, I–23–3 ff
Harmonics, I–50–1 ff
Heat, I–1–3, I–13–3
Heat conduction, II–3–6 ff
Heat diffusion equation, II–3–8
Heat energy, I–4–2, I–4–6, I–10–7,
  I–10–8
Heat engines, I–44–1 ff
Heat flow, II–2–8 f, II–12–2 ff
Heisenberg, W., I–6–10, I–37–1, I–37–9,
  I–37–11, I–37–12, I–38–9, III–1–1,
  III–1–9, III–1–11, III–1–12,
  III–2–9, III–20–17
Helium, III–19–14
Helmholtz, H., I–35–7, II–40–10
Henry (unit), I–25–7
Hermitian adjoint, III–20–3
Hess, II–9–2
Hexagonal cell, II–30–7
High-voltage breakdown, II–6–13 f
Hooke's law, I–12–6, II–38–1 f
Huygens, C., I–15–2, I–26–2
Hydrodynamics, II–40–2 ff
Hydrogen, III–19–14
Hydrogen atom, III–19–1 ff
Hydrogen, hyperfine splitting in,
  III–12–1 ff
Hydrogen molecular ion, III–10–1 ff
Hydrogen molecule, III–10–8 ff
Hydrogen wave functions, III–19–12
Hydrostatics, II–40–1 ff
Hyperfine splitting in hydrogen,
  III–12–1 ff
Hypocycloid, I–34–3
Hysteresis curve, II–37–5 ff
Hysteresis loop, II–36–8

Ideal gas law, I–39–10 ff
Identical particles, III–3–9 ff, III–4–1 ff
Illumination, II–12–10 ff
Image charge, II–6–9
Impedance, I–25–8 f, II–22–1 ff
  complex, I–23–7
Impure semiconductors, III–14–4
Incidence, angle of, I–26–3
Inclined plane, I–4–4

INDEX 3

Independent particle approximation, III–15–1 ff
Index of refraction, I–31–1 ff
Induced currents, II–16–1 ff
Inductance, I–23–6, II–16–4 f, II–17–12 ff, II–22–2 f
    mutual, II–17–9 ff, II–22–16
    self-, II–16–4, II–17–11 f
Induction, laws of, II–17–1 ff
Inductor, I–23–6
Inertia, I–2–3, I–7–11
    moment of, I–18–7, I–19–5 ff
    principle of, I–9–1
Infeld, II–28–7
Infrared radiation, I–23–8, I–26–1
Integral, I–8–7 f
Integral calculus, II–3–1 ff
Insulator, II–1–2, II–10–1
Interference, I–28–6, I–29–1 ff
    two-slit, III–3–5 ff
Interfering amplitudes, III–5–10 ff
Interfering waves, I–37–4, III–1–4
Interferometer, I–15–5
Internal reflection, II–33–12
Ion, I–1–6
Ionic bond, II–30–2
Ionic conductivity, I–43–6 f
Ionic polarizability, II–11–8
Ionization energy, I–42–5
Ionosphere, II–7–5, II–9–3
Irrotational flow, II–40–5
Isotherm, I–2–3
Isothermal atmosphere, I–40–2
Isothermal compression, I–44–5
Isothermal expansion, I–44–5
Isothermal surfaces, II–2–3
Isotopes, I–3–4 ff

Jeans, J., I–40–9, I–41–6 f, II–2–6
Johnson noise, I–41–2, I–41–8
Josephson junction, III–21–14 ff
Joule (unit), I–13–3
Joule heating, I–24–2
Junction, III–14–8 ff

Kármán vortex street, II–41–9
Kepler, J., I–7–1
Kepler's laws, I–7–1 f, I–9–1, I–18–6
Kerr cell, I–33–5
Kilocalorie (unit), II–8–5
Kinematic momentum, III–21–5
Kinetic energy, I–1–7, I–4–2, I–4–5 f, I–39–4
    rotational, I–19–7 ff
Kinetic theory, I–42–1 ff
    of gases, I–39–1 ff
Kirchhoff's laws, I–25–9, II–22–7 ff
Kronecker delta, II–31–6
Krypton, III–19–17

Lamb, II–5–6
Lamé elastic constants, II–39–6
Landé g-factor, II–34–4
Laplace, P., I–47–7
Laplace equation, II–6–1, II–7–1
Laplacian operator, II–2–10
Larmor frequency, II–34–7
Larmor's theorem, II–34–6 f

Laser, I–32–6, I–42–10, III–9–13
Laughton, II–5–6
Laws, of electromagnetism, II–1–5 ff
    of induction, II–17–1 ff
    of quantum mechanics, III–13–1
Least action, principle of, II–19–1 ff
Least time, principle of, I–26–3 ff, I–26–8
Legendre functions, III–19–9
Leibnitz, G. W., I–8–4
Lens formula, I–27–6
Lenz's rule, II–16–4, II–34–2
Leverrier, U., I–7–5
Liénard-Wiechert potentials, II–21–11
Light, II–21–1 f
    absorption of, III–9–14 f
    momentum of, I–34–10 f
    polarized, I–32–9
    scattering of, I–32–5 ff
    speed of, I–15–1, II–18–8 f
Light waves, I–48–1
Lightning, II–9–10 f
Line of charge, II–5–3 f
Line integral, II–3–1
Linear momentum, conservation of, I–4–7, I–10–1 ff
Linear systems, I–25–1 ff
Liquid helium, III–4–12
Lithium, III–19–14
Lodestone, II–1–10
Logarithms, I–22–4
Lorentz, H. A., I–15–3
Lorentz condition, II–25–9
Lorentz contraction, I–15–7
Lorentz force, II–13–1, II–15–14
Lorentz formula, II–21–12 f
Lorentz gauge, II–18–11
Lorentz transformation, I–15–3, I–17–1, I–34–8, I–52–2, II–25–1
    of fields, II–26–1 ff

McCullough, II–1–9
Mach number, II–41–6
Magenta, III–10–12
Magnetic dipole, II–14–7 f
Magnetic dipole moment, II–14–8
Magnetic energy, II–17–12 ff
Magnetic field, I–12–9 f, II–1–2, II–1–3, II–13–1, II–14–1 ff
    relativity of, II–13–6 ff
    of steady currents, II–13–3 f
Magnetic force, II–1–2, II–13–1
    on a current, II–13–2 f
Magnetic induction, I–12–10
Magnetic lens, II–29–3
Magnetic materials, II–37–1 ff
Magnetic moments, II–34–3 f, III–11–4
Magnetic resonance, II–35–1 ff
Magnetic susceptibility, II–35–7
Magnetism, I–2–4, II–34–1 ff
Magnetization currents, II–36–1 ff
Magnetizing fields, II–36–7
Magnetostatics, II–4–1, II–13–1 ff
Magnetostriction, II–37–6
Magnification, I–27–5
Magnons, III–15–4
Marsden, II–5–3
Maser, I–42–10
    ammonia, III–9–1 ff
Mass, I–9–1, I–15–1
    center of, I–18–1 f, I–19–1 ff

electromagnetic, II–28–3 f
    relativistic, I–16–6 ff
Mass energy, I–4–2, I–4–7
Mass-energy equivalence, I–15–10 f
Matrix, III–5–5
Maxwell, J. C., I–6–1, I–6–9, I–28–1, I–40–8, I–41–7, I–46–5, II–1–8, II–1–11, II–5–6, II–18–1 ff
Maxwell's equations, I–15–2, I–25–3, I–47–7, II–2–1, II–2–8, II–4–1, II–6–1, II–18–1 ff, II–32–3 ff, II–42–14
    currents and charges, II–21–1 ff
    free space, II–20–1 ff
Mayer, J. R., I–3–2
Mean free path, I–43–3 f
Mean square distance, I–6–5, I–41–9
Mechanical energy, II–15–3 ff
Meissner effect, III–21–8 ff
Mendeléev, I–2–9
Metastable atom, I–42–10
Meter (unit), I–5–10
Mev (unit), I–2–9
Michelson-Morley experiment, I–15–3 ff
Miller, W. C., I–35–2
Minkowski, I–17–8
Minkowski space, II–31–12
Modes, I–49–1 ff
Mole (unit), I–39–10
Molecular attraction, I–1–3, I–12–6 f
Molecular crystal, II–30–2
Molecular diffusion, I–43–7 ff
Molecular dipole, II–11–1
Molecular motion, I–41–1
Molecule, I–1–3
Moment, dipole, I–12–6
    of force, I–18–5
    of inertia, I–18–7, I–19–5 ff
Momentum, I–9–1 f, I–38–2 ff, III–2–2 ff
    angular, I–7–7, I–18–5 ff, I–20–1, I–20–5, III–20–14 ff
    dynamical, III–21–5
    of light, I–34–10 f
    kinematic, III–21–5
    linear, I–4–7, I–10–1 ff
    relativistic, I–10–8 f, I–16–1 ff
Momentum operator, III–20–2, III–20–9 ff
Momentum spectrometer, II–29–1
Momentum spectrum, II–29–2
Monatomic gas, I–39–5
Monoclinic cell, II–30–7
Mössbauer, R., I–23–9
Mössbauer effect, II–42–11
Motion, I–5–1, I–8–1 ff
    of charge, II–29–1 ff
    circular, I–21–4
    constrained, I–14–3
    harmonic, I–21–4, I–23–1 ff
    parabolic, I–8–10
    planetary, I–7–1 ff, I–9–6 f, I–13–5
Motors, electric, II–16–1 ff
Moving charge, field momentum of, II–28–2 f
Music, I–50–1
Mutual inductance, II–17–9 ff, II–22–16
mv momentum, III–21–5

Negative carriers, III–14–2
Neon, III–19–16

Nernst heat theorem, I–44–11
Neuman, J. von, II–12–9
Neutral K-meson, III–11–12 ff
Neutral pion, III–10–7
Neutrons, I–2–4
  diffusion of, II–12–6 ff
Neutron diffusion equation, II–12–7
Newton, I., I–8–4, I–15–1, I–37–1,
  II–4–10, III–1–1
Newton · meter (unit), I–13–3
Newton's laws, I–2–6, I–7–3 ff, I–7–11,
  I–9–1 ff, I–10–1 ff, I–11–7 f, I–12–1,
  I–39–2, I–41–1, I–46–1, II–7–5,
  II–42–1, II–42–13
Nishijima, I–2–9, III–11–12
Nodes, I–49–2
Noise, I–50–1
Nonpolar molecule, II–11–1
Nuclear cross section, I–5–9
Nuclear energy, I–4–2
Nuclear forces, I–12–12, III–10–6 ff
Nuclear g-factor, II–34–4
Nuclear interactions, II–8–7
Nuclear magnetic resonance, II–35–10 ff
Nucleon, III–11–3
Nucleus, I–2–4, I–2–8 ff
Numerical analysis, I–9–6
Nutation, I–20–7
Nye, J. F., II–30–9

Oersted (unit), II–36–6
Ohm (unit), I–25–7
Ohm's law, I–25–7, I–43–7
One-dimensional lattice, III–13–1 ff
Operator (s), III–8–5, III–20–1 ff
  curl, II–2–8, II–3–1
  divergence, II–2–7, II–3–1
  gradient, II–2–4, II–3–1
  Laplacian, II–2–10
  vector, II–2–6
Operators, III–20–1 ff
Optic axis, I–33–3
Optic nerve, I–35–2
Optics, I–26–1 ff
  geometrical, I–26–1, I–27–1 ff
Orbital angular momentum, III–19–9
Orbital motion, II–34–3
Orientation polarization, II–11–3 ff
Oriented magnetic moment, II–35–4
Orthorhombic cell, II–30–7
Oscillation, amplitude, of, I–21–3
  damped, I–24–3 f
  frequency of, I–2–5
  period of, I–21–3
  periodic, I–9–4
  phase of, I–21–3
Oscillator, I–5–2
  harmonic, I–10–1, I–21–1, I–21–5 f,
  I–23–3 ff

Pais, III–11–12
Pappus, theorem of, I–19–4
Parabolic antenna, I–30–6 f
Parabolic motion, I–8–10
Parallel-axis theorem, I–19–6
Parallel-plate capacitor, I–14–9,
  II–6–11 ff, II–8–3
Paramagnetism, II–34–1 ff, II–35–1 ff
Paraxial rays, I–27–2

Partial derivative, I–14–9
Particles, Bose, III–4–1 ff
  Fermi, III–4–1 ff
  Identical, III–3–9 ff, III–4–1 ff
  spin-one, III–3–1 ff
  spin one-half, III–6–1 ff, III–12–1 ff
  "strange", II–8–7
Pauli spin exchange operator, III–12–7
Pauli spin matrices, III–11–1 ff
Permalloy, II–37–11
Permeability, II–36–9
Pascal's triangle, I–6–4
Passive circuit element, II–22–5
Pendulum, I–49–6 f
Pendulum clock, I–5–2
Period of oscillation, I–21–3
Periodic table, III–19–13 ff
Periodic time, I–5–1 f
Perpetual motion, I–46–2
Phase of oscillation, I–21–3
Phase shift, I–21–3
Phase velocity, I–48–6
Photon, I–2–7, I–26–1, I–37–8, III–4–7 f
  III–1–8
  polarization states of, III–11–9 ff
Physiochemistry of color vision, I–35–9 f
Piezoelectricity, II–11–8
Pines, II–7–7
Planck, M., I–41–6, I–42–8, I–42–9
Planck's constant, I–5–10, I–6–10,
  I–17–8, I–37–11, III–1–11
Plane lattice, II–30–5
Plane waves, II–20–1 ff
Planetary motion, I–7–1 ff, I–9–6 f,
  I–13–5
Plasma frequency, II–7–6, II–32–12
Plasma oscillations, II–7–5 ff
Plimpton, II–5–6
p-momentum, III–21–5
Poincaré, H., I–15–3, I–15–5, I–16–1
Poincaré stress, II–28–4
Point charge, electrostatic energy of,
  II–8–12
  field energy of, II–28–1 f
Poisson's ratio, II–38–2
Polar molecule, II–11–1, II–11–3 ff
Polarization, I–33–1 ff, II–32–1 ff
Polarization charges, II–10–3 ff
Polarization states of photon, III–11–9 ff
Polarization vector, II–10–2 f
Polarized light, I–32–9
Positive carriers, III–14–2
Potassium, III–19–16
Potential energy, I–4–4, I–13–1 ff,
  I–14–1 ff
  conservation of, III–7–6 ff
Potential gradient of the atmosphere,
  II–9–2 f
Power, I–13–2
Poynting, J., II–27–3
Precession, angle of, II–34–4
  of atomic magnets, II–34–4 f
Pressure, I–1–3
Priestly, J., II–5–6
Principle of equivalence, II–42–8 ff
Principle of least action, II–19–1 ff
Principle of superposition, II–1–3,
  II–4–2
Probability, I–6–1 ff
Probability amplitude (s), III–3–1 ff,
  III–16–1 ff

Probability density, I–6–8 f, III–16–6 ff
Probability distribution, I–6–7 ff,
  III–16–6
Propagation, in a crystal lattice,
  III–13–1 ff
Propagation factor, II–22–14
Proton, I–2–4
Proton spin, II–8–7
Pseudo force, I–12–10 ff
Ptolemy, I–26–2
Purkinje effect, I–35–2
Pyroelectricity, II–11–8
Pythagoras, III–11–1

Quadrupole lens, II–7–4, II–29–6
Quadrupole potential, II–6–8
Quantized magnetic states, II–35–1 ff
Quantum electrodynamics, I–2–7, I–28–3
Quantum mechanical resonance, III–10–4
Quantum mechanics, I–2–2, I–2–6 ff,
  I–6–10, I–10–9, I–37–1 ff, I–38–1 ff,
  III–1–1 ff, III–2–1 ff, III–3–1 ff
Quantum numbers, III–12–14

Rabi, I. I., II–35–4
Rabi molecular-beam method, II–35–4 ff
Radiant energy, I–4–2
Radiation, infrared, I–23–8, I–26–1
  relativistic effects, I–35–1 ff
  synchrotron, I–34–3 ff, I–34–6
  ultraviolet, I–26–1
Radiation damping, I–32–3 f
Radiation resistance, I–32–1 ff
Radioactive clock I–5–3 ff
Radius excess, II–42–6
Radius of electron, I–32–4
Ramsey, N., I–5–5
Random walk, I–6–5 ff, I–41–8 ff
Ratchet and pawl machine, I–46–1 ff
Rayleigh's criterion, I–30–6
Rayleigh's law, I–41–6
Rayleigh waves, II–38–8
Reactance, II–22–11
Reciprocity principle, I–30–7
Rectification, I–50–9
Rectifier, II–22–15
Rectifying function, III–14–11
Reflected waves, II–33–7 ff
Reflection, I–26–2 f
  angle of, I–26–3
  internal, II–33–12
  of light, II–33–1 ff
Refraction, I–26–2 f
  anomalous, I–33–9 f
  index of, I–31–1 ff
  of light, II–33–1 ff
Refractive index, II–32–1 ff
Relative permeability, II–36–9
Relativistic dynamics, I–15–9 f
Relativistic energy, I–16–1 ff
Relativistic mass, I–16–1 ff
Relativistic momentum, I–10–8 f,
  I–16–1 ff
Relativity, of electric field, II–13–6 ff
  Galilean, I–10–3
  of magnetic field, II–13–6 ff
  special theory of, I–15–1 ff
  theory of, I–7–11, I–17–11
Resistance, I–23–5

Resistor, I–23–5, II–22–4
Resonant cavity, II–23–6 ff
Resonant circuits, II–23–10 f
Resonant mode, II–23–10
Resonator, cavity, II–23–1 ff
Resolving power, I–27–7 f, I–30–5 f
Resonance, I–23–1 ff
  electrical, I–23–5 ff
  in nature, I–23–7 ff
Resonance interaction, I–2–9
Retarded time, I–28–2
Retherford, II–5–6
Retina, I–35–1
Reynolds' number, II–41–5 f
Rigid body, I–18–1
  angular momentum of, I–20–8
  rotation of, I–18–2 ff
Ritz combination principle, I–38–8,
  II–2–8
Rods, I–35–1, I–36–6
Roemer, O., I–7–5
Root-mean-square distance, I–6–6
Rotation, of axes, I–11–3 f
  plane, I–18–1
  of a rigid body, I–18–2 ff
  in space, I–20–1 ff
  in two dimensions, I–18–1 ff
Rotation matrix, III–6–4
Rushton, I–35–9
Rutherford, II–5–3
Rutherford-Bohr atomic model, II–5–3
Rydberg (unit), I–38–6, III–2–6
Rydberg energy, III–10–4, III–19–3

Scalar, I–11–5
Scalar field, II–2–2 ff
Scalar product, II–25–3 ff
Scattered amplitude, III–13–13
Scattering of light, I–32–5 ff
Schrödinger, E., I–35–6, I–37–1, I–38–9,
  III–1–1, III–2–9, III–20–17,
  III–21–1 ff, III–3–1
Schrödinger equation, II–15–12,
  III–16–4, III–16–11 ff, III–19–1 f,
  III–21–1 ff
Scientific method, I–2–1 f
Screw dislocation, II–30–9
Screw jack, I–4–5
Second (unit), I–5–5
Seismograph, I–51–5
Self-inductance, II–16–4, II–17–11 f
Semiconductor junction, III–14–8 ff
Semiconductors, III–14–1 ff
  impure, III–14–4
  N-type, III–14–5 f
  P-type, III–14–5 f
Shannon, C., I–44–2
Shear modulus, II–38–5
Shear wave, I–51–4, II–38–8
Sheet of charge, II–5–4
Side bands, I–48–4 f
Sigma electron, III–12–3
Sigma matrices, III–11–2
Sigma proton, III–12–3
Sigma vector, III–11–4
Simultaneity, I–15–7 f
Sinusoidal waves, I–29–2 f
Skin depth, II–32–11
Slip dislocation, II–30–9
Smoluchowski, I–41–8

Smooth muscle, I–14–2
Snell, W., I–26–3
Snell's law, I–26–3, I–31–2, II–33–1
Sodium, III–19–16
Solenoid, II–13–5
Solid-state physics, II–8–6
Sound, I–2–3, I–47–1 ff, I–50–1
  speed of, I–47–7 f
Space, I–8–2
Space-time, I–2–6, I–17–1 ff, II–26–12
Special theory of relativity, I–15–1 ff
Specific heat, I–40–7 f, I–45–2, II–37–4
Speed, I–8–2 ff, I–9–2
  of light, I–15–1, II–18–8 f
  of sound, I–47–7 f
Sphere of charge, II–5–4 f
Spherically symmetric solution, III–19–2 f
Spherical harmonics, III–19–1
Spherical waves, II–20–12 ff, II–21–2 ff
Spinel, II–37–12
Spin one-half particles, III–6–1 ff,
  III–12–1 ff
  precession of, III–7–10 ff
Spin-one particles, III–5–1 ff
Spin orbit, II–8–7
Spin-orbit interaction, III–15–13
Spin waves, III–15–1 ff
Spontaneous emission, I–42–9
Standard deviation, I–6–9
State vector, III–8–1
  resolution of, III–8–3 ff
States of definition energy, III–13–3 ff
Statics, II–4–1 f
Stationary state, III–7–2, III–11–22
Statistical fluctuations, I–6–3 ff
Statistical mechanics, I–3–1, I–40–1 ff
Steady flow, II–40–6 ff
Step leader, II–9–10
Stern, II–35–3
Stern-Gerlach apparatus, III–5–1 ff
Stern-Gerlach experiment, II–35–3 ff
Stevinus, S., I–4–5
Stokes' theorem, II–3–10
Strain, I–38–2
Strain tensor, II–31–11, II–39–1 ff
"Strange" particles, II–8–7
Strangeness, III–11–12
  conservation of, III–11–12
"Strangeness" number, I–2–9
Streamlines, II–40–6
Stress, II–38–2
Stress tensor, II–31–9 ff
Striated muscle, I–14–2
Superconductivity, III–21–1 ff
Supermalloy, II–36–9
Superposition, II–13–11 f
  of fields, I–12–9
  principle of, I–25–2 ff, II–1–3, II–4–2
Surface, equipotential, II–4–11 f
  gaussian, II–10–1
  isothermal, II–2–3
Surface tension, II–12–5
Symmetry, I–1–4, I–11–1 ff
  of physical laws, I–16–3, I–52–1 ff
Synchrotron, I–2–5, I–15–9, I–34–3 ff,
  I–34–6, II–17–5

Tamm, I., I–51–2
Taylor expansion, II–6–7
Temperature, I–39–6 ff

Tensor, II–26–7, II–31–1 ff
Tensor field, II–31–11
Tetragonal cell, II–30–7
Theory of gravitation, II–42–13 f
Thermal conductivity, II–2–8, II–12–2
  of a gas, I–43–9 f
Thermal equilibrium, I–41–3 ff
Thermal ionization, I–42–5 ff
Thermodynamics, I–39–2, I–45–1 ff,
  II–37–4 f
  laws of, I–44–1 ff
Thompson, II–5–3
Thompson atomic model, II–5–3
Thompson scattering cross section,
  II–5–3
Three-body problem, I–10–1
Three-dimensional waves, II–20–8 f
Three-dimensional lattice, III–13–7 f
Thunderstorms, II–9–5 ff
Tides, I–7–4 f
Time, I–2–3, I–5–1 ff, I–8–1, I–8–2
  retarded, I–28–2
  standard of, I–5–5
  transformation of, I–15–5 ff
Time-dependent states, III–13–6 f
Torque, I–18–4, I–20–1 ff
Torsion bar, II–38–5 ff
Total internal reflection, II–33–12 f
Transformation, Fourier, I–25–4
  Galilean, I–12–11
  linear, I–11–6
  Lorentz, I–15–3, I–17–1, I–34–8,
  I–52–2, II–25–1, II–26–1 ff
  of time, I–15–5 ff
  of velocity, I–16–4 ff
Transformer, II–16–4 f
Transforming amplitudes, III–6–1 ff
Transient, I–24–1 ff
  electrical, I–24–5 f
Transient response, I–21–6
Transistor, III–14–11 ff
Translation of axes, I–11–1 ff
Transmission line, II–24–1 ff
Transmitted waves, II–35–7 ff
Travelling field, II–18–5 ff
Triclinic lattice, II–30–7
Trigonal lattice, II–30–7
Triphenyl cyclopropenyl, III–15–13
Twenty-one centimeter line, III–12–9
Twin paradox, I–16–3 f
Two-dimensional field, II–7–2 ff
Two-slit interference, III–3–5 ff
Two-state systems, III–10–1 ff,
  III–11–1 ff
Tycho Brahe, I–7–1

Ultraviolet radiation, I–26–1
Uncertainty principle, I–2–6, I–6–10 f,
  I–37–9, I–37–11, I–38–8 f, II–5–3,
  III–1–9, III–1–11, III–2–8
Unit cell, I–38–5
Unit matrix, III–11–2
Unit vector, I–11–10, II–2–3
Unworldliness, II–25–10

van de Graaff generator, II–5–9, II–8–7
Vector, I–11–5 ff, III–8–1 f
Vector algebra, I–11–6 f

Vector analysis, I–11–5, I–52–2
Vector field, II–1–4 f, II–2–1 ff
    flux of, II–3–2 ff
Vector integrals, II–3–1 f
Vector operator, II–2–6
Vector potential, II–14–1 ff, II–15–1 ff
Vector product, I–20–4
Velocity, I–8–3, I–9–2 f
    components of, I–9–3
    transformation of, I–16–4 ff
Velocity potential, II–12–9
Vinci, Leonardo da, I–36–2
Virtual work, principle of, I–4–5
Viscosity, II–41–1 ff
    coefficient of, II–41–2
Viscous flow, II–41–4 f
Vision, I–36–1 ff
    binocular, I–36–4
    color, I–35–1 ff
Visual cortex, I–36–4
Visual purple, I–35–9
Voltmeter, II–16–1
Volume strain, II–38–3
Volume stress, II–38–3
von Neumann, J., II–40–3
Vortex lines, II–40–10 ff
Vorticity, II–40–5

Wall energy, II–37–6
Wapstra, I–52–10
Watt (unit), I–13–3
Wave, I–51–1 ff, II–20–1 ff
    electromagnetic, II–21–1 f
    light, I–48–1
    packet, III–13–6
    plane, II–20–1 ff
    reflected, II–33–7 ff
    shear, I–51–4, II–38–8
    sinusoidal, I–29–2 f
    spherical, II–20–12 ff, II–21–2 ff
    three-dimensional, II–20–8 f
    transmitted, II–33–7 ff
Wave equation, I–47–1 ff, II–18–9 ff
Wavefront, I–47–3
Wave function, III–16–5 ff
    meaning, III–21–6
Waveguides, II–24–1 ff
Wavelength, I–19–3, I–26–1
Wave nodes, III–7–9
Wave number, I–29–2
Weber, II–16–2
Weber (unit), II–13–1
"Wet" water, II–14–1 ff
Weyl, H., I–11–1
Wheeler, II–28–8

Wilson, C. T. R., II–9–9
Work, I–13–1 ff, I–14–1 ff

X-rays, I–2–5, I–26–1
X-ray diffraction, II–30–1

Young, I–35–7
Young's modulus, II–38–2
Yukawa, H., I–2–8, II–28–13
Yukawa potential, II–28–13, III–10–7
Yustova, I–35–8

Zeeman splitting, III–12–9 ff
Zeno, I–8–3
Zero, absolute, I–1–5
Zero curl, II–3–10 f, II–4–1
Zero divergence, II–3–10 f, II–4–1
Zero mass, I–2–10
Zinc, III–19–16